Trustworthy Cyber-Physical Systems Engineering

CHAPMAN & HALL/CRC
COMPUTER and INFORMATION SCIENCE SERIES

Series Editor: Sartaj Sahni

PUBLISHED TITLES

ADVERSARIAL REASONING: COMPUTATIONAL APPROACHES TO READING THE OPPONENT'S MIND
Alexander Kott and William M. McEneaney

COMPUTER-AIDED GRAPHING AND SIMULATION TOOLS FOR AUTOCAD USERS
P. A. Simionescu

DELAUNAY MESH GENERATION
Siu-Wing Cheng, Tamal Krishna Dey, and Jonathan Richard Shewchuk

DISTRIBUTED SENSOR NETWORKS, SECOND EDITION
S. Sitharama Iyengar and Richard R. Brooks

DISTRIBUTED SYSTEMS: AN ALGORITHMIC APPROACH, SECOND EDITION
Sukumar Ghosh

ENERGY-AWARE MEMORY MANAGEMENT FOR EMBEDDED MULTIMEDIA SYSTEMS: A COMPUTER-AIDED DESIGN APPROACH
Florin Balasa and Dhiraj K. Pradhan

ENERGY EFFICIENT HARDWARE-SOFTWARE CO-SYNTHESIS USING RECONFIGURABLE HARDWARE
Jingzhao Ou and Viktor K. Prasanna

FROM ACTION SYSTEMS TO DISTRIBUTED SYSTEMS: THE REFINEMENT APPROACH
Luigia Petre and Emil Sekerinski

FUNDAMENTALS OF NATURAL COMPUTING: BASIC CONCEPTS, ALGORITHMS, AND APPLICATIONS
Leandro Nunes de Castro

HANDBOOK OF ALGORITHMS FOR WIRELESS NETWORKING AND MOBILE COMPUTING
Azzedine Boukerche

HANDBOOK OF APPROXIMATION ALGORITHMS AND METAHEURISTICS
Teofilo F. Gonzalez

HANDBOOK OF BIOINSPIRED ALGORITHMS AND APPLICATIONS
Stephan Olariu and Albert Y. Zomaya

HANDBOOK OF COMPUTATIONAL MOLECULAR BIOLOGY
Srinivas Aluru

HANDBOOK OF DATA STRUCTURES AND APPLICATIONS
Dinesh P. Mehta and Sartaj Sahni

HANDBOOK OF DYNAMIC SYSTEM MODELING
Paul A. Fishwick

HANDBOOK OF ENERGY-AWARE AND GREEN COMPUTING
Ishfaq Ahmad and Sanjay Ranka

PUBLISHED TITLES CONTINUED

Trustworthy Cyber-Physical Systems Engineering

Edited by
Alexander Romanovsky
Newcastle University, UK

Fuyuki Ishikawa
National Institute of Informatics, Tokyo, Japan

CRC Press
Taylor & Francis Group
Boca Raton   London   New York

CRC Press is an imprint of the
Taylor & Francis Group, an **informa** business

A CHAPMAN & HALL BOOK

CRC Press
Taylor & Francis Group
6000 Broken Sound Parkway NW, Suite 300
Boca Raton, FL 33487-2742

Printed on acid-free paper
Version Date: 20160719

International Standard Book Number-13: 978-1-4987-4245-0 (Hardback)

Visit the Taylor & Francis Web site at
http://www.taylorandfrancis.com

and the CRC Press Web site at
http://www.crcpress.com

Printed and bound in the United States of America by
Edwards Brothers Malloy on sustainably sourced paper

Contents

Foreword, ix

Preface, xiii

Acknowledgments, xix

Editors, xxi

Contributors, xxiii

Foreword

F OR DECADES THE VAST MAJORITY OF COMPUTERS PRODUCED have been embedded processors running software that senses, computes, and actuates as part of a larger system.
Over time, increasing numbers of embedded processors have been placed into various systems to improve performance, reduce operating costs, and provide advanced functionality.
As this has happened, system interdependencies have increased, as has the ability of these
systems to exert an ever greater influence over the physical environment. At some point
these complex systems become Cyber-Physical Systems (CPSs), which in general involve
multiple computer systems interacting to sense and control things that happen in the real
world. While a CPS might sound a lot like just a bigger embedded system, the changes that
come with Internet-level scale and the integration of disparate components into aggregated
systems of systems are profound.

As a longtime researcher, teacher, and practitioner in the areas of embedded systems and
computer dependability, I've witnessed dramatic changes in the cruciality of getting these
systems right. Thirty-five years ago I used a computer to control the lights in a building with
primitive power-line-carrier remote switches. (And people asked me why I'd ever want to do
such a thing.) If the networked lighting system failed back then, the physical wall switches
worked just fine. Today, some buildings soft-link networked wall switches to varying sets of
networked lighting fixtures, and do real-time control on illumination conditions building-
wide based on how much light is coming in through the windows on a minute-to-minute
basis. If part of the computing system fails the wall switches might not do anything, because
they are just input devices on a computer network, and the surviving parts must make sure
enough lights are on to get by until a repair is made. In the future, we might see every lightbulb and thermostat in every house automatically negotiating for a better electricity rate with
the national electricity consortium by offering to reduce consumption during peak loads.
(This is already starting to happen with smart electric meters.) But if such a deeply interconnected system fails, we could easily have a significant outage. In principle, a software
bug in a lightbulb could take down an entire power grid.

In other words, functionality, optimization, interdependencies, and the potential cost of
failure are all increasing over time. The scope and capability of CPSs has increased to the
point that dependability is no longer a luxury—it has become a necessity. This is true for util-
ities, building management, chemical processing, transportation, health care, urban infrastructure, agriculture, telecommunications, and just about every other aspect of modern life.
Even computing server farms are being pulled into a critical role via cloud-connected CPS

services. At this point we're still busy connecting previously isolated embedded systems up to the Internet. Soon enough we'll be dealing with the consequences of creating ad hoc CPS amalgams out of components not originally designed for such purposes, and finding out that "good enough" software wasn't really as good as it needed to be.

Which brings us to the purpose of this book. The chapters in this book address many of the problems that will have to be solved to get from where we are now to where we need to be to attain truly dependable CPS design, deployment, and operations. In other words, its important to make sure we can justifiably trust that these systems will actually work as needed. But, developing a suitable level of trust is difficult because the scope of a modern CPS can be so broad, and the issues that must be addressed so pervasive. Topics in this book cover questions such as

- What does having a trustworthy CPS actually mean for something as pervasive as a global-scale CPS?

- How does CPS trustworthiness map onto existing knowledge, and where do we need to know more?

- How can we mathematically prove timeliness, correctness, and other essential properties for systems that may be adaptive and even self-healing?

- How do we represent, reason about, and ensure the correctness of an inherently discrete system (the computer) that interacts with an inherently continuous system (the real world)?

- How can we better represent the physical reality underlying real-world numeric quantities in the computing system?

- How can we expand the notion of trustworthiness to include system support aspects such as ensuring that a software defect doesn't drain the batteries of a critical component?

- How can we establish, reason about, and ensure trust between CPS components that are designed, installed, maintained, and operated by different organizations, and which may never have really been intended to work together?

- How can we make sure that when we use a new replacement part for an older component, the entire system won't come crashing down around us due to a subtle incompatibility?

- Once we have solutions to these problems, how can we impart that knowledge to future designers?

This is a book primarily about concepts, and as such these chapters emphasize theory, formality, modeling, and rigor. For example, more than a third of the chapters feature discussions of the Event-B approach for modeling and reasoning about CPS functionality.

The authors include a broad and deep sampling of CPS researchers spanning many application areas, with an emphasis on European formal methods-based approaches. The authors met for several days at an NII Shonan meeting, which affords researchers a distraction-free setting structured to permit digging deep into a problem and figuring out what needs to be done to solve it. This book is the follow-up to that retreat.

Getting CPS dependability right is essential to forming a solid foundation for a world that increasingly depends on such systems. This book represents the cutting edge of what we know about rigorous ways to ensure our CPS designs are trustworthy. I recommend it to anyone who wants to get a deep look at concepts that will form a cornerstone for future CPS designs.

Phil Koopman
Pittsburgh, Pennsylvania

Preface

THIS PREFACE BRIEFLY INTRODUCES THE AREA OF TRUSTWORTHY Cyber-Physical Systems (TCPS), outlines challenges in developing TCPS, and explains the structure of the book.

TRUSTWORTHY CYBER-PHYSICAL SYSTEMS

Societal and business activities depend more and more on software-intensive systems. Novel systems paradigms have been proposed and actively developed, notably Cyber-Physical Systems (CPS). The envisioned systems expand target entities and processes handled by the systems, stepping more deeply into human activities as well as real-world entities. There are emerging application areas such as automated driving and smart cities, while the existing areas, such as aviation, railways, business process management, and navigation systems, are also evolving by including a richer set of features. Visions for CPS include or extend a number of systems paradigms, such as Systems of Systems, Ubiquitous Computing, the Internet of Things, Smart Cities, and so on. Obviously, the increased involvement with human activities and real-world entities leads to greater demand for trustworthy systems. Here, and in this book, we treat *system trustworthiness* in a broad sense of a system being worthy of confidence or being dependable, so that it delivers service that can be justifiably trusted.

The developers of CPS face unprecedented complexity, caused not only by expanded application features, but also by combined mechanisms for trustworthiness (self-adaptation, resilience, etc.). Considering this growing complexity the construction of trustworthy CPSs and ensuring their trustworthiness is absolutely the key challenge in systems and software engineering.

The only solution that can help tackle this challenge is the development of new engineering methods for trustworthy CPS. One of the difficulties here is development of abstractions and semantic models for effective orchestration of software and physical processes [1]. There is no doubt that foundational theories and various technological components are essential as the building blocks for the new engineering methods. The building blocks for the engineering of trustworthy systems spread across verification algorithms, probabilistic analysis, fault models, self-adaptation mechanisms, and so on. The challenge of complexity requires further elaboration and integration of such blocks into sound engineering methods. Engineering methods define systematic and reliable ways for modeling analyzing, and verifying the system and its trustworthiness while mitigating the complexity. Recently, there have been yet more active efforts in engineering methods for trustworthy systems, on the basis of various approaches. Formal methods are one of the promising approaches that have been actively explored not only by academia but also by industry. Each approach has different, unique features, but essentially relevant to each other, focusing on modeling of the system, modeling of trustworthiness or faults, and their analysis and verification for complex systems, especially CPSs.

It is clear that in order to speed up the development of novel engineering methods for emerging complex CPSs, it is absolutely necessary to promote active discussions on subjects beyond specific applications, engineering approaches, or even paradigms. On October 27–30, 2014 we organized an NII Shonan Meeting on the Science and Practice of Engineering Trustworthy Cyber-Physical Systems. The NII Shonan Meeting, following the well-known Dagstuhl Seminars, is a scheme for meetings to promote discussions by world-leading researchers and practitioners. Our TCPS meeting aimed at providing this opportunity by inviting world-leading researchers working on engineering methods for TCPS. The meeting was attended by 32 participants from Asia, Europe, and North America and helped the participants to exchange ideas and develop a good common understanding of the area of TCPS. It was an extremely successful event that spawned new ideas and new collaborations. Further information about this meeting can be found at the Web site http://shonan.nii.ac.jp/seminar/055/ and in the final report (No. 2014-14 at http://shonan.nii.ac.jp/shonan/report/).

The idea for this book was conceived and discussed during the meeting. The chapters of the book were written by the participants of the seminar, to whom we are really grateful for their participation in the Meeting, for sharing their ideas, and for supporting the idea of the book.

KEY QUESTIONS

During the Meeting we organized several discussion sessions focusing on the following four key questions:

- What are the essential differences between CPS and traditional software systems?

- What are the technical challenges facing TCPS engineering?

- What are the nontechnical challenges facing TCPS engineering?

- What are the gaps between academia and industry (research and practice) in engineering (so far/for future TCPS)?

Most of the participants are top academic researchers on techniques for trustworthiness, such as formal methods, who have active collaborations with industry. There were also participants from industry. Below we include a brief summary of the issues identified.

1. What are the essential differences between CPS and traditional software systems?

   To understand these differences we need to first identify the essential properties of the systems we wish to develop. CPSs consist of collaborating computational elements controlling physical entities, which interact with humans and their environment. The design method should consider their elements differently from those of traditional software-based systems.

The main characteristics of physical systems are their continuous behavior and stochastic nature. This is why the development of CPS requires the involvement of specialists in multiple engineering disciplines. These systems often use close interaction between the plant and the discrete event system, and their complexity can grow due to interaction among multiple CPSs and the need to deal with time.

These are some of the essential features of CPS beyond conventional control systems: context awareness, cognitive computation, and autonomy.

2. What are the technical challenges facing TCPS engineering?

The technical challenges in TCPS engineering derive from the need to deal with system heterogeneity. TCPS engineering needs sound foundations to deal with modeling and analyzing the parts (e.g., continuous and discrete) and their compositions. These systems are often so complex that the humans in the loop may not completely understand them, which calls for special engineering methods to develop systems to assist humans as much as possible and make sure that the systems are always safe.

These systems have to be modeled/reasoned about together with their environments; the difficulties here are in choosing the right level of detail in representing the environment and in ensuring that all relevant elements of the environment are represented.

There is a wide range of CPS starting from social interaction systems and the Internet (Internet of Things) to safety-critical CPS in medicine/surgery.

Typically the general properties of CPSs can only be deduced from the local behaviors of their parts. The execution of these parts might have global effects and typically result in an emergent behavior of the whole. CPS engineering will need to include methods for compositional verification and compositional certification.

It is crucial to guarantee CPS trustworthiness for the guaranteed as well as the nonguaranteed behavior of the system. The safety, security, and reliability risk analysis of CPS require methods that go beyond the traditional hazard analysis techniques.

3. What are the nontechnical challenges facing TCPS engineering?

Perception of technology by the general public is crucial for deploying CPS. The difficulty lies in the problem that public acceptance of technologies depends on more than technical evidence (e.g., proof) of their correctness. It is challenging to communicate arguments, which include stochastic behavior and risks. We need to better understand how people feel about new technologies and also about collecting data to support analysis of trustworthiness.

There have been discussions on the development culture in some companies in which computer science (CS) and mathematics (e.g., formal methods) are not considered so relevant in practice. However, scientific approaches are inevitable to ensure trustworthiness under higher complexity of CPS. There are also obstacles in changing the culture in large companies. We need to understand how to demonstrate advantages of new technologies, and for this we need to clearly demonstrate business cases.

TCPS development requires interdisciplinary approaches that rely on communication between various experts. The CS people are expected to play a larger role by serving as hubs, as well as the ambassadors of new technologies. They will be able to develop rigorous unambiguous foundations for communicating knowledge and decisions across the domain involved.

In the area of certification and assurance important topics are the liability of companies, the use of specification as contract, and the need to develop a standard for standards. One important issue is to understand when and why the life of CPSs end.

The main challenges in education are as follows: CS engineering education does not cover continuous domain, and there is a need to educate the general public about new technologies.

4. What are the gaps between academia and industry (research and practice) in engineering so far/for future TCPS?

There is a cultural gulf between academia and industry, reified by the career requirements of both groups, but there is also a gulf within industry between research and production. There is a gap between different industries as well as between single-product organizations and heterogeneous CPS, where no one industry dominates. Sometimes companies are aware of the advantages of engagement but encounter difficulties in implementing it.

The goals are different as well. There is an academic need for academic excellence and an industrial need for commercial success. Academics are mainly into science; industry is into engineering. To add to this, there is inflexibility in both, academics and industry; for example, industries keep academics at a distance.

The successes achieved so far include government support for smaller joint projects (as in the EU), large EU projects (e.g., the development of software for avionics has successfully mixed academics and industry in long-term projects and large investment has been key), moving PhDs from industry to academia and vice versa.

The ways forward are as follows. Relevant government policies should be developed, and the governments should be lobbied for support. Unfortunately, information and communication technology and CPS are not given the attention afforded to, for example, high energy physics. In addition we need to lobby academia to promote the value of industrial engagement (the detail here is not trivial!) and industry to show the cost implications of academic engagement. Cross-fertilization between academic disciplines and between academia and industry should be encouraged. There should be a greater academic presence on standards committees. Lastly, industry should be more involved in the teaching of CPS.

OBJECTIVES AND STRUCTURE OF THIS BOOK

It will require joint effort by researchers working in a number of disciplines related to systems engineering in general to ensure that the CPSs that we are using now and will be using in the future are trustworthy. The traditional software-centric approaches are not applicable in this domain. In CPS, the physical part, the environment, the humans, and a variety of other systems are interacting with the increasingly cyber realm traditionally handled by computer scientists and software engineers.

There are already several excellent books that have been published on CPS (e.g., [2–4]). The aim of this book is to give practitioners and researchers a comprehensive introduction to the area of TCPS engineering.

The book covers various topics related to engineering of modern/future TCPS, including modeling, verification, rigorous design, simulation, requirements engineering, dependability, resilience, safety, architectural design, codesign/cosimulation, validation, modeling of constraint resources, safety cases, service-level agreement, and run-time resource optimization/allocation.

The book largely draws on the 4-day Shonan meeting that helped the contributors to analyze the challenges in developing trustworthy CPS, to identify important issues in developing engineering methods for CPS, and to develop a common understanding of the area and its challenges.

The individual chapters are written by teams led by the participants in the Shonan meeting. Joint participation in the meeting allowed the contributors to develop a good common understanding of the domain and of the challenges faced by the developers of TCPS.

The book has wider than usual technical coverage addressing various issues contributing to trustworthiness complemented by the contributions on TCSP roadmapping, taxonomy, and standardization, as well as on the experience in deploying advanced systems engineering methods in industry. Bringing all these topics together is unique and extremely important for the domain.

The book features contributions from the leading researchers in Japan, Canada, the USA, and Europe. The European partners have been and are being involved in a number of major FP7 and Horizon 2020 projects on CPS engineering.

The book consists of 16 chapters. The chapters are self-contained and provide discussions from various points of view; thus the readers can read the chapters independently according to their interests.

Chapters 1 through 3 discuss definitions, visions, and challenges of TCPS. This provides a good starting point to review concepts and difficulties in CPS and trustworthiness of CPS. Different viewpoints are provided, such as comparisons between embedded systems and systems of systems, as well as views of CPS as a type system.

Chapters 4 through 7 present specific approaches to ensuring trustworthiness, namely, proof and refinement. Mathematical proofs make it easier to rigorously ensure the trustworthiness of the target system. Refinement adds a further power onto proofs as an effective way of mitigating complexity through incremental or gradual formal development. The core challenges reside in how to model and verify self-* (adaptive, healing, etc.) and resilient systems.

Chapters 8 through 10 focus on engineering methods for dealing with hybrid aspects, that is, both continuous and discrete aspects. This is one of the key unique challenges in CPS, which requires the developers to go beyond the classical realm of computer science and software engineering.

Chapters 11 and 12 exemplify advances in developing the foundational techniques. Two studies are presented that envision strong support for modeling of domain-specific structures and verification of power consumption.

Chapters 13 through 15 present approaches based on agreements and assurance. Both of these concepts focus on properties that define trustworthiness of the target system and are often discussed beyond organizational boundaries. Different agreements and assurance techniques are presented in the chapters: monitoring and enforcement by the system as well as careful analysis and discussion by the involved engineers.

This book ends with Chapter 16, which discusses transfer of these advanced techniques to a wide range of engineers in the industry. This discussion is based on an intensive, long experience of industry education through industry–academia collaboration.

REFERENCES

1. E. A. Lee. Cyber physical systems: Design challenges. *The 11th IEEE International Symposium on Object Oriented Real-Time Distributed Computing (ISORC 2008)*, Orlando, FL, pp. 363–369, 2008.
2. M. Klein, R. Rajkumar, and D. de Niz. *Cyber-Physical Systems (SEI Series in Software Engineering)*. Addison-Wesley Professional, Boston, MA, 2016.
3. F. Hu. *Cyber-Physical Systems: Integrated Computing and Engineering Design*. CRC Press, Boca Raton, FL, 2010.
4. P. Marwedel. *Embedded System Design: Embedded Systems Foundations of Cyber-Physical Systems*. Springer, Berlin Heidelberg, 2010.

Acknowledgments

T HE EDITORS WOULD LIKE TO THANK THE COMMITTEE of the Shonan Meeting at NII (National Institute of Informatics, Tokyo, Japan) for supporting the TCPS meeting that led to the idea for this book. We thank the attendees of the TCPS meeting for their active and interesting discussions during the meeting as well as their contributions to this book. We are grateful to the book contributors for their help in reviewing the chapters. Our thanks to Stefan Hallerstede, Carolyn Talcott, Thai Son Hoang, and Kenji Tei, who kindly reviewed several book chapters. Last but not least, we thank the staff of CRC Press and Randi Cohen who gave us the opportunity and support for publication of this book.

Editors

Alexander Romanovsky is a professor at the Centre for Software Reliability and the leader of the Secure and Resilient Systems Group at the School of Computing Science, Newcastle University, UK. He earned his PhD from St. Petersburg State Technical University (Russia) in 1987. He joined Newcastle University in 1996 and since then has been working on various EU and UK systems dependability projects. One of these was a major EC FP7 Integrated Project on Industrial Deployment of System Engineering Methods Providing High Dependability and Productivity (DEPLOY), which he led in 2008–2012. Professor Romanovsky is an editorial board member of the Elsevier Journal of Systems Architecture and a visiting professor at the National Institute of Informatics (NII, Tokyo, Japan). He has more than 350 publications on various aspects of dependability. His research interests include resilience engineering, fault tolerance, safety verification, fault modeling, exception handling, system structuring, and formal engineering methods. Professor Romanovsky has been successfully working with industrial partners in the railway, automotive, aerospace, and business information domains.

Fuyuki Ishikawa is an associate professor at the National Institute of Informatics (NII) in Tokyo, Japan. He earned a Bachelor of Science degree in 2002 and a Masters degree in information science and technology in 2004, both from the University of Tokyo. He earned a PhD degree in information science and technology from the University of Tokyo in 2007. He is also a visiting associate professor at the University of Electro-Communications in Japan. His research interests include service-oriented computing and software engineering. He has served as the leader of six funded projects and published more than 100 papers.

Contributors

Yamine Aït-Ameur
Université de Toulouse
IRIT-INPT
Toulouse, France

Manamiary Bruno Andriamiarina
Lorraine Research Laboratory
  in Computer Science and its
  Applications
Université de Lorraine
Nancy, France

Guillaume Babin
Université de Toulouse
IRIT-INPT
Toulouse, France

Szilárd Bozóki
Department of Measurement and
  Information Systems
Budapest University of Technology and
  Economics
Budapest, Hungary

Ben Breimer
McMaster Centre for Software
  Certification
McMaster University
Hamilton, Ontario, Canada

John Fitzgerald
School of Computing Science
Newcastle University
Newcastle upon Tyne,
  United Kingdom

László Gönczy
Department of Measurement and
  Information Systems
Budapest University of Technology and
  Economics
Budapest, Hungary

Alexei Iliasov
School of Computing Science
Newcastle University
Newcastle upon Tyne,
  United Kingdom

Claire Ingram
School of Computing Science
Newcastle University
Newcastle upon Tyne,
  United Kingdom

Fuyuki Ishikawa
Digital Content and Media Sciences
  Research Division
National Institute of Informatics
Tokyo, Japan

Paul Joannou
McMaster Centre for Software
  Certification
McMaster University
Hamilton, Ontario, Canada

John Knight
Department of Computer Science
University of Virginia
Charlottesville, Virginia

Imre Kocsis
Department of Measurement and
    Information Systems
Budapest University of Technology and
    Economics
Budapest, Hungary

Linas Laibinis
Department of Information
    Technologies
Åbo Akademi University
Turku, Finland

Peter Gorm Larsen
Department of Engineering
Aarhus University
Aarhus, Denmark

Mark Lawford
McMaster Centre for Software
    Certification
McMaster University
Hamilton, Ontario, Canada

Thierry Lecomte
ClearSy
France

Thomas S. E. Maibaum
McMaster Centre for Software
    Certification
McMaster University
Hamilton, Ontario, Canada

István Majzik
Department of Measurement and
    Information Systems
Budapest University of Technology and
    Economics
Budapest, Hungary

Tom McCutcheon
Defence Science and Technology
    Laboratory
United Kingdom

Dominique Méry
Lorraine Research Laboratory in
    Computer Science and
    its Applications
Université de Lorraine
Nancy, France

Shin Nakajima
National Institute of
    Informatics
Tokyo, Japan

Marc Pantel
Université de Toulouse
IRIT-INPT
Toulouse, France

András Pataricza
Department of Measurement and
    Information Systems
Budapest University of Technology and
    Economics
Budapest, Hungary

Inna Pereverzeva
Department of Information Technologies
Åbo Akademi University
Turku, Finland

Alexander Romanovsky
School of Computing Science
Newcastle University
Newcastle upon Tyne,
United Kingdom

Ágnes Salánki
Department of Measurement and
 Information Systems
Budapest University of Technology and
 Economics
Budapest, Hungary

Neeraj Kumar Singh
Université de Toulouse
IRIT-INPT
Toulouse, France

Kevin Sullivan
Department of Computer Science
University of Virginia
Charlottesville, Virginia

Yoshinori Tanabe
School of Literature
Tsurumi University
Kanagawa, Japan

Elena Troubitsyna
Department of Information
 Technologies
Åbo Akademi University
Turku, Finland

Sasan Vakili
McMaster Centre for Software
 Certification
McMaster University
Hamilton, Ontario, Canada

Alan Wassyng
McMaster Centre for Software
 Certification
McMaster University
Hamilton, Ontario, Canada

Jim Woodcock
Department of Computer Science
University of York
York, United Kingdom

Jian Xiang
Department of Computer Science
University of Virginia
Charlottesville, Virginia

Nobukazu Yoshioka
Information Systems Architecture Science
 Research Division
National Institute of Informatics
Tokyo, Japan

CHAPTER 1

Concepts of Dependable Cyber-Physical Systems Engineering

Model-Based Approaches

John Fitzgerald, Claire Ingram, and Alexander Romanovsky

CONTENTS

T HE ENGINEERING OF CYBER-PHYSICAL SYSTEMS (CPS) is inherently multidisciplinary, requiring the collaborative effort of engineers from a wide range of backgrounds, often with significantly different models, methods, and tools. In such an environment, shared understanding of common concepts and the points at which terminology differs is essential. This is particularly the case in engineering dependable CPS.

In this chapter, we introduce some key concepts for CPS engineering, with a focus on the delivery of dependable systems and the role of model-based techniques.

1.1 INTRODUCTION

The design, development, deployment, and maintenance of dependable CPSs require collaboration among a variety of disciplines such as software, systems, mechanics, electronics, and system architectures, each with well-established notations, models, and methods. As might be expected in this context, terms and concepts that are well known in one discipline may be unknown or understood differently in another. This assumes particular significance in developing CPSs on which reliance is to be placed, where it is necessary to provide a demonstrably sound integration of diverse models. Here we provide a brief introduction to some key concepts for model-based engineering of dependable CPSs, and a short glossary of terms. It should be stressed that we do not seek to provide a survey of CPS engineering, but rather to provide the reader with a platform for subsequent chapters.

We first distinguish the subclass of systems we call *cyber-physical systems* in Section 1.3. In Section 1.4 we consider some concepts useful for dependability. Approaches to development of dependable CPSs are increasingly underpinned by model-based and simulation techniques, which differ among the disciplines involved. We discuss some basic concepts for CPS modeling in Section 1.5. In Section 1.6 we present our conclusions and a brief glossary.

1.2 DEFINITIONS AND CONCEPT BASES OF CPS

The European Commission (EC), the National Science Foundation, and other U.S. agencies[*] have made significant investments in methods and tools for CPS engineering. Among the efforts to provide a common conceptual basis for this emerging field, perhaps the most comprehensive to date is the NIST draft framework for CPS [1]. Among EC projects, work on the DESTECS[†] [2] and COMPASS[‡] [3] projects developed concept bases of embedded systems and systems of systems, respectively. Among more recent EC projects, several have surveyed the state of the art in CPS and embedded systems engineering. The CyPhERS[§] action produced a report to characterize the CPS domain, including key terms and concepts [4]. The ongoing TAMS4CPS[¶] project has published a definitional framework [5] of key concepts for a transatlantic CPS engineering audience, highlighting key commonalities in usage of terms and concepts in Europe and North America.

---

[*] See, for example, http://www.cps-vo.org.
[†] See http://www.destecs.org/.
[‡] See http://www.thecompassclub.org/.
[§] http://www.cyphers.eu/.
[¶] http://www.tams4cps.eu/.

1.3 TYPES OF SYSTEM

A *system* can be defined as a collection of interacting elements, organized to achieve a given purpose [6]. A system interacts with its *environment*; in model-based design, interactions between the system and its environment are represented as a series of *stimuli* provided by the environment to the system and as *responses* from the system to its environment [7]. There are many subtypes of *system*, and one system may fit simultaneously into several different categories.

1.3.1 Defining Cyber-Physical Systems

In a *cyber-physical system* (CPS), some elements are computational and some involve interactions with the physical environment [8–13], integrating "computation, communication, sensing, and actuation with physical systems to fulfill time-sensitive functions with varying degrees of interaction with the environment, including human interaction" [1,14]. A CPS incorporates multiple connected systems, producing a system capable of developing an awareness of its physical environment and context [15], making decisions based on that information, and enacting work that can effect changes in its physical environment [16].

As an example, consider a traffic management system (TMS). In many jurisdictions, road networks are divided into regions, each controlled by a separate autonomous TMS. The TMS is intended to meet several goals, some of which may conflict. These can include, for example, ensuring optimal throughput with minimum congestion, improving road safety, reducing air pollution and fuel burned, ensuring consistent travel times, etc. The TMS relies on data transmitted by large numbers of monitoring devices that are typically installed roadside or buried under the road surface and connected to a local traffic control center (TCC). The TCC conducts analysis, making predictions based on current data about likely congestion in the near future, identifying current problems or hazards, and suggesting appropriate strategies. Decisions made by the TCC are communicated to a variety of further roadside devices that can influence traffic behavior, such as variable speed limits and message boards, dynamic lane closures, and variable timings on traffic lights.

This is an example of a large-scale CPS; it relies on devices that can observe or affect the real world, gathering data from sensors, analyzing it, and making adjustments as necessary to improve performance. It enables a flexible solution that identifies problems and quickly adapts (e.g., by imposing speed limits or opening extra lanes). However, it is a complex system with an enormous variety of sensor types (and therefore significant heterogeneity), as well as complex analysis and data visualization. The application domain demands a high degree of dependability, which in turn is reliant on the behavior of different participating systems, from sensors to communications systems to analysis algorithms. Dependability includes real-time requirements; the situation on the road can change relatively quickly, and if analysis takes too long, the recommendations produced will be based on out-of-date information.

This traffic management example provides an illustration of a CPS in one domain, but the same principle of combining sensors, actuators, and intelligent analysis can be used to build CPSs that deliver improved performance, flexibility, or efficiency in many other domains. For example, assisted living systems can rely on wearable sensors or nonintrusive devices installed around a building to identify when an elderly person who lives alone needs help. CPSs can be used in manufacturing to monitor quality and make adjustments automatically that improve performance and reduce waste or allow a manufacturing line or other industrial process to adapt dynamically to volatile requirements. CPSs are suitable for these domains and a wide range of others.

CPSs can cross organizational boundaries, with one or more organizations contributing constituent parts toward the whole. In addition, a CPS crosses multiple engineering, computer science, and social science disciplines by incorporating elements that interact with the real world, human systems, and complex software systems capable of intelligently processing the large amounts of data that CPSs may encounter [9,17].

The TAMS4CPS definitional framework [5] points out a variety of definitions that exist for CPSs. For example, some define CPS as "integrations of computation and physical processes" [18] or "smart systems that encompass computational (i.e., hardware and software) and physical components, seamlessly integrated and closely interacting to sense the changing state of the real world" [19]. Other definitions emphasize the "cyber" aspects of CPS engineering, for example, defining CPS as

- "ICT systems (sensing, actuating, computing, communication, etc.) embedded in physical objects, interconnected (including through the Internet) and providing citizens and businesses with a wide range of innovative applications and services" [20].

- "Computation, communication and control components tightly combined with physical processes of different nature, e.g., mechanical, electrical, and chemical. Typically a CPS is defined and understood (evaluated) in a social and organizational context" [4].

- "Large complex physical systems that are interacting with a considerable number of distributed computing elements for monitoring, control and management which can exchange information between them and with human users" [21].

- "Systems with embedded software (as part of devices, buildings, means of transport, transport routes, production systems, medical processes, logistic processes, coordination processes and management processes), which

  o Directly record physical data using sensors and affect physical processes using actuators.

  o Evaluate and save recorded data, and actively or reactively interact both with the physical and digital world.

  o Are connected with one another and in global networks via digital communication facilities (wireless and/or wired, local and/or global).

  o Uses globally available data and services.

  o Have a series of dedicated, multimodal human–machine interfaces" [22].

The TAMS4CPS definition framework report suggests that European definitions tend to place the emphasis on the "smart" or "cyber" aspects of CPS engineering [4,20–22] whereas definitions from North America (e.g., [19]) tend to assign equal importance to each [5], although it is clear from these definitions that the differences in emphasis are small.

All definitions of CPS include aspects of *heterogeneity*. We can describe a CPS as heterogeneous because the different aspects of the CPS (e.g., interacting with a physical environment, conducting complex data processing, or integrating human and organizational processes) are very diverse in terms of the information that must be captured and how it is handled. This has significant ramifications for modeling the CPS; we discuss this further in Section 1.5.

CPSs have some similarities with, but are distinct from, some other types of system. There are several in particular that are often associated with CPSs: systems of systems and embedded systems. We describe these here in the interest of clarifying some important similarities and differences.

1.3.2 Systems of Systems

Systems of systems (SoS) are systems that comprise a number of constituent elements, each of which is an independently owned, complex system in its own right [23]. Such an SoS normally exhibits

- Independently owned and operated constituents

- Distribution between its constituents

- Emergent behavior that can only be produced when the constituents are interoperating together (we discuss *emergent behavior* in Section 1.3.4)

- Long-term evolution [23]

However, there are various definitions of SoS (e.g., [24–28]), some of which emphasize *heterogeneity*. CPSs significantly overlap with SoS [15] (the NIST Cyber-Physical Systems Public Working Group considers CPSs as a subtype of SoS [1]), and many CPSs will also be examples of SoS, and vice versa. We would expect that, like an SoS, a CPS will exhibit emergent global behavior, physical distribution between its constituent parts, and an evolutionary life cycle, with constituent systems removed, added, or replaced whenever necessary.

A key feature of an SoS is that constituent parts may also be owned and operated independently. This may or may not be true in a CPS; although as CPSs grow in scale and complexity, it is increasingly likely that a CPS will incorporate components contributed or operated by separate and independent stakeholders. However, this is not usually considered to be a defining feature, and some CPSs may not exhibit a large degree of independence and autonomy. In contrast, a CPS must by definition interact with the noncomputational world; this may or may not be true for SoS. As SoS also grow in complexity and scale, this is also increasingly likely to form a part of SoS engineering.

Because the fields of SoS engineering and CPS engineering overlap, CPS engineers and SoS engineers can share a significant amount of expertise and best practice. Some CPSs may operate in isolation, but many can contribute significant benefits when they are capable of interoperating in groups [1]. Lessons from SoS engineering are therefore important for CPS engineers who need to manage interactions between independent elements in a CPS, moving toward being able to engineer emergent behavior [1].

When a single independent system contributes to a CPS, we follow the terminology employed in SoS engineering literature and refer to it as a *constituent system* of the CPS.

1.3.3 Embedded Systems

An *embedded system* is a system that, like a CPS, incorporates a digital computing element (e.g., control software) and some physical components that interact with the physical environment [29]. These are tightly coupled together, occupying a dedicated piece of hardware that is usually specialized for the given purpose and has (traditionally) very limited resources [11]. An embedded system has a much more limited scale than a CPS; an embedded system is commonly considered at the level of a device,* incorporating a limited functionality designed for a specific task with associated control logic and hardware elements.

CPSs, in contrast, exist at a much larger scale, with many systems integrated together to provide the aspects of physical interaction and computational power: "Embedded systems are closed 'boxes' that do not expose the computing capability to the outside. The radical transformation that we envision comes from networking these devices" [11].

CPSs often incorporate some embedded systems, and some of the concepts relevant to embedded systems are therefore relevant to CPSs. Like an embedded system, a CPS is concerned with integrating computations and real-world interactions.

Embedded control systems are typically divided into a *controller* and a *plant*. The plant is "the part of the system that will be controlled," and "the controller contains the control laws and decision logic that affect the plant" [7,29].

Actuators and *sensors* are key underlying concepts for embedded systems and CPS. Sensors are devices that collect information about the real world (the exact information collected by a sensor will depend on the CPS's goals, but it could include, for example, temperatures, light/darkness levels, noise levels, quantities of various particulates in air or water, water levels or humidity, pressure readings, etc.). Actuators are devices that are capable of generating some effect on the real world. The type of the actuator will, again, depend on the type of CPS and its goals, but examples of actuators can include lights or display screens used for messaging or signaling or motors that can generate motion in components such as gates and barriers, vehicles, or industrial machinery.

A few concepts from the world of embedded systems are particularly important and relevant for thinking about CPSs, as well. *Control systems* are important for CPSs, as they are in many other aspects of engineering. Control systems are responsible for regulating the behavior of a system (or SoS), usually issuing directions after gathering some form of input. A *closed loop*, or feedback, control system includes some method of monitoring current performance or output and making corrections to the currently selected strategy as a result. A common embedded system design choice sees the control system implemented as software.

---

* Notwithstanding approaches developed in the embedded systems field to work with networked or distributed embedded systems.

Control systems vary substantially in the amount of complexity and precision they are capable of delivering; for example, some may take account of many input streams whereas others have a single source of input data. The types, quality, frequency, and precision of data collected by sensors within a single system are likely to be different. This can increase the complexity of designing and engineering dependable global behavior within a CPS; we discuss problems related to heterogeneity among components in Section 1.5.4.

1.3.4 Some Properties of CPS

CPSs exist at the intersection of different types of system, behaviors, and goals. CPSs form a subclass of systems that exhibit common properties; some of these properties pose particular design challenges for CPS engineers. We discuss common CPS properties here.

Like SoS, CPSs produce *emergent* behavior. The term *emergent* can have different meanings in different engineering disciplines, but we suggest that emergent behavior can be thought of as the global system behavior produced by constituents interoperating together [3,30] through the interaction of components/features; it cannot be produced by any one of those components in isolation. Emergent behavior may be unexpected or even undesirable from the point of view of individual constituent systems participating in the CPS, or from the point of view of an integrator who engineers global CPS behavior.

It is typically impossible to accurately predict all emergent behaviors due to a lack of trust between separate independent constituent systems, and therefore a lack of disclosure, as well as the fact that it is usually too computationally expensive to study all possible interactions between constituent systems within one CPS. The interactions that could arise among the systems can be unpredictable, with physical aspects of some constituents (such as choice of communications) producing side effects on apparently unrelated constituents (e.g., interference from one constituent's communications affecting the data packets received by another). Due to the complexity of CPSs, unexpected emergent behavior is rife. This creates challenges for the CPS engineer, since safety is a global property of the CPS, and emergent behaviors can often undermine efforts to deliver "safe," predictable behaviors. As a result, the ability to compose the constituent parts into a CPS and be confident that the system is safe remains a significant challenge in CPS engineering.

The constituent parts of a CPS may or may not exhibit a significant degree of *autonomy*, which is "the ability to complete one's own goals within limits and without the control of another entity" [25]. This can have a significant effect on the choice of CPS design approach. Autonomous constituent systems may opt, for reasons linked to their own goals, not to continue to participate long term. For example, a constituent in a CPS may unveil changes that mean that it can no longer provide the services required for the CPS. This means that it can be difficult to make the same types of judgment on the dependability of a CPS that can be made on systems that do not have autonomous constituents. We discuss dependability further in the next section.

Within a CPS, both physical and digital information may be crossing the boundaries between the constituent systems. For example, we can communicate information about temperature between two neighboring constituent systems. It is also possible for the physical heat from one constituent system to cross the boundary between the two and have some effect on its neighbor. The absence of temperature data on this phenomenon does not mean that nothing has crossed the boundary.

1.4 DEPENDABILITY

We base this discussion of dependability concepts to a large degree on that of Avizienis et al. [31]. A *dependable* system is one on which reliance may justifiably be placed. The requirement for justification that a given CPS is dependable therefore implies the need for some evidence to be assembled with which to justify this claim. Because of the nature of a CPS, such evidence will include analysis of both computational and physical elements (and potentially other types of system) and how they interact. Avizienis et al. suggest that dependability encompasses the following "attributes" [31]:

- *Availability*, or "readiness for correct service"

- *Reliability*, or "continuity of correct service"

- *Safety*, or the "absence of catastrophic consequences on the user(s) and environment"

- *Integrity*, or the "absence of improper system alterations"

- *Maintainability*, or the "ability to undergo modifications and repairs"

1.4.1 Achieving Dependability

A range of techniques are proposed to achieve some measure of dependability, including fault prevention, tolerance, removal, and forecasting. It is beyond the scope of this chapter to introduce all of these techniques, but we provide some discussion of fault tolerance in a CPS here in order to illustrate how fault tolerance techniques, which are well established in software engineering practice, can be adapted when deployed in a CPS context. Firstly, though, we introduce some important underlying concepts: faults, errors, and failures.

- A *failure* describes the situation in which a system deviates from its expected, externally visible service.

- When a system's state or service state deviates from correct behavior the deviation is called an *error*.

- A *fault* is the adjudged or hypothesized cause of an error and can be internal or external [31].

These definitions are provided with the assumption that the fault, error, and failure are associated with a single system. In an SoS situation it is necessary to consider multiple levels of abstraction: the global level, which presents SoS-level views, and the single-system level, which presents the viewpoint of one of the SoS constituents [32,33]. This multiple-level view is also appropriate for CPSs, which frequently comprise a global view of all constituent parts as well as the view of just one constituent of the CPS.

Faults may be present in a system without necessarily leading to an immediate failure [34]. On occasion a system may enter a state in which there is a fault, but this does not result in a service failure immediately (described as an *activated* fault). It may remain *latent* while the system transitions through a number of system states, before entering a state that is capable of manifesting the fault as a service failure. When this happens, the fault is described as *effective* [34].

1.4.2 Fault Tolerance in a CPS

Fault tolerance is the practice of attempting to "detect and correct latent errors before they become effective" [34], or the ability to avoid service failures even if faults are present within the system [31]. Systems relying on hardware to interact with the environment are highly likely to experience random component failures, depending on the properties and performances of the devices in the system. As a result of the close relationship between computation systems and interactions with the real world, CPS devices need to cope with *transient faults*, which are faults that appear temporarily due to an adverse environmental condition, such as electromagnetic interference. The CPS can then continue to operate as normal (although the effects may still be present in the form of corrupted data). These types of faults are believed to be very common in physical devices but not in software, resulting in different approaches and emphases adopted by the various engineering and computer science disciplines [34].

Fault tolerance is implemented via *error detection and recovery*. Instead of recovery techniques for *fault masking* or *error compensation* may be applied. Error recovery may be *forward* (repairing the erroneous state, as is done in exception handling) or *backward*, in which the erroneous state is replaced by a previous state that is believed to be correct, as is done in transaction-based or recovery block approaches. There are many ways to implement fault masking as well. One approach, for example, relies on *modular redundancy*, which may be applied at physical or computational levels in a CPS. If varying types of hardware or combinations of hardware and software are deployed, it allows designers to increase their confidence that randomly arising transient faults will not affect all modules. Without this independence, the voting system does not achieve a high probability of successfully compensating or masking different types of error.

1.4.3 Security in a CPS

There is a clear link between a dependable system (a system we consider reliable) and the notion of *trust*. Trust is a complex concept, applicable in a wide variety of domains, both technical and nontechnical, and in addition is highly context-dependent [35]. Xinlan et al. suggest, "Trust is a psychological state comprising the intention to accept vulnerability based upon positive expectations of the intentions or behavior of another" [36]. Trust has been defined within the dependability community as "accepted dependence" [31].

The concept of *security* is often linked to trust. Security on its own does not guarantee that a system is trusted [37,38], but a system that is believed to be secure is likely to be more trusted. Like trust, the concept of security is heavily context-dependent and differs among disciplines and domains. Computer scientists have traditionally decomposed the notion of computer security into *confidentiality* (prevention of undisclosed information), *integrity* (prevention of unauthorized modification of information), and *availability* (prevention of unauthorized withholding of information or resources) [39]. There are apparent overlaps with the dependability attributes defined above.

Security in a CPS context needs to encompass the CPS's holistic behavior, which includes the CPS's computational activities as well as its real-world interactions. Security violations within a CPS, for example, may include attacks or failures that cause hardware components to deliver unacceptable behavior or performance, which can have undesired or unexpected side effects, as well as attacks or failures that affect the availability of information. Anwar and Ali suggest a two-tiered approach to assess the trustworthiness of a CPS, classifying trust into two subdivisions:

- *Internal trust is* "responsible for different internal trusted entities such as sensors, actuators, communication networks" [38].

- *External trust is* "responsible for physical environment or architecture of cyber-physical system" [38].

1.5 MODELING AND SIMULATION

Modeling is of key importance to modern engineering design. A *model* is a partial description of a system, where the description is limited to those components and properties of the system that are pertinent to the current goal. Details that are not needed for the current modeling goal are not captured in the model.

During the design process, models are constructed to represent relevant aspects of the finished system. If they are sufficiently accurate, the models can be used to make predictions about the behavior or characteristics of the finished system so that engineers can analyze and optimize the design. Models are typically used to confirm the presence of desired behavior and performance in the finished system and the absence of undesired behavior and performance [7]. Previously, models consisted of physical prototypes that could be subjected to testing, but in the modern engineering design process rigorous mathematical models are created and analyzed in software.

To be effective, models should omit unnecesssary information [7]. On the other hand, a model must contain enough information to make realistic and useful predictions about the finished system. A model that includes all the necessary information and permits adequate analysis can be described as *competent* [7]. Some models are *executable*, which means that they can be interpreted and carried out as a sequence of steps on a computer if appropriate tools are available. The execution tool normally includes predictions of the likely effects of each step on the modeled system and its modeled environment. If a model is executable, the activity of executing these steps is called a *simulation*. Simulating a model allows the design engineer to examine the predicted behavior of the finished CPS and its environment in a variety of scenarios. Simulation, along with testing techniques in general, is never complete.

Modeling approaches can be divided into different types. For example, structural or architectural models allow an engineer to consider the *architecture* of a CPS, whereas behavioral modeling approaches allow an engineer to make predictions about the behavior of a CPS.

Analysis models are models created or generated for the purpose of performing an analysis of a system characteristic; they perform an important role during the engineering design process.

1.5.1 CPS Architectural Modeling

The *architecture* of a CPS describes its key constituent parts and how they are connected. The concept of *architecture* is important for a number of domains (e.g., organizational and business studies, software, mechanics, systems, and electronics), and each discipline conceptualizes architecture differently. Software architecture, for example, has been defined as "relationships among major components of the program" [40]. However, many different sources provide very similar definitions, centering on the notion of key constituents and connections between them (e.g., [41–43]). Some also emphasize a system's environment as part of its architectural definition ("The fundamental organisation of a system, embodied in its components, their relationships to each other and the environment, and the principles governing its design and evolution" [44]).

The COMPASS project dealt with the challenges of integrating disparate independent systems into a shared context that delivers some global behavior. Many CPSs also work within this shared context, so one observation made in a COMPASS deliverable addressing architectural approaches is worth considering here:

> The term architecture in the world of software engineering is limited to software elements (including interfaces) that make up a software application. The term architecture in the world of systems engineering may apply to any sort of system, whether it is technical, social, political, financial, or anything else. In the context of SoS engineering it is essential that these broader definitions of architecture are used [45].

Architectural modeling techniques that are often used to describe complete systems include graphical notations such as SysML* and UML.† Models created using these types of notations can be helpful for considering how the CPS is organized, as well as how the separate constituent parts interact and share data.

1.5.2 CPS Behavior Modeling

Engineered systems that incorporate elements for interacting with the physical environment rely heavily on modeling to optimize designs before prototypes are built. Modeling is cheaper than physical prototyping and much quicker; models can produce predictions of the behavior of the finished system, allowing engineers to study the effects of design decisions immediately. As we pointed out earlier in this chapter, there are a variety of disciplines involved in the design of a CPS, and each of these disciplines has its own modeling approaches and notations. The two most widely known modeling approaches are *continuous-time* modeling paradigms and *discrete-event* modeling paradigms.

---

* http://www.omgsysml.org.
† http://www.uml.org/.

When analyzing how an engineered system interacts with the real physical world, modeling techniques that rely on mathematical notations capable of representing *continuous-time* behaviors are used. A continuous-time formalism is one in which "the state of the system changes continuously through time" [46]. Continuous-time modeling is necessary for producing a physics model that can accurately predict the interactions between a system and its physical environment. It uses differential equations and iterative integration methods to capture dynamic behavior in a system [7]. Continuous-time techniques are used for analogue circuits and physical processes [47].

Modeling techniques also exist that allow computer scientists to make predictions about the behavior and properties of software. These modeling techniques also rely on mathematical modeling, but typically the mathematical models used by computer scientists are underpinned by *discrete-event* mathematics. In a discrete-event formalism, "only the points in time at which the state of the system changes are represented" [46]. Discrete-event modeling is typically used for digital hardware [47], communication systems [48], and embedded software [7].

Discrete-event formalisms allow for a rich representation of system behaviors and characteristics that are importance to software engineers, such as the software state or the synchronization among independent processes, but are poor at capturing the *real-time* information necessary for modeling the physical environment interactions that can be captured by continuous-time modeling techniques. On the other hand, a continuous-time formalism offers poor support for the types of analysis that are made possible by discrete-event modeling approaches. This is the crux of a challenge imposed on CPS engineers by the heterogeneity or diversity of the CPS constituent parts. Each discipline's preferred modeling notation has a very limited ability to capture and represent information of key importance to another discipline. There are a variety of approaches that could be used to tackle this challenge (for example, by integrating models into a multiparadigm modeling approach or considering standards for interchange); we discuss these in Section 1.5.4. Lee [11] explains this challenge by considering the example of a software program's successful behavior compared to the successful behavior of a CPS:

> The fact is that even the simplest C program is not predictable and reliable in the context of CPS because *the program does not express aspects of the behavior that are essential to the system...* Since timing is not in the semantics of C, whether a program misses deadlines is in fact irrelevant to determining whether it has executed correctly. But it is very relevant to determining whether the *system* has performed correctly [11].

1.5.3 Real-Time Concepts

Continuous-time techniques are necessary for creating realistic physics models that can accurately predict the behavior of a variety of physical phenomena. These techniques permit accurate representation of *real-time* information; systems including real-world interactions (such as CPSs) usually need to take real-time constraints into account. A real-time system is one in which "the correctness of the system behaviour depends not only on the logical results of the computations, but also on the physical time when these results are produced" [49,50].

The traditional techniques for representing time in models of physical systems uses real numbers (time is represented here as a continuum) or integers (representing time as discrete units) [10]. Software models, and software itself, typically abstract away details of time passing until there is no way to predict how long an operation will take when executed on a real system [10,11].

Systems with real-time performance requirements are often categorized into two types. *Soft real-time* systems do take physical time into account when considering correctness but do not have severe consequences for failure modes. In contrast, *hard real-time* systems do have severe consequences for an inability to meet real-time requirements, so a different amount of rigor is typically adopted to ensure that the correct system behavior is produced [49,50]. Hard real-time requirements might be associated with, for example, vehicles or machinery where short delays in the system's ability to deliver the "correct" result can result in threats to the safety of human passengers or operators.

1.5.4 Modeling Complete Heterogeneous Systems

As described above, the *heterogeneity* of the CPS presents a challenge for coping with different types of system [14]. Models that adequately capture software and computational behavior and models that capture interactions with the physical environment are difficult to integrate, because they capture different information and represent system behavior differently. Our inability to integrate discrete-event and continuous-time modeling paradigms, in fact, has been identified as a key challenge for improving our ability to deliver trustworthy and dependable CPSs in the future [7,10–13,15]. Because of the anticipated complexity of CPSs, a number of researchers have even called for better techniques for integrating results from many different modeling approaches, including human and organizational models in addition to software and hardware models [15,51].

The fact that these two major modeling paradigms (discrete-event and continuous-time) are so difficult to integrate encourages design teams working on CPSs to partition their designs for physical system elements and software elements, and reduces the possibility of working collaboratively. The prevalence of separate design teams, trained in separate disciplines, increases the potential for miscommunications and misunderstandings to arise during the design process [29,52]. To combat this, a number of different modeling techniques have been suggested to create holistic CPS models that integrate heterogeneous elements of a CPS in order to help engineers improve the system design [13,53].

One strategy is the use of hybrid system modeling approaches. This includes hybrid statecharts [54,55], hybrid [56] or timed automata [57], as well as hybrid [58] or differential Petri nets [59]. These approaches take existing techniques (e.g., automata [60] or Petri nets [61]) that are capable of describing systems without capturing time constraints and then extend the approach into a hybrid technique [13].

In an alternative approach, languages that have been designed for modeling holistic embedded systems and CPSs include Stateflow/Simulink, Modelica [62], hybrid CSP [63], and HyVisual [64]. Usually these types of notations attempt to capture discrete-event and continuous-time aspects of a system in a single language. There can be some disadvantages to this approach; for example, [53] note that more abstractions and simplifications may be required than is optimal.

Comodeling (collaborative modeling) is an approach that focuses on creating system models composed of separate models: a discrete-event model (e.g., of a controller) and a continuous-time model (e.g., of a plant) [7,53]. This approach has been deployed previously in embedded systems [7,52,53].

In one such approach, a cosimulation engine called Crescendo[*] [7] is used, which permits two separately created models to be executed simultaneously. Information is shared between models during the simulation, and the overall simulation result takes into account both discrete-event and continuous-time aspects of the system [7]. A collaborative simulation with two submodels like this is called a *cosimulation* [7,29]. In the Crescendo approach, two simulators are used—one for the discrete-event model and another for the continuous-time model—and each is responsible for its own constituent model, while coordination between the two is the responsibility of a *cosimulation engine* [7,29] (this is Crescendo itself).

Crescendo is not the only comodeling approach. For example, Ptolemy II [65] also supports heterogeneous simulation. Using Ptolemy II, a computation mode is specified for each model integrated into a simulation [66]. Another example of a comodeling approach supported by tools can be found in the automotive domain, in the Vanderbilt model-based prototyping tool chain [67]. This approach relies on Functional Mockup Interface,[†] which is an increasingly adopted protocol for exchanging data between models for simulation.

By considering the techniques employed by comodeling approaches like this, we can introduce some useful concepts for considering modeling of holistic CPSs. We will concentrate on the Crescendo environment as an example.

A CPS model needs to adopt the concept of an *event*. An event is a change in the state of the system. The system state changes continuously in a continuous-time notation. This can only be approximated in a discrete-event notation, by taking small discrete-time steps [29]. Events can be scheduled to occur at a specific time, which Crescendo refers to as *time events*. Events may also be responses to a change in a model, in which case they are referred to as *state events* [29].

During a cosimulation, the two simulation environments exchange information via *shared variables* and *shared design parameters*. A design parameter is a property of a model that affects the behavior described by the model but remains constant during the simulation. On the other hand, a variable is a property or piece of information that may change during simulation [29].

The terminology here provides us with a good example of the ease with which misunderstandings can arise between two different disciplines engaged in collaborative design of a CPS as well as the importance of a careful and continual process to identify and describe key common concepts. The term "parameter" is in widespread use in both computer science

---

[*] http://crescendotool.org/.

[†] https://www.fmi-standard.org/.

and various engineering disciplines, but, like many other technical terms, it has different meanings in different disciplines as do many other technical terms.

1.5.5 Modeling and Simulation for Validation and Verification

Modeling and simulation are often conducted as part of the *verification* and *validation* of a system.

- The validation process usually includes the systematic amassing of evidence to demonstrate that the system will meet the needs of users when it is deployed to its operational environment, particularly ensuring "that the right system has been built" [6] and that the requirements that have been created are accurate.

- Verification is concerned with the collection of evidence that the created system fulfills these requirements.

The use of models is important for both of these activities; testing and simulations of models of the system in its intended environment, under varying conditions, can check that the finished design will have the desired effects, that is, that the requirements are the right ones, and, moreover, that the proposed system design will satisfy them.

1.5.6 Fault Modeling

In Section 1.4 we introduced some concepts relevant for fault tolerance and dependability in a system. Here we discuss some concepts for modeling and simulation of faults and fault-tolerant behaviors.

Fault simulation is "the simulation of the system in the presence of faults" [7]. In order to achieve this, a model should capture faulty behavior (this is referred to as *fault modeling*) then the *activation* of the fault. Finally, the model must be capable of analyzing the predicted system behavior when a fault is present [7].

Fault modeling techniques can be very useful, for example, for evaluating the effectiveness of *error detection* and *error recovery* mechanisms that have been designed into the system (see Section 1.4). They are important in CPS engineering, in which there are usually a number of independent and heterogeneous components that must collaborate in order to deliver some global fault-tolerant behaviors. This is because fault modeling allows engineers to test that a proposed design that has been partitioned between these components will deliver the desired global behavior when the separate parts are integrated.

1.6 CONCLUSIONS

In this chapter we have briefly surveyed some key concepts for understanding the basics of dependability in CPSs. This includes examining some definitions of CPS, as well as surveying some of the key approaches for modeling and simulating aspects of CPS technology.

Because CPSs, by definition, involve interaction with the physical world, there are some important concepts to consider that are drawn from related areas such as embedded systems, including the need to be able to capture and reason about real-time requirements. CPSs also have some overlaps with SoS, which are collections of independent systems that contribute toward a global behavior. Many CPSs share features with embedded systems and/or SoS, so concepts from these domains are appropriate. We provide definitions for a few key concepts in our glossary.

Modeling and simulating CPSs that are cross-disciplinary is not a trivial problem, since the modeling paradigms that are usually employed to reason about interactions with the real world and complex software are not readily compatible. However, in the future, as CPSs grow more and more complex, there will be an increasing need to develop tools and techniques that allow engineers from a variety of disciplines, each with their own peculiarities, to contribute their advanced knowledge and skills to CPSs together. This will require methods and tools for integrating information and models from very different perspectives.

GLOSSARY

Below is our short glossary of selected terms. Words in italics appear in the glossary with their own definitions.

Activated fault: A *fault* can be said to be activated when the system has entered an erroneous state. The fault may remain *latent* throughout a sequence of system states before it becomes *effective* [34].

Architecture: "The fundamental organisation of a system, embodied in its components, their relationships to each other and the environment, and the principles governing its design and evolution" [44].

Autonomy: "The ability to complete ones own goals within limits and without the control of another entity" [25].

Availability: "Readiness for correct service" [31].

Backward error recovery: This is a technique used to implement *error recovery*. This technique concentrates on replacing the erroneous state with a copy of a previous system state that is believed to be correct [34].

Comodel: A collaborative modeling approach used in embedded systems. The approach uses two separate simulators, one to manage a *discrete-event* model that represents a *controller* and another to manage a *continuous-time* model that represents a *plant*. A comodeling engine is necessary to coordinate the two [7].

Cosimulation: A *simulation* involving a *comodel* [7].

Competent: "We regard a model as being competent for a given analysis if it contains sufficient detail to permit that analysis" [7].

Confidentiality: "Prevention of undisclosed information" [39].

Constituent system: An independent system that contributes toward a CPS; a CPS may comprise many constituent systems.

Continuous-time: A system in which "the state of the system changes continuously through time" [46]. In some systems (e.g., systems that focus on the behavior and properties of components that interact with the real world) this is an important property, and so in order to be useful, models of the system must also be capable of capturing and representing this property.

Controller: In embedded systems, the controller contains the control laws and decision logic that affect the plant directly by means of actuators and receive feedback via sensors [7].

Cyber-physical system (CPS): A system in which some elements are computational and some elements involve interactions with the physical environment [8–10].

Dependable: A dependable system is one on which reliance may justifiably be placed [31].

Design parameter: A property of a model that affects the behavior described by the model, but which remains constant throughout the simulation [29].

Discrete-event: A system, or a model of a system, in which "only the points in time at which the state of the system changes are represented" [46].

Effective fault: A *fault* that has resulted in a *failure*. The system may have entered a system state that was incorrect (erroneous) some time earlier without showing any evident failures in service. The fault is not described as effective until it results in a failure [34].

Embedded system: A single, self-contained system incorporating some software for control logic and some interaction with the real world. An embedded system typically has a very specific function and often very limited computational resources.

Emergent: Emergent behavior is produced when all the *constituent systems* of a CPS are collaborating; it is the global CPS behavior. Emergent behavior may be expected or unexpected.

Environment: Entities outside the boundary of a CPS that interact with it. This can include hardware, software, humans, and the physical world.

Error: The part of the CPS state that can lead to its subsequent service *failure*.

Error detection: The first stage of *fault tolerance*; it involves detection of *latent errors* in the CPS [34].

Error compensation: Also known as fault masking [34].

Error recovery: Error recovery is one of several steps involved in implementing *fault tolerance* and can only arise following *error detection*. During error recovery, the CPS's erroneous state must be "replaced by an acceptable valid state" [34].

Event: A change in the state of a system [29].

Executable: A model is executable if it can be interpreted by an appropriate tool on a computer as a sequence of instructions [7].

Failure: A deviation of the service provided by a CPS (or at a different level of abstraction, a single *constituent system* within a CPS) from expected (correct) behavior. This results in incorrect behavior visible at the boundary of the CPS (or at the boundary of the constituent system).

Fault: The adjudged or hypothesized cause of an *error*. Note that a *failure* when viewed at the level of a constituent system (i.e., a failure in one single *constituent system*) becomes a *fault* when viewed from the perspective of the CPS.

Fault forecasting: A family of techniques to "estimate the present number, the future incidence, and the likely consequences of faults" [31].

Fault masking: Also known as *error compensation*. This approach uses techniques to identify potential faults and avoid entering erroneous states. *Modular redundancy* is one technique that can be used for this.

Fault modeling: The process of capturing faulty behavior in a model. This can be used to analyze proposed *error detection* and *error recovery* strategies in a given design [7].

Fault prevention: A family of techniques to "prevent the occurrence of introduction of faults" [31].

Fault tolerance: A fault tolerant system attempts to "detect and correct *latent errors* before they become *effective*" [34].

Fault removal: A family of techniques to "reduce the number and severity of faults" [31].

Fault simulation: The simulation of the system in the presence of faults [7].

Fault tolerance: The avoidance of service *failures* in the presence of *faults* [31].

Forward error recovery: This is a technique used to implement *error recovery*. It concentrates on "repairing the erroneous state; exception handling is an example of this" [34].

Hard real-time: A system that has severe consequences for *real-time* failure modes, and consequently is rigorous in ensuring that the correct system behavior is produced [50].

Heterogeneity: A *system* is heterogeneous if it comprises a range of very diverse *constituent systems*. Models of these constituent systems capture very different system aspects and are difficult to integrate.

Integrity: "Absence of improper system alterations" [31].

Latent fault: A fault that has been *activated* but not yet resulted in a service *failure* [34].

Maintainability: "Ability to undergo modifications and repairs" [31].

Model: A partial description of a system, where the description is limited to those components and properties of the system that are pertinent to the current goal.

Modular redundancy: A technique for implementing *fault masking*, in which multiple independent modules are used to generate each computation and the final result selected after taking a vote between all modules' results.

Plant: In embedded systems, the plant "is the part of the system that will be controlled" [7].

Real-time system: A system in which "the correctness of the system behaviour depends not only on the logical results of the computations, but also on the physical time when these results are produced" [50].

Reliability: "Continuity of correct service" [31].

Response: The CPS produces responses to the environmental *stimuli*.

Safety: "Absence of catastrophic consequences on the user(s) and environment" [31].

Simulation: Some models are *executable*. An execution of such a model is called a simulation; the simulation normally involves making predictions about the likely behavior of the finished CPS and its environment.

Soft real-time: A system which has real-time requirements, but does not have severe consequences for real-time failure modes [50].

State event: An *event* that occurs in response to a change in the system's state [29].

Stimuli: The CPS's environment provides stimuli as inputs to the CPS.

System: A system is a collection of interacting elements organized in order to achieve a given purpose [41].

System of systems: A system of systems (SoS) is a system comprising some number of constituent parts, each of which is an independently owned, complex system in its own right [23].

Time event: An *event* that is scheduled to occur at a specific time interval [29].

Transient fault: A temporary *fault* afflicting a system, for example, due to an environmental factor such as electromagnetic interference [34].

Trust: "Trust is a psychological state comprising the intention to accept vulnerability based upon positive expectations of the intentions or behaviour of another" [36].

Validation: "The purpose of the Validation Process is to provide objective evidence that the services provided by a system when in use comply with stakeholders requirements, achieving its intended use in its intended operational environment" [6].

Variable: A piece of information that may change during a simulation [29].

Verification: "The Verification Process confirms that the system-of-interest and all its elements perform their intended functions and meet the performance requirements allocated to them (i.e., that the system has been built right)" [6].

REFERENCES

1. Cyber Physical Systems Public Working Group NIST. *Draft Framework for Cyber-Physical Systems*. Technical Report Draft Release 0.8, National Institute of Standards and Technology, September 2015.
2. DESTECS. *D2.3: Methodological Guidelines 3*. Technical Report [Online] http://www.destecs.org/images/stories/Project/Deliverables/D23MethodologicalGuidelines3.pdf [Last visited August 2015], DESTECS Project, 2012.
3. COMPASS. *Compass Deliverable D11.3: Convergence Report 3*. Technical Report [Online] http://www.compass-research.eu/deliverables.html [Last visited August 2015], COMPASS project, 2014.
4. CyPhERS. *D2.2: Structuring of CPS Domain: Characteristics, Trends, Challenges and Opportunities Associated with CPS*. Technical Report [Online] http://cyphers.eu/project/deliverables [Last visited June 2015], CyPhERS (Cyber-Physical European Roadmap and Strategy), 2014.
5. TAMS4CPS. *D1.1: Definitional Framework*. Technical Report [Online] http://www.tams4cps.eu/wp-content/uploads/2015/04/TAMS4CPS_D1-1_Definitional-Framework.pdf [Last visited August 2015], TAMS4CPS project, 2015.
6. INCOSE. *Systems Engineering Handbook: A Guide for System Life Cycle Processes and Activities, Version 3.2.2*. Technical Report INCOSE-TP-2003-002-0.3.2.2, International Council on Systems Engineering (INCOSE), October 2011.
7. J. Fitzgerald, P.G. Larsen, and M. Verhoef. *Collaborative Design for Embedded Systems*. Springer-Verlag, Berlin Heidelberg, 2014.
8. M. Broy. Engineering: Cyber-physical systems: Challenges and foundations. In *Proceedings of the Fourth International Conference on Complex Systems Design & Management CSD&M 2013*. Paris, France, pp. 1–13. 2013.
9. J.S. Fitzgerald, K.G. Pierce, and C.J. Gamble. A rigorous approach to the design of resilient cyber-physical systems through co-simulation. In *DSN Workshops*. Boston, MA, IEEE, 2012.
10. E.A. Lee. CPS Foundations. In *Proceedings of the 47th Design Automation Conference, DAC'10*, Anaheim, CA, pp. 737–742, 2010.
11. E.A. Lee. *Cyber Physical Systems: Design Challenges*. Technical Report UCB/EECS-2008-8, EECS Department, University of California, Berkeley, January 2008.
12. NIST. *Strategic Vision and Business Drivers for 21st Century Cyber-Physical Systems*. Technical Report, National Institute of Standards and Technology, January 2013.
13. X. Zheng, C. Julien, M. Kim, and S. Khurshid. *On the State of the Art in Verification and Validation in Cyber Physical Systems*. Technical Report TR-ARiSE-2014-001, The University of Texas at Austin, The Center for Advanced Research in Software Engineering, 2014.

14. NIST. *Foundations for Innovation in Cyber-Physical Systems; Workshop Report.* Technical Report, National Institute of Standards and Technology, January 2013.

15. M. Broy, M.V. Cengarle, and E. Geisberger. Cyber-physical systems: Imminent challenges. In R. Calinescu and D. Garlan, editors, *Large-Scale Complex IT Systems. Development, Operation and Management,* volume 7539 of *Lecture Notes in Computer Science,* pp. 1–28. Springer, Berlin Heidelberg, 2012.

16. E. Geisberger and M. Broy. *Living in a Networked World: Integrated Research Agenda Cyber-Physical Systems (agendaCPS).* Technical Report [Online] http://www.acatech.de/de/publikats ionen/empfehlungen/acatech/detail/artikel/living-in-a-networked-world-integrated-research-agenda-cyber-physical-systems-agendacps.html [Last visited August 2015], acatech National Academy of Science and Engineering, 2015.

17. H. Giese, B. Rumpe, B. Schätz, and J. Sztipanovits. Science and engineering of cyber-physical systems (Dagstuhl Seminar 11441). *Dagstuhl Reports,* 1(11):1–22, 2011.

18. E. Lee. *Computing Foundations and Practice for Cyber-Physical Systems: A Preliminary Report.* Technical Report No. UCB/EECS-2007-72, Electrical Engineering and Computer Sciences, University of California at Berkeley, May 2007.

19. Inc. Energetics. *Foundations for Innovation in Cyber-Physical Systems.* Technical Report US: National Institute of Standards and Technology, US Dept. Commerce, Washington DC, 2013.

20. European Commission. Cyber-physical systems: Uplifting Europes innovation capacity. In *Report from the Workshop on Cyber-Physical Systems: Uplifting Europes Innovation Capacity.* INCOSE, October 2013.

21. CPSoS. *Definitions Used Throughout the Project.* Technical Report [Online] http://www.cpsos. eu/project/what-are-cyber-physical-systems-of-systems/ [Last visited April 2015], CPSoS (Towards a European Roadmap on Research and Innovation in Engineering and Management of Cyber-Physical Systems of Systems), 2015.

22. acatech. *Driving Force for Innovation in Mobility, Health, Energy and Production.* Technical Report [Online] http://www.acatech.de/fileadmin/user_upload/Baumstruktur_nach_Websi te/Acatech/root/de/Publikationen/Stellungnahmen/acatech_POSITION_CPS_Englisch_WEB. pdf [Last visited June 2015], acatech Position Paper, December 2011.

23. M.W. Maier. Architecting principles for systems-of-systems. *Systems Engineering,* 1(4):267–284, 1998.

24. R. Abbott. Open at the top; open at the bottom; and continually (but slowly) evolving. In *System of Systems Engineering, 2006 IEEE/SMC International Conference on,* Los Angeles, CA, pp. 41–46, IEEE, April 2006.

25. W.C. Baldwin and B. Sauser. Modeling the Characteristics of System of Systems. In *System of Systems Engineering, 2009. SoSE 2009. IEEE International Conference on,* Albuquerque, NM, pp. 1–6. IEEE, 2009.

26. J. Boardman and B. Sauser. System of Systems—the meaning of "of". In *Proceedings of the 2006 IEEE/SMC International Conference on System of Systems Engineering,* pp. 118–123, Los Angeles, CA, April 2006. IEEE.

27. D. Cocks. How should we use the term "system of systems" and why should we care? In *Proceedings of the 16th INCOSE International Symposium 2006.* Orlando, FL, pp. 427–438, INCOSE, July 2006.

28. D.A. Fisher. *An Emergent Perspective on Interoperation in Systems of Systems.* Technical Report: CMU/SEI-2006-TR-003. Technical Report, Software Engineering Institute, Carnegie Mellon University, Pittsburgh, PA, March 2006.

29. J.S. Fitzgerald, P.G. Larsen, K.G. Pierce, and M.H.G. Verhoef. A formal approach to collaborative modelling and co-simulation for embedded systems. *Mathematical Structures in Computer Science,* 23(04):726–750, 2013.

30. J.W. Sanders and G. Smith. Emergence and refinement. *Formal Aspects of Computing,* 24(1): 45–65, 2012.

31. A. Avizienis, J.-C. Laprie, B. Randell, and C. Landwehr. Basic concepts and taxonomy of dependable and secure computing. *IEEE Transactions on Dependable and Secure Computing*, 1(1): 11–33, 2004.

32. Z. Andrews, J. Fitzgerald, R. Payne, and A. Romanovsky. Fault modelling for systems of systems. In *Proceedings of the 11th International Symposium on Autonomous Decentralised Systems (ISADS 2013)*, Mexico City, Mexico, pp. 59–66, March 2013.

33. Z. Andrews, C. Ingram, R. Payne, A. Romanovsky, J. Holt, and S. Perry. Traceable engineering of fault-tolerant SoSs. In *Proceedings of the INCOSE International Symposium 2014*, Las Vegas, NV, pp. 258–273, June 2014.

34. J. Rushby. Critical system properties: Survey and taxonomy. *Reliability Engineering and System Safety*, 2(43):189–219, 1994.

35. M. Blaze, J. Feigenbaum, and J. Lacy. Decentralized trust management. In *Proceedings of IEEE Symposium on Security and Privacy*, Oakland, CA, pp. 164–173, 1996.

36. Z. Xinlan, H. Zhifeng, W. Guangfu, and Z. Xin. Information security risk assessment methodology research: Group decision making and analytic hierarchy process. In *Second WRI World Congress on Software Engineering*, pp. 157–160, 2010.

37. S. Massoud. Amin cyber and critical infrastructure security: Toward smarter and more secure power and energy infrastructures. In *Canada-US Workshop on Smart Grid Technologies*, Vancouver, Canada, March 2010.

38. R.W. Anwar and S. Ali. Trust based secure cyber physical systems. In *Workshop Proceedings: Trustworthy Cyber-Physical Systems. Technical Report Series: Newcastle University, Computing Science, No: CS-TR-1347*, pp. 1–10, 2012.

39. D. Gollman. *Computer Security*. John Wiley, New York, NY, 1999.

40. R.S. Pressman. *Software Engineering—A Practitioner's Approach*. Computer Science Series. McGraw-Hill International Editions, 2nd edition, New York, NY, 1998.

41. Department of Defense. DoD Architecture Framework, Version 1.5, 04 2007.

42. J. Sanders and E. Curran. *Software Quality: Framework for Success in Software Development and Support*. ACM Press Series, Addison Wesley, 1994.

43. S.R. Schach. *Software Engineering with Java*. McGraw-Hill International editions, 1997.

44. IEEE. *IEEE Recommended Practice for Architectural Description of Software-Intensive Systems*. Technical Report [Online] IEEE std 1471. IEEE, NY, USA, 2000, The Institute of Electrical and Inc. Electronics Engineers, 2000.

45. COMPASS. *COMPASS Deliverable D22.1: Initial Report on SoS Architectural Models*. Technical Report [Online] http://www.compass-research.eu/deliverables.html [Last visited June 2015], COMPASS project, 2012.

46. S. Robinson. *Simulation: The Practice of Model Development and Use*. John Wiley and Sons, 2004.

47. J. Liu. *Continuous Time and Mixed-Signal Simulation in Ptolemy II*. Technical Report No. UCB/ERL M98/74, EECS Department, University of California, Berkeley, 1998.

48. J. Banks, J. Carson, B.L. Nelson, and D. Nicol. *Discrete-Event System Simulation*. Prentice Hall, 4th edition, 2004.

49. H. Kopetz. Software Engineering for Real-Time: A Roadmap. In *ICSE'00 Proceedings of the Conference on the Future of Software Engineering*, pp. 201–211, Limerick, Ireland, 2000.

50. H. Kopetz. *Real-Time Systems: Design Principles for Distributed Embedded Applications*. Springer, USA, 2 edition, 2011.

51. COMPASS. *COMPASS Deliverable D11.4: Roadmap for Research in Model-Based SoS Engineering*. Technical Report [Online] http://www.compass-research.eu/Project/Deliverables/D11.4.pdf [Last visited June 2015], COMPASS project, 2014.

52. X. Zhang and J.F. Broenink. A concurrent design approach and model management support to prevent inconsistencies in multidisciplinary modelling and simulation. In *19th European*

Concurrent Engineering Conference, pp. 21–28. EUROSIS, EUROSIS-ETI Publication, June 2013.

53. Y. Ni and J.F. Broenink. A co-modelling method for solving incompatibilities during co-design of mechatronic devices. *Advanced Engineering Informatics*, 28(3):232–240, 2014.

54. Y. Kesten and A. Pnueli. Timed and hybrid statecharts and their textual representation. In J. Vytopil, editor, *Formal Techniques in Real-Time and Fault-Tolerant Systems, Second International Symposium, Nijmegen, The Netherlands, January 8–10, 1992, Proceedings*, volume 571 of *Lecture Notes in Computer Science*, pp. 591–620. Springer, 1992.

55. O. Maler, Z. Manna, and A. Pnueli. From timed to hybrid systems. In J.W. de Bakker, C. Huizing, W.P. de Roever, and G. Rozenberg, editors, *Real-Time: Theory in Practice, REX Workshop, Mook, The Netherlands, June 3–7, 1991, Proceedings*, volume 600 of *Lecture Notes in Computer Science*, pp. 447–484. Springer, 1992.

56. R. Alur, C. Courcoubetis, N. Halbwachs, T.A. Henzinger, P.-H. Ho, X. Nicollin, A. Olivero, J. Sifakis, and S. Yovine. The algorithmic analysis of hybrid systems. *Theoretical Computer Science*, 138(1): 3–34, 1995.

57. R. Alur and D. Dill. The theory of timed automata. *Theoretical Computer Science*, 126:183–235, 1994.

58. J. Le Bail, H. Alla, and R. David. Hybrid Petri nets. In *European Control Conference*, ECC 91, Grenoble, France, 1991.

59. I. Demongodin and N.T. Koussoulas. Differential Petri nets: Representing continuous systems in a discrete-event world. *IEEE Transactions on Automatic Control*, 43(4):573–579, 1998.

60. J.E. Hopcroft, R. Motwani, and J.D. Ullman. *Introduction to Automata Theory, Languages, and Computation*. Pearson Education, USA, 2001.

61. T. Murata. Petri nets: Properties, analysis and applications. *Proceedings of the IEEE*, 77(4):541–580, 1989.

62. P. Fritzson and V. Engelson. Modelica—A unified object-oriented language for system modelling and simulation. In *ECCOP '98: Proceedings of the 12th European Conference on Object-Oriented Programming*, pp. 67–90. Springer-Verlag, 1998.

63. H. Jifeng. From CSP to hybrid systems, In: *A Classical Mind*, edited by A W. Roscoe. pp. 171–189. Prentice Hall, Hertfordshire, UK, 1994.

64. E.A. Lee and H. Zheng. Operational semantics of hybrid systems. In M. Morari and L. Thiele, editors, *Hybrid Systems: Computation and Control, 8th International Workshop, HSCC 2005, Zurich, Switzerland, March 9–11, 2005, Proceedings*, volume 3414 of *Lecture Notes in Computer Science*, pp. 25–53. Springer, 2005.

65. J. Eker, J.W. Janneck, E.A. Lee, J. Liu, X. Liu, J. Ludvig, S. Neuendorffer, S. Sachs, and Y. Xiong. Taming heterogeneity—The ptolemy approach. *Proceedings of the IEEE*, 91(1):127–144, January 2003.

66. J. Davis, R. Galicia, M. Goel, C. Hylands, E.A. Lee, J. Liu, X. Liu et al. *Ptolemy-II: Heterogeneous Concurrent Modeling and Design in Java*. Technical Memorandum UCB/ERL No. M99/40, University of California at Berkeley, July 1999.

67. P.J. Mosterman, J. Sztipanovits, and S. Engell. Computer-automated multiparadigm modeling in control systems technology. *IEEE Transactions on Control Systems Technology*, 12(2):223–234, 2004.

CHAPTER 2

Pathways to Dependable Cyber-Physical Systems Engineering

John Fitzgerald, Claire Ingram, and Tom McCutcheon

CONTENTS

C YBER-PHYSICAL SYSTEMS (CPS) have become the subject of a growing body of research, alongside an increasing awareness of the challenges posed by their technological and semantic foundations and the impact of their deployment in a wide range of new applications. In such a large emergent field, there is a need to consider a wide variety of potential areas for research and development. We suggest that research road maps that can identify possible relevant research directions and their impacts will prove to be important for CPS research across Europe, in order to facilitate coordination between the disparate disciplines and fields touched by CPS engineering, and allowing the maximum research impact to be achieved.

There are many approaches to the development of road maps for both research and technology development. In this chapter, we review road-mapping activities relevant to CPS research to date, identifying stepping-stones on the way to methods and tools for the design of CPSs that merit reliance being placed on their correct functioning.

2.1 INTRODUCTION

Cyber-physical systems (CPS) have been attracting increasing levels of interest in the research community in recent years. Definitions for the term *cyber-physical system* vary, however. We provide here some simple explanations intended to bring clarity to the discussion in this chapter.

2.1.1 Some Definitions

Firstly, we suggest that a *system* can usefully be defined as a collection of interacting elements, organized in order to achieve some explicit purpose [1]. A CPS is a system in which some elements are computational and some elements involve interactions with the physical environment [2–4]. This goes beyond the traditional definition of a single embedded system, which, although it also incorporates an element of interaction with the real world, has a much more limited scale. We would suggest that an embedded system is traditionally conceived at the level of a single device, whereas a CPS incorporates multiple connected systems, potentially at the scale of a total enterprise, or even across multiple enterprises:

Cyber-Physical Systems enable a variety of novel functions, services and features that far exceed the current capabilities of embedded systems with controlled behaviour. Powerful Cyber-Physical Systems are able to directly register the situation of their current distributed application and environment, interactively influence it in conjunction with their users and make targeted changes to their own behaviour in accordance with the current situation [5].

Typically, CPSs are large scale, cross-domain, and can include diverse constituent systems, such as networked embedded systems, control systems, communication networks, and physical systems, involving expensive deployment costs and complex network interactions [3,6]. A CPS couples together (potentially many) sensors of different types with significant computational power, and as a result a CPS is able to gather information about complex environmental events, analyze it, and take some intelligent action based on that analysis. This goes further than embedded systems, that conventionally do not possess significant computational ability that can be used for developing knowledge sources in this way. In contrast, a CPS can, for example, monitor some manufacturing or industrial process continually, automatically making precise adjustments to reduce wastage and improve efficiency. A CPS can go even further, identifying unexpected situations or changing requirements and adapting dynamically, allowing engineers to build flexible, adaptable systems. This functionality can be deployed to a wide range of application domains to improve flexibility, performance, and efficiency.

We consider in this chapter road maps suitable for dependable systems. A *dependable* system is one on which reliance may justifiably be placed [7]. Assuring ourselves that a given CPS is dependable or trustworthy therefore implies the need for some evidence to be assembled, with which to justify this claim. Because of the nature of a CPS, such evidence will include analysis of computational elements as well as physical elements (and potentially other types of systems such as human systems) and how they interact.

2.1.2 Chapter Structure

In this chapter we consider recent and ongoing activities relevant to the development of research road maps and strategies in CPS, with the intention of delivering an introduction to key research road maps for CPS. We introduce some road-mapping approaches and concepts in the next section, followed by discussion of road maps themselves in Sections 2.3 through 2.5.

We are unable to provide a complete overview of all road-mapping activities that are relevant for CPS engineering due to space limitations. We have chosen therefore to narrow our focus to examine primarily European research projects studying CPS and related subjects. This allows us to cover a rich range of relevant material. There is recognition within Europe that CPS research benefits from strategies to consciously identify collaboration opportunities across national boundaries and across disciplines, and as a result research projects have been funded by the European Commission in order to achieve just this task. Such projects have generally made their recommendations and output road maps publicly available, and we have drawn on these in this chapter. This does not imply that there is not a rich and important range of research outputs in CPS engineering published in other regions.

A number of research road maps relevant to CPS engineering are currently under development at time of writing; we consider here the small number of CPS-specific road maps and research strategies that have been published to date. There is also a rich collection of relevant road maps from a variety of recently concluded projects that have concentrated on technologies and disciplines connected to CPSs. This is particularly true of systems of systems (SoS), an area that has received much research interest in recent years. Projects studying aspects of SoS development and modeling have already recently delivered road maps and research strategies.

Finally, we also survey some research recommendations and reports that, while not actually road maps *per se*, have already had some influence on research directions.

2.2 ROAD MAPPING APPROACHES

Road mapping is a technique for planning research and development activities in a variety of organizations, including commercial industries, public sector government bodies, research funders, and universities, among others. Road mapping is often used for identifying key technologies and pace-setting, or for planning application systems in addition to planning outputs such as products [8]. Road maps may also be used to coordinate research activities (including research taking place outside the current organization) or for planning marketing and competitive strategies [8].

The techniques and processes used to generate the road map, and the format for its presentation, can vary significantly, but in general the final road map that is output at the end of the process is a plan that identifies those subjects that should receive research and development attention.

2.2.1 Road Mapping Concepts

Recommendations for research and development activities are often associated with a timescale on a road map—for example, activities can be prioritized for short-term, medium-term, or long-term investments. Recommendations in the final road map are usually also accompanied by a structured rationale that links recommended research activities to their likely useful outcomes. An industrial manufacturer might use road mapping to plan possible future products by predicting what future gaps in the market might be exploitable, what products will be needed to fill those gaps, and therefore what research will be needed to develop those products. On the other hand, a research funder might use similar road mapping techniques to identify future societal needs and then identify research activities that will be required to deal with these future needs.

This structured rationale, which links research to likely outputs, is the major strength of a road map this structure allows decision-makers to identify the likely future benefits and opportunities that will be enabled by current research activities, and the likely future investments that will be necessary to recoup the costs.

2.2.2 Customizing the Road Mapping Process

There are many effective ways to develop a road map and each organization should customize the road mapping process to ensure that it adequately fits that organization's needs and resources [9]. Organizations employing road mapping as a planning activity need to ensure that road maps are appropriately customized to suit specific domain requirements and technological and socio-technical challenges.

As an example of a structured road mapping approach, we describe briefly here the "Fast-Start" approach [10,11]. This documented approach advocates a recognizable road map format with five "layers" of information:

1. Target markets that have been identified for the future

2. Products that can be aimed at these target markets

3. Enabling technologies that must be developed in order to implement these products

4. Research and development programs that will be necessary in order to develop these enabling technologies

5. Funding programs that will be necessary to run these research and development programs [10]

In the list above, "technology" can be interpreted as process, technique, or knowledge base [12].

Although this approach is documented as a specific recommended process (the "Fast-Start" approach), many other road mapping approaches are in use, often resulting in similar final road map structures.

A typical road mapping activity should span multiple sessions (e.g., in a series of half-day and full-day workshops) and often relies on gathering a wide variety of inputs from domain experts and practitioners, perhaps using "brainstorming" style interactive sessions that are structured to a greater or lesser degree. Multiple sessions will ensure that there are opportunities both to generate ideas, as well as opportunities in later sessions to validate, reflect upon, and structure those ideas. Naturally, the quality of a road map depends to a certain extent on gathering a wide range of inputs from people with varied perspectives, who have expertise in the desired areas.

2.2.3 Road Mapping Methodologies

In terms of perspective, there are two major types of road mapping methodology [13].

A "technology push" road mapping approach begins by examining the current state of the art and building on this to consider future technologies. For example, a research funder may begin by surveying the state of the art in a particular technology, and then use road mapping techniques to identify what further technologies can be developed based on these, and how these could be employed in a specific domain.

In contrast, a "requirements-pull" road mapping approach begins by examining some supposed future requirements, and working backward to identify technologies that will be needed. For example, a technology manufacturer may identify some future market requirements that should be satisfied by a new product, and then work backward to identify the features that will be needed, and the enabling technologies that need to be developed, in order to make the new product a reality.

2.3 SURVEYING EXISTING ROAD MAPS

CPSs are currently the subject of significant research attention, and a number of public sector bodies and research bodies have begun to consider key research priorities in this area. A small number of organizations have already published some indications of priorities or their own road maps, but for many organizations CPS research planning is still ongoing. We briefly survey here extant road maps examining technologies relevant for trustworthy CPS.

Because there are relatively few road maps that have already been published explicitly addressing CPS, we also expand our survey to take into account road maps from related areas, such as systems of systems (SoS), a field that has some common overlap with CPS. These two fields see many common challenges, such as dynamic reconfiguration, substantial heterogeneity in modeling approaches, and coping with distribution and communications [2].

2.3.1 Road Maps and Research Priorities in SoS

An SoS is generally regarded as a system composed of constituent systems (CSs), each of which is independently owned or operated in its own right: "an SoS is an integration of a finite number of constituent systems which are independent and operable, and which are networked together for a period of time to achieve a certain higher goal" [14]. Another well-known definition [15] describes five properties associated with SoSs: operational independence; managerial independence; evolutionary development; emergence (the SoS performs functions that do not reside in any individual CS); and distribution.

The general concepts of independence, continuous evolution, and emergence have been widely acknowledged, and variations of these properties feature in subsequent definitions of SoS (e.g., Fisher [16], Abbott [17], Boardman and Sauser [18], and Baldwin and Sauser [19]). The latter two also reference heterogeneity typically seen between CSs within an SoS and the fact that a CS must accept the need to adapt.

2.3.2 Relationship between CPS and SoS

The challenges addressed by SoSs thus share some overlap with the concerns of CPS engineers. The EU-funded AMADEOS* project, for example, identifies emergence, interoperation, representation of time, evolution, dependability and security, governance, and quality metrics as key areas of CPS and SoS engineering [20]. Much could be written about the relationship between CPSs and SoSs, but we present some simple, brief comparisons of SoSs and CPSs in Table 2.1; this is intended to provide a brief indications only and is not intended as a comprehensive survey of the two fields.

---

* http://amadeos-project.eu/.

TABLE 2.1 Comparison of CPS and SoS

Cyber-Physical System	System of Systems
Definition	
A system incorporating computation and interaction with physical environment, allowing flexibility and complex analysis of surroundings	A system composed of constituents, which are also independent systems in and complex analysis of surroundings their own right
Multiple Domains	
Diverse selection of systems and domains	Diverse selection of systems and domains
Global Behavior	
Global behavior is composed from the behaviors of constituents, which may or may not be independent	Global behavior is from the behaviors of independent constituent systems
Distribution	
Systems are distributed, including devices that interact with physical environment, networked together	Systems are distributed as they are independently owned and operated
Independence of Constituents	
Constituents may be owned and operated independently, or may be directed/owned by one organization	Constituents are owned and operated independently, and have their own goals
Evolution	
Evolves continuously	Evolves continuously
Physical Environment	
Includes interaction with physical environment, via sensors and/or actuators	May or may not include interaction with physical environment
Heterogeneity	
Heterogeneity of constituents is a major design challenge, particularly where design crosses disciplines—for example, designing hardware alongside software	Heterogeneity is commonly found in SoSs, particularly where constituents are drawn from different domains

Both CPSs and SoSs may present challenges relating to

- Modeling a diverse selection of systems and domains

- Designing for emergence or global behavior

- Distribution

- Coping with the fact that many of the constituents in a CPS are independently owned and/or operated

Although these concerns do not apply universally to all CPSs, a subset of CPSs clearly could be defined as SoSs, and a subset of SoSs can legitimately be regarded as CPSs. For this reason, we feel it is appropriate to consider some of the outputs from recent SoS research here, since many aspects of SoS engineering grapple with the same challenges as CPS engineering, and because many high-quality road mapping reports have been produced in recent years from across the SoS community.

2.3.3 Modeling and Simulation in SoS: The COMPASS Project

The COMPASS[*] project was an international EU-funded research consortium that ran from 2011 to 2014. COMPASS primarily developed tools and techniques for modeling and simulation of systems of systems, but also produced a road map [21] document to accompany the project's final deliverables. The road map specifically considered issues in modeling and simulation of SoSs. The COMPASS road map identified 11 research goals related to modeling and simulation; they are

1. *Heterogeneity and tool interoperability:* Tools are needed to provide support for heterogeneous cross-domain and cross-discipline analysis techniques, and integration of tools across sectors.

2. *Accessibility of verification methods:* Tools and methods are needed to make formal approaches available and accessible to engineers from a range of disciplines.

3. *SoS architectures:* Architectures suitable for SoSs and systems of interconnected devices need further development [22,23]. There is a clear need for improved architectural guidance for SoSs [24,25].

4. *Verification of emergent behavior:* Tools and techniques are needed to verify the emergent behavior and nonfunctional properties of an SoS (COMPASS defines "emergent" behavior as the behavior that is produced by collaborations of CSs, and cannot be reproduced by one CS working alone [21]).

5. *Fault tolerance:* Methods and tools to support fault modeling across an SoS are needed.

6. *Dynamic configuration:* Tools and techniques to support reasoning about dynamic reconfiguration and/or autonomous decision-making are necessary for SoSs. Many SoSs have significant requirements for fault tolerance, resilience, and robust communications, but at the same time inhabit challenging physical environments where connectivity problems or lack of full disclosure can lead to the sudden withdrawal of a CS, or degraded service levels [26].

7. *Process mobility:* Modeling and simulation techniques are needed for engineering and reasoning about emergent behavior in SoSs that exhibit mobility. Both processes and processing in SoSs that orient around embedded devices may be highly distributed [23].

---

[*] http://thecompassclub.org.

8. *Security:* There is a need for guidance on ensuring privacy and security in wirelessly connected systems. There must be frameworks and techniques supported by modeling tools for ensuring that data remains secured even when handled, gathered, and processed by independent constituent systems.

9. *Performance optimization:* Methods and guidelines for verifying optimization strategies will be needed in the future. A key requirement for most SoSs is to achieve optimized performance. Furthermore, many SoSs operate in domains with highly complex environments where an "optimal" level of SoS service is required, but where the definition of "optimal" can vary substantially for different stakeholders.

10. *Stochastic behaviors:* Tools and techniques for modeling dependable human-centred SoSs of the future need to model stochastic events and human behavior.

11. *Evolution:* Tools and methods for reasoning about and analyzing SoSs through a series of configurations or versions over a period of time are needed. Many researchers have noted that SoSs constantly evolve (e.g., [15,17]). This poses challenges relating to traceability and reasoning about correctness. There is a need to model and plan migration strategies and possibilities for a phased transition [21].

COMPASS key research priorities are relevant for a consideration of research directions for trustworthy CPS, in part because of the overlap of CPS and SoS research fields, and in part because modeling and simulation are key tools relied on in engineering and computer science disciplines to ensure correctness and dependability.

2.3.4 Fault Tolerance in an SoS

One of the research priorities identified by COMPASS in the list above is the problem of guaranteeing fault tolerant behavior in an SoS. COMPASS is not alone in emphasizing dependability issues as key challenges for SoSs; for example, the EU-funded AMADEOS project[*] also calls for attention to be paid in this area:

> Whereas the occurrence of a fault is considered an exceptional event in classical system designs, in an SoS design the occurrence of a fault must be regarded as a normal event, since any CS may decide to leave the SoS for whatever reason [20].

The road map produced by COMPASS emphasized the difficulty of delivering fault tolerance in an SoS:

> Techniques for reasoning about the identification of faults, errors and failures from the point of view of the SoS are needed (noting that a system failure at the level of a CS becomes a fault at the level of the SoS). Detection of faulty behaviour and the design and implementation of recovery behaviour can involve the collaboration of

---

[*] http://amadeos-project.eu.

multiple CSs, making fault-tolerance a global emergent property that can be difficult to verify for correctness [21].

While the text here refers to the particular situation of an SoS, which is always comprised of independent constituent systems, many CPSs will also experience similar difficulties for verifying fault tolerance in cases where global behavior is a composition of behaviors from separate devices or systems.

The reference to system/SoS faults and failures draws upon a well-known "dependability taxonomy" [7] that identifies important dependability concepts, and is widely employed in the field of dependable software. Reasoning about properties or behaviors within an SoS is complicated by the multiple views, all important and valid, that apply at different levels of granularity within an SoS. This arises because we can apply our analysis to the global level to consider the SoS's globally emergent behavior, and we can also—equally importantly—apply analysis to individual single systems within the SoS.

In traditional dependability taxonomy [5], a "fault" is described as being located within a component part of a system; a fault may eventually result in a "failure," that is visible at the external boundary of the system [5]. The COMPASS project distinguishes between a failure of a constituent system; when viewed at the constituent system level, this is a failure, but viewed from the point of view of the SoS, this is a fault [27]. Reasoning about fault tolerance in distributed, heterogeneous and independent systems has been studied previously (for example, see [27–30]), although more can be done in this area.

The COMPASS road map recommends additional work to develop further the range of tools and methods available to support reasoning about fault tolerance in this area, as well as tools and methods for incorporating stochastic techniques into SoS modeling approaches [21]. This would allow for the modeling of human behavior (SoSs typically rely on substantial inputs or stimuli from humans) as well as the ability to identify, model, and plan for problems. As a project that concentrated on formal modeling practices that are widely used for development of trustworthy and dependable systems, many of the future research directions called for by COMPASS are important for further developing our ability to build trustworthy SoSs and CPSs.

2.3.5 The Road2SoS Project

COMPASS is not the only project in the field of SoS engineering that has published road maps or recommendations for future research directions relevant for CPS engineering. The Road2SoS project* was an EU-funded project that concentrated on development of road maps and recommendations for future research into SoSs; recommendations were made after surveys that crossed a number of SoS domains [31]. Priority research themes identified by Road2SoS include (among others) [31]:

---

* http://road2sos-project.eu/.

- The need for improved communications infrastructures that can support high speed, high capacity data exchange—including novel and wireless technologies, operating under low power requirements and with potentially large numbers of nodes.

- Interoperability between heterogeneous systems. Like COMPASS, Road2SoS also identifies diverse ranges of constituent systems as a challenge.

- Smart sensors and data fusion techniques, which can improve the apparent performance of inexpensive, low-energy sensors and deliver more dependable information.

- Improvements in big data handling and processing. This is important in an SoS context, since an SoS is capable of generating large quantities of data in real time (or close to real time), which often is needed in order to make decisions about operations and performance.

- Improvements in the ability to conduct forecasting activities and present the resulting information in an understandable way.

- Autonomous decision-making, allowing limitation of the amount of human intervention required with safe decision-making, reasoning about an SoS environment, and coping with uncertainty.

- Platforms for coordinated planning and decision-making.

- Modeling and simulation tools supporting aspects of SoS engineering and design.

- Understanding emergent behavior.

- Elicitation of appropriate metrics and measurement of SoSs for assessing operating performance.

- Elicitation of SoS architectural patterns, to reduce effort and provide transparency.

- Engineering for resilience, adaptability, and flexibility, including techniques to deliver flexibility at runtime.

- Business models reflecting multiple ownership and distributed governance.

2.3.6 The T-AREA-SOS Project

The T-AREA-SOS[*] EU-funded project considered a range of SoS domains under the general aim of developing a collaborative transatlantic strategic research agenda.

The final agenda comprised 12 important SoS themes, including [25]

- Characterization and description of SoS, necessary in order to ensure clarity and support the move for businesses to consider an SoS engineering environment.

---

[*] https://www.tareasos.eu/.

- *Theoretical foundations for SoS*: This includes selecting and identifying a generally agreed framework for SoS theories that supports interoperability; identifying fundamental principles for SoS at varying levels of abstraction.

- *Emergence*: As with other projects (such as COMPASS [21] and Road2SoS [31]), T-AREA-SoS identifies emergence as a key challenge. The causes of emergence need to be better understood and strategies for identifying and coping with both expected and unexpected emergent behavior are needed.

- *Multilevel modeling of SoS*: Techniques are needed that span disciplines and can cope with varying levels of abstraction needed for the architectural level.

- *Measurement and metrics for SoS*: There is a need for predictive measures and use of real-time measurements, including domain-specific measurements and identification of metrics that can span domains.

- *Evaluation of SoS*: Tools and techniques are needed to allow SoSs to be assessed against expected behavior.

- *Definition and evolution of SoS architecture*: Architecture for SoSs is poorly understood; techniques are needed to improve the SoS architectural approach and reduce its cost.

- *Prototyping SoS*: Prototyping is a standard way to reduce risk and optimize design, but the evolutionary nature of SoSs complicates this since SoSs are rarely "blank-slate" developments. Methods are needed to reduce risk in this situation.

- *Trade-offs in SoS*: Evaluating decisions regarding selection and configuration of constituent systems is difficult for an SoS due to the complexity of the design space. Tools to support these types of decisions are needed.

- *Security in SoS*: The "physical, ethical and cyber aspects of safety and security of the SoS, its people, its end-users, and its social and physical environment" need to be addressed [25].

- *Human aspects of SoS*: Human error and unpredictable interactions with SoSs are a significant factor in SoS operation. This requires an ability to understand human psychology, such as how humans gather information from an SoS and make decisions, as well as means for incorporating predictions on human behaviors.

- *Energy efficient SoS*: There is a need to optimize SoSs and minimize their effect on the environment.

In general, there is substantial overlap in the recommendations produced by these SoS engineering research projects, with dependability forming a major theme, and all calling for further work to improve security and/or privacy, improve our ability to reason about reliable emergent behavior, assess SoS performance, and process large amounts of data. We will return to these themes in Section 2.6.

2.3.7 Road Maps and Research Priorities in CPS

In the previous section we considered some road maps published in the field of SoS. Here we examine road maps specifically addressing CPS topics.

2.3.8 The CyPhERS Project

The EU-funded CyPhERS project[*] collected an analysis of various SoS domains into an overall SoS road map and strategy report [32]. The project adopted an holistic view of CPS engineering, considering a wide range of views, from market structure to human behavior and engineering design. We concentrate here on discussing challenges identified by CyPhERS in the development of dependable and trustworthy CPSs, which included the following [32]:

- *Multiparadigm challenges*: This particular challenge is repeated elsewhere in other road maps (e.g., [21]). CPSs need to cope with a diverse range of systems, each bringing a diverse range of concepts, which can complicate attempts to reason about the whole. CyPhERS identifies cross-disciplinary concepts; cross-domain concepts (e.g., use of models or partial models at varying levels of detail suitable for different domain requirements); and theories and techniques for creating models to capture aspects of human behavior. "Foundational theories" are needed to support cross-domain and cross-discipline CPS engineering: the problem is that "individual theories are not combined in a common integrating theory" [32]. Such a theory could express the links between separate domain- or discipline-specific reasoning, facilitating reasoning that spans process chains.

- Challenges of complexity, resulting from size and heterogeneity as well as reconfiguring and optimizing efforts. Modeling must cope with independent constituent parts that are not motivated to make full disclosures, as well as ensuring concepts such as dependability and privacy even when CPSs are capable of self-configuring dynamically, or self-optimizing dynamically. Approaches for dealing with uncertainty without comprising important nonfunctional properties like this are needed.

- *Challenges to competence*: This is important because CPS engineering typically crosses domains and disciplines, whereas many advanced training courses at the undergraduate and postgraduate level emphasize separated disciplines. There is a lack of platforms for sharing knowledge on CPSs.

Having elucidated these challenges, CyPhERS builds a collection of recommendations for future research, including the following, which we suggest are pertinent to the development of dependable CPSs [32]:

- More research in key scientific fields, including big data, knowledge sharing, CPS systems, and physical technologies

---

[*] http://www.cyphers.eu/.

- More research into human–machine interaction

- Efforts designed to encourage and support cross-disciplinary research

- Support for projects that can mature the emergent technologies designed to support CPS engineers

- Recommending appropriate reference platforms to encourage interoperability

- Harmonizing interoperability standards

- Creating systems-levels design methodologies

Trustworthiness is also a key pillar of CyPhERS recommendations [32], which is relevant here. Recommendations for this aspect of CPS engineering include: improving the security of existing infrastructures, including the internet; creating legal frameworks to protect data ownership; and prioritizing research programs that facilitate automated operations in CPSs (such as dynamic reconfiguration) and live updates that do not compromise dependability [32].

2.3.9 The CPSoS Project

CPSoS* (Cyber-Physical Systems of Systems) is a project funded by the European Union that aims to foster sharing of best practices and state of the art for a number of communities and sectors interested in CPS and systems of systems [33]. At time of writing this project was engaged in establishing the state of the art and future challenges in cyber-physical SoSs, noting that many domains had already encountered, and were trying to tackle, key challenges in the area.

The CPSoS project emphasizes the need to integrate modeling and simulation techniques and engineering expertise from a variety of domains. This includes continuous-time modeling as well as discrete-event modeling (continuous-time modeling approaches are traditionally used to model interactions with the physical environment, whereas discrete-event modeling approaches are sometimes used to model software systems) [33]. CPSoS is not the only project to suggest this as a key challenge—for example, the COMPASS project also noticed this [21].

CPSoS identifies some common future research challenges across a range of varying cyber-physical/SoS domains, as follows [33]:

1. Engineering support for the design-operation continuum of CPSoS. This requires frameworks and modeling and simulation tools that can: support collaborations between separate organizations where not all information is revealed; support asynchronous development cycles; and can cope with the fact that SoS development projects rarely begin with a "blank slate": engineering methodologies need to take into account that systems are ready deployed and operational while updates and changes are being designed. Models from different domains need to be managed and integrated,

---

* http://www.cpsos.eu/.

and engineers need modeling tools to evaluate optimization strategies that straddle collections of complex and heterogeneous SoSs.

2. Distributed, reliable, and efficient management of CPSoSs. This task is made more complex as many CPSoSs typically see responsibility for management shared by distributed constituent systems; tasks that were previously carried out in a centralized way are now carried out by subsets of systems with partial local autonomy. This situation means that there is considerable uncertainty in the running and design of SoS-level (i.e., global) behavior, so SoS design tools need to be able to incorporate stochastic reasoning [33].

3. Cognitive CPSoSs. This challenge addresses the problem of building a systems-wide view, when management mechanisms and processes are distributed and the quantities of data generated are very large.

As is the case with COMPASS, researchers working with CPSoS note that the ability to perform verification at the SoS level is an important research challenge facing developers of engineering tools [33]. Also like COMPASS, CPSoS researchers see SoS-level fault analysis and stochastic reasoning as key components for SoS modeling; the COMPASS road map identifies stochastic modeling as a tool for analyzing human behavior and fault tolerance [21] while CPSoS reports add that stochastic modeling is needed to cope with the considerable uncertainty imposed by an SoS's distributed authority and independent constituents [33].

CPSoS researchers also consider the importance and challenges of fault tolerance in an SoS (or CPSoS) environment, noting that because of the complexity and scale of these systems, fault tolerance is the expected norm [33]. Advanced monitoring capabilities presented by increased use of sensor technology and the ability to process large quantities of data in real time will become important for fault tolerance and dependability of these systems. Together these capabilities will enable detection of system-level faults before they become SoS-level faults that can propagate from constituent to constituent and progress to an SoS-level failure, paving the way toward fault prevention behaviors at the enterprise level.

CPSoS reports note that engineering tools (e.g., modeling and simulation tools) suitable for SoSs and/or CPSs may result in considerable investment of effort—perhaps more than is justified by the visible returns [33]. This may result in models only coarsely describing the behavior of constituent systems; more effort must be directed into the processing of large and complex data, efficient computing, and handling and presentation of highly heterogeneous models, as well as model management and development of control strategies and algorithms [33]. This is a challenge also flagged by a number of other SoS-related research projects.

2.4 OTHER RESEARCH RECOMMENDATIONS

The previous section described the recommendations of road maps produced by structured European research projects. These are complemented by published research strategies that are not structured in the same way as a traditional road map, but which are nevertheless still useful pointers for future research planning.

One such report, which has been influential within Europe, is the report Agenda Cyber-Physical Systems [5] published by the German National Academy of Science and Engineering (acatech). This report introduces CPS and discusses potential contributions to be made by CPS industries to the German economy as well as the potential impact on German society. Although the report concentrates on Germany, of course many of the report's recommendations and observations can be generalized to other national and international engineering industries and social concerns.

The social and economic impact of CPS adoption is also illustrated through the use of scenarios depicting facets of everyday life for citizens and businesspeople in Germany; this includes, for example, integrated autonomous vehicles that provide high capacity, high speed, carefully distributed traffic routing; dynamic travel recommendations based on current road conditions; and smart systems installed around the home to provide automatic, nonintrusive support for elderly persons living alone.

While outlining the potential impacts of CPS adoption, the report identifies challenges in a number of areas, including technological challenges as well as sociological and economic challenges that currently hinder widespread adoption of effective CPSs. These include technological challenges such as

- Domain models are needed to capture and reason about user behavior, including methods for recognizing user intentions, strategies to govern interactions between humans and machines, and detecting conflicts and deploying resolutions.

- Techniques and models are needed for providing support for self-analysis of CPS.

- There is a need for "multimodal interfaces, enhanced sensor technologies and networks, processing and semantic aggregation of large data sets in real time and semantic annotation of the data that have been captured and made available" [5].

- Methods and models are also needed to support reasoning about cooperative behaviors, including autonomously cooperating CPS systems or humans and CPS systems. This will require techniques for: establishing and validating complex environmental data and high-speed processing and communications; complex requirements models; techniques for reasoning about quality of service, interoperability, architecture, and composition; and techniques for reasoning about risk, among many others [5].

- Methods to support development of systems that can learn and plan, organize themselves, and identify and design strategies for cooperating.

- Methods are also needed to support development of systems that operate in environments with considerable uncertainty, and with uncertain knowledge available to them.

Modeling techniques that are currently separated into different fields need to be integrated, so that, for example, techniques suitable for reasoning about real-time control requirements associated with the physical world are integrated with models that capture information about human behaviors, system goals and requirements, contexts, and so on. This includes models from a wide variety of different fields from engineering, mathematics, computer science, science, and social sciences.

There is a need to ensure that CPSs are socially acceptable and accepted; this requires methods and techniques to take into account privacy requirements—supported, if necessary, by new legal frameworks or self-regulation guidelines—as well as security protection, dependability, reliability, and so on.

2.5 MULTIDISCIPLINARITY IN CPS

Cyber-Physical Systems are intrinsically multidisciplinary. This is well reflected from an engineering standpoint as described throughout this chapter. However, CPSs exist to be used by and for humans. They must therefore be studied in a socio-technical context. While this need is expressed in the CyPhERS program and in a wide range of documents [5,13,21,25,31], there is still a lack of research in this area.

Some examples of the need for further interdisciplinary research are presented below.

2.5.1 Socio-Technical Aspects of Safety and Security

Unlike more bounded systems, large-scale CPS may be compromised by human interactions. First, the capricious nature of the human user is difficult to anticipate and constrain. Second, it may not be possible to bound TCPS in a rational way. The automotive industry provides a good exemplar. While CPS vehicles may behave safely according to their specification, other road users will behave unpredictably from time to time. Such users will include drivers, pedestrians, cyclists, and others. This complexity, combined with deliberate rule breaking, specialist requirements, environmental factors, and unplanned outages, dramatically complicates the road situation and could easily undermine required levels of safety.

2.5.2 Data

As mentioned previously in this chapter, there is also a research need to consider the very large quantities of data produced and consumed by CPSs in their day-to-day operation. An example of this would be in the smart cities sector where interlocking CPSs would enable a great deal of information to be amassed about individuals and groups, This could pose a serious privacy threat. Further, the potential backlash against such a threat could impact on the wider acceptability (and uptake) of a range of CPSs and other systems. There is a limited amount of research being sponsored into the Privacy and Trust aspects of CPS by EPSRC[*] and the UK Digital Catapult future cities initiative,[†] and the nascent Alan Turing Institute[‡] is considering such issues for a major research thrust, but these are just initial steps in a very large topic.

---

[*] https://www.epsrc.ac.uk/funding/calls/compatriotsresearchhub.
[†] https://futurecities.catapult.org.uk.
[‡] https://turing.ac.uk.

2.5.3 Legal Aspects

Currently, advances in CPS technologies and services are outstripping existing legal frameworks. This trend is common to many new technologies, seems set to increase, and is augmented by the complex relationship among CPS, CPS-generated data, and the mixing of CPS systems with traditional, human-centric systems, often legacy. The use of CPS in a mixed CPS/traditional environment, especially if a degree of autonomous control is involved, raises a number of currently unexplored questions. For example, in an automotive accident involving a CPS, where does responsibility lie? Traditionally if a vehicle is involved then the person in charge is largely liable, but if this person is merely using a CPS service then the service provider, infrastructure owner, or some other individual or organization may be liable. There is currently little understanding and less consistency in such areas, with a concomitant need for further study.

2.5.4 Ethical and Cultural Aspects

While CPSs under human control may pose few societal challenges, autonomous or semi-autonomous systems could lead to a wide range of concerns. These could range from safety concerns (autonomous vehicles) through privacy (ubiquitous sensing and data management) to challenging the role of humans in society (autonomous decision-making in CPS and mixed cyber-physical/human systems). These challenges would be culturally dependent and have real impact on individuals and societies. Accordingly, some initial research into culture and CPS would seem appropriate.

2.6 CONCLUSIONS

In this section we survey the road maps and research strategies discussed in this chapter, identifying some common consensus topics as well as a few potential gaps in their recommendations. Many of the projects and strategies addressing CPS and the related field of SoS share a few key concerns in their recommendations. In Table 2.2 we present a tabularized indication of key research recommendations from previous relevant road maps, showing where projects overlap, and in this section we identify and briefly summarize some key recommendations shared by multiple road maps.

2.6.1 Heterogeneity and Tool Interoperability

The need to deal with heterogeneity and/or varying levels of abstraction in modeling approaches for CPS and SoS has been identified by many CPS-related road maps (e.g., [5,21,25,31–33]). Heterogeneity is a key problem for CPS, which by their very nature span several engineering disciplines. Although an ability to incorporate modeling outputs from all of the engineering teams in CPS would be enormously advantageous, this is not a trivial problem. Each engineering discipline (as well as human and business modeling approaches) has developed its own notations and tools, that typically capture only the information needed and leave no room for capturing information that is only of interest to other disciplines. This means that, for example, while the discrete-event notations employed for reasoning about software excel at analysis of the concepts and data needed for software, such notations

TABLE 2.2   Summary of Road Map Recommendations for CPS and SoS

Recommendation	COMPASS	Road2SoS	T-AREA-SOS	CyPhERS	CPSoS
Heterogeneity/cross-disciplinary development and tool interoperability	*	*	*	*	*
Tool support, accessible verification	*	*		*	*
SoS architectures and patterns	*	*	*		
Understanding and engineering emergence	*	*	*		
Fault tolerance, resilience, adaptability	*	*			*
Process mobility	*				
Security	*		*		
Performance optimization	*				*
Human aspects of SoSs	*		*	*	
Continuous evolution and design-operation continuum	*				*
Improved communications infrastructures for high speed, high-capacity data exchange		*		*	
Smart sensors and data fusion		*			
Big data and scalability		*		*	*
Forecasting and visualization of results		*			
Autonomous decision-making		*			
Metrics and measurement, evaluation and monitoring		*	*		
Business models		*			
Characterization and description of SoS			*		
Theoretical foundations for SoS			*		
Understanding trade-offs			*		
Energy efficient SoSs			*		
Systems-levels design methodologies				*	*
Competence and skills				*	
Distributed control and coordinated planning		*			*
Efficient processing					*

frequently do not provide any support for the real-time functionality that is needed to reason about aspects of real-world interactions.

2.6.2 Engineering Emergent (Global) Behavior and the Need for Tool Support

The problems of coping with emergent behavior are familiar to both SoS and CPS communities, and are referred to in different ways in many of the road maps we have mentioned here. Emergence is behavior that is produced when separate systems in an SoS/CPS collaborate. The emergent behavior cannot be reproduced by the constituent systems acting in isolation. Sometimes the emergent behavior can be undesired, unexpected behavior against which we want to guard.[*]

---

[*] In fact, in some disciplines, the term emergence is reserved solely for behaviors that are unexpected.

There is a need to be able to reason about emergent behaviors, firstly in order to ensure that the global CPS behavior can be relied upon to produce the functionality we expect, and secondly that no additional behaviors are produced. However, this is not a trivial problem. Because CPSs are often implemented by many independently operated systems, CPS engineers are often required to design with only limited information available about the constituent parts. It is difficult to apply conventional analysis techniques used for reasoning about system behavior to the analysis of global behavior which is a composition of separate behaviors contributed by a variety of participants in a CPS. Analysis tools and techniques used in different disciplines for ensuring reliable system behavior typically are constrained to representing the concepts and aspects important in their respective disciplines (for example, analysis techniques used to ensure verifiable software behavior do not capture real-time behaviors well). And the challenge of analyzing all possible interactions is typically computationally prohibitive, in any case.

There is a need for methods and tools that can provide some ability to analyze these behaviors in the face of incomplete system information, hidden internal dependencies, and considerable environmental uncertainty.

2.6.3 Resilience, Fault Tolerance, and Adaptability

CPSs need dynamic methods at runtime to deal with uncertainty, including techniques that allow the CPS to continue delivering key services while coping with changing situations flexibly. This can be a useful tool in the CPS designer's toolkit, one of many that can be deployed to help achieve resilience or fault tolerance. It can also lead to CPSs that can deliver improved quality, consistency, or efficiency as they adapt swiftly to new requirements or scenarios.

Adaptability and dynamic reconfiguration are mentioned by the COMPASS road map [21], whereas flexibility is mentioned by Road2SoS [31] and CyPhERS [32], for example. This, in turn, requires significant research into enabling technologies, such as frameworks and techniques that will allow CPSs to self-monitor [5] and make decisions (including autonomously [31]) to improve performance that is mentioned by further road maps. Self-monitoring is particularly pertinent for CPSs, since a major feature of a CPS is its ability to collect environmental data and leverage some computational processing ability in order to deliver intelligent behavior, including adaptability for coping with volatile requirements or operational environments, self-monitoring and self-healing behaviors for implementing aspects of resilience and security, and coping with challenges such as continual evolution.

2.6.4 Big Data

Challenges will be posed by big data—for example, in terms of our ability to cope with the expected quantities of data that CPS can generate, transmit it sufficiently quickly, and use it to produce predictions or simulations in adequate, useful timescales [5,21,31–33]. An inability to cope with big data in these ways will limit the potential for some future CPSs (e.g., transportation systems that gather data from many thousands of nodes will need to be able to exchange this data sufficiently quickly and process it fast enough to produce helpful recommendations in soft real time; likewise smart grids may be required to cope with data from many thousands of nodes).

2.6.5 Privacy and Data Security

Connecting many different devices and systems together naturally poses some questions relating to data privacy, security, and other, similar issues, all raised by several road maps discussed here (e.g., [5,21,25,32]). These issues are of key importance in ensuring that the wider public can accept the widespread move toward ubiquitous CPSs [5].

2.6.6 Architectures

One area where the CPS community clearly believes much progress can be made is that of architectures to support SoSs and CPSs. Architectural decisions made by CPS and SoS designers are often constrained. In some systems, a subset of constituents may be already fixed in place and the engineer is unable to alter them to suit an optimal architecture that might have been the choice given a "blank slate." This is common in large-scale SoSs, where some constituents are operated by independent third parties, or in systems that have significant longevity, where legacy constituents are included. However, there is very little in the way of guidance to help engineers and designers develop optimal choices for these types of systems, including patterns for delivering security, privacy, or performance in a system that straddles several engineering disciplines. Many road mapping projects surveyed here have suggested that patterns could be a useful tool for collecting and organizing "lessons learned" about architectures for CPSs and SoSs.

2.6.7 Human Aspects

The behavior of humans who interact with CPS needs to be considered in a variety of ways, as recommended by a number of previous road maps in these areas. A CPS can potentially gather large quantities of information for analysis and forecasting in real time (or near real time). Sometimes analysis involves humans to authorize automatic decision-making, where skilled human operators are presented with data gathered by the CPS and recommendations for future action. Human performance in this scenario can vary, depending on the methods we use to present complex data to humans.

Humans also form part of the operational environment of CPSs. Since humans are capable of an almost infinite variety of highly complex and sometimes unpredictable actions, this places considerable demands on the decision-making abilities of a CPS. For example, autonomous vehicles are likely to come into contact with pedestrians and cyclists, whereas assisted living systems may be responsible for identifying subtle and complex signs that all is not well with an elderly person who lives alone. Coping with these challenges requires an ability to reason about stochastic events and to capture and predict human actions, as well as highly efficient processing techniques to support this level of analysis.

2.6.8 The Overall Picture

There is considerable consensus, therefore, in the nature of the topics that require some future attention. Most of the road maps surveyed in this chapter connect these topics and our future ability to deliver useful, reliable CPSs. The impact of this has been linked to a number of domains, including: transportation (including mass transit systems as well as road networks); health and medical systems; assisted living at home for the elderly or infirm; smart grids; intelligent and efficient homes and offices; highly flexible and efficient production systems including large-scale customizable manufacturing; and environmental monitoring systems. In order to deliver on their promises, all of these domains will require reliable CPSs that are secure and protect user data, but can also deliver on their nonfunctional guarantees of predictable, reliable behavior.

In addition to the consensus, there are some area where the road maps currently available could go further. It is clear, for example, that many road maps regard social acceptance of CPSs as key; this will require significant investment in ensuring that data protection, privacy guarantees, certifiable safety, and highly secure CPSs are available [5,21,25,31,32]. In order to be useful, many domains will also need to make significant investments in considering how complex data should be presented to humans required to interoperate with the systems (e.g., human–computer interaction issues are mentioned in References 5 and 32). The Agenda Cyber-Physical Systems [5] provide some possible concrete suggestions for building an initial high-level framework to safeguard security and/or privacy, with the concept of *unlinkability* used to control the ability to amass user data.

Many CPS and SoS road maps recommend the need to take a variety of modeling approaches into account, including human and organizational modeling approaches [5,21, 31,32], and to develop appropriate legal frameworks to safeguard data [5,21,32].

However, although there is general agreement among researchers producing CPS engineering road maps that further work is needed in these areas, strategies and road maps that plan out potential research paths, specifically in these areas as they relate to CPS engineering, are lacking at the moment. This is a gap in the CPS research road maps currently available publicly, and should be addressed by future investigation.

REFERENCES

1. INCOSE. *Systems Engineering Handbook: A Guide for System Life Cycle Processes and Activities, Version 3.2.2.* Technical Report INCOSE-TP-2003-002-0.3.2.2, International Council on Systems Engineering (INCOSE), October 2011.
2. M. Broy. Engineering cyber-physical systems: Challenges and foundations. In M. Aiguier, Y. Caseau, D. Krob, and A. Rauzy, editors, *Proceedings of the Third International Conference on Complex Systems Design & Management CSD&M 2012*, pp. 1–13. Springer-Verlag, Berlin, Heidelberg, 2013.
3. J.S. Fitzgerald, K.G. Pierce, and C.J. Gamble. A rigorous approach to the design of resilient cyber-physical systems through co-simulation. In *IEEE/IFIP International Conference on Dependable Systems and Networks Workshops (DSN 2012)*, pp. 1–6, Boston, MA, 2012.
4. E.A. Lee. CPS Foundations. In *Proceedings of the 47th Design Automation Conference, DAC'10*, Anaheim, CA, pp. 737–742, June 13–18, 2010.
5. E. Geisberger and M. Broy. *Living in a Networked World: Integrated Research Agenda Cyber-Physical Systems (agendaCPS).* Technical Report [Online] http://www.acatech.de/de/publikationen/empfehlungen/acatech/detail/artikel/living-in-a-networked-world-integrated-research-agenda-cyber-physical-systems-agendacps.html [Last visited August 2015], acatech National Academy of Science and Engineering, 2015.

6. H. Giese, B. Rumpe, B. Schätz, and J. Sztipanovits. Science and engineering of cyber-physical systems (Dagstuhl Seminar 11441). *Dagstuhl Reports*, 1(11):1–22, 2011.
7. A. Avizienis, J.-C. Laprie, B. Randell, and C. Landwehr. Basic concepts and taxonomy of dependable and secure computing. *IEEE Transactions on Dependable and Secure Computing*, 1(1):11–33, 2004.
8. M.G. Moehrle, R. Isenmann, and R. Phaal. Basics of technology roadmapping. In M.G. Moehrle, R. Isenmann, and R. Phaal, editors, *Technology Roadmapping for Strategy and Innovation*, pp. 1–9. Springer, Berlin Heidelberg, 2013.
9. S. Lee and Y. Park. Customization of technology roadmaps according to roadmapping purposes: Overall process and detailed modules. *Technological Forecasting and Social Change*, 72(5):567–583, 2005.
10. R. Phaal, C. Farrukh, and D.R. Probert. Fast-start roadmapping workshop approaches. In M.G. Moehrle, R. Isenmann, and R. Phaal, editors, *Technology Roadmapping for Strategy and Innovation*, pp. 91–106. Springer, Berlin Heidelberg, 2013.
11. K. Vlaanderen. *Technology Roadmapping—Fast Start*. Technical Report [Online] http://www.cs.uu.nl/wiki/Spm/FastStart [Last visited May 2014], Utrecht University, Faculty of Science: Information and Computing Sciences, 2010.
12. R. Phaal, C.J.P. Farrukh, and D.R. Probert. Technology roadmapping—A planning framework for evolution and revolution. *Technological Forecasting and Social Change*, 71(1–2):5–26, 2004.
13. Road2SoS. *Road2SoS Deliverable D3.3g: Roadmap for the Domain of Distributed Energy Generation and Smart Grids*. Technical Report [Online] http://road2sos-project.eu/cms/front_content.php?idcat=12 [Last visited June 2015], Road2SoS project, 2013.
14. Ed. Jamshidi, M. *System-of-Systems Engineering – Innovations for the 21st Century*. J. Wiley and Sons, 2009.
15. M.W. Maier. Architecting principles for systems-of-systems. *Systems Engineering*, 1(4):267–284, 1998.
16. D.A. Fisher. *An Emergent Perspective on Interoperation in Systems of Systems. Technical Report: CMU/SEI-2006-TR-003*. Technical Report, Software Engineering Institute, Carnegie Mellon University, Pittsburgh, PA, March 2006.
17. R. Abbott. Open at the top; open at the bottom; and continually (but slowly) evolving. In *System of Systems Engineering, 2006 IEEE/SMC International Conference on*. Los Angeles, CA, pp. 41–46. IEEE, April 2006.
18. J. Boardman and B. Sauser. System of systems – The Meaning of "of". In *Proceedings of the 2006 IEEE/SMC International Conference on System of Systems Engineering*, pp. 118–123, Los Angeles, CA, April 2006. IEEE.
19. W.C. Baldwin and B. Sauser. Modeling the characteristics of system of systems. In *System of Systems Engineering, 2009. SoSE 2009. IEEE International Conference on*, Albuquerque, NM, pp. 1–6. IEEE, 2009.
20. AMADEOS. *D1.1: SoSs, Commonalities and Requirements*. Technical Report [Online] http://amadeos-project.eu/documents/public-deliverables/ [Last visited June 2015], AMADEOS14 (Architecture for Multi-criticality Agile Dependable Evolutionary Open System-of-Systems), 2014.
21. COMPASS. *COMPASS Deliverable D11.4: Roadmap for Research in Model-Based SoS Engineering*. Technical Report [Online] http://www.compass-research.eu/Project/Deliverables/D11.4.pdf [Last visited June 2015], COMPASS project, 2014.
22. RFID Resource Network (RFID-RNET). *Rfid Contribution to Strategic Research Agenda Smart Systems Integration*. Technical Report [Online] http://www.rfid-rnet.com/documents/Contribution\%20to\%20Strategic\%20Research\%20Roadmap\%20RFID-RNET\%202009_V01.pdf [Last visited June 2014], RFID Resource Network (RFID-RNET), 2009.

23. O. Vermesan and P. Friess. *Internet of Things: Converging Technologies for Smart Environments and Integrated Ecosystems*. River Publishers, Algade 42, 9000 Aalborg, 2013.

24. R.S. Kalawsky, D. Joaanou, Y. Tuan, and A. Fayoumi. Using architecture patterns to architect and analyze systems of systems. In *Proceedings of the Conference on Systems Engineering Research (CSER'13)*, Atlanta, GA, pp. 283–292, 2013.

25. T-AREA-SoS. *T-AREA-SOS Deliverable D3.1: The Systems of Systems Engineering Strategic Research Agenda*. Technical Report [Online] https://www.tareasos.eu/results.php [Last visited June 2014], T-AREA-SOS Project (Trans-Atlantic Research and Education Agenda in Systems of Systems Project), 2013.

26. COMPASS. *D34.2: Specialised Test Strategies*. Technical Report [Online] http://www.compass-research.eu/deliverables.html [Last visited June 2014], COMPASS, 2014.

27. Z. Andrews, C. Ingram, R. Payne, A. Romanovsky, J. Holt, and S. Perry. Traceable engineering of fault-tolerant SoSs. In *Proceedings of the INCOSE International Symposium 2014*, pp. 242–257, Las Vegas, NV, June 2014. John Wiley & Sons, Boston, MA.

28. Z. Andrews, J. Fitzgerald, R. Payne, and A. Romanovsky. Fault modelling for systems of systems. In *Proceedings of the 11th International Symposium on Autonomous Decentralised Systems (ISADS 2013)*, pp. 59–66, Mexico City, Mexico, March 2013.

29. DSoS. *Architecture and Design Initial Results on Architectures and Dependable Mechanisms for Dependable SoSs*. Technical Report DSoS Deliverable BC2, DSoS (Dependable Systems of Systems), 2001.

30. C. Ingram, S. Riddle, J. Fitzgerald, A.H.L. Al-Lawati, and A. Alrbaiyan. SoS fault modelling at the architectural level in an emergency response case study. In R. Payne, Z. Andrews, and U. Schulze, editors, *Proceedings of the Workshop on Engineering Dependable Systems of Systems (EDSoS) 2014*, p. 5, Technical Report 1416, School of Computing Science, Newcastle University, Newcastle upon Tyne, UK, July, 2104.

31. Road2SoS. *Road2SoS Deliverables D5.1 and D5.2: Report on Commonalities in the Four Domains and Recommendations for Strategic Action*. Technical Report [Online] http://road2sos-project.eu/cms/upload/documents/Road2SoS_D5.1_D5.2_Commonalities_and_Recommendations.pdf [Last visited June 2015], Road2SoS project, 2013.

32. CyPhERS. *CyPhERS Deliverables D6.1 and D6.2: Integrated CPS Research Agenda and Recommendations for Action*. Technical Report [Online] http://cyphers.eu/project/deliverables [Last visited June 2015], CyPhERS (Cyber-Physical European Roadmap and Strategy), 2015.

33. CPSoS. *D2.4: Analysis of the State-of-the-Art and Future Challenges in Cyber-physical Systems of Systems*. Technical Report [Online] http://cpsos-project.eu [Last visited June 2015], CPSoS (Towards a European Roadmap on Research and Innovation in Engineering and Management of Cyber-Physical Systems of Systems), 2013.

CHAPTER 3

A Rigorous Definition of Cyber-Physical Systems

John Knight, Jian Xiang, and Kevin Sullivan

CONTENTS

3.1 INTRODUCTION

In this chapter, we present a rigorous definition of Cyber-Physical Systems (CPS) based on the three essential components of such a system: (1) a computing platform, (2) a set of physical entities with which the computing platform interacts, and (3) a connection between the first two components. These components seem familiar, and the third probably seems trivial. But the third component is crucial because it defines how logical values read and produced by the computing platform will be affected by and will affect the various physical entities (for example, whether an integer input of "73" that is meant to be a heading is measured relative to true North or magnetic North and whether an output with logic value of "1" is meant to cause a pump to start or stop).

Thus, correctly and precisely defining the connection between the computing platform and the physical entities is crucial, because defects in the definition can lead to undesired and possibly dangerous behavior of the CPS.

We characterize the details of the three components precisely, placing particular emphasis on the connection between the computing platform and the physical entities. We use these characterizations to construct a rigorous definition of CPS. From this definition, we derive various validation, verification, and specification properties that can serve to support the design, analysis, and assurance of CPS.

3.1.1 Cyber and Physical

The fundamental concept underlying CPS is that a logic function defined by a digital computer (cyber) interacts with the real world (physical) in order to sense and affect entities in the real world. The details of the implementation of CPS, such as the use of particular sensors and/or actuators, do not affect this underlying concept. Similarly, the functionality of a system, whether the system is in the healthcare, transportation, or some other domain, does not affect the underlying concept. In other words, the role of CPS is to implement an explicit *interaction* between the two parts of the cyber-physical combination.

With these informal notions in mind, an intuitive definition of CPS might be

> *Cyber-physical system:* A system that includes a discrete logic function in its design such that either: (a) the state of the logic function could be changed by a change in the state of the real world, or (b) the state of the real world could be changed by a change in the state of the logic function, or (c) both.

Based on this intuitive definition, most systems that involve a computer are CPS. And, in the most general sense, this is the case. The only computer systems that are not CPS are those that do not affect and are not affected by the real world, that is, software that is pure logic—a compiler, for example. Such software is unusual.

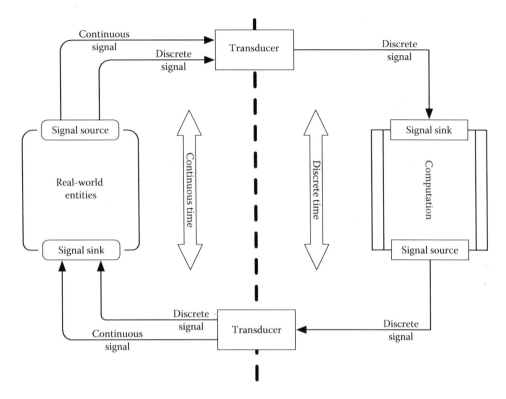

FIGURE 3.1   The basic elements of CPS.

For practical purposes, a rigorous definition and the associated ability to undertake system fault detection is more important for certain types of CPS than others. Systems for which a rigorous definition is likely to be important are those that have serious consequences of failure, such as medical systems that operate a control loop aiding a patient, automobile engine controllers that execute complex algorithms to manage pollution control, and nuclear shutdown systems that include complex decision-making algorithms. In such systems, failure could lead to equipment damage, human injury or death, or environmental pollution, so required levels of assurance of correct operation are necessarily high.

In practice, CPS have the general form shown in Figure 3.1. An important aspect of CPS is the distinction between continuous functions and discrete functions. Simple two-state ("on" and "off") discrete functions arise in the real world with switches, but many signals in the real world that are intended to change the state of the logic function are continuous. Signals to which the logic function might react, and the signals that it generates, are discrete. Transducers process continuous signals in the real world to produce discrete signals that are input to the logic function. By making various state transitions, the logic function affects a "computation" that produces digital signals designed to affect real-world entities. Transducers process these digital signals to produce continuous signals that are inputs to the real world.

An important element of the distinction between continuous and discrete functions is *time*. Time is continuous for both real-world entities and logic. But logic models the progress of time as a discrete function and changes state only at regular, discrete-time intervals.

Since the implementation of CPS logic functions are almost exclusively in software, we will use the term "software" in the remainder of this chapter. Nevertheless, we note that the definition presented in this chapter also applies to CPS in which the logic functions are implemented in hardware or in FPGAs.

The importance of the notion of interaction between real-world entities and software can be seen when one notes that many CPS are safety critical, and assurance of their correct operation depends on the interaction between real-world entities and software being complete and correct. Unless the interaction is precisely that which is intended *and* known to be so, doubt in the effects of such systems is inevitable.

Software failing to interact with real-world entities with sufficient fidelity has been a causal factor in various accidents and incidents. In 1999, the Mars Climate Orbiter was lost because different parts of the software system used different units of measurement [1]. More recently, in 2013 a delay in berthing Orbital's Cygnus spacecraft with the International Space Station (ISS) occurred because Cygnus and the ISS used different forms of GPS time data—one based on the original 1980 ephemeris and the other based on an ephemeris designed in 1999 [2]. In both of these examples, real-world entities were affected by operations defined in software in ways that made no real-world sense, that is, the systems failed because the essential interaction was improperly defined.

The software in CPS should respect invariants that derive from the real world the system senses, models, and affects. For example, software that computes physical quantities should respect measurement units and physical dimensions, for example, as defined by the ISO/IEC 80,000 standards. Frequently, however, the interaction between real-world entities and the system's software is undocumented or is documented incompletely, informally, or implicitly.

Even more frequently, the completeness and correctness of the interaction is not checked comprehensively. Invariants derived from the real world are stated and enforced either in ad hoc ways or not at all. Crucial, nontrivial relationships between software representations and real-world entities remain under-specified, and programs treat machine-world representations as if they were the real-world entities themselves. As a result, faults are introduced into systems due to unrecognized discrepancies, and executions end up violating constraints derived from the real world. The results are software and system failuresas well as adverse downstream consequences.

3.1.2 Logic Interpretation

The need to define the interaction between the software in CPS and entities in the real world explicitly arises because software is a logic function with no *interpretation*. The notations that are used for defining software are formal languages. High-level languages, assembly languages, and machine languages are *all* formal and, as formal languages, have no inherent real-world meaning, that is, the logic is *uninterpreted*.

For a statement in a formal language to be anything other than a purely syntactic entity, an *interpretation* has to be added to the logic. The interpretation defines the intended *meaning* in the real world of elements of the logic. In doing so, the interpretation exposes the logic

(a)

```
/* Returns the year after 1900. */
  public int getYear() {
  if (!expanded)
      expand();
  return tm_year;}


/* Sets the year. */
  public void setYear(int v) {
  tm_year = v;
  valueValid = false;}


/* Returns the month. ie: 0-11 */
  public int getMonth() {
  if (!expanded)
      expand();
  return tm_mon;}
```

(b)

```
public int lnmgtyu() {
  if (!wpou88kj)
      xcvbbhu71();
  return tm_lnmgtyu;}

public void tyugfds(int v) {
  tm_ugfdsrew = v;
  ascboi9jjk  = false;}

public int sdgtyu() {
  if (!wpou88kj)
      xcvbbhu71();
  return tm_sdgtyu;}
```

FIGURE 3.2 Source code with (a) and without (b) an ad hoc interpretation.

to constraints and invariants that derive from the real world, such as the laws of physics. The logic must conform to these constraints and invariants to provide a rich opportunity for error checking of the software.

Surprisingly, the interpretation of a software system is *always* present in practice but usually documented in an ad hoc, informal, and sometimes implicit manner using casual techniques, such as "descriptive" comments, "meaningful" identifiers, and design documents. Figure 3.2a shows a block of Java text taken from an open-source library. The meanings of the values used by the parameters of the various functions are documented in part by the names of the parameters and in part by the comments. The block of Java text with the identifiers replaced with random strings and the comments removed is shown in Figure 3.2b. This version of the Java text compiles correctly, and execution of the software is unaffected by the changes. Although the logic function is unaffected, human understanding of what the logic does is almost completely destroyed. In the function originally named getmonth(), the comment explaining that the encoding of the months of the year used by the function is 0—January, 1—February, 2—March, etc. is *essential*. That particular encoding is unusual and impossible to discern from the code.

As noted previously, for CPS that are safety critical, assurance of their correct operation depends on the interpretation of the logic being complete and correct. Dealing with this dependence begins with the interpretation being precisely and comprehensively documented. Without such documentation, doubt in the correctness of the effects of such systems is inevitable.

The explicit and systematic documentation of the interpretation of software logic provides three major advantages:

- The interpretation informs the design of the software of the actual entities that the software will affect, allowing better design choices.

- The interpretation documents essential reference material in a centralized and well-defined form, allowing rigorous examination for correctness and completeness by human inspection.

- The real-world constraints and invariants that the interpretation exposes can be checked providing a new mechanism for detecting software faults.

All three of these advantages are valuable, but the provision of a significant new capability for detecting software faults is especially important. Static analysis of a system's software where the analysis derives from the interpretation allows the detection of faults that result from misuse of real-world entities or violate real-world constraints. In case studies, such analyses revealed both unrecognized faults and faults that had been reported as bugs in real systems after deployment [3]. Examples of this new fault detection capability are given in Section 3.7.

3.2 CONCEPT OF INTERPRETATION

The role of interpretation can be thought of as an enhanced version of *abstraction functions* in logic that map concrete representations, such as variables, to abstract representations, such as abstract data types. A stack, for example, is an abstract data type that has a concrete implementation as an array and an integer index into the array.

In a similar way, an interpretation defines the real-world meaning of an element of logic. For example, an integer variable in an avionics program might be used to represent the altitude of an aircraft. Within the logic of the software, the variable is merely an integer. The role of the interpretation is to reveal everything about the *actual* altitude of the aircraft.

Figure 3.3 illustrates this idea. The system design process starts with a problem to be solved in the real world and develops a concrete solution in logic. The interpretation provides the abstract (real world) details of a concrete entity (logic). With an explicit interpretation, important characteristics of real-world entities, such as units and dimensions, and associated real-world constraints, such as not mixing units, can be stated and enforced. In addition, crucial relationships between logic representations and real-world entities, such as accuracy of sensed values, can be fully specified. The introduction of faults due to unrecognized discrepancies resulting from violating rules inherited from the real world can be significantly reduced or even eliminated.

As an example, consider again the altitude of an aircraft and the representation of altitude in avionics software. Aircraft altitude is not just a number, even though it might be represented as such in software. Altitude has many important attributes that impact the way that CPS, such as an aircraft's autopilot, compute using altitude. A partial list of those attributes include:

- *Measurement units*: The altitude value will be measured in prescribed units (feet, meters, etc.).

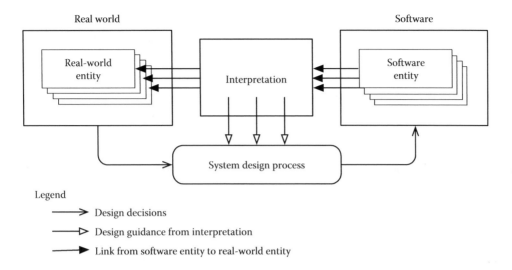

FIGURE 3.3 The role of interpretation in CPS.

- *Physical dimensions*: Altitude has the fundamental physical dimension of length.

- *Frame of reference*: Altitude is defined based on an origin and the direction, that is, a frame of reference.

- *Sensor performance*: A sensor, that is, a transducer, will have determined the value to be supplied to the software and so that value will be of limited precision and accuracy because sensors are imperfect transducers.

- *Sensing schedule*: Transducers supply values according to a discrete-time schedule. The value supplied to the software is the value obtained when the sensor sample was taken, not the "current" value, that is, the altitude of the aircraft "now."

With an explicit interpretation that documents details of a quantity (like altitude), a wide variety of checks of the software in CPS (like autopilot) are possible, such as

- *Mixed measurement units*: Expressions that mix units of measurement are probably erroneous unless an appropriate conversion is provided. For example, adding an altitude measured in feet to a displacement measured in meters is an error if no conversion factor is included in the computation.

- *Mixed physical dimensions*: Dimensional analysis is a standard error detection mechanism in physics. Thus, for example, assigning the result of dividing altitude with physical dimension length by a time to a variable that is not a speed (speed has dimensions length/time) is an error.

- *Mixed frames of reference*: Altitude is measured in a frame of reference with an origin and an orientation. A distance calculation between two points is probably erroneous if the two points are from different frames of reference.

The necessity of defining an interpretation for software explicitly indicates the need for a composite entity that includes both software and its interpretation. A candidate entity is the *interpreted formalism*.

3.3 INTERPRETED FORMALISMS

Having established the role and value of an interpretation of the logic in CPS, we turn to the structure needed to incorporate an explicit interpretation into the engineering artifacts that are needed for CPS. The structure that we introduce is called an *interpreted formalism* and has the form shown in Figure 3.4.

The logic in an interpreted formalism is defined in whatever manner is appropriate for the system of interest, that is, the choice of programming language, programming standards, compiler, and so on are unaffected by the interpreted formalism structure. The key difference, of course, is the addition of the explicit interpretation.

As discussed in Section 3.1.2, in practice an interpretation is always present for any CPS. The interpreted formalism combines the interpretation and the software in a manner that makes the interpretation a first-class entity.

In the development of any particular CPS of interest, the task is no longer to develop software. The task is, in fact, to develop an interpreted formalism for the systems of interest. Without the explicit interpretation, whatever would be developed as "software" runs the risk of failing to correctly define the desired interaction with the real world, where the implementation of that interaction is the entire purpose of CPS.

The switch from developing software to developing an interpreted formalism is a *paradigm shift*. Such a change should not be undertaken lightly, but with the number,

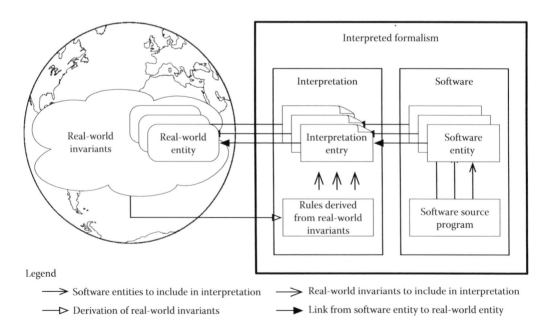

FIGURE 3.4 An interpreted formalism combines logic with an explicit interpretation.

criticality, and consequences of failure of CPS increasing, the shift needs to be seriously considered.

3.4 RIGOROUS DEFINITION OF CYBER-PHYSICAL SYSTEMS

With the concept, role, and importance of interpretation defined and with the interpreted formalism structure in place, we have the elements needed to define CPS.

CPS are triple {Rw, Hw, If}, where

- Rw is a set of entities in the real world with which the system interacts.

- Hw is a hardware platform that (a) executes the system's software and (b) provides the physical connections between the set of real-world entities and the system's software.

- If is an interpreted formalism that combines the definitions of (a) the software of the system and (b) the interpretation of that software.

For the most part, Rw and Hw are familiar. The set Rw can be enumerated based on the problem to be solved by theCPS. In the case of a drug-infusion pump, for example, the pump has to interact with the patient's physiology; the patient and medical professionals as "users" through a human–machine interface; and the environment, including gases, radiation fields, and physical entities that might cause damage.

Hw for a drug-infusion pump is a collection of specialized hardware items including a pump, a drug reservoir, multiple valves, switches, a keyboard, one or more computers, a network interface, and so on.

If is a new concept, and so we focus on the details of the interpreted formalism in the remainder of this chapter.

With this definition, we need to modify the basic elements of CPS shown in Figure 3.1. Figure 3.5 shows the original elements but includes an explicit interpretation and an explicit indication that the design of the computation derives fundamentally from the real-world context within which the system will operate.

3.5 STRUCTURE OF AN INTERPRETED FORMALISM

The concept of logic interpretation is well established, but defining the content and structure of an effective and complete interpretation for *practical* use is a significant challenge. In this section, we present a preliminary design. In later sections we present an implementation of the design and the results of experiments undertaken to assess the feasibility and performance of the concept.

The design we present is based on the concept of a *real-world type* system [3]. The design of the interpretation is a set of real-world types defined within the framework of a real-world type system.

3.5.1 Real-World Types

A real-world type is the real-world analog of a type in a formal language. A real-world type defines the values that an object in the real world of that type can have and the operations

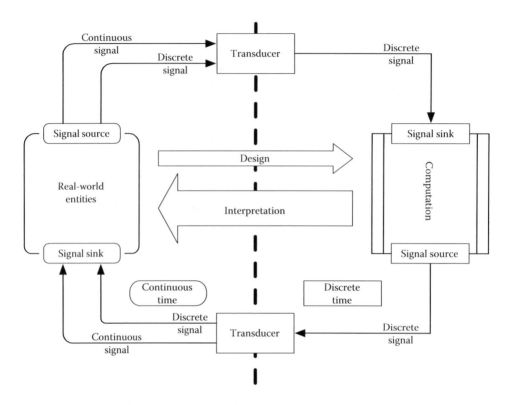

FIGURE 3.5   The basic elements of CPS based on the rigorous definition.

in which the object can engage. The real-world type for the physical concept of an aircraft's altitude would include all the details noted in Section 3.2.

More generally, a real-world type *system* documents:

- The real-world attributes associated with the type. These attributes are properties possessed by an entity of the associated type. For the aircraft altitude example discussed earlier, the attributes include measurement units, physical dimensions, sensor precision, sensed value delay, frame of reference, and so on.

- The type rules that define allowable operations on entities of the various types. The type rules derive from what constitutes meaningful operations in the real world. The examples given earlier for aircraft altitude (measurement unit compatibility, etc.) will all be included. Rules that derive from application-specific details will be included also. For example, an aircraft's altitude might be measured by radar and by barometric pressure, but for managing altitude in an autopilot, the system developers might require that the radar measurement be used. Type rules can be established to identify incorrect uses of altitude derived from barometric pressure.

- The machine representations for entities of the real-world types. This allows analysis of the implication of limitations within that representation. For example, a particular precision might be defined, but analysis of the computations in which that item is used

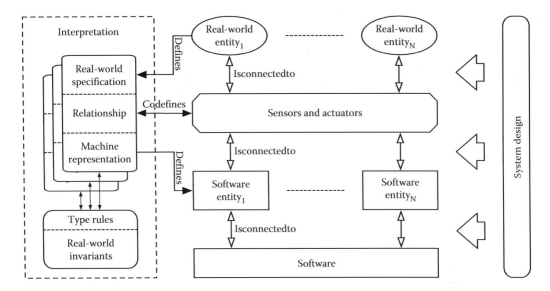

FIGURE 3.6   Real-world type system.

or expected range values derived from the associated real-world entity might reveal the possibility of overflow.

- The relations between real-world entities and their machine representation. The relation defines the mapping of values to meaning in the real world. In the aircraft altitude example, the machine representation might be feet, meters, or flight level (altitude in hundreds of feet). In addition, for engineering reasons, values might be offset by a fixed increment.

- An explication. All of the above provide reference information. An explication is a careful, detailed, and precise statement in natural language that provides the reader with a detailed explanation of the various terms and concepts in use.

Our preliminary structure for an explicit interpretation is a set of real-world types together with a set of bindings between entities in the real world and entities in the logic.

A real-world type system and its connections to an application system of interest are shown in Figure 3.6. In the figure, real-world entities are sensed and affected by sensors and actuators. In the system software, there are software entities that model the associated real-world entities. The relationship between a real-world entity and a software entity documents information such as sensing precision and sensing delay. The details of both, and their relationships, are documented in the real-world types shown on the left of the figure.

Real-world types are intended to facilitate design in computer engineering *from* the real world *to* the machine world so as to enable all relevant aspects of the real world to be considered in developing a computer system. Within an interpretation defined using real-world types is a set of type rules (see Figure 3.6) that are derived from real-world invariants. For example, rules about unit conversion, valid and invalid expressions using variables with units, and types resulting from expressions have to be defined in their entirety to enable suitable checking.

3.5.2 Example Real-World Type

An example of a real-world type is a point in 3D space. Measurements designed to locate a point are only relevant if the associated coordinate system is defined completely. If multiple coordinate systems are in use in a program, they must be distinguished. Thus, the real-world type information needs to document all of the different aspects of the coordinate system.

A possible definition for a point in 3D space, including all of the coordinate system information, is shown in Figure 3.7. The field names in this definition are the attributes of interest. Many other formats and sets of attributes are possible. In this definition, the explications are omitted for simplicity. Note that this type definition is created just to distinguish coordinate systems. Separately, we need the types of the three fields that will be used for a point in the coordinate system.

For the types of the three fields, one, two, or three different type definitions might be needed. For this example, we assume that the x and y variables can share a type definition and a second definition is used for z. All three dimensions are actually distances within the coordinate system. As such, the real-world physical dimensions are length. In this example, the encoding of physical dimensions is based on the standard set of seven dimensions from physics: length, mass, time, electric current, temperature, amount of substance, and luminosity [52]. For convenience in many CPS, we have added angle as a basic dimension, bringing the total to eight. In the real-world type definitions, the actual physical dimensions of an entity are specified as a vector of powers of these dimensions. Thus, for example, velocity, which is length per unit time and has physical dimensions LT^{-1}, is encoded as $(0,1,-1,0,0,0,0,0)$.

```
geographic_cartesian_coord_sys:
    Specification
        explication              : <text>
        real_world_semantics
            coordinate_sys_type  : cartesian
            target_space         : Earth
            origin               : center of mass of Earth
            dimensionality       : 3
            earth_model          : spheroid
            x_axis_orientn       : positive toward 0 degrees longitude
            y_axis_orientn       : positive toward 90 degrees E longitude
            z_axis_orientn       : positive northward
    Representation
        machine_semantics
            representation       : record structure - (float, float, float)
    Relationship                 : <null>
```

FIGURE 3.7 A real-world type definition for horizontal axis (a) and for vertical axis (b) for a coordinate in a frame of reference.

```
(a)
horizontal_cartesian_axis:
   Specification
      explication            :  <text>
      real_world_semantics
         linear_units        :  mile
         physical dimensions :  (1,0,0,0,0,0,0,0)
         technology          :  GPS
         geometry_plane      :  horizontal
   Representation
      machine_semantics
         representation      :  float
         mutable             :  no
   Relationship
         value_error < delta1 and delay < tau1

(b)
vertical_cartesian_axis:
   Specification
      explication            :  <text>
      real_world_semantics
         linear_units        :  feet
         physical dimensions :  (1,0,0,0,0,0,0,0)
         technology          :  radar
         geometry_plane      :  vertical
         offset_origin       :  mean sea level
   Representation
      machine_semantics
         representation      :  float
         mutable             :  no
   Relationship
         value_error< delta2 and delay < tau2
```

FIGURE 3.8   Real-world type definitions for lengths in a coordinate system.

For x and y, we define the type shown in Figure 3.8a. In this example, altitude is part of a complete reference frame with an origin at the center of mass of the Earth but with a presumed offset to mean sea level. The appropriate type definition is shown in Figure 3.8b. Such a type might be used to hold data in any space of interest. For example, the type could be used to hold location information for aircraft, climbers, balloons, etc.

3.5.3 Real-World Type Rules

Given the real-world type definition for a coordinate, example type rules that could be stated, checked, and indicate operations that are probably erroneous if violated include:

- The units of an angle and a latitude must match if they are added. The result is of type latitude measured in the same units.

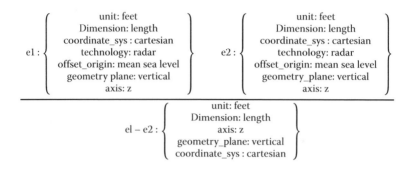

FIGURE 3.9   Example type rule definition.

- A velocity, dimensions $(0,1,-1,0,0,0,0,0)$, cannot be added to a distance, dimensions $(0,1,0,0,0,0,0,0)$.

- A latitude or a longitude cannot be added to a latitude or a longitude.

- An x coordinate in one frame of reference cannot be used in any arithmetic operation with a coordinate from a different frame of reference.

- A variable of type *magnetic* heading cannot be used in an expression expecting a variable of type *true* heading, even if both are represented as integers and are commensurable.

- A variable of type *geodetic* latitude cannot be used in an expression expecting a variable of type *geocentric* latitude, even if both are represented as floating point and are commensurable.

As an example of type-rule definition, consider the semantics of the result of subtracting two operands of type `vertical_cartesian_axis`, for example, for calculating the altitude difference between two points in the same Cartesian coordinate system. The definition is illustrated in Figure 3.9.

The notation e:T denotes a type judgment (e is of type T), and the overall construct defines an inference rule defining the type of the result of applying a specific operator, here subtraction, to operands, e1 and e2, of the specified types.

3.6 AN IMPLEMENTATION OF INTERPRETATIONS FOR JAVA

3.6.1 Design of the Implementation

If explicit interpretations are to be used in the development of realistic CPS, an approach to integrating them into widely used languages and development methods is needed. We have developed a prototype that implements interpretations as sets of real-world types for Java. The implementation supports

- The definition of a set of real-world types for a Java program of interest.

- The definition of set real-world type rules by system experts based on real-world and application invariants.

- Creation of bindings between the real-world type definitions and entities in the Java source program.

- Static type checking of the Java program based on the set of real-world type rules.

- Synthesis of assertions as Java fragments that can be inserted into the subject program to implement runtime checking of type rules that cannot be checked statically.

- Synthesis of a checklist of locations in the subject program at which human inspection is required to check type rules that cannot be checked statically or dynamically.

To illustrate the analysis capabilities that are possible with an explicit interpretation, we present case studies in Section 3.7 in which real-world types were applied to a set of pertinent projects, including a project with 14,000 LOC. The real-world type checking revealed both unreported faults and faults that had previously been reported as bugs.

An important design choice made for the implementation was that the system should operate without requiring changes to the subject Java program. This choice provides four major advantages:

- The interpretation does not obscure the basic structure of the Java program.

- An interpretation can be added to an existing Java program without having to modify (and possibly break) the original source text.

- An interpretation can be added to a Java program asynchronously, thereby not impeding development of the Java program itself and permitting an interpretation to be added to an existing legacy program.

- An interpretation can be added incrementally so that developers can experiment with the technology to determine its efficacy for their application.

Motivated by this design choice, the implementation operates separately from the compiler via its own user interface.

The structure of the Java prototype system is shown in Figure 3.10. The subject Java source program is shown on the left toward the top of the figure, and the interpretation is shown on the right. The development of the interpreted formalism, the combination of the source of the Java program and the interpretation, is shown at the top of the figure.

3.6.2 Use of the Implementation

In the prototype implementation, the interpretation is accessed via the user interface, enabling:

- The establishment and display of bindings between items in the Java program and real-world type definitions in the interpretation.

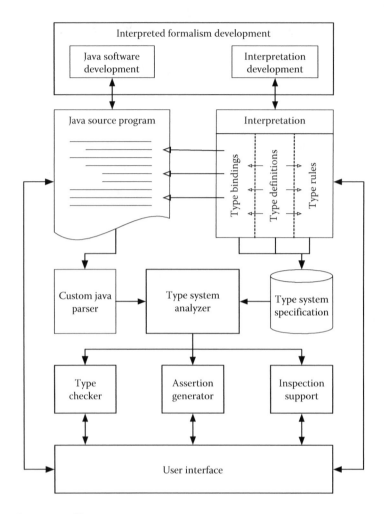

FIGURE 3.10   Structure of Java prototype.

Selecting an entity in the Java program that is to have a real-world type (clicking on the text) and selecting the particular real-world type to be used (clicking on the type name) establishes a binding.

- Reference to the details of the interpretation.

All definitional aspects of the real-world types and all bindings to Java entities can be displayed. The set of bindings can be displayed in various ways, for example, all bindings, binding of a given Java entity, all Java entities bound to a particular real-world type, etc.

To support analysis of the system, a custom parser produces a representation of the subject Java program as an abstract syntax tree, and the implementation of the interpretation produces a database that documents all of the details of the interpretation. The abstract syntax tree and the details of the interpretation are processed by an analyzer shown in the center of the figure, which supports three types of analyses:

- *Type checking*: The type rules define invariants that are derived from the real world and from the application. All of the type rules are checked throughout the Java program and diagnostics displayed for the user identifying violations.

- *Assertion generation*: Some type rules cannot be checked statically. Real-world invariants often document relationships between real-world entities of constraints on values. For example, an aircraft's altitude should lie between approval flight limits, and an autopilot should not command a pitch-up angle beyond a safe limit. Such restrictions are common, and assertions are sometimes included in software to check them at run-time. The analysis system synthesizes assertions as Java code fragments and optionally will insert them (with explanatory comments) into the Java source program.

- *Inspection checklist generation*: Some aspects of real-world type checking cannot be performed either statically or dynamically. For these type rules, the analysis develops simple checklists that can be used to guide an inspection. An example of this type of rule is that the maximum difference between the value of a physical entity available to the system's software should be less within some prescribed tolerance of the actual value. An aircraft's altitude as represented in an autopilot might be required to be within some fraction of the aircraft's true altitude in order to establish stability in an autopilot. The difficulty that arises in checking derives from the sensor inaccuracy, sensor sampling delay, delay in processing of the sampled value prior to its being available to the autopilot, and so on. The only reasonable way to have confidence in the required type rule is to conduct human inspection. The analysis system synthesizes inspection checklists based on the use of identified relevant software entities so as to provide focus for human inspectors.

3.6.3 Type Conversion

An important issue in the type system is type *conversion*. For example, a length variable whose real-world type indicates that the measurement unit is feet could be switched to a different measurement unit, say meters, by calling a function that effects the switch, multiplying by a constant, or multiplying by a variable. Each of these mechanisms could be implemented as stand-along assignment statements, within other expressions, as expressions stated as actual parameters, as return values, and so on.

The analysis system deals with explicit type conversion simply by including type rules associated with whatever special operator or function is used. For example, a conversion function is documented as taking one real-world type as its input parameter and a second real-world type as its return value.

Implicit type conversion is more difficult. Conversions between real-world types can be syntactically simple. For example, a conversion from feet to inches requires multiplying a variable storing a value in feet by 12, and the constant might not be identified specifically to support analysis. The difficulty lies in locating such conversions automatically without generating false negatives.

Implicit type conversion is dealt with in the prototype implementation by requiring that the programmer investigate each diagnosed type error and mark implicit type conversions as such. Thus, diagnostics of which the type system was unaware will be generated for type conversion because the mismatch appears to be a violation of the type rules. In those cases, the programmer suppresses the diagnostic by indicating that there is an expected implicit type conversion. By doing so, the programmer indicates that the diagnostic has been investigated and the code found to be as desired.

3.6.4 Typed Java Entities

The Java entities that require binding to real-world types are (a) local variables, (b) fields in classes, (c) method parameters, (d) method return parameters, and (e) class instances. In order to make the development of the prototype tractable, the current version imposes some restrictions on the use of real-world types in Java, specifically

- *Fields*: Fields in classes are assumed to be monomorphic, that is, a field in a class is assumed to have the same real-world type in all class instances. Fields are bound to real-world types inside the class declaration body.

- *Class instances*: Different instances of a class might have different real-world meanings and so the type definition is of the *instance*, not the class. For example, suppose a class `Point` has three fields `x, y, z`. Further, suppose that `pt1` and `pt2` are both instances of `Point` but are from different coordinate systems. Writing a statement that involves both `pt1.x` and `pt2.x` such as `pt1.x + pt2.x` might be an error, so the two instances need to be distinguished.

- *Method return parameters*: If a particular method is not bound to a real-world type, the analysis treats the method as polymorphic. For a polymorphic method, at each invocation site, the type checker examines all the expressions in the method declaration body and determines the real-world type for the return statement. That will be the real-world type for the method invocation ultimately. If the method contains multiple return statements, the real-world type for the return value will be the one with no type errors. Also, if real-world types for return statements are inconsistent, the type checker issues a warning message.

- *Arrays*: Since individual array elements cannot be typed with real-world types, all objects inside an array are treated as having the same real-world type.

Type checking can only rely on limited type inference because many of the result types of operations are defined by real-world circumstances. For example, adding a variable of type angle measured in degrees to a variable of type latitude also measured in degrees (commonly but not necessarily) yields a result of type latitude because the purpose of the addition is (probably but necessarily) to compute a new heading in some sort of navigation system. Without a comprehensive type inference system, checking type rules leads to three difficulties:

- *Determining the types of constants*: Variables are given a real-world type when declared, but constants are used as needed. Constants are dealt with simply by associating each one with a hidden variable and associating a real-world type with the hidden variable.

- *Function-return types*: Defining a function essentially introduces a new type rule. Function return types are dealt with by associating a real-world type with each function that returns a value with a real-world type.

- *Defining the types of compound objects*: Classes introduce the possibility of nesting real-world types because the class might have a real-world type and the fields within the class might have real-world types. In that case, the type of a qualified name is the union of the attributes of all the elements in the path to a specific item of interest in an expression. This same rule applies to method invocation where fields are retrieved, such as `cs2.get_x();`.

3.7 PERFORMANCE/EXPERIMENTAL RESULTS

Informal arguments, as presented in this chapter, suggest that the rigorous definition of CPS and the associated analysis might have value, but that is insufficient reason to change current development methods. Recall from Section 3.3 that the change is of the magnitude of a paradigm shift. Empirical evidence to show feasibility and utility is essential.

To gain some insight into the feasibility and utility of interpreted formalisms, we conducted a two-part study in which we developed interpretations structured as sets of real-world types for several open-source projects with which we have no association. In the first part, a complete set of real-world types were defined for a project called the Kelpie Flight Planner [54]. Various elements of the software were given real-world types, a set of type rules were defined, and type checking was performed.

In the second part, we reused real-world types and type rules created in part one on a set of projects that access the same real-world entities. For these projects, type checking has only been applied to pertinent packages and files to detect errors.

3.7.1 Kelpie Flight Planner

The Kelpie Flight Planner is an open-source Java project based on Flightgear [5]. The program uses the airport and navaid databases of Flightgear to determine routes between airports based on user inputs. The program is 13,884 lines long, is organized as 10 packages, and is contained in 126 source files.

A critical element of the data used by the Kelpie Flight Planner in modeling aircraft movement is the *velocity surface,* a two-element vector consisting of the horizontal velocity (motion across the Earth's surface) and the vertical velocity (climb or sink rate) of the aircraft. The details of the velocity surface are shown in Figure 3.11.

Various models of the Earth's geometry have been created, including a sphere and an ellipsoid. For the ellipsoid, different models have been developed for special purposes; for

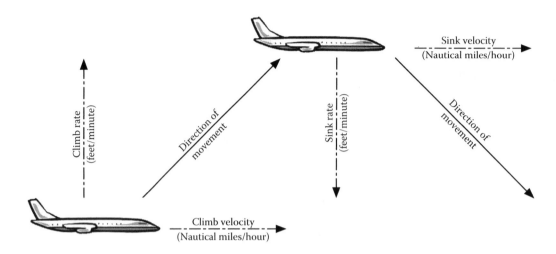

FIGURE 3.11   The velocity surface.

example, the International Geomagnetic Reference Field (IGRF) [6] and the World Magnetic Model (WMM) [7]. In order to undertake useful calculations, programs like the Kelpie Flight Planner have to operate with a model of the Earth's geometry, and details of the model need to be included in the real-world type system in order to allow appropriate checking.

For the Kelpie Flight Planner, 35 different real-world types were defined along with 97 type rules. The total number of type bindings for the project was 255. For illustration, we summarize three of six faults that were identified by the real-world type analysis. None of these faults had been reported in the project error log.

The program source code containing the first fault is

```
alt -= legTime * plan.getAircraft().getSinkSpeed()/60;.
```

The expression references the wrong data. `getSinkSpeed()` returns a quantity measured horizontally and `alt` is measured vertically.

The source code containing the second fault is

```
alt += legTime * plan.getAircraft().getClimbRate()/60;.
```

`plan.getAircraft().getClimbRate()` returns the climb rate in feet/minute, the variable `legTime` is time in hours, and `alt` is altitude in feet. The conversion factor is 60, but the conversion requires multiplication by 60, not division.

The source code containing the third fault is

```
alt -= legTime * plan.getAircraft().getSinkSpeed()/60;.
```

The expression references the wrong data. As in the first fault, `getSinkSpeed()` returns a quantity measured horizontally and `alt` is measured vertically. Correcting this fault yields code with the same units issues as arose in the second fault, requiring an additional fix.

3.7.2 Other Java Applications

We reused the real-world types and type rules created for the Kelpie Flight Planner project to check packages and source files in other applications with pertinent functions. We chose applications for which a log was available for defects reported from the field after release. We include examples here for illustration.

OpenMap is a Java Beans-based tool kit for building applications and applets that access geographic information [8]. The code for the first fault we detected is

```
lon2 = l12.getY();.
```

The variable `lon2` is a longitude, but the method `getY()` returns a latitude. The code for the second fault is

```
double[] llpoints = GreatCircle.greatCircle(startLLP.getY(),
    startLLP.getX(), endLLP.getY(), endLLP.getX(), numPoints, true);.
```

The arguments to `greatCircle()` should be in radians, but in this call the arguments are measured in degrees.

Geoconvertor is a Java API that converts latitude and longitude to points in the Universal Transverse Mercator coordinate system [9]. The faulty code detected is

```
if (datum.name().equals("WGS84")) {
    e1=eccentricityOne(WGS84.MAJOR_AXIS, SAD69.MINOR_AXIS);
    e2=eccentricityTwo(WGS84.MAJOR_AXIS, SAD69.MINOR_AXIS)};}.
```

The constructors `eccentricityOne()` and `eccentricityTwo()` each expect two arguments of the same Earth model, SAD69 for the first and WGS84 for the second. The code has the argument Earth models confused.

3.8 RELATED WORK

A variety of previous research results are relevant to the ideas presented in this chapter. In this section, we summarize the results in various related areas and provide references for more detailed information.

3.8.1 Modeling the Relationships between the Real World and the Machine World

Interpretation of logic is about binding the real world to the machine world. Research efforts on requirements and specification have also modeled the connections between the real world and the machine world [10–14,47].

Jung and Saglietti defined a language for describing interfaces between system components to support fault detection in component interfaces [53]. The language facilitates the definition of system details (including those that we refer to as real-world attributes, such as units). From interface descriptions, software wrappers can be derived to align interfaces, thereby enabling the reuse of reusable components. The approach does not address the issue of real-world information and analysis within software source code.

Johnson et al. developed an approach to detection of undocumented assumptions in Simulink/Stateflow models in which traces are used to develop system invariants [4]. Since the source for the approach is a high-level system model, the approach can detect undocumented assumptions at the level of real-world entities. The approach does not include mechanisms to define real-world properties over and above those available in Simulink/Stateflow.

Representing multiple domains, including both physical and cyber, and determining inconsistencies between those domains has been addressed by Bhave et al. using a set of architectural views [15]. The views are derived from various models and a base architecture for subject CPS. The approach is based on typed graph models of the subject system, and consistency is defined by morphisms from views to the base architecture.

The four-variable model proposed by Parnas and Madey introduces relationships that the authors labeled IN and OUT [14]. The IN and OUT relationships define the connections between mathematical variables available to the software and the environmental variables in the real world. The relationship between real-world entities and the machine world is described mathematically. The hardware and software requirements are intertwined in the REQ relation. Miller and Tribble introduced an extension to the four-variable model that isolates the virtual versions of the monitored and controlled values in subsystems [13]. The relationships emerge as relations between virtual and real versions of variables.

The work of Zave and Jackson characterized phenomena of interest to the system and separated world phenomena from machine phenomena [11,12,16]. The reference model of Gunter et al. gave a detailed explanation of different classes of phenomena and the relationship between environment and system [10]. These results model the picture of the connection between machines and the real world. In contrast to these results, real-world types provide a comprehensive set of real-world semantic attributes and emphasize imposing constraints on software inherited from the real world.

3.8.2 Conventional Types

Conventional types as used in formal languages are closely related to the machine context, and as a result they cannot comprehensively describe real-world information. For example, most real-world semantic attributes are associated with compile-time values; such attributes and associated values should not be represented as variables. Some real-world semantic attributes might be represented as variables, fields, or other structures. However, these program elements can only convey limited real-world meanings through identifier names or unstructured comments. In addition to real-world semantic attributes, the relationships between real-world entities and their machine representations are also ignored or vaguely expressed in conventional types. The discrepancies caused by sensors and timing differences are frequently neglected. As a result, real-world constraints are incompletely documented and are thereby enforced in conventional type systems in ad hoc ways or not at all.

3.8.3 Enhanced Type Systems

A real-world type system is an enhanced and extended version of the concept underlying conventional type systems with the goal of supporting checking constraints inherited from the real world in addition to default type rules.

Powerful extensions to the basic notion of type have been developed, in particular in the form of pluggable type systems [17–20]. Pluggable type systems [20] enhance the built-in type systems in applicable formal languages and provide support for additional checking capabilities. The Checker framework [17,20] and JAVACOP [18] implement the idea of pluggable type system for Java. These frameworks refine built-in type systems to allow users to define additional types and check additional type rules.

Dependent types [21,48,50,56] are another powerful type system concept that allows programmers to specify and enforce rich data invariants and guarantee that unwanted program behaviors are detectable by analysis. They are important in computing environments where users must certify and check properties of untrusted programs [22]. Dependent type systems, such as Coq [23] and Agda [24], provide formal languages to write mathematical definitions, executable algorithms, and theorems and then support development of proofs of these theorems.

Pluggable type systems and dependent type systems are designed to provide greater flexibility in type mechanisms. Increased flexibility is valuable in supporting language expressivity. However, the resulting flexibility remains within the mathematical framework of machine logic and does not address the notion of deriving and exploiting type information from the real world.

3.8.4 Real-World Constraints

An interpretation based on real-world types supports analyzing real-world constraints. As special kinds of real-world constraints, dimensional analysis and unit checking have been explored in many programming languages [25–27,51]. Previous research focused on extending programming languages to allow checking that constraints on dimensions of equations are not broken. Extensions to support dimensional and unit analysis have been developed for several programming languages. For the most part, previous research focused on checking dimensions of equations and validating unit correctness [26,28–32]. Nevertheless, these efforts are limited to basic rules derived from dimensions or combinations of entities with different units.

3.8.5 Improved Program Understanding

Real-world semantic attributes in real-world types can improve the understanding of programs. Other researchers have attempted to improve program understanding by linking structured semantic information in real-world contexts to programs [33–35,49]. Ontology is a widely used structure for documenting domain concepts and relationships among the concepts. Ratiu et al. developed techniques to improve the understanding of program elements by making explicit mappings between ontology classes and program elements [33,34].

3.8.6 Concept Location

Concept location, that is, identifying parts of a software system that implement some aspect of the problem, is related to the idea of mapping real-world entities to software and has been studied extensively [35–40,55,57]. Information-retrieval-based approaches have been developed to reduce the effort required to understand and to locate source code that requires change. Poshyvanyk attempted to further reduce the efforts by producing a concept lattice using the most relevant attributes (terms) selected from the top-ranked documents (methods) [41]. Grant et al. proposed approaches that identify statistically independent signals that could lead to concepts [39].

The use of parts of speech of terms in identifiers has been investigated as a means of extracting information from the source code. Binkley et al. improved identifier name tagging using templates and defined rules to improve the structure of field names [42]. Hill et al. generated noun, verb, and prepositional phrases from the signatures of program elements [40].

Parts of speech have also been used to extract domain models, such as ontologies. Abebe and Tonella [36,37] have used the parts of speech of terms and the natural-language dependencies to extract ontologies from source code. Raitu et al. have proposed an approach to extract domain-specific ontologies from APIs [33,34]. Other researchers have used WordNet to automatically extract semantics and relationships between the semantics [43,44].

3.8.7 Context Representation and Reasoning

The notion of *context* is important in many areas of computing. In ubiquitous computing, for example, context is needed to enable suitable processing of inputs received [45,46]. This notion of context is related to the basic functionality of the system and is closely linked to machine learning and other aspects of artificial intelligence. Context representation and reasoning is related to the rigorous definition developed in this chapter only to the extent that it helps to identify the real-world entities with which the system of interest interacts.

3.9 SUMMARY

In this chapter, we have presented a rigorous definition of CPS based on the idea that such a system is composed of a computing platform, a set of physical entities, and a connection between the two. The emphasis in the definition is on a precise and complete definition of the connection between the physical entities and the computing platform.

The connection has been defined using the notion of logical interpretation. The interpretation precisely defines the meaning of items in the logic in terms of physical entities and their associated properties. Many important invariants can be derived from physical, that is, real-world, entities in a comprehensive and systematic way. Defining the connection between physical entities and the computing platform in this way enables a new class of fault detection mechanisms for the logic, that is, for CPS software.

REFERENCES

1. Mars Climate Orbiter Mishap Investigation Board Phase I Report, 1999. National Aeronautics and Space Administration, Washington DC, November 10, 1999.
2. Bergin, C. and P. Harding. 2013. Cygnus delays ISS berthing following GPS discrepancy. http://www.nasaspaceflight.com/2013/09/cygnus-cots-graduation-iss-berthing/
3. Xiang, J., J. Knight, and K. Sullivan. 2015. Real-world types and their application. In *Proceedings of the 34th International Conference on Computer Safety, Reliability and Security (SAFECOMP)*. Delft, 2015, 471–484. Springer, 2015.

4. Johnson, T., B. Stanley, and D. Steven. 2015. Cyber-physical specification mismatch identification with dynamic analysis. In *Proceedings of the ACM/IEEE Sixth International Conference on Cyber-Physical Systems* (ICCPS). 208–217. ACM Press, Seattle, WA, 2015.

5. FlightGear. http://www.flightgear.org/

6. International Geomagnetic Reference Field. http://www.ngdc.noaa.gov/IAGA/vmod/igrf.html

7. World Magnetic Model. http://www.ngdc.noaa.gov/geomag/WMM/DoDWMM.shtml

8. OpenMap. https://code.google.com/p/openmap/

9. Geoconvertor. https://code.google.com/p/geoconvertor/

10. Gunter, C. A., E. L. Gunter, M. Jackson, and P. Zave. 2000. A reference model for requirements and specifications. *IEEE Softw.* 17(3):37–43. IEEE, 2000.

11. Jackson, M. and P. Zave. 1993. Domain descriptions, In *Proceedings of the Second IEEE International Symposium on Requirements Engineering.* Los Alamitos, CA, 56–64. IEEE, 1993.

12. Jackson, M. and P. Zave. 1995. Deriving specifications from requirements: An example. In *Proceedings of the 17th International Conference on Software Engineering (ICSE).* 15–24. ACM Press, Seattle, WA, 1995.

13. Miller, S. P. and A. C. Tribble. 2001. Extending the four-variable model to bridge the system-software gap. In *Proceedings of the 20th Digital Avionics System Conferences.* Daytona Beach, 14–18. IEEE Computer Society, 2001.

14. Parnas, D. L. and L. Madey. 1995. Functional documents for computer systems. In *Sci. Comput. Program.* 25:41–61. Amsterdam: Elsevier North-Holland, Inc., 1995.

15. Bhave, B., B. H. Krogh, D. Garlan, and B. Schmerl. View consistency in architectures for cyber-physical systems. In *Proceedings of the 2011 IEEE/ACM International Conference on Cyber-Physical Systems (ICCPS)*, Chicago, 2011, 151–160. IEEE Computer Society, 2011.

16. Jackson, M. 2000. *Problem Frames: Analyzing and Structuring Software Development Problems.* Boston, Addison-Wesley Longman Publishing Co., Inc., 2000.

17. Dietl, W., S. Dietzel, M. Ernst, K. Mužlu, and T. Schiller. 2011. Building and using pluggable type-checkers. In *Proceedings of the 33rd International Conference on Software Engineering (ICSE).* Waikiki, Honolulu, 681–690. ACM Press, 2011.

18. Markstrum, S., D. Marino, M. Esquivel, T. Millstein, C. Andreae, and J. Noble. 2010. JavaCOP: Declarative pluggable types for java. In *ACM Trans. Program. Lang. Syst.* 32:1–37. ACM Press, New York, 2010.

19. Milanova, A. and W. Huang. 2012. Inference and checking of context-sensitive pluggable types. In *Proceedings of the ACM SIGSOFT 20th International Symposium on the Foundations of Software Engineering (FSE).* Article 26. ACM Press, North Carolina, USA, 2012.

20. Papi, M., M. Ali, C. T. Luis, J. H. Perkins, and M. D. Ernst. 2008. Practical pluggable types for Java. In *Proceedings of the ACM SIGSOFT International Symposium on Software Testing and Analysis (ISSTA)*, edited by B. G. Ryder and A. Zeller. 201–212. ACM Press, Seattle, WA, 2008.

21. Bove, A. and P. Dybjer. 2009. Dependent types at work. In *Language Engineering and Rigorous Software Development*, edited by Ana Bove, Luís Soares Barbosa, Alberto Pardo, and Jorge Sousa Pinto. 57–99. Springer, Berlin, Heidelberg, 2009.

22. Ou, X., G. Tan, Y. Mandelbaum, and D. Walker. 2004. Dynamic typing with dependent types. In *Exploring New Frontiers of Theoretical Informatics.* 155:437–450. Springer, 2004.

23. Coq. https://coq.inria.fr/

24. Agda. http://wiki.portal.chalmers.se/agda/pmwiki.php

25. Chen, F., G. Rosu, and R. P. Venkatesan. 2003. Rule-based analysis of dimensional safety. In *Proceedings of the 14th International Conference on Rewriting Techniques and Applications (RTA)*, edited by Robert Nieuwenhuis. 197–207. Springer, Berlin, Heidelberg, 2003.

26. Grein, C., D. Kazakov, and F. Wilson. 2003. A survey of physical unit handling techniques in ada. In *Proceedings of the 8th Ada-Europe international Conference on Reliable Software Technologies (Ada-Europe)*, edited by Jean-Pierre Rosen and Alfred Strohmeier. 258–270. Springer, Berlin, Heidelberg, 2003.

27. Wyk, E. and Y. Mali. 2007. Adding dimension analysis to Java as a composable language extension. In *Generative and Transformational Techniques in Software Engineering II*. E Ralf Lämmel and Joost Visser (Eds), Lecture Notes in Computer Science, Vol. 5235. 442–456. Springer, 2007.

28. Antoniu, T., P. A. Steckler, S. Krishnamurthi, E. Neuwirth, and M. Felleisen. 2004. Validating the unit correctness of spreadsheet programs. In *Proceedings of the 26th International Conference on Software Engineering (ICSE)*, Edinburgh, Scotland, 2004, 439–448. IEEE Computer Society, 2004.

29. Delft, V. 1999. A Java extension with support for dimensions. In *Softw. Pract. Exper.* 29:605–616. John Wiley & Sons, Inc., 1999.

30. Jiang, L. and Z. Su. 2006. Osprey: A practical type system for validating dimensional unit correctness of C programs. In *Proceedings of the 28th International Conference on Software engineering (ICSE)*. Shanghai, 262–271. ACM Press, 2006.

31. Kennedy, A. 1999. Dimension types. In *Proceedings of the 5th European Symposium on Programming (ESOP)*. Edinburg, UK. Lecture Notes in Computer Science volume 788. Springer, 1999.

32. Petty, G. 2001. Automated computation and consistency checking of physical dimensions and units in scientific programs. In *Softw. Pract. Exper.* 31:1067–1076. New York: John Wiley & Sons, Inc., 2001.

33. Ratiu, D. and F. Deissenboeck. 2007. From reality to programs and (Not Quite) back again. In *Proceedings of the 15th IEEE International Conference on Program Comprehension (ICPC)*. Banff, Alberta, 91–102. IEEE Computer Society, 2007.

34. Ratiu, D., M. Feilkas, and J. Jurjens. 2008. Extracting domain ontologies from domain specific APIs. In *Proceedings of the 2008 12th European Conference on Software Maintenance and Reengineering (CSMR)*. Athens, 203–212. IEEE Computer Society, 2008.

35. Yang, L., H. Yang, and W. Chu. 2000. Generating linkage between source code and evolvable domain knowledge for the ease of software evolution. In *Proceedings International Symposium on Principles of Software Evolution*, Kanazawa, 2000. 196–205. IEEE Computer Society, 2000.

36. Abebe, S. and P. Tonella. 2010. Natural language parsing of program element Names for Concept Extraction. In *Proceedings of the 18th International Conference on Program Comprehension (ICPC)*, Braga, 2010, 156–159. IEEE Computer Society, 2010.

37. Abebe, S. and P. Tonella. 2011. Towards the extraction of domain concepts from the identifiers. In *Proceedings of the 18th Working Conference on Reverse Engineering (WCRE)*, Limerick, 2011, 77–86. IEEE Computer Society, 2011.

38. Gay, G., S. Haiduc, A. Marcus, and T. Menzies. 2009. On the use of relevance feedback in IR-based concept location. In *Proceedings of the 25th International Conference on Software Maintenance (ICSM)*. Edmonton, 2009, 351–360. IEEE Computer Society, 2009.

39. Grant, S., J. R. Cordy, and D. Skillicorn. 2008. Automated concept location using independent component analysis. In *Proceedings of the 15th Working Conference on Reverse Engineering (WCRE)*, Antwerp, 2008, 138–142. IEEE Computer Society, 2008.

40. Hill, E., L. Pollock, and K. Vijay-Shanker. 2009. Automatically capturing source code context of nl-queries for software maintenance and reuse. In *Proceedings of the 31st International Conference on Software Engineering (ICSE)*, Washington, DC, 2009, 232–242. IEEE Computer Society, 2009.

41. Poshyvanyk, D. and A. Marcus. 2007. Combining formal concept analysis with information retrieval for concept location in source code. In *Proceedings of the 15th IEEE International Conference on Program Comprehension (ICPC)*, Banff, Alberta, 2007, 37–48. IEEE Computer Society, 2007.

42. Binkley, D., M. Hearn, and D. Lawrie. 2011. Improving identifier informativeness using part of speech information. In *Proceedings of the 8th Working Conference on Mining Software Repositories (MSR)*, Waikiki, 2011, 203–206. ACM Press, 2011.

43. Ruiz-Casado, M., E. Alfonseca, and P. Castells. 2005. Automatic extraction of semantic relationships for wordnet by means of pattern learning from Wikipedia. In *Proceedings of the 10th International Conference on Natural Language Processing and Information Systems (NLDB)*, edited by A. Montoyo, R. Muñoz, and E. Métais. 67–79. Springer, Berlin Heidelberg, 2005.

44. Wordnet. http://wordnet.princeton.edu

45. Kofod-Petersen, A. and M. Mikalsen. 2005. Context: Representation and reasoning: Representing and reasoning about context in a mobile environment. Special issue of the Revue d'Intelligence Artificielle on Applying Context-Management. 2005.

46. Perttunen, M., J. Riekki, and O. Lassila. 2009. Context representation and reasoning in pervasive computing: A review. *International Journal of Multimedia and Ubiquitous Engineering*. 4(4): 1–28. 2009.

47. Ait-Ameur, Y., J. P. Gibson, and D. Méry. 2014. On implicit and explicit semantics: Integration issues in proof-based development of systems. In *Leveraging Applications of Formal Methods, Verification and Validation. Specialized Techniques and Applications*, edited by Tiziana Margaria and Bernhard Steffen, 604–618. Springer, Berlin Heidelberg, 2014.

48. Brady, E. 2011. Idris: Systems programming meets full dependent types. In *Proceedings of the 5th ACM Workshop on Programming Languages Meets Program Verification (PLPV)*, Austin, 2011, 43–54. ACM Press, 2011.

49. Castro, J., M. Kolp, and J. Mylopoulos. 2001. A requirements-driven development methodology. In *Advanced Information Systems Engineering*, edited by K.R. Dittrich, A. Geppert, M. Norrie, CAiSE 2001. LNCS, vol. 2068, 108–123. Springer, Berlin Heidelberg, 2001.

50. Christiansen, D. 2013. Dependent type providers. In *Proceedings of the 9th ACM Workshop on Generic Programming*, 23–34. ACM Press, Boston, MA, 2013.

51. Hangal, S. and M. S. Lam. 2009. Automatic dimension inference and checking for object-oriented programs. In *Proceedings of the 31st International Conference on Software Engineering (ICSE)*. Washington, DC, 155–165. IEEE Computer Society, 2009.

52. The International System of Units (SI). National Institute of Standards Technology, Washington, DC, 2008.

53. Jung, M. and F. Saglietti. 2005. Supporting component and architectural re-usage by detection and tolerance of integration faults. In *Proceedings of the 9th IEEE International Symposium on High-Assurance Systems Engineering (HASE)*. Washington, DC, 47–55. IEEE Computer Society, 2005.

54. Kelpie Flight Planner for Flightgear. http://sourceforge.net/projects/fgflightplanner/

55. Marcus, A., V. Rajlich, J. Buchta, M. Petrenko, and A. Sergeyev. 2005. Static techniques for concept location in object-oriented code. In *Proceedings of the 13th International Workshop on Program Comprehension (IWPC)*. St. Louis, MO, 33–42. IEEE Computer Society, 2005.

56. McKinna, J. 2006. Why dependent types matter. In *Proceedings of the 33rd ACM SIGPLAN-SIGACT Symposium on Principles of Programming Languages* (POPL). 1–1. ACM Press, 2006.

57. Rajlich, V. and N. Wilde. 2002. The role of concepts in program comprehension. In *Proceedings of the 10th International Workshop on Program Comprehension (IWPC)*. Washington, DC, 271–278. IEEE Computer Society, 2002.

CHAPTER 4

A Generic Model for System Substitution

Guillaume Babin, Yamine Aït-Ameur, and Marc Pantel

CONTENTS

4.1 INTRODUCTION

One of the key properties studied in Cyber-Physical Systems (CPS) [1] is the capability of a system to react to changes (e.g., failures, quality of service change, context evolution, maintenance, resilience, etc.). The development of such systems needs to handle explicitly, and at design time, the reactions to changes occurring at runtime. One of the solutions consists in defining another system that substitutes the original one. This substitute system may be a new system or an improvement of the original one.

System substitution can be defined as the capability to replace a system by another one that preserves the specification of the original one. It may occur in different reconfiguration situations like failure management or maintenance. When substituting a system at runtime, a key requirement is to identify the restoring state and to correctly restore the identified state of the substituted one. The correctness of the state restoration relies on the definition of safety properties expressed by invariants and their preservation.

The objective of the work presented in this paper is to present a generic framework defining a system substitution operator. We rely on formal methods, more precisely the proof and refinement-based formal method Event-B [2] to describe both the studied systems and the introduced substitution operation. We have chosen to describe systems as state transition systems. The substitution operation is defined using specific state restoration that preserves system invariants. The choice of state transition systems is motivated by the fact that several CPS models use such state transition systems for monitoring, controling, etc.

Moreover, the proposed framework supports different substitution features. Indeed, depending on the chosen state restoration operation and on the defined invariants, degraded, upgraded, or equivalent system substitution can be handled. The framework also supports the capability to define substitution at design time (cold start) or at runtime (warm start). Furthermore, this framework is applicable to the design of adaptive, resilient, dependable, self-∗ trustworthy CPSs that require the capability to substitute running systems.

We have structured this paper as follows. Next section reviews some of the main approaches for system reconfiguration. Section 4.3 gives an overview of the Event-B method. Then, Section 4.4 defines the systems we are addressing as state transition systems, and gives the associated formal Event-B models. The main contribution of this chapter is presented in Section 4.5. It describes a stepwise methodology for the design of a correct system substitution. The properties of the substitution operator we have defined are discussed in Section 4.6. A case study illustrating the approach is presented in Section 4.7. The possible applications of the defined substitution operator are discussed in Section 4.8. Finally, a conclusion summarizes our contribution in the last section.

4.2 STATE OF THE ART

CPSs are strongly linked to their environment, which can change significantly. Thus, there is a strong requirement for these systems to adapt to these changes to ensure their correctness. System substitution is a key element to implement adaptation. There are various techniques and tools that are proposed by several authors.

First, various formal tools are used to ensure the correctness of dynamic reconfiguration. In Reference 3, π-calculus and process algebra are used to model the system and exploit behavioral matching based on bisimulation to reconfigure the system appropriately. An extended transaction model is presented to ensure consistency during reconfiguration of distributed systems in Reference 4. The B method is applied for validating dynamic reconfiguration of the component-based distributed systems using proofs techniques for consistency checking and model-checking for timing requirements [5]. A high-level language is used to model architectures (with categorical diagrams) and for operating changes over a configuration (with algebraic graph rewriting) [6].

Second, dynamic reconfiguration can be used as part of a fault-tolerance mechanism that is a major concern for designing the dependable systems [7,8]. Rodrigues et al. [9] presented the dynamic membership mechanism as a key element of a reliable distributed storage system. Event-B is demonstrated in the specification of cooperative-error recovery and dynamic reconfiguration for enabling the design of a fault-tolerant multiagent system, and to develop dynamically reconfigurable systems to avoid redundancy [10–12]. This approach enables the discovery of possible reconfiguration alternatives that are evaluated through probabilistic verification.

Third, dynamic reconfiguration is used to meet several objectives of autonomic computing [13,14] and self-adaptive systems [15,16] such as self-configuration and self-healing. The self-configuring systems require dynamic reconfiguration that allows the systems to adapt automatically. Similarly, the dynamic reconfiguration allows to correct faults in self-healing systems. Finally, note that some approaches dealing with adaptive systems address nonfunctional requirements [17–19].

4.3 EVENT-B: A CORRECT-BY-CONSTRUCTION METHOD

An Event-B model [2] (see Model 4.1) is defined in a machine. It encodes a state transition system that consists of the variables declared in the variables clause to represent the state; and the events declared in the events clause to represent the transitions (defined by a Before-After predicate BA) from one state to other.

The model contains also invariants and theorems to represent relevant properties of the defined model. A variant may introduce convergence properties when needed. An Event-B machine is related, through the sees clause, to a context that contains the relevant sets, constants, axioms, and theorems needed for defining an Event-B model. The refinement [20],

```
Context ctxt_id_2
Extends ctxt_id_1
Sets s
Constants c
Axioms A(s,c)
Theorems T_c(s,c)
End
```

```
Machine machine_id_2
Refines machine_id_1
Sees ctxt_id_2
Variables v
Invariants I(s,c,v)
Theorems T_m(s,c,v)
Variant V(s,c,v)
Events
 Event Initialisation =
  Begin
   v :| D(s,c,x,v')
  End
 Event evtr =
  Refines evt
  Any x
  Where G(s,c,v,x)
  Then v :| BA(s,c,v,x,v')
  End
End
```

Model 4.1   Structures of Event-B contexts and machines.

TABLE 4.1   Some Proof Obligations Generated for an Event-B Model

Theorems	$A(s,c) \Rightarrow T_c(s,c)$
	$A(s,c) \land I(s,c,v) \Rightarrow T_m(s,c,v)$
Invariant preservation	$A(s,c) \land I(s,c,v) \land G(s,c,v,x) \land BA(s,c,v,x,v') \Rightarrow I(s,c,v')$
Event feasibility	$A(s,c) \land I(s,c,v) \land G(s,c,v,x) \Rightarrow \exists v'.BA(s,c,v,x,v')$
Variant progress	$A(s,c) \land I(s,c,v) \land G(s,c,v,x) \land BA(s,c,v,x,v')$
	$\Rightarrow V(s,c,v') < V(s,c,v)$

introduced by the refines clause, decomposes a model (thus a transition system) into another transition system with more design decisions while moving from an abstract level to a concrete one. New variables and new events may be introduced at a new refinement level. In the new refinement level, the given invariants must link the variables of the refined machine and the abstract machine. Gluing invariants can be introduced for this purpose. The obtained refinement preserves the proved properties and supports the definition of new ones.

Once an Event-B machine is defined, a set of proof obligations is generated. They are passed to the prover embedded in the RODIN platform. Proof obligations associated to an Event-B model are listed in Table 4.1. The prime notation is used to distinguish between pre (x) and post (x') variables. More details on proof obligations can be found in Reference 2.

4.4 STUDIED SYSTEMS

The systems addressed by our approach are formalized by state transition systems [21], that have proven to be useful to model various kinds of systems and particularly hybrid systems [22] or CPSs [1]. In particular, controllers are modeled within state transitions systems.

A system is characterized by a state that may change after a transition occurs. A state is defined as a set of pairs (*variable*, *value*). The values of a given variable are taken in a set of values satisfying safety properties expressed within invariants (Kripke structure). A transition characterizes a state change, through updating of variable values.

Below, we show how such transition systems are modeled with the Event-B method.

4.4.1 Specification of Studied Systems

When the studied systems are described as state transition systems, they are modeled using Event-B as follows:

- A set of variables, in the Variables clause is used to define system states. The Invariants clause describes the relevant properties of these variables.

- An Initialisation event determines the initial state of the described system by assigning initial values to the variables.

- A set of (guarded) events defining transitions is introduced. They encode transitions and record variable changes.

A state transition system (where the variables clause defines states and the events clauses define transitions, see Model 4.3) is described in an Event-B machine Spec. This machine sees the context C0 (see Model 4.2) from which it borrows relevant definitions and theories.

4.4.2 Refinement of Studied Systems

The previously defined transition system may be defined at a given abstraction level. It constitutes a system specification. Several candidate systems S_i may implement the same specification *Spec*. These implementations are more concrete state transition systems that refine an abstract one. Model 4.4 shows such a refinement. A new set of variables and events is introduced that refines the abstract model.

Refinement relies on the definition of a gluing invariant. The verification of the correctness of this refinement ensures that the refined system is a correct implementation of the specification it refines.

We have chosen to use the refinement relationship in order to characterize all the substitute systems. If we consider a system characterized by an original specification, then all the systems that refine this specification are considered as a potential substitute. Obviously, we are aware that these refining systems are different and may behave differently, but we are

```
Context C0
Sets s
Constants c
Axioms A(s, c)
End
```

Model 4.2   Context C0.

sure that these behaviors include the one of the refined systems. The definition of the horizontal invariant introduced by the repairing event introduced later is the key property ensuring properties preservation after system substitution.

```
Machine Spec
Sees C0
Variables v_A
Invariants I_A(s, c, v_A)
Events
 Event Initialisation =
  Begin
   v_A :| D_A(s, c, v'_A)
  End
 Event Evt =
  Any
   x_A
  Where
   G_A(x_A, s, c, v_A)
  Then
   v_A :| BA_A(x_A, s, c, v_A, v'_A)
  End
End
```

Model 4.3   Machine *Spec*.

```
Machine Sys_S
Refines Spec
Sees C0
Variables v_S
Invariants I_S(s, c, v_S, v_A)
Variant VN_S
Events
 Event Initialisation =
  Begin
   v_S :| D_S(s, c, v'_S)
   VN_S :| VN_S_InitValue
  End
 Event s_evt =
  Any x_S
  Where
   G_S(x_S, s, c, v_S)
  Then
   v_S :| BA_S(x_S, s, c, v_S, v'_S)
  End
  . . .
 Event Evt Refines Evt = . . .
End
```

Model 4.4   Machine *SysS*.

4.5 SYSTEM SUBSTITUTION

The availability of several refinements for a given specification means that several systems may implement a single specification. Each of these systems behaves like the defined specification. The systems that refine the same specification can be gathered into a class of systems. The availability of such a class makes it possible to address the problem of system substitution or system reconfiguration. The stepwise methodology for system substitution considers one system of this class as a running system, and substitutes it with another system belonging to the same class. Indeed, when a running system is halted (in case of failure or loss of quality of service, etc.), a system of this class can be chosen as a substitute. Here, we describe a formal method allowing system developers to define correct-by-construction system substitution or system reconfiguration. By "correct," we intend the preservation of safety properties expressed by the invariants that also include deadlock freeness invariants.

4.5.1 A Stepwise Methodology

Our approach to define a correct system substitution setting is given in several steps. This stepwise methodology leads to the definition of a system substitution operator whose properties are discussed later.

- *Step 1. Define a system specification*: A state transition system characterizing the functionalities and the suited behavior of the specification system is defined.

- *Step 2. Characterize candidate substitute systems*: All the refinement of the specification represent substitutes of the specified system that preserve the invariants properties expressed at the specification level. A class of substitutes contains all the systems refining the same specification.

- *Step 3. Introduce system modes*: Modes are introduced to identify which system is running, those that have been halted, and the remaining available systems for substitution. A mode is associated to each system and at most one system is running.

- *Step 4. Define system substitution as a composition operator*: When a running system is halted, the selected substitute system becomes the new running system. During this substitution, the state of the halted system shall be restored in the substitute system. Restoring the state of the halted system consists in copying the values of the state of the halted system to the variables of the state of the substitute system. To formalize this operation, a sequence of two specific Event-B events is introduced. The first event, named Fail, consists in halting the running system and switching it to a failure mode. The second one, namely Repair, restores the system state and switches the control to the substitute system. Because Repair depends on the modeling of the internal state of both systems, it has to be explicitly defined for each pair of systems (parameter of the substitution operator). Here, only pairs of systems where the relation between each internal state can be explicitly defined are considered. When substituting a system, the cause leading to halting the running system and the process of determining which system is selected as the next substitute are out of scope of this paper.

4.5.2 An Event-B Model for System Substitution

In this section, we give the Event-B models corresponding to the stepwise methodology presented above. First a system specification *Spec* is given, then we show how a source system S_S defined as a refinement *SysS* of the machine *Spec* can be substituted by a target system S_T defined as a refinement *SysT* of the same machine *Spec*. Two events Fail and Repair for halting a system S_S and for transferring the control to the target system S_T are introduced.

4.5.2.1 Step 1. Define a System Specification

The specification of the system is given by an abstract description of its functionalities and its behavior. An Event-B machine *Spec*, corresponding to the one in Model 4.3, defines the system specification. More events may be introduced to define this behavior; we have just limited our description to one single event.

4.5.2.2 Step 2. Characterize Candidate Substitute Systems

As stated above in Section 4.5.1, a class of substitute systems is defined as the set of the systems that are described as an Event-B refinement of the original Event-B machine *Spec*. Two systems *SysS* and *SysT* described by the Event-B refinements Models 4.5 and 4.6 are substitute systems for the system described by the specification *Spec*. Note that several refinement steps may be required before the final models of the substitute systems are obtained.

```
Machine SysS
Refines Spec
Sees C0
Variables vS
Invariants IS(s, c, vS, vA)
Variant VNS
Events
 Event Initialisation =
  Begin
     vS  :|  DS(s, c, v'S)
     VNS :|  VNS_InitValue
  End
 Event s_evt =
  Any xS
  Where
     GS(xS, s, c, vS)
  Then
     vS  :|  BAS(xS, s, c, vS, v'S)
  End
     . . .
 Event Evt Refines Evt = . . .
End
```

Model 4.5   Machine *SysS*.

```
Machine SysT
Refines Spec
Sees C0
Variables vT
Invariants IT(s, c, vT, vA)
Variant VNT
Events
 Event Initialisation =
  Begin
     vT  : |  DT(s, c, v'T)
     VNT : |  VNT_InitValue
  End
 Event t_evt =
  Any xT
  Where
    GT(xT, s, c, vT)
  Then
    vT  : |  BAT(xT, s, c, vT, v'T)
  End
    . . .
 Event Evt Refines Evt = . . .
End
```

Model 4.6   Machine *SysT*.

On these two refinements *SysS* and *SysT*, we note the presence of

- New sets of variables

- An invariant describing the properties of the system and gluing the variables with the ones of the abstraction in the *Spec* machine

- New events may be either added or refined in order to describe the behavior of the new variables or define behaviors that were hidden in the specification

We consider that both *SysS* and *SysT* see the context *C0* of the specification *Spec*, and we assume that no new specific element is needed for their own contexts.

4.5.2.3 Step 3. Introduce System Modes

The introduction of modes is a simple operation consisting of defining a new variable m (standing for *mode*). The values of the variable mode may be either the system identifier or the value F to represent a halted system in a failure mode. Moreover, the invariant related to each substitute system S shall be valid when the variable m equals the system identifier S. Models 4.7 and 4.8 show the description of the systems S and T with introduced mode. Again, each of the machines *SysS** and *SysT** refine the original specification *Spec*. At this step, we also anticipate any name clashes by renaming some elements.

```
Machine  Sys*_S
Refines  Sys_S  Sees  C0
Variables  v_S, m
Invariants  m = S ⇒ I_S(s, c, v_S, v_A)
Variant  VN_S
Events
 Event  Initialisation =
  Begin
   m := S
   v_S  :|  D_S(s, c, v'_S)
   VN_S  :|  VN_S_InitValue
  End
 Event  s_evt =
  Any  y_S
  Where
   m = S ∧ G_S(y_S, s, c, v_S)
  With
   y_S : x_S = y_S
  Then
   v_S  :|  BA_S(y_S, s, c, v_S, v'_S)
  End
  . . .
 Event  Evt  Refines  Evt = . . .
End
```

Model 4.7   Machine $SysS^*$.

```
Machine  Sys*_T
Refines  Sys_T  Sees  C0
Variables  v_T, m
Invariants  m = T ⇒ I_T(s, c, v_T, v_A)
Variant  VN_T
Events
 Event  Initialisation =
  Begin
   m := T
   v_T  :|  D_T(s, c, v'_T)
   VN_T  :|  VN_T_InitValue
  End
 Event  t_evt =
  Any  y_T
  Where
   m = T ∧ G_T(y_T, s, c, v_T)
  With
   y_T : x_T = y_T
  Then
   v_T  :|  BA_T(y_T, s, c, v_T, v'_T)
  End
  . . .
 Event  Evt  Refines  Evt = . . .
End
```

Model 4.8   Machine $SysT^*$.

4.5.2.4 Step 4. Define System Substitution as a Composition Operator

The machines *SysS** and *SysT** are composed into a single Event-B machine with two new events Fail and Repair. The role of the substitution operation is to enable the following sequence of events.

1. The source system S is the first running system. The variable mode m is initialized to the value S in order to transfer the control to the events of the system S.

2. When a halting event occurs, the Fail event (Model 4.9) is triggered. This event changes the value of the variable mode m to the value F. At this state, the system S is stopped and the invariant I_S is valid at the current state. Note that the event Fail can be triggered for any reason in the current formalization. Furthermore, this event can be refined in order to introduce failure conditions like loss of quality of service.

3. At this stage, the Repair event (Model 4.10) is triggered because its guard ($m = F$) becomes true. This event serves two purposes. On the one hand it restores the state of the halted system by defining the values of the variables v_T of the substitute system S_T and on the other hand it sets up the variable VN_T used to express the variant, to allow the restart of the system S_T at the suited state (or an approximate state). Finally, the mode is changed to T so that the control transfers to the substitute system S_T.

```
Event fail =
  Where
    m = S
  Then
    m := F
  End
```

Model 4.9   Extract: Event fail.

```
Event repair =
  Where
    m = F
  Then
    -- New values for state variables
    vS, vT := ...
    -- New values for variants
    VNT := ...
    -- Change mode
    m := T
  End
```

Model 4.10   Extract: Event repair.

The definition of the Repair event implies the definition of state restoration. The new values of the variables of system S_T must fulfill safety conditions in order to move the control to S_T in order for the invariant T_T to hold in the recovery state. In other words, specific proof obligations are associated to the Repair event.

4.5.3 Substitution as a Composition Operator

As stated above, the Repair event shall be defined so that the state restoration preserves the safety properties described in the invariants. The definition of this event is completed in Model 4.11.

At this level, two predicates are defined.

1. The *Recover* predicate characterizes the new values of the variables v_T such that the invariant I_T holds in the next state. It represents *horizontal* invariant that glues the state variables of system S_S with the variables of system S_T.

2. The *Next* predicate describes the next value of the variant. It determines which state in the system S_T, is used as the new restoring state.

4.5.4 The Obtained Composed System with Substitution

Once the Fail and Repair events have been defined, the obtained model is composed of the two systems S_S and S_T. The sequence described above is encoded using a predefined sequence of assignments of the mode variable m in the corresponding events.

Moreover, the invariant of the final system is defined by cases depending on the value of the mode variable. When the system S_S is running, then invariant I_S holds, when the system S_T is running, then invariant I_T holds and finally, as stated previously, the invariant I_S holds when the system S_S is halted and substituted. The obtained invariant is then a conjunction of three implications.

The obtained global system is again described as a refinement of the original specification. It is formalized by Event-B machine Sys_G as shown in Model 4.12.

$$S_G = S_S \underset{(Recover, Next)}{\circ} S_T \; Refines \; Spec \tag{4.1}$$

```
Event repair =
  Where
    m = F
  Then
    v_S, v_T  :|  Recover(v_S, v_T, v'_S, v'_T)
    VN_T      :|  Next(V_S, V'_T)
    m := T
  End
```

Model 4.11   Extract: Event repair.

```
Machine SysG                            . . .
Refines Spec                            Event Evt Refines Evt = ...
Sees C0                                 Event fail
Variables vS, vT, m                       Where
Invariants m = S ⇒ IS(s, c, vS)             m = S
            ∧ m = F ⇒ IS(s, c, vS)        Then
            ∧ m = T ⇒ IT(s, c, vT)          m := F
Variant VNS + VNT                         End
Events                                  Event repair =
 Event Initialisation =                   Where
  Begin                                     m = F
   m := S                                 Then
   vS :| DS(s, c, v'S)                      vS, vT :| Recover(vS, vT, v'S, v'T)
   vT :| ⊤                                  VNT :| Next(VS, V'T)
   VNS :| VNS_InitValue                      m := T
   VNT :| 0                               End
  End                                     Event t_evt =
 Event s_evt =                            Any xT
  Any xS                                  Where
  Where                                     m = T ∧ GT(xT, s, c, vT)
   m = S ∧ GS(xS, s, c, vS)               Then
  Then                                       vT :| BAT(xT, s, c, vT, v'T)
   vS :| BAS(xS, s, c, vS, v'S)           End
  End                                   End
```

Model 4.12   Machine Sys_G.

Finally, as defined in Equation 4.1, the obtained composition operator is parameterized by the *Recover* and *next* predicates.

4.6 PROOF OBLIGATIONS FOR THE SYSTEM SUBSTITUTION OPERATOR

The proof obligations resulting from the definition of our substitution operator concern invariant preservation by the different events of the Event-B machine Sys_G. Let us analyze these proof obligations.

- For the initialization and the events of system *SysS*, the preservation of the invariant is straightforward. The proofs are those that have been performed for the refinement introducing modes in the previous step.

- The same situation occurs for the events of system *SysT*. Again, the associated proof obligations are those obtained and proved when introducing modes in the previous step.

- The Fail event preserves the invariant since it does not modify any variable except the mode, it maintains the invariant I_S with $m = S \Rightarrow I_S(s, c, v_S) \land m = F \Rightarrow I_S(s, c, v_S)$.

- Finally, the Repair event considers that I_S holds before substitution and it must ensure that the invariant I_T holds after substitution.

So, the introduction of the Repair event entails specific proof obligations that need to be discharged in order to ensure the correctness of the substitution. The definition of the Recover predicate is the key point to obtain a correct system substitution. The proof obligations associated to the Repair event consists first in preserving the invariants and second in restoring the correct variant value.

4.6.1 Invariant Preservation Proof Obligation

Invariant preservation for the repair event requires to establish that the invariant I_T of system S_T holds in the recovery state. In other words, under the hypotheses given by the axioms $A(s, c)$, the guard $m = F$, the invariant $m = S \Rightarrow I_S(s, c, v_S) \land m = T \Rightarrow I_T(s, c, v_T) \land m = F \Rightarrow I_S(s, c, v_S)$ and the new variable values $Recover(v_S, v_T, v'_S, v'_T) \land m' = T$, the invariant $m' = S \Rightarrow I_S(s, c, v'_S) \land m' = T \Rightarrow I_T(s, c, v'_T) \land m' = F \Rightarrow I_S(s, c, v'_S)$ hold for the variables in the next state. The sequent in Equation 4.2 describes this proof obligation.

$$A(s, c),$$
$$m = S \Rightarrow I_S(s, c, v_S) \land m = T \Rightarrow I_T(s, c, v_T)$$
$$\land m = F \Rightarrow I_S(s, c, v_S),$$
$$m = F,$$
$$Recover(v_S, v_T, v'_S, v'_T) \land m' = T$$
$$\vdash$$
$$m' = S \Rightarrow I_S(s, c, v'_S) \land m' = T \Rightarrow I_T(s, c, v'_T)$$
$$\land m' = F \Rightarrow I_S(s, c, v'_S) \qquad (4.2)$$

After simplification, the previous proof obligation leads to the definition of the final proof obligation of Equation 4.3 associated to invariant preservation.

$$\boxed{A(s, c) \vdash I_S(s, c, v_S) \land Recover(v_S, v_T, v'_S, v'_T) \Rightarrow I_T(s, c, v'_T)} \qquad (4.3)$$

4.6.2 Variant Definition Proof Obligation

The introduction of the new variant value determines the restoring state in the target system S_T. The predicate *Next* needs to be defined so that the variant $VN_S + VN_T$ of the global system decreases. It is required to establish that $VN'_S + VN'_T < VN_S + VN_T$. The next value of VN'_T determines the restoring state in system S_T. Since the value of the variant VN_S does not change, only the variant VN_T decreases. The associated proof obligation is given by the sequent of Equation 4.4.

$$A(s, c),$$

$$m = S \Rightarrow I_S(s, c, v_S) \wedge m = T \Rightarrow I_T(s, c, v_T)$$

$$\wedge\, m = F \Rightarrow I_S(s, c, v_S),$$

$$m = F,$$

$$Next(VN_S, VN_T') \wedge m' = T \wedge VN_S' = VN_S$$

$$\vdash$$

$$VN_S' + VN_T' < VN_S + VN_T \tag{4.4}$$

After simplification, the previous proof obligation leads to the definition of the final proof obligation of Equation 4.5 associated to variant definition.

$$\boxed{A(s, c) \vdash Next(VN_S, VN_T') \wedge VN_S = VN_S' \Rightarrow VN_T' < VN_T} \tag{4.5}$$

4.6.3 About Restored States

As shown on the proof obligations obtained in Equations 4.3 and 4.5, the definition of the *Recover* and *Next* predicates is identified as the fundamental characteristic for the correct substitution operation.

The *Recover* predicate defines the *horizontal invariant*. This invariant defines the properties needed to restore the state variables of the original halted system in the substitute state variables. It also describes the safety property of the substitute system. According to the definition of this predicate, as discussed in Section 4.8, different substitution cases are identified.

Regarding the *Next* predicate, one can note that any value of the variant that decreases the variant VN_T is accepted. For instance, one could set up the variant to the final state of system S_T meaning that the substitution has been done in the final state. The only condition concerns the *Recover* predicate that shall restore the correct values of the variables in this final state.

4.7 A CASE STUDY

In order to illustrate our approach, we have selected a simple case study borrowed from E-commerce. In this example, a simple web application consisting of enabling a purchase of a set of products from a single supplier is considered. A user selects some products in a cart, pays the corresponding amount of money, receives an invoice, and then the products are delivered by the logistics part of the system. As depicted by a simple state transition system in Figure 4.1, this application can be described as a sequence events leading to the purchase of a set of products.

The state transition system depicted in Figure 4.1 represents the high-level specification for the considered system.

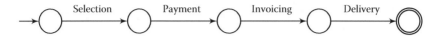

FIGURE 4.1 A simple state transition system describing a sequence of events for purchasing products.

We suppose that during the *selection* activity, a failure occurs, for instance due to an error on the supplier site. To restore this selection activity, we suppose that two other suppliers are available on two different sites. The defined substitution mechanism consists in transferring the selection activity to these two suppliers in order to complete the purchasing of products. Moreover, we require that all the previously selected products are transferred to the substitute system. In other words, the substitute system is equivalent to the original system since the set of products to be purchased remains unchanged.

4.7.1 Step 1. Specification

The Event-B context given in Model 4.13 defines the set of products *P* to be purchased and the set *SITE* of suppliers. Other axioms relevant for the proof process are given as well.

The specification of the considered system is given by a set of five events (Initialisation, Selection, Payment, Invoicing, and Delivery) corresponding to the state transition system of Figure 4.1.

The Event-B machine *Spec* (Model 4.14) describes the whole specification of the studied system. In the remainder of the paper, we provide the details on Selection event relevant to illustrate the behavior of the system substitution operation. The carts variable represents the set of purchased products. It is an empty set at initialization, and it is filled, with an arbitrary set of products (someCarts) for one single site (grd4) in the Selection event. A variable seqCount initialized to value 4 defines the sequencing of the events. Two relevant safety invariant have been defined. The first one (prop1) states that the Selection event completes the purchasing ($seqCount < 4$) of all the products of the set *P* and the second states that

```
Context
 ContProd
Sets
 PRODUCTS  -- all the products that can be purchased
 SITES  -- all the sites where products may be purchased
Constants
 STOCKS
 P  -- the set of products to purchase
Axioms
 axm1:  finite(PRODUCTS)
 axm2:  finite(SITES)
 axm3:  card(SITES) ≥ 3 -- at least three suppliers are available
 axm4:  STOCKS = SITES ×PRODUCTS
 axm5:  P ⊆ PRODUCTS
End
```

Model 4.13 The context ContProd defining relevant sets and properties for products and suppliers.

```
Machine
 Spec
Sees
 ContProd
Variables
 seqCount
 carts
Invariants
 type1: seqCount ∈ ℕ
 type3: carts ⊆ STOCKS
 prop1: (seqCount < 4) ⇒ ran(carts) = P
 prop2: ∀p · p ∈ ran(carts) ⇒ card(carts⁻¹[{p}]) = 1
Theorems
 type2: P ⊆ PRODUCTS
Variant
 seqCount
Events

 Event Initialisation =
  Then
   act1: seqCount := 4
   act3: carts := ∅
  End

 Event selection =
  Any
   someCarts
  Where
   grd1: seqCount = 4
   grd2: someCarts ⊆ SITES ×P
   grd3: ran(someCarts) = P
   grd4: ∀p · p ∈ ran(someCarts) ⇒ card(someCarts⁻¹[{p}]) = 1
  Then
   act1: seqCount := seqCount −1
   act2: carts := someCarts
  End

 Event payment := ...
 Event billing := ...
 Event delivery := ...
End
```

Model 4.14   The Event-B machine Spec defining the high-level specification.

each product of P is purchased from a single supplier (for any purchased product p, we have $card(carts^{-1}[p]) = 1$).

4.7.2 Step 2. Candidate Systems

In order to illustrate the substitution mechanism, we consider two systems that refine the original specification. The first one uses one single supplier and the second uses two suppliers for the same set of products.

4.7.2.1 The First Substitute System

The first substitute system considers a selection process encoded by a loop iterating on the set of products. Each loop step adds one product in the set variable *carts_ref*. Once the number of suited products *P* is reached, the variable *carts_ref* is assigned to the *carts* abstract variable of the specification. This assignment ensures that the set of products *P* has been selected. A counter variable *loopCount* is introduced, and it is initialized to the number of products (act2). Events Initialisation, addItemToCart_loop, and Selection are defined to encode this loop.

The details of the main events of this system SysS are shown in Model 4.15. The whole model is described in the appendix of this chapter in Model 4.21 (see later in text).

4.7.2.2 The Second Substitute System: Two Suppliers

The second substitute system considers a process that selects products from two suppliers represented by two variables *site_A* and *site_B*. At the initialization, the products of *P* are split on both suppliers *site_A* and *site_B*.

This is shown on the first excerpt in Model 4.16. The whole model is available in the appendix of this chapter in Model 4.22 (see later in text).

Then, the selection process (Model 4.17) allows to select products from each of the two suppliers *site_A* and *site_B*. For each supplier, the used loop schema is identical to the one presented for the first substitute system *SysS*. Moreover, the described system authorizes for selecting the products from both suppliers in parallel. The interleaving activity of the events

```
Machine SysS Refines Spec
. . .
 Event addItemToCart_loop :=
  Any someProduct
  Where . . .
   grd2: loopCount > 0
   grd3: someProduct ∈ P \ran(carts_ref)
  Then
   act1: loopCount := loopCount −1
   act2: carts_ref := carts_ref ∪{site ↦ someProduct}
  End
 Event selection :=
  Refines selection
  Where . . .
   grd2: loopCount := 0
   grd3: ran(carts_ref) := P
   grd4: ∀p · p ∈ ran(carts_ref) ⇒ carts_ref⁻¹[{p}] := {site}
  With someCarts: someCarts := carts_ref
  Then . . .
   act2: carts := carts_ref
  End
 . . .
End
```

Model 4.15   The Event-B refinement of the specification Spec.

```
Machine SysT Refines Spec Sees ContProd
Variables carts, carts_ref, varPar_A, site_A, site_B, loopCount_A, ...
Invariants ...
 type1: varPar_A ∈ ℕ
 type3: carts_ref ⊆ SITES ×P
 type4: loopCount_A ∈ ℕ
 type6: site_A ∈ SITES
 type7: site_B ∈ SITES
Variant
 seqCount + varPar_A + varPar_B + loopCount_A + loopCount_B
Events
 Event Initialisation =
  Begin ...
   act1: seqCount := 4
   act2: loopCount_A, loopCount_B :|
         loopCount_A' + loopCount_B' = card(P)
         ∧ loopCount_A' ∈ ℕ∧loopCount_B' ∈ ℕ
   act3: carts := ∅
   act4: varPar_A := 1
   act6: carts_ref := ∅
   act7: site_A :∈ SITES
   act8: site_B :∈ SITES
  End
```

Model 4.16   Excerpt 1 of the system SysT, refinement of Spec.

```
Event selection_A_loop =
 Any someProduct
 Where ...
  grd2: varPar_A = 1
  grd3: loopCount_A > 0
  grd4: someProduct ∈ P \ran(carts_ref)
 Then
  act1: loopCount_A := loopCount_A −1
  act2: carts_ref := carts_ref ∪{site_A ↦ someProduct}
 End
Event selection_A_loop_end =
 Where ...
  grd2: varPar_A = 1
  grd3: loopCount_A = 0
 Then act1: varPar_A := 0
 End
```

Model 4.17   Excerpt 2 of the system SysT, refinement of Spec.

is ensured by the introduction of the control variables $varPar_A$ and $varPar_B$. These variables are set to 0 when the selection for each supplier is completed.

Again, the set of products variable *carts_ref* is filled with products issued from both suppliers *site_A* and *site_B*. Once all the products of *P* are in the set *carts_ref*, the Selection event can be refined by assigning the *carts* variable in the refined Selection event (Model 4.18).

```
Event selection =
 Refines selection
 Where ...
  grd2: ran(carts_ref) = P
  grd3: ∀p·p ∈ ran(carts_ref) ⇒
        (carts_ref⁻¹[{p}]={site_A} ∨carts_ref⁻¹[{p}]={site_B})
  grd4: varPar_A = 0
  grd5: varPar_B = 0
  grd6: loopCount_A = 0
  grd7: loopCount_B = 0
 With someCarts: someCarts = carts_ref
 Then ...
  act2: carts := carts_ref
 End
 ...
End
```

Model 4.18   Excerpt 3 of the system SysT, refinement of Spec.

Note that the variant of Event-B machine *SysS* defined as $seqCount + varPar_A + varPar_B + loopCount_A + loopCount_B$ decreases.

4.7.3   Step 3. Introducing System Modes

Once the two systems *SysS* and *SysT* are defined, the mode variable *m* is introduced and the invariants of each Event-B machine are modified accordingly (implicative invariants). We do not provide the Event-B models produced at this step because they are similar to our previous models obtained in step 2.

4.7.4   Step 4. Define System Substitution

Finally, to obtain the composed system, we introduce the Fail and Repair events formalizing the substitution of *SysS* by *SysT*.

4.7.4.1   The Fail Event

The Fail event (Model 4.19) corresponds to the definition given in Model 4.9. Its guard is strengthened so that it allows to the event to be triggered only during the selection activity (grd1: seqCount = 4).

As defined, it sets the variable mode *m* to the value *F*.

```
Event Fail =
 Where
  grd1: seqCount = 4
  grd2: m = S
 Then
  act1: m := F
 End
```

Model 4.19   Introduction of the failure during selection.

4.7.4.2 The Repair Event

The Repair event (Model 4.20) also follows the definition given in Model 4.11. The predicate *Recover* and *Next* are defined so that the substitution restores the halted state of system *SysS* in an equivalent state in system *SysT*. The mode is then turned to *T* so that the system *SysT* switches to the correct running state of the substitute system.

The *Recover* predicate is defined as follows. The restored state variable *carts_ref* recording the already purchased products is defined with

- A range containing the same set of products as the one at the halted state ($ran(carts\_ref') = ran(carts\_ref)$)

- A domain representing the two suppliers corresponding to the two parallel selections ($dom(carts\_ref') \subseteq site\_A, site\_B$)

- And a property requiring that an already purchased product appears, in the restored state, only once in the set of products and is thus not purchased more than once ($\forall p.p \in ran(carts\_ref') \Rightarrow card(carts\_ref'^{-1}[p]) = 1$)

In this case, the *Recover* predicate corresponds to a substitution of a system by another system that behaves equivalently to the original system.

The *Next* predicate is defined so that the restored state in *SysT* corresponds to the state where the set of products remaining to be purchased is exactly the same as in the halted system *SysS*. This is achieved by using the property $loopCount\_A' + loopCount\_B' = loopCount$. This property guarantees that the number *loopCount* of purchased products in *SysS* is exactly the same as the number of purchased products $loopCount\_A' + loopCount\_B'$ in the restored state of *SysT*.

```
Event Repair =
 Where
  grd1: seqCount = 4
  grd2: m = F
 Then
  act1: m := T
  act2: carts_ref :|
           carts_ref' ⊆ SITES ×P
         ∧ ran(carts_ref') = ran(carts_ref)
         ∧ dom(carts_ref') ⊆ {site_A, site_B}
         ∧ ∀p p ∈ ran(carts_ref') ⇒ card(carts_ref'⁻¹[{p}]) = 1
  act3: loopCount_A, loopCount_B :|
           loopCount_A' + loopCount_B'= loopCount
         ∧ loopCount_A' ∈ ℕ∧loopCount_B' ∈ ℕ
 End
```

Model 4.20 Definition of the Repair event restoring the equivalent state.

4.8 SUBSTITUTION CHARACTERISTICS

Many substitution, reparation, or maintenance approaches are currently used in system engineering. This section introduces some properties of the system substitution operation defined in Section 4.5.

4.8.1 Cold and Warm Start

One first characteristic is the persistence of the state after substitution, usually named cold or warm start.

Cold start means that the substitute system will start from its initial state without any data nor state variables values originated from the state where the original system was halted.

Warm start means that the substitute system will recover as much data and state variable values as possible originated from the state where the original system was halted. We will tag the substitution as static for a cold start and dynamic for a warm start.

In the approach we have sketched in Section 4.5, this characteristic is handled by the correct definition of the *Recover* and *Next* predicates. Indeed, according to the definition on these predicates the restored state may be the initial state or any other state of the substitute system.

4.8.2 Identical, Included, or Disjoint Sets of State Variables

In the framework presented in Section 4.5, v_S and v_T represent the set of state variables for the original and substitute systems. According to the properties linking these two sets in the Repair event using the *Recover* predicate, different substitution cases occur.

- The sets of variables are identical, that is, $v_S = v_T$. This situation means that the original and the substitute systems are the same system. The effect of the Repair event is to restore a new state (correct with respect to the given invariants) after substitution. This situation usually occurs in case of maintenance.

- The sets of variables are partially shared, that is, $v_S \cap v_T \neq \varnothing$. In this case, part of the original system state variables are restored in the substitute system, and the substitute system introduces new state variables that describe new behaviors.

- Finally, the sets of variables are disjoint, that is, $v_S \cap v_T = \varnothing$. Disjointness implies that the substitute system is independent of the original system and it is a new system. The Repair event transfers the control to a completely new substitute system.

4.8.3 Equivalence, Upgrade, and Degradation

There is another characteristic that relates to the behavior of the substitute system and the associated quality. Depending on the expressed invariants, several substitute systems, refining a system specification may be available, they may offer different functionalities and have different behavior. In general, these differences relate on the one hand to the availability of more or less functionalities and on the other hand to quality of service.

Within the provided framework three cases can be considered. The substitute system *SysT* may be equivalent to the original system *SysS*, upgrade it (enhance it), or degrade it.

- Equivalence means that the original system properties are preserved, that is, the substitute system offers the same functionalities, but may differ from quality of service point of view.

- Upgrade is stronger than equivalence. The substitute system provides the same functionalities as the original system, but it also provides more functionalities. Quality of service may be impacted.

- Degradation is weaker than equivalence. The substitute system provides the less functionalities than the original system, and quality of service may be impacted.

As quality of service is out of scope of our framework, the three previous cases can be described with adequate definitions of the *Recover* and *Next* predicates. In fact, the definition of each case relies on the provided invariants to be preserved during substitution, that is, by the Repair event.

Let us assume that there exists two predicates Φ and Ψ ($\Phi \neq$ *False* $\wedge \Psi \neq$ *False*) such that $I_S \wedge \Phi \Longleftrightarrow I_T \wedge \Psi$, then the three identified cases can be expressed.

- *Equivalence* is obtained when $I_S \Longleftrightarrow I_T$. It means that the substitute preserves the same invariant properties as the original system since $\Phi \Longleftrightarrow$ *True* and $\Psi \Longleftrightarrow$ *True*. The case study presented in Section 4.7 illustrates this case. The set of products purchased with the substitute system *SysT* is identical to the original system *SysS*.

- *Upgrade* occurs when $I_S \wedge \Phi \Longleftrightarrow I_T$. Here, the substitute system offers more functionalities characterized by the invariant part Φ than the original system.

- *Degradation* is dual to upgrade and it occurs when $I_S \Longleftrightarrow I_T \wedge \Psi$. Here, the substitute system looses some of the functionalities characterized by the invariant part Ψ of the original system.

4.8.4 Static or Dynamic Set of Substitutes

In the framework presented in the previous section, we have assumed that the set of substitute systems is known and does not change (static). Modes have been introduced to identify the running system and the selected substitute system is known by the Repair event. One can imagine that the set of substitutes may evolve dynamically. A substitute can be added or removed from the set of substitutes. To handle such a mechanism, event managing (adding or removing substitutes) a set of modes corresponding to substitute systems (that refine a common specification) must be added, and the repair event must select a substitute in this set.

4.9 CONCLUSION

This paper addressed the problem of correct system substitution, where systems are described as state transition systems. It provides a stepwise correct-by-construction approach based on refinement and proof supported by the Event-B method.

This approach relies on (1) the definition of a class of systems that implement (i.e., refine) the same specification and (2) a system substitution operator parameterized by a recovery property, namely a *horizontal invariant*. This composition operator combines two or more systems that refine the same specification. It is parameterized by the *substitution* or *reparation property* ensuring that the current state (the state where the source system is halted) is correctly restored in the substitute system.

The defined framework for substitution ensures that, when a system is halted (a failure occurs for instance), the state of the source system is correctly restored to the state of the target system. Depending on the definition of the horizontal invariant, the composition operator entails three types of substitution: equivalent, degraded, or upgraded substitute systems can be obtained.

Two different substitution relationships are studied. The first one is a static substitution (corresponding to a *cold start*) that relies on refinement to characterize the set of systems that conforms to the same specification. A class of potential implementation systems are thus characterized by refinement. The second one addresses the dynamic substitution (substitution at runtime or *warm start*) that uses state restoration by transferring the control to the adequate state in the substitute system.

Finally, the proposed framework has been deployed on a didactic case study. However, this framework is generic and can be set up for any kind of systems provided that they are expressed as state transition systems. The whole proposed approach has been modeled within the Event-B method. Refinement and proof have been extensively used to obtain the whole model.

APPENDIX

```
Machine
 SysS
Refines
 Spec
Sees
 ContProd
Variables
 seqCount
 carts
 carts_ref
 loopCount
 site
Invariants
 type1: carts_ref ⊆ SITES ×P
 type2: loopCount ∈ ℕ
 type3: site ∈ SITES
Variant
 seqCount + loopCount
Events

 Event Initialisation =
  Begin
   act1: seqCount := 4
```

```
  act2: loopCount := card(P)
  act3: carts := ∅
  act4: carts_ref := ∅
  act5: site :∈ SITES
 End

Event addItemToCart_loop =
 Any
  someProduct
 Where
  grd1: seqCount = 4
  grd2: loopCount > 0
  grd3: someProduct ∈ P \ran(carts_ref)
 Then
  act1: loopCount := loopCount −1
  act2: carts_ref := carts_ref ∪{site ↦ someProduct}
 End

Event selection =
 Refines
  selection
 Where
  grd1: seqCount = 4
  grd2: loopCount = 0
  grd3: ran(carts_ref) = P
  grd4: ∀p · p ∈ ran(carts_ref) ⇒ carts_ref⁻¹[{p}] = {site}
 With
  someCarts: someCarts = carts_ref
 Then
  act1: seqCount := seqCount −1
  act2: carts := carts_ref
 End

Event payment = ...
Event billing = ...
Event delivery = ...
End
```

Model 4.21   The Event-B refinement SysS of the specification Spec.

```
Machine
 SysT
Refines
 Spec
Sees
 ContProd
Variables
 seqCount
 carts
 carts_ref
 varPar_A
 varPar_B
 site_A
```

```
 site_B
 loopCount_A
 loopCount_B
Invariants
 type1: varPar_A ∈ ℕ
 type2: varPar_B ∈ ℕ
 type3: carts_ref ⊆ SITES ×P
 type4: loopCount_A ∈ ℕ
 type5: loopCount_B ∈ ℕ
 type6: site_A ∈ SITES
 type7: site_B ∈ SITES
Variant
 seqCount + varPar_A + varPar_B + loopCount_A + loopCount_B
Events

 Event Initialisation =
 Begin
  act1: seqCount := 4
  act2: loopCount_A, loopCount_B :|
           loopCount_A' + loopCount_B' = card(P)
         ∧ loopCount_A' ∈ ℕ
         ∧ loopCount_B' ∈ ℕ
  act3: carts := ∅
  act4: varPar_A := 1
  act5: varPar_B := 1
  act6: carts_ref := ∅
  act7: site_A :∈ SITES
  act8: site_B :∈ SITES
 End

 Event selection_A_loop =
 Any
  someProduct
 Where
  grd1: seqCount = 4
  grd2: varPar_A = 1
  grd3: loopCount_A > 0
  grd4: someProduct ∈ P \ran(carts_ref)
 Then
  act1: loopCount_A := loopCount_A −1
  act2: carts_ref := carts_ref ∪{site_A ↦ someProduct}
 End

 Event selection_A_loop_end =
 Where
  grd1: seqCount = 4
  grd2: varPar_A = 1
  grd3: loopCount_A = 0
 Then
  act1: varPar_A := 0
 End

 Event selection_B_loop =
 Any
```

```
    someProduct
  Where
    grd1: seqCount= 4
    grd2: varPar_B = 1
    grd3: loopCount_B > 0
    grd4: someProduct ∈ P \ran(carts_ref)
  Then
    act1: loopCount_B := loopCount_B −1
    act2: carts_ref := carts_ref ∪{site_B ↦ someProduct}
  End

Event selection_B_loop_end =
  Where
    grd1: seqCount = 4
    grd2: varPar_B = 1
    grd3: loopCount_B = 0
  Then
    act1: varPar_B := 0
  End

Event selection =
  Refines
    selection
  Where
    grd1: seqCount = 4
    grd2: ran(carts_ref) = P
    grd3: ∀p · p ∈ ran(carts_ref) ⇒
          (carts_ref⁻¹[{p}]={site_A} ∨carts_ref⁻¹[{p}]={site_B})
    grd4: varPar_A = 0
    grd5: varPar_B = 0
    grd6: loopCount_A = 0
    grd7: loopCount_B = 0
  With
    someCarts: someCarts = carts_ref
  Then
    act1: seqCount := seqCount −1
    act2: carts := carts_ref
  End

Event payment = ...
Event billing = ...
Event delivery = ...
End
```

Model 4.22   The Event-B refinement SysT of the specification Spec.

REFERENCES

1. Edward Ashford Lee and Sanjit Arunkumar Seshia. *Introduction to Embedded Systems—A Cyber-Physical Systems Approach*. LeeSeshia.org, Edition 1.5, 2014.
2. Jean-Raymond Abrial. *Modeling in Event-B: System and Software Engineering*. Cambridge University Press, New York, NY, USA, 1st edition, 2010.

3. Anirban Bhattacharyya. Formal modelling and analysis of dynamic reconfiguration of dependable systems. PhD thesis, Newcastle University School of Computing Science, January 2013.

4. Noël de Palma, Philippe Laumay, and Luc Bellissard. Ensuring dynamic reconfiguration consistency. In *6th International Workshop on Component-Oriented Programming (WCOP 2001)*, at ECOOP 2001, Budapest, Hungary, pp. 18–24, 2001.

5. Arnaud Lanoix, Julien Dormoy, and Olga Kouchnarenko. Combining proof and model-checking to validate reconfigurable architectures. *Electronic Notes in Theoretical Computer Science*, 279(2):43– 57, 2011. *Proceedings of the 8th International Workshop on Formal Engineering approaches to Software Components and Architectures (FESCA).*

6. Michel Wermelinger, Antónia Lopes, and José Luiz Fiadeiro. A graph based architectural (re)configuration language. In *Proceedings of the 8th European Software Engineering Conference Held Jointly with 9th ACM SIGSOFT International Symposium on Foundations of Software Engineering*, ESEC/FSE-9, pp. 21–32, New York, NY, USA, 2001. ACM.

7. Rogério de Lemos, Paulo Asterio de Castro Guerra, and Cecília Mary Fischer Rubira. A fault-tolerant architectural approach for dependable systems. *Software, IEEE*, 23(2):80–87, March 2006.

8. Ilya Lopatkin and Alexander Romanovsky. *Rigorous Development of Fault-Tolerant Systems Through Co-Refinement*. Technical Report, School of Computing Science, University of Newcastle upon Tyne, January 2014.

9. Rodrigo Rodrigues, Barbara Liskov, Kathryn Chen, Moses Liskov, and David Schultz. Automatic reconfiguration for large-scale reliable storage systems. *Dependable and Secure Computing, IEEE Transactions on*, 9(2):145–158, March 2012.

10. Inna Pereverzeva, Elena Troubitsyna, and Linas Laibinis. Development of fault tolerant MAS with cooperative error recovery by refinement in Event-B. Fuyuki Ishikawa and Alexander Romanovsky, editors, *DS-Event-B 2012: Workshop on the Experience of and Advances in Developing Dependable Systems in Event-B*, in Conjunction with ICFEM 2012, Kyoto, Japan, Proceedings: http://arxiv.org/html/1211.2259, November 13, 2012.

11. Inna Pereverzeva, Elena Troubitsyna, and Linas Laibinis. A refinement-based approach to developing critical multi-agent systems. *International Journal of Critical Computer-Based Systems*, 4(1):69–91, Jan 2013.

12. Anton Tarasyuk, Inna Pereverzeva, Elena Troubitsyna, Timo Latvala, and Laura Nummila. Formal development and assessment of a reconfigurable on-board satellite system. In Frank Ortmeier and Peter Daniel, editors, *Computer Safety, Reliability, and Security*, volume 7612 of *Lecture Notes in Computer Science*, pp. 210–222. Springer, Berlin Heidelberg, 2012.

13. Xin An, Gwenaël Delaval, Jean-Philippe Diguet, Abdoulaye Gamatié, Soguy Gueye, Hervé Marchand, Noël de Palma, and Eric Rutten. Discrete control-based design of adaptive and autonomic computing systems. In Raja Natarajan, Gautam Barua, and Manas Ranjan Patra, editors, *Distributed Computing and Internet Technology*, volume 8956 of *Lecture Notes in Computer Science*, pp. 93–113. Springer International Publishing, 2015.

14. Manish Parashar and Salim Hariri. Autonomic computing: An overview. In Jean-Pierre Banâtre, Pascal Fradet, Jean-Louis Giavitto, and Olivier Michel, editors, *Unconventional Programming Paradigms*, volume 3566 of *Lecture Notes in Computer Science*, pp. 257–269. Springer, Berlin Heidelberg, 2005.

15. Rogério Lemos, Holger Giese, Hausi A. Müller, Mary Shaw, Jesper Andersson, Marin Litoiu, Bradley Schmerl et al. Software engineering for self-adaptive systems: A second research roadmap. In Rogério Lemos, Holger Giese, Hausi A. Müller, and Mary Shaw, editors, *Software Engineering for Self-Adaptive Systems II*, volume 7475 of *Lecture Notes in Computer Science*, pp. 1–32. Springer, Berlin Heidelberg, 2013.

16. Danny Weyns, M. Usman Iftikhar, Didac Gil de la Iglesia, and Tanvir Ahmad. A survey of formal methods in self-adaptive systems. In *Proceedings of the Fifth International C* Conference on*

Computer Science and Software Engineering, C3S2E '12, pp. 67–79, New York, NY, USA, 2012. ACM.

17. Antonio Filieri, Carlo Ghezzi, and Giordano Tamburrelli. A formal approach to adaptive software: Continuous assurance of non-functional requirements. *Formal Aspects of Computing*, 24(2):163–186, 2012.

18. Raffaela Mirandola, Pasqualina Potena, and Patrizia Scandurra. Adaptation space exploration for service-oriented applications. *Science of Computer Programming*, 80, Part B(0):356–384, 2014.

19. Pasqualina Potena. Optimization of adaptation plans for a service-oriented architecture with cost, reliability, availability and performance tradeoff. *Journal of Systems and Software*, 86(3):624–648, 2013.

20. Jean-Raymond Abrial and Stefan Hallerstede. Refinement, decomposition, and instantiation of discrete models: Application to Event-B. *Fundamenta Informaticae*, 77(1):1–28, 2007.

21. Roel J. Wieringa. *Design Methods for Reactive Systems: Yourdon, Statemate, and the UML*. Morgan Kaufmann Publishers, 1st edition, 2003.

22. Rajeev Alur. Formal verification of hybrid systems. In Samarjit Chakraborty, Ahmed Jerraya, Sanjoy K. Baruah, and Sebastian Fischmeister, editors, *Proceedings of the 11th International Conference on Embedded Software, EMSOFT 2011, part of the Seventh Embedded Systems Week, ESWeek 2011*, Taipei, Taiwan, October 9–14, 2011, pp. 273–278. ACM, 2011.

CHAPTER 5

Incremental Proof-Based Development for Resilient Distributed Systems

Manamiary Bruno Andriamiarina, Dominique Méry, and
Neeraj Kumar Singh

CONTENTS

D ISTRIBUTED SYSTEMS AND APPLICATIONS require efficient and effective techniques (e.g., self-(re)configuration and self-healing) for ensuring safety, security, and, more generally, dependability properties, including stabilization and resilience. The complexity of these systems is increased by several factors—for example, dynamic topology, interconnection of heterogeneous components, and automatic failure detection. This chapter presents a methodology for developing protocols satisfying safety and convergence requirements of the distributed self-★ systems. The self-★ systems are based on the idea of managing complex infrastructures, software, and distributed systems, with minimal user interactions. *Correct-by-construction* and *service-as-event* paradigms are used for formalizing the system requirements, where the formalization process is based on incremental refinement in EVENT-B. We describe a fully mechanized proof of correctness of self-★ systems along with an interesting case study related to P2P-based self-healing protocols. Moreover, we also address the formal reasoning steps to introduce resilience in the distributed networks.

5.1 INTRODUCTION

Nowadays, our lives are affected by advanced technologies, including computers and smartphones. These technologies are integrated into distributed systems with different kinds of complexities such as mobility, heterogeneity, security, fault tolerance, and dependability. Distributed systems are largely used in many applications to provide required functionalities from the interactions between a large collection of possibly heterogeneous and mobile components (nodes and/or agents). Within the domain of distributed computing, there is an increasing interest in the self-stabilizing systems that can automatically recover from a failure state [1]. The autonomous property of the self-★ systems tends to take a growing importance in the analysis and development of the distributed systems. It is important to get a better understanding of the self-★ systems (emergent behaviors, interactions between agents, etc.) if we want to reason about their security, correctness, and trustworthiness. The formal methods community has been analyzing a similar class of systems for many years, namely distributed algorithms.

In this chapter, we use a *correct-by-construction* approach [2] for modeling the distributed self-★ systems and the resilient distributed systems in particular, which allow recovery from a failure state or the ability to adapt new changing states. Moreover, we also emphasize the use of the *service-as-event* [3] paradigm, which identifies the phases of a *self-stabilization* mechanism simplified into more stable and simple coordinated steps.

We consider that a given system S (see Figure 5.1) is characterized by a set of *events*, also called *procedures*, modeling either *phases* or basic actions according to an abstraction level. *Legal states* satisfying a *safety property P* are defined by a subset CL of possible events of the system S. The events of CL represent the possible big or small computation steps of the system S introducing the notion of *closure* [4], where any computation starting from a *legal state* satisfying the *property P* leads to another *legal state* that also satisfies the property P. The occurrence of a fault f leads the system S into an *illegal state* (incorrect state) that violates the property P. The fault f is defined as an event f that belongs to a subset $\mathcal{F}$ of events. When considering the hypothesis of having a self-★ system, we assume that there are *procedures* (*protocols* or *actions*) that *implement* an identification of the *current illegal*

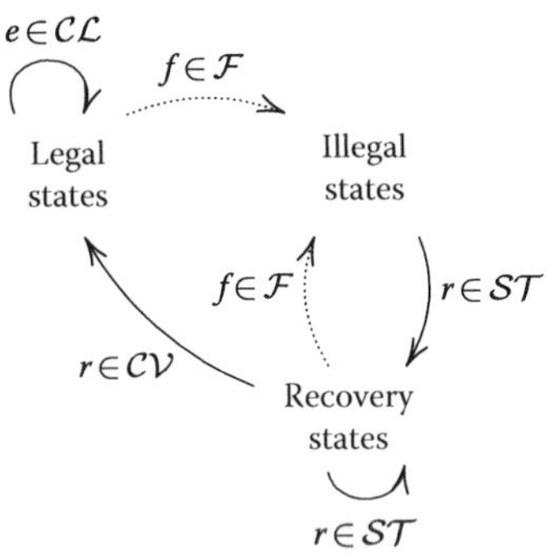

FIGURE 5.1   Abstract view for a self-stabilizing system $\mathcal{S}$.

states and their recovery for *legal states*. There is a subset $\mathcal{ST}$ of events modeling *recovery* phases to demonstrate the *stabilization* process. The system recovers using a finite number of *stabilization* steps (r). The process is modeled as an event r of $\mathcal{CV} \subseteq \mathcal{ST}$, eventually leading to the legal states (*convergence* property) from *recovery* states. During the *recovery* phase, a fault may occur (see dotted transitions in Figure 5.1).

The system $\mathcal{S}$ is represented by a set of events $\mathcal{M} = \mathcal{CL} \cup \mathcal{ST} \cup \mathcal{F}$ where $\mathcal{M}$ contains a set ($\mathcal{CL}$) of events representing the *computation steps* of the system $\mathcal{S}$. When a fault occurs, a set ($\mathcal{ST}$) of events *simulates* the *stabilization* process that is performed by $\mathcal{S}$. The formal representation expresses a *closed* model, but we do not know what is the complete set of events for modeling faults/failures. We characterize a fault model in a very abstract way, and it may be possible to develop the fault model according to the assumptions about the given environment, but we do not consider this in the current study. We restrict our study by making explicit the events of $\mathcal{ST}$, modeling the *stabilization* of the system from illegal/failed states. We ensure that the *convergence* is always possible: a subset $\mathcal{CV}$ of $\mathcal{ST}$ *eventually* leads $\mathcal{S}$ into the *legal states* satisfying the invariant P of the system. Whenever the system $\mathcal{S}$ is in a *legal state*, we consider that the events of $\mathcal{ST}$ are either nonoperative or simply preserving the invariant P of the system.

We formalize the system $\mathcal{S}$ using the EVENT-B modeling language [5], dealing with *events* and *invariant* properties, including *convergence* properties, by using temporal logic. The *service-as-event* paradigm [3] helps express the process for developing the concrete models: the procedures (1) *leading* from an *illegal state* to a *recovery state* and (2) *leading* from a *recovery state* to a *legal state* stated by (abstract) events. The next step is to unfold each (abstract procedure) event, by refinement, to a set of concrete events that form a body of the procedure.

In this chapter we extend our previous work [6], a preliminary study on the design and verification of self-★ systems using refinement-based modeling in EVENT-B. The chapter is organized as follows: Section 5.2 summarizes existing modeling works related to the resilient systems. Section 5.3 introduces the EVENT-B modeling framework including the *service-as-event* paradigm, as well as a graphical notation for helping the formal process. Section 5.4 illustrates the proposed methodology by studying the self-healing P2P-based protocol [7]. Finally, Section 5.5 concludes the chapter with future work.

5.2 FORMAL MODELING FOR RESILIENT SYSTEMS

Systems usually run in a complex environment with frequent and unexpected changes. This aspect increases interest of autonomous, self-★, or resilient architectures. These systems are able to adapt to new changes automatically according to new system states (i.e., faults) or the changing environment. Applying formal methods to resilient systems originates from understanding how these systems behave and how they meet their specifications. A resilient system relies on *emergent behaviors* resulting from interactions between components of the system as, for instance, self-★ systems [8].

Traditionally, the correctness of self-★, autonomous, and resilient systems is validated through simulation and testing [9,10]. However, simulation and testing are not sufficient to cover a complete set of possible states of a system. Therefore, formal methods are an alternative approach for validating those systems. Formal techniques can assert the correctness of those systems and certify the target properties, such as trustworthiness, security, efficiency, etc., under the rigorous mathematical reasoning [11–13].

Smith et al. [8] have applied the stepwise refinement using Z [14] to study a case of self-reconfiguration, where a set of autonomous robotic agents is able to assemble and reach a global configuration. They do not validate models using an adequate tool (e.g., proof checker and proof assistant), and models are not localized. Calinescu et al. [11] have used Alloy [15] to demonstrate the correctness of the autonomic computing policies. However, Alloy does not provide a mechanism for expressing the *correct-by-construction* paradigm. Méry et al. [16,17] have also investigated self-reconfiguring systems using the EVENT-B framework.

State exploration approaches, such as model-checking, are also used to study self-★ systems. Model-checkers like PRISM, NUSMV, and UPPAAL are used for property specification and acquiring evidence that properties, such as flexibility and robustness of self-★ systems, hold [11–13,18]. Moreover, these tools allow users to obtain the metrics for the self-★ systems, such as performance and quantitative evaluations [11–13,18]. Model-checking and state-space evaluation can be used for analyzing the self-★ systems, but they are especially used for runtime verification [13,18]. The limit of model-checking is clearly the size of models.

Other formal techniques like static analysis and design by contract are also applied for the formal specification of self-★ systems [19]. These techniques are mainly used for *runtime verification*. Graphical approaches, such as Petri nets, are used to model the temporal aspects and communication flows among different components of a self-★ system and help study the cases like self-reconfiguration (replacement of a component, removal of a link between two components, etc.) [13].

Finally, visual notations (e.g., UML) help represent self-★ systems with comprehensible figures [20]. Their general purpose is to provide users with insight of a self-★ system by describing its architecture and the relationships between agents of the system (ADELFE [9]) or by presenting the system as a composition of extendable/instantiable primitives (FORMS [20]). These notations are generally graphical front-ends for more complex representations of the self-★ systems, where the source code [9] and formal models [20] can be generated from the notations.

Our proposed methodology integrates the EVENT-B method and elements of temporal logic. Using the refinement technique, we gradually build a formal model of a self-★ system in the EVENT-B framework. Moreover, we use the *service-as-event* paradigm to describe the *stabilization* and *convergence* from *illegal* states to *legal* ones. Self-★ and resilient systems require the expression of trace properties like liveness, and we borrow a minimal set of inference rules for deriving liveness properties. The concept of a *refinement diagram* generalizes predicate diagrams [21] and intends to capture intuition of designer for progressively deriving the target self-★ system.

5.3 MODELING FRAMEWORK

5.3.1 EVENT-B

We advocate the use of the *correct-by-construction* paradigm for modeling the self-★ systems. The key concept is the incremental refinement, which relates discrete models by preserving properties. The EVENT-B modeling language designed by Abrial [5,22] is based on *set theory* and *refinement* of models. An abstract model expresses requirements of a given system; it can be verified and validated easily. A concrete model corresponding to the actual system is *constructed* progressively by *refining* abstraction. EVENT-B is supported by the complete toolset RODIN [23], providing features like refinement, proof obligations generation, proof assistants, and model-checking.

5.3.1.1 Modeling Actions Over States

The EVENT-B modeling language can express *safety properties*, which are either *invariants* or *theorems* in a model corresponding to a system.

Two main structures are available in EVENT-B: (1) context describes the static part of a model (for instance, graph properties like connectivity) and (2) a machine describes the dynamic part of a system and safety properties. An EVENT-B model is defined by a context and a machine. A machine organizes events (or actions) modifying state variables and uses static properties defined in contexts. An EVENT-B model is characterized by a list x of *state variables* possibly modified by a list of *events*.

An invariant $I(x)$ states the properties that must always be satisfied by the variables x and *maintained* by the firing of the events. The general form of an event e is as follows:

$$\text{ANY } t \text{ WHERE } G(t, x) \text{ THEN } x : |P(t, x, x') \text{ END.}$$

It corresponds to a transformation of state described by a *before-after* predicate $BA(e)(x, x')$. The predicate is semantically equivalent to $\exists t \cdot G(t, x) \wedge P(t, x, x')$ and expresses the relationship linking the values of the state variables before (x) and just after (x') the *execution* of the event e. $grd(e)(x)$ is equivalent to $\exists t \cdot G(t, x)$. Proof obligations can be produced (see Table 5.1) by RODIN [24] prover: INV1 and INV2 state that an assertion $I(x)$ is inductively invariant for events. The general form of these proof obligations is defined using the before-after predicate $BA(e)(x, x')$ of each event e of a machine. FIS expresses the feasibility of an event e with respect to the invariant I. The proof of feasibility demonstrates that

TABLE 5.1   Proof Obligations

INV1	INV2	FIS
$Init(x) \Rightarrow I(x)$	$I(x) \wedge BA(e)(x, x') \Rightarrow I(x')$	$I(x) \wedge grd(e)(x) \Rightarrow \exists\, z \cdot BA(e)(x, z)$

the before-after predicate $BA(e)(x, z)$ provides the next state whenever the guard $grd(e)(x)$ holds. It means that the guard is an enabling condition of the event e.

5.3.1.2 Model Refinement

Refinement of machines or, equivalently, refinement of models defines a binary link between an abstract machine and a concrete machine. This feature allows us to develop EVENT-B models for self-⋆ systems gradually and validate each decision step using the proof tool. The refinement relationship is expressed as follows: A model AM is refined by a model CM, when CM simulates AM (i.e., when a concrete event ce occurs in CM, there must be a corresponding enabling abstract event ae in AM). The final concrete model is closer to the behavior of a real system that observes events.

The refinement of a formal model enriches the model gradually following the *correct-by-construction* approach [2]. Refinement provides a way to strengthen invariants and add details to a model. It is also used to transform an abstract model into a more concrete version by modifying the state description. This is done by transforming a list of state variables (possibly suppressing some of them), refining each abstract event to a set of possible concrete versions, and adding new events. We consider that an abstract model AM with variables x and an invariant $I(x)$ is refined by a concrete model CM with variables y. The abstract state variables x and the concrete variables y are linked together by means of a so-called gluing invariant $J(x, y)$.

Suppose that the event ae is in the abstract model AM and event ce is in the concrete model CM. The event ce refines the event ae. $BA(ae)(x, x')$ and $BA(ce)(y, y')$ are predicates of the events ae and ce, respectively. We have to discharge the following proof obligation:

$$I(x) \;\wedge\; J(x, y) \;\wedge\; BA(ce)(y, y') \;\Rightarrow\; \exists x' \cdot (BA(ae)(x, x') \;\wedge\; J(x', y')).$$

More details are available in References 5 and 22. The refinement-based development of EVENT-B requires the integration of possibly *tough* interactive proofs. To assist the development of the self-⋆ systems, we use the *service description* and *decomposition* that are provided by the *service-as-event* [3] paradigm (derived from the *call-as-event* approach [25]).

5.3.2 Temporal Logic of Actions (TLA)

Liveness properties as well as fairness properties cannot be expressed in an explicit way in EVENT-B. Therefore, we interpret EVENT-B events as TLA [26] actions, and we define, from EVENT-B states, traces over states using TLA formulas. This technique has already been used by Méry and Poppleton [27] for developing population protocols and extending the EVENT-B refinement. Let us recall that a TLA action is a relation over unprimed and primed variables, whereas a primed variable states the next state of the variable. In TLA, we can express safety properties using a notation $\Box P$ (*always P*) or liveness properties using a notation $P \rightsquigarrow Q$ (*P eventually leads to Q*). These formulas can be defined over traces generated from states of EVENT-B models. Particularly, a system is defined by a set of traces satisfying a stutter-free formula Φ defined as follows:

$$\Phi \;\widehat{=}\; Init_\Phi \wedge [N]_f \wedge WF_f(F_1) \wedge SF_f(F_2)$$

where $N \;\widehat{=}\; N_1 \vee N_2 \vee \ldots$ is a logical disjunction of system actions (i.e., *the next* transition), denoting progress subject to possible stuttering over f, the list of variables. Stuttering is required to allow us to specify and prove refinements. The *WF* and *SF* constraints are the weak and strong fairness constraints possibly required by the system actions in order to progress, and F_1 and F_2 are two logical combinations of actions. We say that an action A is weakly fair, when if it is eventually enabled, then it is guaranteed to fire infinitely often. A is strongly fair, when if it is infinitely often enabled, then it is guaranteed to fire infinitely often. We have the following definitions for *WF* and *SF*:

$$WF_f(A) \;\widehat{=}\; \Diamond\Box Enabled\langle A\rangle_f \Rightarrow \Box\Diamond\langle A\rangle_f$$
$$SF_f(A) \;\widehat{=}\; \Box\Diamond Enabled\langle A\rangle_f \Rightarrow \Box\Diamond\langle A\rangle_f.$$

TLA [26] has been designed by Lamport for modeling reactive systems, especially distributed algorithms. A specification language TLA$^+$ [28] proposes a set-theoretical framework for modeling data structures. We consider only the temporal part, namely TLA, and we use it within the set-theoretical language of EVENT-B. We express trace properties from EVENT-B models using the *service-as-event* paradigm. We recall the inference rules from Reference 26 that are used for deriving proofs as follows:

- Implication (impl):

$$\frac{[}{(}impl)]P \rightsquigarrow QP \Rightarrow Q$$

- Transitivity (trans):

$$\frac{[}{(}trans)]P \rightsquigarrow QP \rightsquigarrow R \quad R \rightsquigarrow Q$$

- Disjunction (disj):

$$\frac{[}{(}disj)]P \vee R \rightsquigarrow QP \rightsquigarrow Q \quad R \rightsquigarrow Q$$

- **LATTICE:**

$$\frac{\succ \text{ a well-founded partial order on a set } S\ (S, \succ)}{F \wedge c \in S \Rightarrow (H(c) \rightsquigarrow (G \vee \exists d \in S \cdot (c \succ d) \wedge H(d)))}$$
$$F \Rightarrow ((\exists e \in S \cdot H(e)) \rightsquigarrow G)$$

LATTICE is an induction rule over $(S, \succ)$. The given indexed predicate $H(c)$ leads to either the goal G or $H(d)$ for some d strictly smaller than c, then $\exists e \in S \cdot H(e)$ eventually leads to G under assumption F.

- **FAIRNESS:**

  - WF1:

$$\frac{\begin{array}{c} P \wedge [N]_x \Rightarrow (P' \vee Q') \\ P \wedge \langle N \wedge A \rangle_x \Rightarrow Q' \\ P \Rightarrow \text{ENABLED}\langle A \rangle_x \end{array}}{\Box[N]_x \wedge WF_x(A) \Rightarrow (P \rightsquigarrow Q).}$$

WF1 states conditions under which weak fairness of an action A is enough to guarantee that $P \rightsquigarrow Q$. A stuttering progress step produces either P or Q in the next state. A nonstuttering action $\langle N \wedge A \rangle_x$ (equivalent to $N \wedge A \wedge x \neq x'$) leads Q from P. From P an action A is always enabled ($\text{ENABLED}\langle A \rangle_x$ stands for $\exists y.A(x, y) \wedge x \neq y$). Therefore, under WF1 assumption and $\Box[N]_x$, P leads to Q.

  - SF1:

$$\frac{\begin{array}{c} P \wedge [N]_x \Rightarrow (P' \vee Q') \\ P \wedge \langle N \wedge A \rangle_x \Rightarrow Q' \\ \Box P \wedge \Box[N]_x \wedge \Box F \Rightarrow \Diamond \text{ENABLED}\langle A \rangle_x \end{array}}{\Box[N]_x \wedge SF_x(A) \wedge \Box F \Rightarrow (P \rightsquigarrow Q).}$$

SF1 states conditions under which strong fairness of an action A is enough to guarantee that $P \rightsquigarrow Q$. The first two premises state that from P either the next transition reaches P or Q and that $\langle N \wedge A \rangle_x$ helps to reach Q. However, $\text{ENABLED}\langle A \rangle_x$ is possibly not holding from P, and the third premise states that under the computation process ($\Box[N]_x$), the stability of P ($\Box P$) and some auxiliary assumptions F ($\Box F$) and $\langle A \rangle_x$ are enabled. Finally, under SF1, $\langle A \rangle_x$ is infinitely often enabled and will be executed. Reference 29 details the use of this rule for proving the population protocols.

5.3.3 The *Service-as-Event* Paradigm

This section presents the *service-as-event* paradigm and introduces the *refinement diagram* [3,25], which is a graphical notation supporting liveness properties while refining. A brief overview on using these formalisms for modeling the self-⋆ systems is given below. We are considering the self-⋆ systems as a class of resilient distributed system. EVENT-B may deal with liveness properties and traces for modeling the self-⋆ systems, as long as one reasons at an abstract level. The first objective of this paradigm is to identify the services/functionalities provided by the studied systems and express them using temporal logic [3,25,26,30] and liveness properties, such as $P \rightsquigarrow Q$. Then, we use the inference rules of TLA [26], related to the liveness properties, to guide the refinement steps. The *service-as-an-event* paradigm is a way to integrate both languages EVENT-B and TLA and exploit the tools and techniques of EVENT-B as much as possible.

5.3.3.1 The Self-⋆ Mechanism

We characterize a self-stabilizing system $\mathcal{S}$ (more generally a self-⋆ system) by its ability to recover autonomously from an *illegal* (faulty) state (violating the invariant P of the system) to a *legal* (correct) state statisfying the invariant property P of system $\mathcal{S}$. Temporal logic [3,25,26,30] can be used to describe such a mechanism, using the liveness properties: We represent the *stabilization* (especially the *convergence*) property as a *service* where the system $\mathcal{S}$, in an *illegal* state (characterized by $\neg P$), *eventually* leads to a *legal* state (satisfying P). This service is expressed using the *leads to* ($\rightsquigarrow$) operator, as follows: $(\neg P) \rightsquigarrow P$. This *leads to* property (equivalently $\Box((\neg P) \Rightarrow \Diamond P)$) states that every *illegal* state (satisfying $\neg P$) will *eventually* (at some point in the future) lead to a *legal* state (satisfying P).

We define a temporal extension for an EVENT-B model M modeling a system $\mathcal{S}$ by the following TLA specification:

$$\mathcal{S}pec(M):\ Init(y) \wedge \Box[Next]_y \wedge L,$$

where $Init(y)$ is a predicate specifying the initial states; $Next \equiv \exists e \in E \cdot BA(e)(y, y')$ is an action formula representing the next-state relation, in which E is the set of events of M; L is a conjunction of fairness assumptions. We list *fairness assumptions* for some events e (or some combinations of events) that are modeling steps of the recovery process (we do not add any fairness on events leading to *illegal states* (*faults*), since these events are due to the environment and cannot be constrained). $\mathcal{S}pec(M)$ defines a set of traces characterizing the system under consideration, and it gives a semantical framework for interpreting formulas: M satisfies Φ means that $\mathcal{S}pec(M) \Rightarrow \Phi$ or equivalently Φ holds for M or equivalently Φ characterizes the set of possible fair traces of M. In the next section, we introduce *refinement diagrams*, which are a methodological support for designing both model and properties like fairness assumptions.

5.3.3.2 Refinement Diagrams

The self-⋆ mechanism is modeled using EVENT-B together with liveness properties and fairness assumptions. Figure 5.1 presents the concrete view of the system recovering from *illegal states*, and we have identified graphical notations called *refinement diagrams* that support the refinement process by helping choose fairness asumptions over events of EVENT-B models. Back [31] has defined refinement diagrams for supporting a diagrammatic reasoning over models for developing large software systems, and we use the same name for our diagrams, which are close to the *predicate diagrams* [21].

Refinement diagrams (see Figure 5.2), introduced by Méry et al. [3,25,29], allow the development of EVENT-B models and the addition of a control flow inside these models. Refinement diagrams state liveness properties. A refinement diagram D for a machine M (with variables x and invariant $I(x)$) is defined as follows: $D \triangleq (A, G, E, V, F, M)$, where A is a set of nodes (which are assertions); G is a set of assertions for M called conditions/guards $g(x)$; E is a set of events of machine M; V is a finite set of k labels $\{(v_1, \prec_1), (v_2, \prec_2), \ldots, (v_k, \prec_k)\}$ (each term v_i is paired with a well-founded relation $\prec_i$); and F is a partial function that assigns a fairness condition (WF, SF) with event or a set of events of E.

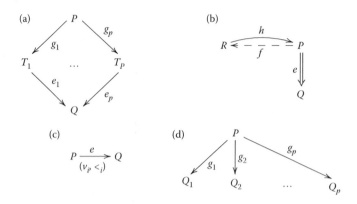

FIGURE 5.2   Examples of parts of refinement diagrams.

The diagram D is a labeled directed graph over A, with labels from $G \cup E \cup E \times V$, and a set of edges is a subset of $A \times (G \cup E \cup E \times V) \times A$. We say that an event e is observable at node n labeled with $P \in A$, if the node n is connected to another node using an edge labeled with e. Méry [25] initially defined refinement diagram as an acyclic structure, because it was related to the development of recursive sequential algorithms. Refinement diagrams intend to be useful for supporting and guiding the refinement of models when considering liveness properties and fairness constraints of resilient distributed systems. By applying the refinement diagrams in several case studies, we have identified a set of patterns for dealing with fairness assumptions; for instance, Figure 5.2b is a special case for using the strong fairness assumption together with the rule SF1 of TLA. As we have already mentioned, a set of edges is a subset of $A \times (G \cup E \cup E \times V) \times A$, and there are different kinds of labeled transitions as $P \xrightarrow{e} Q$, $P \overset{e}{\Rightarrow} Q$, or $P \dashrightarrow^{e} Q$. Each kind of transition has a meaning related to fairness, assumptions: $P \xrightarrow{e} Q$ means that e is under weak fairness, $P \overset{e}{\Rightarrow} Q$ means that e is under strong fairness and $P \dashrightarrow^{e} Q$ means that e has no fairness assumption and may lead to Q from P. We provide a nonexhaustive list of conditions and properties of refinement diagrams and introduce assumptions and meaning over refinement diagrams.

Property 5.1. *If two assertions P and Q are related by one of the e-labeled links ($P \xrightarrow{e} Q$, $P \overset{e}{\Rightarrow} Q$, and $P \dashrightarrow^{e} Q$), then the following properties hold:*

$$\begin{cases} (1) \forall x, x' \cdot P(x) \wedge I(x) \wedge BA(e)(x, x') \Rightarrow Q(x') \\ (2) \forall x \cdot P(x) \wedge I(x) \Rightarrow (\exists x' \cdot BA(e)(x, x')). \end{cases}$$

Property 5.2. *Suppose that $P \xrightarrow{e} Q$ is a unique link from P. Then, if other event f ($f \neq e$) is observed, when P is satisfied, it maintains P or may lead to Q: $\forall f \neq e. \forall x, x' \cdot P(x) \wedge I(x) \wedge BA(f)(x, x') \Rightarrow (P(x') \vee Q(x'))$, $F(e)= WF$ (or e is under weak fairness assumption) and $P \rightsquigarrow Q$ for M under weak fairness assumption over e.*

Property 5.3. *Suppose that $P \xrightarrow{e_1} Q_1$ and $P \xrightarrow{e_2} Q_2$ are the two unique links from P. Then, if other events f ($f \neq e$) are observed, when P is satisfied, they maintain P or may lead to Q_1 or Q_2: $\forall f \notin \{e1, e2\}. \forall x, x' \cdot P(x) \wedge I(x) \wedge BA(f)(x, x') \Rightarrow (P(x') \vee Q_1(x') \vee Q_2(x'))$,*

$F(e_1 \lor e_2) = WF$ *(or $e_1 \lor_2$ is under weak fairness assumption) and $P \rightsquigarrow (Q_1 \lor Q_2)$ for M under weak fairness assumption over $e_1 \lor e_2$.*

Two transitions start from P, and two events $e1$ and $e2$ are possibly observed. Each event $e1$ and $e2$ may lead to Q and is always enabled from P. Weak fairness is over the event defined as the disjunction of $e1$ and $e2$. By WF1 rule, we infer $P \rightsquigarrow (Q_1 \lor Q_2)$. This property shows that the use of fairness is tricky and that refinement diagrams help when they have specific form.

Property 5.4. *Suppose that $P \overset{e}{\Rightarrow} Q$ is the unique link from P to Q and there is a link from P to R as in Figure 5.2b and no link from R to other nodes, but P. Then $F(h) = WF$ and $P \rightsquigarrow Q$ for M under strong fairness assumption over e.*

Property 5.5. *If P is related to Q using an arrow labeled by $e \in E$ and $(v_i, \prec_i) \in V$ in form of $P \xrightarrow[(v_i, \prec_i)]{e} Q$, then M satisfies the following property: $\forall x, x' \cdot P(x) \land I(x) \land BA(e)(x, x') \Rightarrow (v_i(x') \prec_i v_i(x))$.*

Property 5.6. *If P is related to $Q_1, \ldots, Q_p$, as seen in Figure 5.2d (each arrow from P to Q_i is labeled by a guard $g_i \in G$.), then*

$$
\begin{cases}
(1) \forall i \in 1..p. \begin{cases} \forall x \cdot P(x) \land I(x) \land g_i(x) \Rightarrow Q_i(x) \\ \forall x, j.j \in 1..p \land j \neq i \land P(x) \land I(x) \land g_i(x) \Rightarrow \neg g_j(x) \end{cases} \\
(2) \forall x \cdot P(x) \land I(x) \Rightarrow (\exists i \cdot i \in 1..p \land g_i(x)).
\end{cases}
$$

We have nonexhaustively listed possible patterns for refinement diagram properties, which are constraining our diagrams according to the derivation of liveness properties under correct fairness assumptions. For instance, under a suitable pattern in a refinement diagram, using the transition labeled by decreasing events, one can derive the termination of an iterative process. From the diagram D, one can extract properties [25] that are satisfied by M:

Property 5.7.

1. *If M satisfies $P \rightsquigarrow Q$ and $Q \rightsquigarrow R$, it satisfies $P \rightsquigarrow R$.*

2. *If M satisfies $P \rightsquigarrow Q$ and $R \rightsquigarrow Q$, it satisfies $(P \lor R) \rightsquigarrow Q$.*

3. *If I is an invariant for M and if M satisfies $P \land I \rightsquigarrow Q$, then M satisfies $P \rightsquigarrow Q$.*

4. *If I is an invariant for M and if M satisfies $P \land I \Rightarrow Q$, then M satisfies $P \rightsquigarrow Q$.*

These *refinement diagrams* are hidden in EVENT-B models for deriving liveness properties. For example, Figure 5.2a represents a model of a self-stabilizing system: the diagram relates a pair of assertions $(\neg P, P)$, where $\neg P$ is a precondition stating that the studied system is in an *illegal* state (P does not hold) and P is the postcondition for describing the *desired legal* state. We observe that the *leads to* property $(\neg P) \rightsquigarrow P$, demonstrating the stabilization and convergence, is satisfied by the diagram.

5.3.3.3 Applying the Service-as-Event Paradigm

We apply the *service-as-event* [3] paradigm for modeling the self-⋆ systems. An application of the *service-as-event* paradigm is described as follows:

1. *Describing stabilization and convergence as a service*: Let AM be an abstract model with variables x and invariant $I(x)$. This model contains the following events:

    i. An abstract event (e) models the service of stabilization and convergence expressed by $(\neg P) \rightsquigarrow P$: $(\neg P) \xrightarrow{e} P$, where $(\neg P)$ is a *precondition* for triggering the event (e) and P is a *postcondition* defined by the event (e) that should be satisfied after the *observation* of the event (e). The point is to interpret an invocation of the service/procedure in EVENT-B as a guarded event, which is a defensive manner of modeling the idea. Consequently, the event (e) is as follows:

        EVENT e $\widehat{=}$
          WHEN
            $\neg P(x)$
          THEN
            $x : |P(x')$
          END

        The guard of event (e) is $(\neg P)$: event (e) is observed when the values of the variables x satisfy $(\neg P)$. The new values x' of the variables x satisfy the postcondition P.

    ii. An abstract event (f) models possible faults, and it is not a part of the previous events of the underlying EVENT-B model. The EVENT-B model is open to include a fault and model the existence of a possible fault. The system S enters an illegal state satisfying $(\neg P)$ from a legal state satisfying property P or stays in an illegal state $(\neg P)$ (if the previous state was already illegal).

    iii. An abstract event (c) models possible actions of the system S: event (c) is triggered when a state of the system S satisfies P and the action of (c) produces a new state satisfying P. This event does not lead system S into an illegal state satisfying $(\neg P)$ (*closure* property of self-⋆ systems).

    If the system S is leading to an illegal state satisfying $(\neg P)$, neither the event (f) nor the event (c) modifies the state of the system S (it still satisfies $(\neg P)$). $\neg P$ is the condition of observation of the event (e): it means that the other events $(f$ and $c)$ of the model AM do not **interfere** with event (e). The fairness hypothesis L_0 on the model AM is as follows: $L_0 \widehat{=} WF(e) \wedge WF(c)$. We add some weak fairness assumptions over the events (e and c). We do not add any fairness assumption on event f. We summarize our discussion with a refinement diagram and by defining the next-state relation NEXT as follows:

    $$\text{NEXT} \widehat{=} BA(e)(x,x') \vee BA(f)(x,x') \vee BA(c)(x,x').$$

    We prove that AM satisfies $(\neg P(x)) \rightsquigarrow P(x)$.

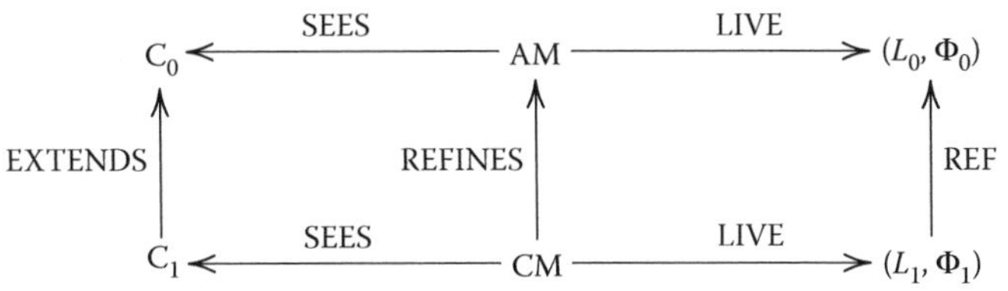

$$C_0 \xleftarrow{\quad\text{SEES}\quad} AM \xrightarrow{\quad\text{LIVE}\quad} (L_0, \Phi_0)$$

$$\text{EXTENDS} \qquad\qquad \text{REFINES} \qquad\qquad \text{REF}$$

$$C_1 \xleftarrow{\quad\text{SEES}\quad} CM \xrightarrow{\quad\text{LIVE}\quad} (L_1, \Phi_1)$$

FIGURE 5.3   *Service-as-event* paradigm: Refinement.

2. *Decomposing stabilization and convergence into simple steps*: The refinement process is driven by the use of the inference at each refinement step for deriving detailed and concrete liveness (*leads to*) properties from the *abstract* ones.

   Let us denote by Φ_0 the abstract properties satisfied by AM under fairness hypothesis L_0, and Φ_1 the more detailed and concrete properties satisfied by the refinement CM of AM under fairness hypothesis L_1. The refinement process is described in Figure 5.3:

   Relating the EVENT-B models and the properties of liveness and fairness, we introduce two relationships called LIVE and REF. The diagram does not state that the relations LIVE, REF, and REFINES can be combined. In other words, the diagram does not commute. The relationship LIVE expresses the following properties:

   a. $\mathcal{S}pec(\text{AM}) \Rightarrow L_0$.

   b. $\mathcal{S}pec(\text{CM}) \Rightarrow L_1$.

   c. For any liveness property $P \rightsquigarrow Q$ of Φ_0, AM satisfies $P \rightsquigarrow Q$.

   d. For any liveness property $P \rightsquigarrow Q$ of Φ_1, CM satisfies $P \rightsquigarrow Q$.

   The relationship REF expresses that any liveness property $P \rightsquigarrow Q$ of Φ_0 is derivable from Φ_1 and CM:

$$\forall P, Q \cdot (P \rightsquigarrow Q \in \Phi_0) \Rightarrow (\mathcal{S}pec(\text{CM}), \Phi_1 \vdash (P \rightsquigarrow Q)).$$

   REF is close to the notion of deduction and proof. The refinement step is achieved by defining a list of liveness properties Φ_1. Then one uses Φ_1 to define the new EVENT-B model CM. This last point is difficult and tricky, since one should determine the fairness condition L_1 and define carefully how fairness is preserved by the new events. We illustrate here the refinement process with an example (see proof tree below) showing the use of inference rules (here *transitivity*) to detail an abstract service stated by $(\neg P) \rightsquigarrow P$. We decompose the abstract service stated by $(\neg P) \rightsquigarrow P$ into simple *subprocedures/steps*, using the *inference rules* [26] related to the *leads to* properties:

$$\cfrac{\cfrac{\cdots\quad\cdots}{(\neg P) \rightsquigarrow R_0}\,trans_2 \qquad \cfrac{\cfrac{\cdots\quad\cdots}{R_0 \rightsquigarrow R_1}\,trans_4 \qquad \cfrac{\cdots\quad\cdots}{R_1 \rightsquigarrow P}\,trans_5}{\cfrac{R_0 \rightsquigarrow P}{}\,trans_3}\,trans_1}{(\neg P) \rightsquigarrow P}$$

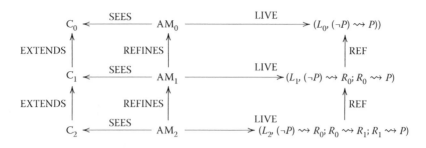

FIGURE 5.4 Decomposition and refinement.

This process is similar to refinement (see Figure 5.4), since we add a new state R_k ($0 \leq k \leq n$) leading from ($\neg P$) to P at each level of the proof tree. The initial property ($\neg P$) $\rightsquigarrow P$ is decomposed until the identification of the *stabilization* steps is satisfactory. The *stabilization* phase is expressed by the property ($\neg P$) $\rightsquigarrow R_0; R_0 \rightsquigarrow R_1; \ldots; R_{n-1} \rightsquigarrow R_n; R_n \rightsquigarrow P$, which states the *convergence* leading to the *desired legal* states. Each level of the proof tree corresponds to a level of refinement (see Figure 5.4) in the formal development. A *leads to* property demonstrates the *service* of *stabilization* defined by an event in the model.

5.4 STEPWISE FORMAL DEVELOPMENT OF SELF-HEALING P2P-BASED APPROACH

5.4.1 Self-Healing P2P-Based Approach

P2P-based self-healing protocols are proposed by Marquezan et al. [7], where *system reliability* is the main concern. The self-healing process ensures the maintenance of proper functioning of the system services. If a service fails, then it switches from a *legal* state to a *faulty* state. The self-healing/recovery procedure ensures that the service switches back to the *legal* state. The services run in a distributed (P2P) system composed of *agents/peers* executing instances of tasks. The *services* and *peers* notions are introduced as follows:

1. *Management services:* Tasks/services are executed by the peers

2. *Instances of management services:* Peers execute a certain type of management service

3. *Management peer group (MPG):* Instances of the same management service

The self-healing property can be described as follows:

1. *Self-identification* is triggered to detect a service failure. This mechanism identifies running or failed instances of a management service.

2. *Self-activation* is started whenever self-identification detects that a management service has failed. The *self-activation* evaluates when the management service recovers from failure. If there are still enough safe instances of service, the recovery procedure

does not start. Otherwise, the *self-configuration* mechanism is triggered for repairing the service.

3. *Self-configuration* is activated if a service failure is critical; the role of this mechanism is to instantiate the failed management service and return the service into a *legal* state.

5.4.2 Stepwise Formal Development

The formal development of the *self-healing P2P-based approach* includes EVENT-B machines organized by the refinement relationship. The machine M0 abstracts the self-healing approach. The refinement machines M1, M2, and M3 introduce, step-by-step, the *self-detection*, *self-activation*, and *self-configuration* phases, respectively. The remaining refinement machines, from M4 to M20, localize the system model: each step of the algorithm is made *local* to a node. The last refinement machine M21 presents a local model that describes a set of procedures for the recovery process of the P2P system.

5.4.2.1 Abstracting the Self-Healing Approach (M0)

This section presents an abstraction of the self-healing procedure using services. Each service (s) is described by two states: *RUN* (*legal/running* state) and *FAIL* (*illegal/faulty* state). A variable *serviceState* is defined as $s \mapsto st \in serviceState$, where s denotes a service and st denotes a possible state. A property $P_0(s)$ expresses that a service (s) is in a *legal running* state that is formalized as $P_0(s) \; \widehat{=} \; (s \mapsto RUN \in serviceState)$. An event FAILURE models a faulty behavior, where service (s) enters into a faulty state (*FAIL*) that satisfies the property $\neg P_0(s)$. The *self-healing* of management service (s) is expressed as $(\neg P_0(s)) \rightsquigarrow P_0(s)$. The *recovery* procedure is stated by an event $HEAL_0$ $((\neg P_0(s)) \xrightarrow{HEAL_0} P_0(s))$, where service ($s$) recovers from an *illegal faulty* state (*FAIL*) to a *legal running* state (*RUN*). The refinement diagram* (see Figure 5.5a) and the events below sum up the abstraction of a *recovery* procedure.

EVENT $HEAL_0$
ANY
$\quad s$
WHERE
$\quad grd1 \; : s \in SERVICES$
$\quad grd2 \; : s \mapsto FAIL \in serviceState$
THEN
$\quad act1 \; : serviceState :=$
$\quad\quad (serviceState \setminus \{s \mapsto FAIL\})$
$\quad\quad \cup \{s \mapsto RUN\}$

EVENT FAILURE
ANY
$\quad s$
WHERE
$\quad grd1 \; : s \in SERVICES$
$\quad grd2 \; : s \mapsto RUN \in serviceState$
THEN
$\quad act1 \; : serviceState :=$
$\quad\quad (\{s\} \lhd serviceState) \cup \{s \mapsto FAIL\}$

The *macro/abstract view* of the *self-healing* is detailed in the further refinement levels, using intermediate steps. The diagram Figure 5.5a states that $HEAL_0$ is constrained by weak fairness assumption, since it is an arrow with a label $HEAL_0$. A set of new variables is introduced to capture the system requirements in the specification, and the new variables are declared in form of *NAME_{Refinement Level}*.

---

* The assertions ($s \mapsto st \in serviceState$), describing the state (st) of a service (s), are shortened into (st) in the diagrams.

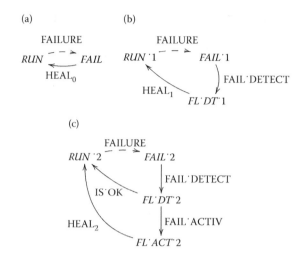

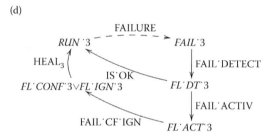

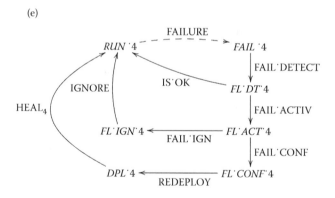

FIGURE 5.5 Refinement diagrams for M0, M1, M2, M3, and M4. (a) Abstraction, (b) self-detection, (c) self-activation, (d) self-configuration, and (e) self-healing steps.

5.4.2.2 Abstracting the Self-Detection (M1)

In this refinement step, the abstract variable *serviceState* is replaced with a new variable *serviceState*_1. The state *RUN* is decomposed into two states: *RUN_1* and *FL_DT_1*. The state *FAIL* is refined into a state *FAIL_1*. We formalize these new states as follows: $P_1(s) \triangleq (s \mapsto RUN\_1 \in serviceState\_1)$ defines that a service *s* is in a *running* state; $(\neg P_1(s)) \triangleq (s \mapsto FAIL\_1 \in serviceState\_1)$ defines a *failed* state of the service (*s*); and $R_1(s) \triangleq (s \mapsto FL\_DT\_1 \in serviceState\_1)$ defines that a service (*s*) is in a state where a failure has been detected. Gluing invariants of the machine M1 describe the relationships between the abstract states and new defined states by preserving the required properties:

$$inv1 \ : \ \forall s \cdot s \in SERVICES \wedge P_0(s) \Rightarrow (P_1(s) \vee R_1(s))$$
$$inv2 \ : \ \forall s \cdot s \in SERVICES \wedge (P_1(s) \vee R_1(s)) \Rightarrow P_0(s)$$
$$inv3 \ : \ \forall s \cdot s \in SERVICES \wedge (\neg P_0(s)) \Rightarrow (\neg P_1(s))$$
$$inv4 \ : \ \forall s \cdot s \in SERVICES \wedge (\neg P_1(s)) \Rightarrow (\neg P_0(s)).$$

We decompose the healing process into two steps: Before triggering, a service (s) can **suspect** and **identify** (1) a failure state (*FAIL_1*) and (2) a recovery phase (HEAL$_1$). It means that the healing process is characterized by a liveness property that can be decomposed into two liveness properties following the *trans* rule.

$$\frac{(\neg P_1) \rightsquigarrow R_1 \qquad R_1 \rightsquigarrow P_1}{(\neg P_1) \rightsquigarrow P_1} \ trans.$$

We detail the self-healing procedure, defined by $(\neg P_1) \rightsquigarrow P_1$, by introducing the property R_1 between $(\neg P_1)$ and P_1 (inference rule **transitivity**). The application of inference rules and the refinement are described in Figure 5.5b.

The event FAIL_DETECT, refining HEAL$_0$, is introduced to show the *self-detection*: the failure state (*FAIL_1*) of a service (s) is detected (*FL_DT_1*). The property $(\neg P_1) \rightsquigarrow R_1$ is described by the event FAIL_DETECT. $R_1 \rightsquigarrow P_1$ is defined by a new event HEAL$_1$, where the service (s) restores to a *legal running* state (*RUN_1*) after the failure detection.[*]

```
EVENT FAILURE
REFINES FAILURE
    . . .
WHERE
    ⊖ s ↦ RUN ∈ serviceState
    ⊕ s ↦ RUN_1 ∈ serviceState_1
THEN
    ⊖ act1 : . . .
    ⊕ serviceState_1 :=
        (serviceState_1 \ {s ↦ RUN_1})
            ∪{s ↦ FAIL_1}


EVENT FAIL_DETECT
REFINES HEAL₀
ANY
    s
WHERE
    grd1 : s ∈ SERVICES
    grd2 : s ↦ FAIL_1 ∈ serviceState_1
THEN
    act1 : serviceState_1 :=
        (serviceState_1 \ {s ↦ FAIL_1})
            ∪{s ↦ FL_DT_1}
```

---

[*] ⊕: to add elements to a model, ⊖: to remove elements.

EVENT HEAL$_1$
ANY
  s
WHERE
  $grd1 : s \in SERVICES$
  $grd2 : s \mapsto FL\_DT\_1 \in serviceState\_1$
THEN
  $act1 : serviceState\_1 := (serviceState\_1 \setminus \{s \mapsto FL\_DT\_1\}) \cup \{s \mapsto RUN\_1\}$.

We prove that the model M1 satisfies the abstract liveness property $(\neg P_0) \rightsquigarrow P_0$.

Proof 5.1. *The model M1 satisfies the liveness property* $(\neg P_1) \rightsquigarrow P_1$. *The invariant inv2 and the inference rule* **implication** *allow us to deduce that* $P_1 \rightsquigarrow P_0$. *The invariant inv3 and the inference rule* **implication** *allow us to deduce that* $(\neg P_0) \rightsquigarrow (\neg P_1)$. *The inference rule* **transitivity** *allows us to prove that* M1 *satisfies* $(\neg P_0) \rightsquigarrow P_0$.

The same method is applied to identify other steps of the *self-healing* algorithm. Due to limited space, we focus only on the interesting parts of models and liveness properties. The complete formal development of this case study is available on the web.[*]

5.4.2.3 Abstracting the Self-Activation (M2)

The *self-activation* mechanism is modeled in this refinement M2 (see Figure 5.5c), where a service failure is evaluated in terms of critical or noncritical. We introduce new states and properties: $P_2(s) \,\hat{=}\, (s \mapsto RUN\_2 \in serviceState\_2)$ defines that a service (*s*) is in a *running/legal* state; $(\neg P_2(s)) \,\hat{=}\, (s \mapsto FAIL\_2 \in serviceState\_2)$ defines that a service (*s*) is in a *failed/illegal* state; $R_2(s) \,\hat{=}\, (s \mapsto FL\_DT\_2 \in serviceState\_2)$ shows that a failure is detected by a service (*s*); and $S_2(s) \,\hat{=}\, (s \mapsto FL\_ACT\_2 \in serviceState\_2)$ defines that the criticality of a detected failure of a service (*s*) is evaluated using a new state *FL_ACT_2*. Gluing invariants of this model M2 state the relationships between the abstract states and new states, including properties:

$$inv1 : \forall s \cdot s \in SERVICES \wedge P_1(s) \Rightarrow (P_2(s) \vee S_2(s))$$
$$inv2 : \forall s \cdot s \in SERVICES \wedge (P_2(s) \vee S_2(s)) \Rightarrow P_1(s)$$
$$inv3 : \forall s \cdot s \in SERVICES \wedge (R_1(s)) \Rightarrow (R_2(s))$$
$$inv4 : \forall s \cdot s \in SERVICES \wedge (R_2(s)) \Rightarrow (R_1(s))$$
$$inv5 : \forall s \cdot s \in SERVICES \wedge (\neg P_1(s)) \Rightarrow (\neg P_2(s))$$
$$inv6 : \forall s \cdot s \in SERVICES \wedge (\neg P_2(s)) \Rightarrow (\neg P_1(s)).$$

The invariants express how the abstract states of the model *M1* are decomposed into coresponding concrete states in the model *M2*. The state *RUN_1* (defined by $P_1(s)$) is decomposed (*inv1* and *inv2*) into two concrete states *RUN_2* (defined by $P_2(s)$) and *FL_ACT_2* (defined by $S_2(s)$). *FL_DT_1* (defined by $R_1(s)$) is equivalent to (*inv3* and *inv4*) *FL_DT_2* (defined by $R_2(s)$). *FAIL_1* (defined by $(\neg P_1(s))$) is equivalent to *FAIL_2* (defined by $(\neg P_2(s))$) (*inv5* and *inv6*).

---

[*] http://eb2all.loria.fr/html_files/files/selfhealing/report.pdf; http://eb2all.loria.fr/html_files/files/selfhealing/self-healing.zip.

The refinement intends to detail what happens to a service (s) after the self-detection. The healing phase taking place after the self-detection is defined by a liveness property $R_2 \rightsquigarrow P_2$. We distinguish here two cases

- If the detected failure is a false alarm, then the service (s) returns to a legal running state. This case is stated by the liveness property as $R_2 \rightsquigarrow P_2$. An event IS_OK (refinement of $HEAL_1$) models this case.

  EVENT IS_OK REFINES $HEAL_1$
  ANY
    s
  WHERE
    $grd1$: $s \in SERVICES \wedge s \mapsto FL\_DT\_2 \in serviceState\_2$
  THEN
    $act1$: $serviceState\_2 := (serviceState\_2 \setminus \{s \mapsto FL\_DT\_2\}) \cup \{s \mapsto RUN\_2\}$.

- If a failure is real, then its criticality must be evaluated before starting an actual self-healing process. This case is stated by the two liveness properties:

  - $R_2 \rightsquigarrow S_2$ shows that the failure of service (s) is evaluated in terms of critical or noncritical. This self-activation is modeled by an event FAIL_ACTIV (refinement of $HEAL_1$):

    EVENT FAIL_ACTIV REFINES $HEAL_1$
    ANY
      s
    WHERE
      $grd1$: $s \in SERVICES \wedge s \mapsto FL\_DT\_2 \in serviceState\_2$
    THEN
      $act1$: $serviceState\_2 := (serviceState\_2 \setminus \{s \mapsto FL\_DT\_2\}) \cup \{s \mapsto FL\_ACT\_2\}$.

  - $S_2 \rightsquigarrow P_2$ shows the healing procedure that occurs after the self-activation, and it is modeled by a new event $HEAL_2$:

    EVENT $HEAL_2$
    ANY
      s
    WHERE
      $grd1$: $s \in SERVICES \wedge s \mapsto FL\_ACT\_2 \in serviceState\_2$
    THEN
      $act1$: $serviceState\_2 := (serviceState\_2 \setminus \{s \mapsto FL\_ACT\_2\}) \cup \{s \mapsto RUN\_2\}$.

The diagram in Figure 5.5c states that the disjunction of events FAIL_ACTIV and IS_OK is under weak fairness assumption.

5.4.2.4 Abstracting the Self-Configuration (M3)

The *self-configuration* step is introduced in the refinement M3 (see Figure 5.5d), which expresses that if a failure of service (s) is critical, then the *self-configuration* procedure for a service (s) will be triggered (state *FL_CONF_3*); otherwise, the failure will be ignored (state *FL_IGN_3*). A set of new properties $P_3(s)$, $S_3(s)$, and

$T_3(s)$ is defined as $P_3(s) \ \widehat{=}\ (s \mapsto RUN\_3 \in serviceState\_3)$, $S_3(s) \ \widehat{=}\ (s \mapsto FL\_ACT\_3 \in serviceState\_3)$, and $T_3(s) \ \widehat{=}\ ((s \mapsto FL\_IGN\_3 \in serviceState\_3) \vee (s \mapsto FL\_CONF\_3 \in serviceState\_3))$, respectively.

This refinement details the self-activation procedure after triggering a service (s). The goal of this refinement is to detail the healing process defined by $S_3 \rightsquigarrow P_3$. We introduce T_3 (the results of self-activation) in between S_3 and P_3:

- $S_3 \rightsquigarrow T_3$ states that a failure service (s) can be either evaluated as critical (state FL_CONF_3) or ignored (state FL_IGN_3). This liveness property is expressed by an event FAIL_CF_IGN (refinement of $HEAL_2$):

  EVENT FAIL_CF_IGN REFINES $HEAL_2$
  ANY
    s, st
  WHERE
    $grd1\ : s \in SERVICES$
    $grd2\ : st \in \{FL\_CONF\_3, FL\_IGN\_3\}$
    $grd3\ : s \mapsto FL\_ACT\_3 \in serviceState\_3$
  THEN
    $act1\ : serviceState\_3 := (serviceState\_3 \setminus \{s \mapsto FL\_ACT\_3\}) \cup \{s \mapsto st\}$.

- $T_3 \rightsquigarrow P_3$ states that a service (s), after reconfiguration or ignoring the failed state, will eventually return to a *running/legal* state. This liveness property is expressed by a new event $HEAL_3$:

  EVENT $HEAL_3$
  ANY
    s, st
  WHERE
    $grd1\ : s \in SERVICES$
    $grd2\ : st \in \{FL\_CONF\_3, FL\_IGN\_3\}$
    $grd3\ : s \mapsto st \in serviceState\_3$
  THEN
    $act1\ : serviceState\_3 := (serviceState\_3 \setminus \{s \mapsto st\}) \cup \{s \mapsto RUN\_3\}$.

5.4.2.5 Abstracting the Global Behavior of the System (M4)

The developed models are refined explicitly introducing the following steps (see Figure 5.5e): (1) *self-detection*, (2) *self-activation*, and (3) *self-configuration*. The self-detection phase is used to detect any failure in the autonomous system using two events FAIL_DETECT and IS_OK. The event FAIL_DETECT models a process for failure detection. The event IS_OK states that if the detected failure of a service (s) is a *false alarm*, then the service (s) returns to a *legal* state (RUN_4). The self-activation process is used to evaluate when an actual failure is detected using the following events: FAIL_ACTIV, FAIL_IGN, IGNORE, and FAIL_CONF. The events FAIL_IGN and IGNORE are used to ignore a failure of service (s) when the detected failure is not in the critical state (FL_IGN_4), and the event FAIL_CONF is used to evaluate a failure of service (s) when the detected failure is critical (FL_CONF_4). The last phase, *self-configuration*, presents the healing procedure of a *failed* service using an event REDEPLOY. In this refinement, we clearly formalize the self-activation process: the self-activation phase leads a service (*s*) to the self-configuration if a failure of the service (*s*) is critical; otherwise, the service (*s*) simply returns to a *legal/running* state.

5.4.2.6 Localizing Models (from M4 to M20)

From model M4 to M20, we localize the events (we switch from a *service* point of view to the instances/peers point of view) and detail the macro (global) steps by adding new events, variables, and constraints. The refinements M5, M6, and M7 introduce the running (*run_peers(s)*), faulty (*fail_peers*[{*s*}]), suspicious (*susp_peers(s)*), and deployed peers/instances (*dep_inst*[{*s*}]) for a service (*s*). A function (*min_inst*) associates each service (*s*) with a minimal number of instances that is required for running a service (*s*) and helps to detail the *self-activation* process; if the number of running instances of service (*s*) is below the minimum, then the failure is critical. The models M8, M9, and M10 detail the *self-detection* and *self-configuration* steps to introduce the *token owners* for the services. Models from M11 to M20 gradually localize the events (to switch from a *service* point of view to the instances/peers point of view). The detailed formal development of various steps from M5 to M20 are described in our technical report. In the following section, we present only the local model M21.

5.4.2.7 The Local Model (M21)

This model details locally the *self-healing* procedure of a service (*s*). The peers instantiating management service (*s*) are introduced, including the notion of *token owner*. The *token owner* is a peer instance of service (*s*) that is marked as a *token owner* for the MPG. It performs the *self-healing* procedure using the *self-detection*, *self-activation*, and *self-configuration* steps, which are now local to the *token owner*, including more details of new events and/or refinement of abstract events.

1. *Self-Detection* introduces an event SUSPECT_INST where a *token owner* for a service (*s*) is able to *suspect* a set (*susp*) of unavailable peers of service (*s*). Other events RECONTACT_INST_OK and RECONTACT_INST_KO are used to model a success or a failure to recontact the unavailable peers for ensuring the failed states. Moreover, the *token owner* is able to monitor the status of service (*s*) using two events FAIL_DETECT and IS_OK. If there are unavailable instances after the recontacting procedure, the *token owner* informs the safe members of MPG of failed instances, using the event FAIL_DETECT. Otherwise, the *token owner* indicates that service is running properly.

2. *Self-Activation* introduces an event FAIL_ACTIV, where the *token owner* evaluates if the failure is critical, if there are failed instances of the service (*s*). Another event FAIL_IGNORE specifies that the failure is not critical. An event IGNORE can ignore the failure if several instances (more than minimum) are running correctly. If the number of instances for the running service (*s*) is lower than the minimum required services, then the failure will be declared critical. The *self-healing* process will be triggered using an event FAIL_CONFIGURE.

```
MACHINE M21 ...
  EVENT SUSPECT_INST
    ANY
      s, susp
    WHERE
      grd1 : s ∈ SERVICES
      grd2 : susp ⊆ PEERS
      grd3 : susp = run_inst(token_owner(s) ↦ s) ∩ unav_peers
      grd4 : suspc_inst(token_owner(s) ↦ s) = ∅
      grd5 : inst_state(token_owner(s) ↦ s) = RUN_4
      grd6 : susp ≠ ∅
    THEN
      act1 : suspc_inst(token_owner(s) ↦ s) := susp
    END
  EVENT FAILURE ...
  EVENT RECONTACT_INST_OK ...
  EVENT RECONTACT_INST_KO ...
  EVENT FAIL_DETECT ...
  EVENT IS_OK ...
  EVENT FAIL_ACTIV ...
  EVENT FAIL_IGNORE ...
  EVENT IGNORE ...
  EVENT FAIL_CONFIGURE ...
  EVENT REDEPLOY_INSTC ...
  EVENT REDEPLOY_INSTS ...
  EVENT REDEPLOY ...
  EVENT HEAL₄ ...
  EVENT MAKE_PEER_UNAVAIL ...
  EVENT UNFAIL_PEER ...
  EVENT MAKE_PEER_AVAIL ...
```

3. *Self-Configuration* introduces three events REDEPLOY_INSTC, REDEPLOY_INSTS, and REDEPLOY that specify that if the failure of service (s) is critical, then the new instances of running service (s) are deployed until the minimum number of instances is reached. Then, the event $HEAL_4$ can be triggered, corresponding to the *convergence* of the self- healing process.

It is worth noting that the *architectural decomposition* of the self-healing process is emphasized in this model in the form of events related to an algorithm. There is also a set of events describing actions related to the environment. MAKE_PEER_UNAVAIL: a set of peers (*prs*) becomes unavailable (cannot be contacted); MAKE_PEER_AVAIL: a formerly unavailable instance (*p*) becomes available; and UNFAIL_PEER: a failed instance reenters into a *legal running* state. The model M21 describes locally the *self-healing P2P-based approach*, where we have formalized *hypotheses*, by events, for ensuring the correct functioning of the self-healing process. (1) Event MAKE_PEER_AVAIL: if the token owner of a service (s) becomes unavailable, at least one peer with the same characteristics as the disabled token owner (state, local information about running, failed peers, etc.), can become the new token owner of the service (s) and (2) Event REDEPLOY_INSTC: there is always a sufficient number of available peers that can be deployed to reach a legal running state of a service (s).

In a nutshell, we say that our methodology allows users to understand the self-★ mechanisms and gain insight into their architectures (components, coordination, etc.) and provides evidence of the correctness of self-★ systems under some assumptions/hypotheses.

5.5 DISCUSSION, CONCLUSION, AND FUTURE WORK

In this chapter, we presented a methodology based on liveness properties and *refinement diagrams* for modeling the self-★ systems using EVENT-B. We characterize the self-★ systems by three modes (abstract states): (1) *legal (correct) state*, (2) *illegal (faulty) state*, and (3) *recovery* state. We have proposed a generic pattern for deriving the correct self-★ systems (see Figure 5.1). The *service-as-event* and *call-as-event* paradigms provide a way to express the relationships between modes for ensuring required properties as convergence. The *correct-by-construction* principle gives us the possibility to refine procedures from events and link it to the modes. The key idea is to identify the modes (considered as abstract states) and the required abstract steps to allow the navigation between modes, and then to gradually enrich abstract models, using refinement to introduce the concrete states and events. We have illustrated our methodology by the *self-healing approach* [7]; we have stated the self-healing mechanism with a simple abstract liveness property $(\neg P) \rightsquigarrow P$. Then we have used inference rules [26] to detail the liveness properties by guiding and expressing the refinement steps.

TABLE 5.2 Summary of proof obligations (RODIN 3.0)

Models	Total	Auto (%)		Interactive (%)	
M0	3	3	100	0	0
M1	21	21	100	0	0
M2	46	46	100	0	0
M3	68	68	100	0	0
M4	142	142	100	0	0
M5	46	41	89.13	5	10.87
M6	83	59	71.08	24	28.92
M7	35	31	88.57	4	11.45
M8	56	55	98.21	1	1.79
M9	60	59	98.33	1	1.67
M10	60	50	83.33	10	16.67
M11	124	101	81.45	23	18.55
M12	63	51	80.95	12	19.05
M13	33	27	81.82	6	18.18
M14	113	80	70.8	33	29.2
M15	48	34	70.83	14	29.17
M16	201	36	17.91	165	82.09
M17	69	44	63.77	25	36.23
M18	60	29	48.33	31	51.67
M19	29	17	58.62	12	41.38
M20	37	21	56.76	16	43.24
M21	13	11	84.62	2	15.38
Total	1410	1026	72.77	384	27.53

The complexity of the development is measured by the number of proof obligations (PO) that are automatically/manually discharged (see Table 5.2). It should be noted that a large majority (72.77%) of the 1410 generated PO are automatically discharged by the proof engine of Rodin using SMT solvers (veriT [32] and CVC3 [33]). Examples of interactive PO are related to proving the *finiteness* of MPG during the *redeployment operation* of the *self-configuration phase*.

Furthermore, our refinement-based formalization allows us to produce final local models close to the *source code*. Our future work includes the development of techniques for generating applications from the resulting model, extending tools like EB2ALL [34,35]. Moreover, further case studies will help us discover new patterns; these patterns will be added to a catalog of patterns that could be implemented in the Rodin platform.

REFERENCES

1. Shlomi Dolev. *Self-Stabilization*. MIT Press, 2000.
2. Gary T. Leavens, Jean-Raymond Abrial, Don S. Batory, Michael J. Butler, Alessandro Coglio, Kathi Fisler, Eric C. R. Hehner et al. Roadmap for enhanced languages and methods to aid verification. In Stan Jarzabek, Douglas C. Schmidt, and Todd L. Veldhuizen, editors, *GPCE*, pp. 221–236. ACM, 2006.
3. Manamiary Bruno Andriamiarina, Dominique Méry, and Neeraj Kumar Singh. Integrating proved state-based models for constructing correct distributed algorithms. In *IFM*, pp. 268–284, 2013.
4. Andrew Berns and Sukumar Ghosh. Dissecting self-* properties. In *Proceedings of the 2009 Third IEEE International Conference on Self-Adaptive and Self-Organizing Systems*, SASO '09, pp. 10–19, Washington, DC, USA, 2009. IEEE Computer Society.
5. J.-R. Abrial. *Modeling in Event-B: System and Software Engineering*. Cambridge University Press, 2010.
6. Manamiary Bruno Andriamiarina, Dominique Méry, and Neeraj Kumar Singh. Analysis of self-* and p2p systems using refinement. In Yamine Ait-Ameur and Klaus-Dieter Schewe, editors, *ABZ*, volume 8477 of *Lecture Notes in Computer Science*, pp. 117–123. Springer, 2014.
7. Clarissa Cassales Marquezan and Lisandro Zambenedetti Granville. *Self-* and P2P for Network Management—Design Principles and Case Studies*. Springer Briefs in Computer Science. Springer, 2012.
8. Graeme Smith and Jeffrey W. Sanders. Formal development of self-organising systems. In *Proceedings of the 6th International Conference on Autonomic and Trusted Computing*, ATC '09, p. 90–104, Berlin, Heidelberg, 2009. Springer-Verlag.
9. Mariachiara Puviani, Giovanna Di Marzo Serugendo, Regina Frei, and Giacomo Cabri. A method fragments approach to methodologies for engineering self-organizing systems. *ACM Trans. Auton. Adapt. Syst.*, 7(3):33:1–33:25, October 2012.
10. Jan Sudeikat, Jan-Philipp Steghöfer, Hella Seebach, Wolfgang Reif, Wolfgang Renz, Thomas Preisler, and Peter Salchow. Design and simulation of a wave-like self-organization strategy for resource-flow systems. In *MALLOW'10*, pp. –1–1, 2010.
11. Radu Calinescu, Shinji Kikuchi, and Marta Kwiatkowska. Formal methods for the development and verification of autonomic IT systems. In Cong-Vinh, P. editor, *Formal and Practical Aspects of Autonomic Computing and Networking: Specification, Development and Verification*, IGI Global, pp. 90–104. 2011.

12. Matthias Güdemann, Frank Ortmeier, and Wolfgang Reif. Safety and dependability analysis of self-adaptive systems. In *Proceedings of the Second International Symposium on Leveraging Applications of Formal Methods, Verification and Validation*, ISOLA '06, pp. 177–184, Washington, DC, USA, 2006. IEEE Computer Society.

13. Danny Weyns, M. Usman Iftikhar, Didac Gil de la Iglesia, and Tanvir Ahmad. A survey of formal methods in self-adaptive systems. In *Proceedings of the Fifth International C* Conference on Computer Science and Software Engineering*, C3S2E '12, pp. 67–79, New York, NY, USA, 2012. ACM.

14. Jonathan Bowen. *Formal Specification and Documentation Using Z: A Case Study Approach*. International Thomson Computer Press (ITCP) Thomson Publishing, 2003.

15. Daniel Jackson. *Software Abstractions, Revised Edition Logic, Language, and Analysis*. MIT Press, 2011.

16. Manamiary Bruno Andriamiarina, Hayat Daoud, Mostefa Belarbi, Dominique Méry, and Camel Tanougast. Formal verification of fault tolerant NoC-based architecture. In *First International Workshop on Mathematics and Computer Science (IWMCS2012)*, p. 7, Tiaret, Algérie, December 2012.

17. Dominique Méry and Neeraj Kumar Singh. Analysis of DSR protocol in Event-B. In *Proceedings of the 13th International Conference on Stabilization, Safety, and Security of Distributed Systems*, SSS'11, pp. 401–415, Berlin, Heidelberg, 2011. Springer-Verlag.

18. M. Usman Iftikhar and Danny Weyns. A case study on formal verification of self-adaptive behaviors in a decentralized system. In *FOCLASA'12*, pp. 45–62, 2012.

19. Davide Tosi. *Research Perspectives in Self-Healing Systems*. Technical Report, DISE LTA, 2004.

20. Danny Weyns, Sam Malek, and Jesper Andersson. Forms: Unifying reference model for formal specification of distributed self-adaptive systems. *ACM Trans. Auton. Adapt. Syst.*, 7(1):8:1–8:61, May 2012.

21. Dominique Cansell, Dominique Mery, and Stephan Merz. Diagram refinements for the design of reactive systems. *J. Univers. Comput. Sci.*, 7(2):159–174, Feb 2001. http://www.jucs.org/jucs_7_2/diagram_refinements_for_the.

22. Dominique Cansell and Dominique Méry. The Event-B modelling method: concepts and case studies. In Dines Bjørner and Martin C. Henson, editors. *Logics of Specification Languages*. EATCS Textbook in Computer Science, pp. 33–140. Springer, 2007.

23. Project RODIN. Rigorous open development environment for complex systems. http://www.eventb.org/, 2004–2010.

24. Jean-Raymond Abrial, Michael J. Butler, Stefan Hallerstede, Thai Son Hoang, Farhad Mehta, and Laurent Voisin. Rodin: An open toolset for modelling and reasoning in Event-b. *STTT*, 12(6):447–466, 2010.

25. Dominique Méry. Refinement-based guidelines for algorithmic systems. *Int. J. Softw. Inform.*, 3(2–3):197–239, June/September 2009.

26. Leslie Lamport. The temporal logic of actions. *ACM Trans. Program. Lang. Syst.*, 16(3):872–923, 1994.

27. Dominique Méry and Michael Poppleton. Formal modelling and verification of population protocols. In *IFM*, pp. 208–222, 2013.

28. Leslie Lamport. *Specifying Systems: The TLA+ Language and Tools for Hardware and Software Engineers*. Addison-Wesley, 2002.

29. Dominique Méry and Mike Poppleton. Towards an integrated formal method for verification of liveness properties in distributed systems—With application to population protocols. *Int. J. Softw. Syst. Model.*, 1–33, 2016.

30. I. S. Wishnu, B. Prasetya, and S. Doitse Swierstra. Formal design of self-stabilizing programs: Theory and examples, 2000.

31. Ralph-Johan Back. Incremental software construction with refinement diagrams. In Michael Johnson and Varmo Vene, editors, *Algebraic Methodology and Software Technology*, volume 4019 of *Lecture Notes in Computer Science*, pp. 1–1. Springer, Berlin Heidelberg, 2006.

32. Thomas Bouton, Diego Caminha B. de Oliveira, David Déharbe, and Pascal Fontaine. Verit: An open, trustable and efficient smt-solver. In Renate A. Schmidt, editor, *Proc. Conference on Automated Deduction (CADE)*, Lecture Notes in Computer Science. Springer-Verlag, 2009.

33. Clark Barrett and Cesare Tinelli. CVC3. In Werner Damm and Holger Hermanns, editors, *Proceedings of the 19th International Conference on Computer Aided Verification (CAV '07)*, volume 4590 of *Lecture Notes in Computer Science*, pp. 298–302. Springer-Verlag, July 2007. Berlin, Germany.

34. Dominique Méry and Neeraj Kumar Singh. Automatic code generation from Event-b models. In *Proceedings of the Second Symposium on Information and Communication Technology*, SoICT'11, pp. 179–188, New York, NY, USA, 2011. ACM.

35. Neeraj Kumar Singh. *Using Event-B for Critical Device Software Systems*. Springer, 2013.

CHAPTER 6

Formalizing Goal-Oriented Development of Resilient Cyber-Physical Systems

Inna Pereverzeva and Elena Troubitsyna

CONTENTS

D EVELOPMENT OF RESILIENT CYBER-PHYSICAL SYSTEMS (CPS), that is, the systems
   that can deliver trustworthy services despite changes, is a complex engineering task.
Resilience can be considered as either a system's ability to achieve its goals despite negative
changes, such as failures of components, or as a system's capability to achieve its desired
goals more efficiently, for instance by increasing component utilization. In this work, we
present a formal goal-oriented approach to developing resilient CPS in Event-B. We define
the main abstractions required for reasoning about system goals and introduce the specification patterns explicitly defining architectural reconfiguration mechanisms allowing the
system to achieve resilience. We demonstrate how formal goal-oriented development in
Event-B facilitates the structuring of component interdependencies and derivation of the
overall distributed architecture of CPS.

6.1 INTRODUCTION

Cyber-Physical Systems are examples of complex distributed systems that carry on their
functions in constant interaction with the physical world [1]. Due to continually changing
operating conditions, achieving resilience—a system's ability to provide trustworthy services
despite changes [2]—becomes one of the main objectives of CPS development.

Essentially, resilience means that the system can absorb the effect of changes, for example,
tolerating a component or scale failure to handle an increased load. Resilience is typically
implemented via reconfiguration. To cope with a component failure, the system should
reconfigure, that is, reallocate responsibilities for providing different functionality from
failed components to the healthy ones. To cope with an increased load, either caused by
increased rate of external service requests or degradation of internal conditions, the system
should optimize resource allocation, for example, by employing idle components.

In this work, we adopt a goal-oriented style of resilient CPS development. Goals [3,4] are
the functional and nonfunctional objectives of a system. In software engineering, goals have
been recognized as useful primitives for capturing system requirements [3,4]. In particular,
resilience can be seen as a property that allows the system to progress toward achieving its
goals [5]. Reasoning in terms of goals promotes structuring the top-down system design.

We formalize goal-oriented development of CPS in Event-B. Event-B [6] is a state-based
framework that relies on abstract modeling, refinement, and theorem-proving to create and
verify specifications of complex systems. Development in Event-B starts with an abstract
model that captures only the most essential properties and behaviors of the system. In a
number of correctness-preserving steps—refinements—the abstract model is transformed
into a detailed specification of the overall system. The resultant specification is decomposed into the specifications of the independent subsystems that, through their interactions,
are guaranteed to preserve the system-level properties. The Rodin platform [7] supports
development and verification in Event-B.

In this work, we demonstrate how formal Event-B development of CPS in a goal-oriented style facilitates reasoning about resilience. In our development, we start by decomposing system-level goals into subgoals (tasks) and establishing a correspondence between tasks and components' functional capabilities. We introduce a hierarchical structure of the components that essentially corresponds to the distributed supervisory control over the lower level service-provisioning components. Then we explicitly define the communication between the system components. Our modeling allows us to systematically derive complex reconfiguration mechanisms, ensuring that the system correctly adapts to changing operating conditions.

We argue that the goal-oriented style employed in our formal development facilitates modeling of component interactions, structuring of intercomponent communication, and derivation of complex reconfiguration mechanisms.

The chapter is structured as follows. In Section 6.2 we introduce the main concepts of our goal-oriented development style, discuss the link between goal reachability and resilience, as well as define the basic architectural and functional properties of resilient CPS. In Section 6.3 we describe our formal modeling framework—Event-B. In Sections 6.4 and 6.5 we present our formal development in a goal-oriented style. We define generic specification patterns addressing goal decomposition, modeling reconfiguration, and component interdependencies. In Section 6.6 we describe the last refinement step, which allows us to arrive at a decentralized model by decomposition. Finally, in Section 6.7 we review the related work and discuss the achieved results.

6.2 GOALS AND ARCHITECTURE OF RESILIENT CPS

6.2.1 Goal-Oriented Development and Reconfiguration

In this chapter, we use the notion of a goal as a basis for reasoning about resilient CPS. Goals are the functional and nonfunctional objectives that the system should achieve [3,4]. Reasoning in terms of goals facilitates structuring requirements and supports the top-down system design. In particular, resilience can be seen as a property that allows a system to progress toward achieving its goals [5].

Usually, a system has different types of interdependent goals. Thus, goals can be structured, for example, to form a hierarchy. Generally, they can be formulated at different levels of abstraction. The process of goal detalization (i.e., decomposition into subgoals) is performed until a certain level of granularity is reached, that is, when a subgoal can be assigned to and consecutively realized by the system components [3,4].

In order to achieve their individual or common goals, system components *interact* with each other. Component interactions might be simple (for example, information exchange) or complex (for example, involving requests for service provisioning from one component to another). The system components interact with each other and behave cooperatively to implement system-level goals. Component *interactions* are achieved by communication. Communication allows the individual components to share their local information with other components to facilitate goal achievement.

Since we focus on studying the functional aspect of resilience of CPS behavior in this work, we should explicitly represent off-nominal situations such as component or communication failures and design mechanisms allowing the system to cope with them. As a result of failure, components usually lose the ability to perform their predefined tasks. Therefore, to ensure resilience, the system needs to *reconfigure* and guarantee that the goals remain reachable. The reconfiguration is based on reallocation of responsibilities between components either to ensure that the healthy components can substitute the failed ones or to enable more efficient utilization of components. In both cases, the system components should collaborate to ensure system resilience.

6.2.2 Generic Requirements for Goal-Oriented Resilient Systems

The goal-oriented framework enables reasoning about the system behavior at different levels of abstraction. At the same time, the goal decomposition process facilitates incremental unfolding of the system architecture. It also helps us to build a hierarchy of components according to their responsibilities in achieving goals. Moreover, the goal-oriented framework allows us to formulate reconfigurability as an ability of components to redistribute their responsibilities to ensure goal reachability.

In our formal goal-oriented development, we rely on the following assumptions about system architecture and behavior:

(PR1) There is the main system *goal* that the system aims at accomplishing during its execution.

(PR2) The main system goal can be decomposed into a subset of corresponding *subgoals* (tasks) that can be executed by the system components.

(PR3) The system consists of a number of autonomic software components that are organized hierarchically, that is, one component may be a *manager* of one or a group of other components called *workers*.

(PR4) The manager components *coordinate* the activities related to task achievement; they can be viewed as a distributed implementation of the supervisory control.

(PR5) The worker components *perform* activities to achieve the task.

(PR6) Every task is associated with a specific manager that supervises its execution.

(PR7) A manager component can also be associated with a number of worker components that it is supervising at the moment.

(PR8) The manager components interact with the corresponding associated workers in order to assign tasks to them and receive confirmation about task completion.

(PR9) A worker cannot perform more than one task simultaneously.

(PR10) The manager components interact with other managers in order to distribute their local information about the statuses of the tasks assigned to them.

(PR11) The system components are unreliable, that is, they might fail during system execution.

(PR12)   In the case of failures of components, the system should, if possible, dynamically reconfigure itself to achieve the main system goal.

(PR12.1)   In the case of a worker failure during a task execution, its associated manager reassigns the task of the failed worker component to another available worker.

(PR12.2)   In the case of a manager failure, other available manager components become responsible for the tasks and the workers of the failed one.

(PR12.3)   The system can also reconfigure itself to achieve some of its goals more efficiently, for example, by means of deploying the idle components of both types.

The aim of our modeling is to derive the distributed architecture of CPS described previously. We demonstrate how to formally define the system goal and, in a stepwise manner, derive a detailed specification of the system architecture. While refining the system specification, we gradually introduce a representation of the main elements of the architecture—managers and workers—as well as their communication. Moreover, we explicitly model failures and define the reconfiguration mechanisms for coping with them. We rely on the aforementioned properties (PR1)–(PR12) to derive a distributed architecture of resilient CPS.

6.2.3   General Approach of Developing Distributed Systems in Event-B

To facilitate the development of a distributed architecture in Event-B, we define a set of specification and refinement patterns that reflects the main concepts of the goal-oriented development. Namely, we start by abstractly defining system goals; then we perform goal decomposition by refinement. We define and prove the relevant gluing invariants, establishing a formal relationship between goals and the corresponding subgoals. Next we introduce a representation of system components and the required reconfiguration mechanisms to ensure that the system progresses toward achieving its goals.

We consider dynamic reconfiguration to be a powerful technique for achieving system resilience because it allows the system to adapt to changes by modifying its structure, intercomponent relationships, and dependencies. However, ensuring correctness of the incorporated reconfiguration mechanisms is a complex task. To address this issue, we formalize the possible interdependencies between goals and components as well as formulate the conditions for ensuring correctness of the design of the reconfiguration mechanism. In the refinement process, we also explicitly introduce component failures. To introduce a possibility of reconfiguration, we define the conditions on the information exchange between the components, that is, determine to which part of the local state the components should communicate in order to provide a sufficient basis for reconfiguration. Finally, we decompose the obtained specification into a number of independent components, that is, arrive at the distributed architecture.

The proposed formalization facilitates a systematic, disciplined development of CPS and can be seen as an example of generic guidelines for designing reconfigurable systems.

6.3   MODELING AND REFINEMENT IN EVENT-B

In this section, we give an overview of our formal development framework—Event-B. Event-B is a state-based formal approach that promotes the correct-by-construction development paradigm and formal verification by theorem-proving. In Event-B, a system model is specified using the notion of an *abstract state machine* [6]. An abstract state machine encapsulates the model state, represented as a collection of variables, and defines operations on the state, that is, it describes the dynamic behavior of a modeled system. The important system properties to be preserved during the execution are defined as model invariants. A machine usually has the accompanying component, called context. A context may include user-defined carrier sets and constants. Their properties are formulated as model *axioms*.

The dynamic behavior of the system is defined by a collection of atomic *events*. An event is essentially a *guarded command* that, in the most general form, can be defined as follows:

$$e \;\widehat{=}\; \textbf{any } a \textbf{ where } G_e \textbf{ then } R_e \textbf{ end,}$$

where e is the event's name, a is the list of local variables, and (the event *guard*) G_e is a predicate over the model state. The body of an event is defined by a *multiple* (possibly nondeterministic) assignment to the system variables. In Event-B, this assignment is semantically defined as the next-state relation R_e. The event guard defines the conditions under which the event is *enabled*, that is, its body can be executed. If several events are enabled at the same time, any of them can be chosen for execution nondeterministically. If none of the events are enabled, then the system deadlocks. The occurrence of events represents the observable behavior of the system.

6.3.1 Event-B Refinement

Event-B employs a top-down, refinement-based approach to system development. A development starts from an abstract specification that nondeterministically models the most essential functional system behavior. In a sequence of refinement steps, we gradually reduce nondeterminism and introduce detailed design decisions. A machine can be refined in two possible ways: either using *data refinement* or *superposition refinement*. In particular, we can replace abstract variables with their concrete counterparts, that is, perform *data refinement*. In this case, the invariant of the refined machine formally defines the relationship between the abstract and concrete variables. Via such a *gluing* invariant—"refinement relation"—we mathematically establish a correspondence between the state spaces of the refined and the abstract machines.

During *superposition refinement*, new implementation details are introduced into the system specification by means of *new* events and *new* variables. The new events cannot affect the variables of the abstract specification but only define computations on the newly introduced variables.

The new events correspond to the stuttering steps that are not visible at the abstract level, that is, they refine implicit *skip* (the null action). To guarantee that the refined specification preserves the global behavior of the abstract machine, we should demonstrate that the newly introduced events *converge*. To prove it, we have to define a *variant*—an expression over a finite subset of natural numbers—and show that the execution of new events decreases it. Sometimes, convergence of an event cannot be proved due to a high level of nondeterminism. In that case, the event obtains the status *anticipated*. This obliges the designer to prove, at some later refinement step, that the event indeed converges.

The correctness and consistency of Event-B models, that is, verification of the model well-formedness, invariant preservation, deadlock-freeness, and correctness of the refinement steps, are demonstrated by proving the relevant verification theorems—*proof obligations*. Proof obligations are expressed as logical sequences, ensuring that the transformation is performed in a correctness-preserving way [6].

Modeling, refinement, and verification of Event-B models are supported by an automated tool—the Rodin platform [7]. The platform provides the designers with an integrated modeling environment as well as supports automatic generation and proving of the necessary proof obligations by means of a wide range of automated provers. Moreover, various plug-ins created for the Rodin platform allow a modeler to transform models from one representation to another, for example, from UML to Event-B language [8,9] or from Event-B specification to programming languages C/C++ [10,11], ADA [12,13], etc.

The Event-B refinement process allows us to gradually introduce implementation details and ensure adherence to the abstract specification. Such an approach seamlessly weaves verification with model development and allows us to construct detailed models of complex systems in a highly automated, incremental manner. Providing immediate feedback on the correctness of the model transformations helps to cope with the complexity of system development. Another important mechanism for handling complexity of formal development is *decomposition*.

Model decomposition helps the designers to separate component development from the overall system model but ensure that the components can be recomposed into the overall system in a correctness-preserving way [14]. Event-B is equipped with three forms of decomposition: shared-variable [15–17], shared-event [16], and modularization [18], all of which are supported by the corresponding Rodin plug-ins [19,20]. In this work we rely on a modularization extension of Event-B [18].

6.3.2 Modularization

Modularization extension allows the designers to decompose a system into *modules*. Modules are components containing groups of callable atomic operations [18,19]. Modules can have their own (external and internal) state and invariant properties. An important characteristic of modules is that they can be developed separately and, when needed, composed with the main system. Since decomposition is a special kind of refinement, such a model transformation is also a correctness-preserving step that has to be verified. Hence, a number of corresponding proof obligations are generated and should be proved.

A module description consists of two parts: the *module interface* and the *module body*. A module interface is a separate Event-B component that consists of the external module variables, the module invariants, and a collection of module operations characterized by their pre- and postconditions. In addition, a module interface may contain a group of standard Event-B events. These events model autonomous module threads of control, expressed in terms of their effect on the external module variables. In other words, they describe how the module external variables may change between operation calls. Development of a module starts with deciding on its interface. Once the interface is defined, it cannot be changed in any manner during development. This ensures that a module body may be constructed independently from a system model that relies on the module interface. A *module body* is an Event-B machine. It implements the interface by providing a concrete behavior for each of the interface operations. To guarantee that each interface operation has a suitable implementation, a set of additional proof obligations is generated.

The modularization extension of Event-B facilitates formal development of complex systems by allowing the designers to decompose large specifications into separate components and verify system-level properties at the architectural level. As a result, proof-based verification, as well as reliance on abstraction and decomposition adopted in Event-B, offers the designers a scalable support for the development of complex distributed systems.

6.4 DERIVING RESILIENT ARCHITECTURE

We start our development in Event-B by creating a high-level abstract specification. In the abstract model, we focus on specifying the overall system behavior. Here we aim at specifying property (PR1). The dynamic behavior of the system is modeled by the abs_behaviour machine. We define a variable $goal \in STATES$ that models the current state of the system goal, where $STATES = \{incompl, compl\}$. The variable $goal$ obtains the value $compl$ when the main goal is achieved. Otherwise $goal$ has the value $incompl$. To abstractly model the process of achieving the goal—specified in the event process—the variable $goal$ may change its value from $incompl$ to $compl$. The system continues its execution until the goal is reached:

Machine abs_behaviour
Variables $goal, finish$
Invariants
  $finish \in BOOL$
  $goal \in STATES$
  $finish = TRUE \Rightarrow goal = compl$
Events
process $\widehat{=}$
  where $goal \neq compl$
  then $goal :\in STATES$
  end
finish $\widehat{=}$
  where $goal = compl \wedge finish = FALSE$
  then $finish := TRUE$
  end

6.4.1 Functional Decomposition by Refinement

The objective of our first refinement step is to elaborate on the structure and the behavior of the abstract model. The overall system execution consists of a number of smaller steps, which we call *tasks* (as defined by (PR2) property). Essentially, a task represents a single functional step that the system has to execute in order to progress toward the goal achievement. In this work, for simplicity, we consider only a simple forward execution scenario. In general, our model permits the definition of any type of a complex execution scenario, for example, combining sequential and parallel task execution, branching, rollbacking, etc. In Event-B, we define the execution scenario via formulating a set of axioms in the model's context.

In our first refinement step, we focus on representing the overall system execution as an iterative execution of a set of tasks. To represent all tasks that constitute system execution, we introduce a new abstract type (set) *TASKS* into the model's context. Naturally, we require that the set *TASKS* is finite and nonempty.

In the machine part of the first refinement, we define the new variable *tasks_state* that stores the current execution status of each task as

$$tasks\_state \in TASKS \rightarrow STATES.$$

Initially, none of the task is completed, that is, the status of each task is *incompl*. After successful execution, the task's status changes to *compl*. The event process is now refined to represent the progress in task executions. Note that this event is parametrized—the parameter *t* designates the id of the task being processed:

```
process ≙ refines process
  any t, res
  where t ∈ TASKS
           tasks_state(t) = incompl
           res ∈ STATES
  then   tasks_state(t) := res
  end
```

To establish the relationship between the main goal and tasks, that is, to model the property (PR2) in our specification, we formulate and prove the following gluing invariant:

$$goal = compl \iff (\forall t.\ t \in TASKS \Rightarrow tasks\_state(t) = compl).$$

The invariant postulates that *the main system goal is achieved if and only if all of the involved tasks are successfully completed.*

Let us note that the proposed refinement step can be repeatedly used to refine tasks into subtasks of finer granularity until the desired level of details is reached.

6.4.2 Second Refinement

6.4.2.1 Introducing System Components

In our previous refinement step, we focused on functional decomposition of the system behavior into execution of atomic tasks. We have deliberately abstracted away from associating these tasks with the specific system components that perform the tasks and orchestrate their execution. The goal of our second refinement step is to (i) introduce representation of the system components into the model, (ii) model their possible activation and deactivation (including both normal and abnormal situations), and (iii) abstractly define system reconfiguration mechanisms.

Our system consists of two types of components: *managers* and *workers*. The manager components orchestrate the tasks execution, while the worker components actually perform the tasks. A manager monitors its associated tasks, assigns these tasks to workers, and receives/processes the results from the workers. Moreover, it interacts with the other manager components to share the information on task status. By integrating into the system design an information-sharing functionality, we ensure that if a manager fails, some other available manager component has sufficient information to take over the responsibilities for the tasks and workers of the failed one.

We assume that both types of system components are unreliable, that is, any component might fail during the system execution. To ensure that all system tasks will be accomplished (and, consequently, the main system goal will eventually be achieved), we should incorporate into the system design some mechanisms that would allow the system to complete its execution despite the component failures.

To model system components in Event-B, we first introduce several new data structures into our model's context. In particular, we introduce two new abstract types (sets)— *MANAGERS* and *WORKERS*—that store all available components of each type. Moreover, we define the constants, *Init_Managers* and *Init_Workers*, which are subsets of the sets *MANAGERS* and *WORKERS*, respectively. These constants define the sets of active components of the corresponding type at the beginning of the system operation.

The set *TASKS* contains all tasks that the system should accomplish to reach its main goal. We assume that each worker is capable of performing a certain subset of tasks. Hence, before giving an assignment to a worker, we have to choose a worker of a suitable type, that is, the one that is able to perform the assigned task. In order to associate the worker components with computational tasks they are able to perform, we define the following functions as axioms in the model's context:

$$wtype \in WORKERS \rightarrow WTYPE,$$

$$WT\_Rel \in WTYPE \rightarrow \mathbb{P}(TASKS).$$

Here *WTYPE* is the set that contains all possible types of workers. Essentially, *wtype* associates each worker with its respective type. In its turn, *WT_Rel* associates each worker type with a subset of specific tasks that the workers of this type are capable of executing. Such mappings allow us to check, in a very straightforward way, whether a worker is able to accomplish a specific task.

In the machine part of the Event-B specification, we define a set of active manager components as the variable *managers*, where *managers* ⊆ *MANAGERS*. Similarly, we introduce the set of currently active worker components *workers* ⊆ *WORKERS*.

To model activation and deactivation of system components, we introduce a number of new model events: m_activation, m_deactivation, w_activation, and w_deactivation. For instance, the event w_activation presented below models activation of the worker component:

```
w_activation ≙
  any w
  where finish = FALSE
        w ∈ WORKERS
        w ∉ workers
  then  workers := workers ∪ {w}
  end
```

6.4.2.2 Modeling Dependencies between System Elements

A manager component can be associated with a number of tasks to be supervised. Moreover, each task should always be associated with some manager. However, during the system execution, a system task can be dynamically reallocated from one manager to another. The association of the tasks to the managers is defined by the function *Responsible*:

$$Responsible \in TASKS \rightarrow managers.$$

The function assigns each task to some manager. The mapping might dynamically change during the system operation. Note that *Responsible* is a *total function* because every task should be associated with some manager. Such an approach allows us to represent the property (PR6).

Furthermore, each manager can be associated with a number of workers to supervise. Again, the manager–worker relationship is dynamic since any worker can be reassigned to another manager during the system execution. To introduce this dynamic characteristic, and formalize property (PR7), we define a new variable *Attached*, which maps managers to workers:

$$Attached \in workers \twoheadrightarrow managers.$$

Here $\twoheadrightarrow$ denotes a partial function, which reflects the fact that some workers may not be attached to a manager.

Finally, the new variable *Assigned* associates the currently executed tasks with the corresponding workers:

$$Assigned \in TASKS \rightarrowtail\mkern-14mu\twoheadrightarrow workers.$$

Here $\rightarrowtail\mkern-14mu\twoheadrightarrow$ denotes a partial *injection*. The function is injective because a worker cannot perform more than one task simultaneously. In such a way, we formulate the property (PR9). Naturally, a manager can assign only incomplete tasks to a worker:

$$\forall t.\, t \in dom(Assigned) \Rightarrow tasks\_state(t) = incompl.$$

Moreover, as we have discussed earlier, only workers of a correct type can be assigned for a task execution. We formulate this property by the following model invariant:

$$\forall t, w.\, (t \mapsto w) \in Assigned \Rightarrow t \in WT\_Rel(wtype(w)).$$

Essentially, all components and tasks defined here are connected to each other via the introduced dependencies. For instance, if some task t has been assigned to the worker w, and m is the manager responsible for this task, then m also supervises w:

$$\forall m, t, w \cdot (t \mapsto w) \in Assigned \wedge (t \mapsto m) \in Responsible \Rightarrow (w \mapsto m) \in Attached.$$

Similarly, if task t is assigned to some worker w, and w is supervised by the manager m, then m becomes also responsible for the task t:

$$\forall m, t, w \cdot (t \mapsto w) \in Assigned \wedge (w \mapsto m) \in Attached \Rightarrow (t \mapsto m) \in Responsible.$$

The new event assign (given below) models assignment of a task t to a worker component w by its manager m. To make such an assignment, a number of conditions should be satisfied. Firstly, we have to be sure that t is not yet accomplished and is not being currently executed by any other worker. Secondly, the type of the worker should be suitable for performing the task t. Moreover, the worker w should be free, that is, not involved in the execution of an already assigned task. Finally, we have to check that both t and w "belong" to the manager m.

```
assign ≘
    any w, t, m
    where tasks_state(t) = incompl   // task t is not yet accomplished
        t ∉ dom(Assigned)            // the task t is not assigned to any other worker
        t ∈ WT_Rel(wtype(w))         // the worker w is able to execute the task t
        w ∉ ran(Assigned)            // the worker w has not already an assigned task
        (w ↦ m) ∈ Attached           // the worker w is attached to the manager m
        Responsible(t) = m           // the manager m is responsible for the task t
    then Assigned(t) := w
    end
```

6.4.2.3 Modeling Task Execution

Upon receiving the assignment from its manager, the worker starts to perform the assigned task. After successfully completing the task, the worker reports this to the manager. However, while performing a given task, a worker might fail, which subsequently leads to the assigned task not being performed. In the case of a worker failure, the manager should choose another available worker to execute the failed task.

To reflect this behavior in our model, we refine the abstract event process by two events, task_compl and task_incompl, which model, respectively, successful and unsuccessful execution of the task. If the task has been successfully performed by the assigned worker w, its supervising manager m changes the status of the tasks t to completed:

```
task_compl ≘ refines process
    any w, t, m
    where t ↦ w ∈ Assigned
        t ↦ m ∈ Responsible.
    then tasks_state(t) := compl
        Assigned := Assigned \ {t ↦ w}
    end
```

The event task_incompl abstractly models the opposite situation when the worker fails to complete the assigned task due to a failure. In this case, the task t can be reassigned to another available worker.

In our model, we assume that in our system, there exists a special component—the *middleware*—that is responsible for detecting the component failures at low level. For example, failure detection can be based on the following heartbeat mechanism. The system components periodically send simple messages to middleware to inform it about their status. These messages are called *heartbeats*. If an expected heartbeat is not received during a specified time period, it is assumed that the corresponding component has failed. In the case of a worker failure, the middleware will notify the supervising manager about the failure of its worker. In the case of a manager failure, the middleware will notify the other managers about that. It would allow them to initiate a system reconfiguration.

6.4.2.4 Toward Modeling System Reconfiguration

As we described earlier, the responsibilities for executing a task and supervising the workers can be reallocated from one manager to another as a part of the system reconfiguration. To model redistribution of tasks and workers between the responsible/supervising managers, we define the new event redistribute. It models the situation when the manager m_2 takes over the tasks ts and workers ws of the manager m_1:

```
redistribute ≙
  any ws, ts, m₁, m₂
  where   m₁ ∈ managers
          ts = dom(Responsible ▷ {m₁})  // all the tasks of the manager m₁
          ws = dom(Attached ▷ {m₁})  // all the workers of the manager m₁
          ts ≠ ∅ ∧ ws ≠ ∅ ∧ m₁ ≠ m₂
          ws ⊄ ran(Assigned)
  then    Responsible := Responsible ⩤ (ts × {m₂})
          Attached := Attached ⩤ (ws × {m₂})
  end
```

Here $⩤$ denotes the overriding relation, that is, $q ⩤ r = r \cup (dom(r) ⩤ q))$, where $⩤$ denotes the domain substraction, that is, $S ⩤ r = \{x \mapsto y \mid x \mapsto y \in r \wedge x \notin S\}$.

We also allow the system to reassign the workers from one manager to another as a part of component cooperation aimed at improving resource utilization. In that case, the workers of the manager that completed all of its tasks can be sent to some other manager that still unfinished tasks. Additionally, we reserve the possibility to cancel all of the current assignments for a group of workers. This functionality will be needed later on, for example, to describe the effect of manager failures.[*]

Let us note that in this refinement step, we have abstracted away from the reasons behind redistribution of responsibilities between manager components. However, later on, when we introduce the failures of manager and worker components, we will refine this behavior by adding the specific conditions imposed on the system reconfiguration.

Since the currently available resources (the active manager and worker components) might be insufficient to accomplish the main goal, the system has the capability to activate idle (spare) components. This behavior has been modeled by the abstract events w_activation and m_activation. Note that the system should be able to integrate the new components into its execution flow. We model this by the new event attach_worker. Finally, to release the excessive system resources, some components might be deactivated. However, while deactivating system components, we should guarantee that the associated intercomponents relationships are preserved.

---

[*] We do not show here the corresponding events to avoid overloading the chapter details of the Event-B specifications.

6.4.3 Representing Manager's Local Knowledge

The variable *tasks_state* represents the *global* knowledge about statuses of all tasks. In our abstract specification, to orchestrate the system operation (assigning uncompleted task to workers, etc.), the active managers access (read) this global knowledge. However, in the decentralized systems, the components typically lack knowledge about the global system state. Thus, in our system, the managers should rely on their local state—*local knowledge*—while coordinating the workers' activities. The local knowledge of each manager should contain accurate information about the statuses of its own tasks, and also the most recent (up to specified delay) information about the statuses of the tasks of all other managers in the system.

To model the local knowledge of the active managers, we introduce the new variable *local_tasks*:

$$local\_tasks \in managers \rightarrow (TASKS \rightarrow STATES).$$

Essentially, the local knowledge is defined for all active managers that have any tasks to coordinate:

$$ran(Responsible) \subseteq dom(local\_tasks).$$

Our abstract variable *tasks_state* is now refined by the newly introduced *local_tasks* variable. To prove data refinement and establish the relationship between the values of concrete and abstract variables, we formulate and prove the following gluing invariant:

$$\forall\ m,\ t \cdot t \in TASKS \wedge m \in managers \Rightarrow$$

$$(local\_tasks(m)(t) = compl \Rightarrow tasks\_state(t) = compl).$$

The invariant defines the restrictions of the consistency of information between global and local state to be maintained by the system. Namely, it states that for any active manager, its local knowledge is consistent with the global one, that is, if a particular task is marked as completed in the local knowledge of this manager, then this task is considered completed in the global knowledge as well. Moreover, for each manager component, the local information about its *own* tasks always coincides with the global knowledge about these tasks:

$$\forall\ m,\ t.\ t \in TASKS \wedge Responsible(t) = m \Rightarrow$$

$$(local\_tasks(m)(t) = compl \Leftrightarrow tasks\_state(t) = compl).$$

A manager component keeps track of the completed and noncompleted tasks, and also periodically receives the information from the other managers about their completed tasks. However, the knowledge is inaccurate for the time span when the information is sent but not yet received. In this refinement step, we abstractly model a manager receiving the information by introducing the new event update_local:

```
update_local ≘
  any m, t, mm
  where m ∈ managers
        local_tasks(m)(t) ≠ compl
        Responsible(t) = mm
        local_tasks(mm)(t) = compl
  then  local_tasks(m) := local_tasks(m) ⩤ {t ↦ compl}
  end
```

The event update_local models updating the local knowledge of a manager m. Here we have to ensure that the obtained information is always consistent with the information stored in the knowledge of the responsible manager. Specifically, the manager m marks the task t as completed only if it has completed status in the knowledge of the manager mm that is responsible for executing this task.

Moreover, in order to accept responsibilities for new tasks, the newly activated manager should also have information about the current situation of the task statuses. Thus, its local knowledge should be initialized according to the current situation. We model this behavior by the new event initialize_local (omitted here).

In our previous refinement step, in the event redistribute we abstractly modeled the possibility of redistributing responsibilities for tasks and workers between two managers. However, a manager can take responsibility for the new tasks only if it has an accurate knowledge about statuses of these tasks. We add this as an additional condition in the event guards, where we check that the local knowledge about reallocated tasks coincides for both involved managers:

```
redistribute ≘ refines redistribute
  any ws, ts, m₁, m₂
  where ...
        ts = dom(Responsible ▷ {m₁})
        ∀ t. t ∈ ts ⇒ local_tasks(m₁)(t) = local_tasks(m₂)(t)
        ...
```

Let us note that at this stage we still rely on the ability of the manager components to "read" the knowledge of the other managers, that is, we are not yet ready to decompose the system into a distributed model. However, in the fifth refinement step, the managers will rely solely on the intercomponent communication to learn about the global system state.

6.4.3.1 Local Knowledge about Attached Workers and Responsible Tasks

During reconfiguration, the manager receiving new responsibilities should have information about the tasks and workers it will accept. There are two ways to model this. The first one is similar to what we used for modeling the local knowledge about the task statuses—in the local knowledge of each manager we can also store the data about the attached workers of all managers in the system. In that case, the managers should periodically exchange this information among themselves.

The second mechanism that can be employed is based on reliance on middleware services. Specifically, middleware might have a responsibility to inform the managers which tasks and workers they should accept. In our modeling, we have chosen the second strategy.

6.4.4 Explicit Model of Component Failures

In our fourth refinement step, we aim at modeling possible component failures. To achieve this, we partition the active components into operational and failed ones. For example, the current set of all operational managers is defined by a new variable *operational_managers*. Initially all active managers are operational, that is, *operational_managers = Init_Managers*. To model possible failure of an active manager, we define the managerFailure event as

```
managerFailure ≙
    any m
    where m ∈ operational_managers
    then
            operational_managers := operational_managers \ {m}
```

The events assign, redistribute, reattach, and attach_worker are now refined to reflect that only the operational managers can give task assignments and participate in the system reconfiguration. Moreover, in the event redistribute we add an additional condition—only if a manager is classified as failed (i.e., $m \in managers \wedge m \notin operational\_managers$) can another manager take over its tasks and workers. Similarly, we introduce worker failures and specify their effects on the task execution and redistribution.

6.5 MODELING OF COMMUNICATION

6.5.1 Communication between Managers

The aim of the fifth refinement step is to define the communication model between the system components. After receiving a notification from a worker about a successful task completion, a manager updates its local knowledge and broadcasts the message about the completed tasks to the other managers. In its turn, upon receiving such a message, each manager component correspondingly updates its own local knowledge. Here we assume that the communication between managers is reliable, that is, no sent message is lost and every base station will eventually receive it. We could also consider communication failures, as we did in, for example, [21]. However, since communication failures and recovery are not the main focus of this work, for the sake of simplicity we decided not to include them in the model.

The middleware is responsible for enabling the intercomponent communication. In particular, the middleware implements a simple point-to-point communication protocol. The middleware observes the outgoing buffer of the manager components. When a manager component puts a message into its outgoing buffer, the middleware delivers the message to all other managers, that is, the middleware puts this message into all corresponding incoming buffers. As soon as a new message is delivered to the manager, the manager updates its local knowledge.

To model communication between manager components, we introduce the following communication structures:

- *taskOutgoingM*: Manager's outgoing (one-place) buffer for its accomplished task id.

- *taskIncomingM*: Manager's incoming buffer for accomplished task ids.

The middleware constantly observes the managers' outgoing buffers and reacts on the appearance of a new message awaiting delivery. Here we assume that a manager can place only one task at a time to its outgoing buffer.

In the Event-B machine created at this refinement step, we model the corresponding buffers as the variables *taskOutgoingM* and *taskIncomingM* with the following properties:

$$taskOutgoingM \in managers \rightarrow \mathbb{P}\,(TASKS),$$

$$ran(Responsible) \subseteq dom(taskOutgoingM),$$

$$\forall m.\, card(taskOutgoingM(m)) \leq 1.$$

and

$$taskIncomingM \in managers \rightarrow \mathbb{P}(TASKS),$$

$$ran(Responsible) \subseteq dom(taskIncomingM).$$

If for any manager *m*, a task *t* belongs to its outgoing buffer, that is, $t \in taskOutcomingM(m)$, then this task *t* has been completed. This property can be formulated as the following model invariant:

$$\forall\, m, t.\, m \in managers \land t \in taskOutgoingM(m)$$

$$\Rightarrow local\_tasks(m)(t) = compl.$$

Moreover, if for any manager *m*, a task *t* belongs to its incoming buffer, that is, $t \in taskIncomingM(m)$, then this tasks *t* has been completed before:

$$\forall\, m, t, mm.\, m \in managers \land t \in taskIncomingM(m)$$

$$\land\, Responsible(t) = mm \Rightarrow local\_tasks(mm)(t) = compl.$$

Additionally, if for any manager *m*, a task *t* belongs to its incoming buffer, that is, $t \in taskIncomingM(m)$, then the task *t* is still considered unfinished in the local knowledge of this manager:

$$\forall\, m, t.\, m \in managers \land t \in taskIncomingM(m) \Rightarrow$$

$$local\_tasks(m)(t) = incompl.$$

A manager component keeps track of the completed and noncompleted tasks and periodically receives the information from the other managers about their completed tasks. However, the knowledge is inaccurate for the time span when the information is sent but not yet received.

6.5.2 Communication between Managers and Workers

Similarly to communication between managers, we introduce the communication between a manager and its supervised workers. The core manager functionality is to control task execution and to assign tasks to its workers. To assign a task, a manager sends a corresponding message to a worker, which is delivered by middleware. In its turn, upon completion of the task, the worker sends a report message to its manager. Once the manager receives the message, it marks the corresponding task as completed. To model this behavior, we introduce the following worker communication structures:

- *taskIncomingW*: Worker's incoming buffer for the assigned task from the manager.

- *taskOutgoingW*: Worker's outgoing buffer for the completed task.

The manager's structures for communication with a worker are

- *taskAssignedM*: Manager's outgoing buffer for an assigned task.

- *taskCompletedM*: Manager's incoming buffer for the accomplished task.

Middleware constantly observes the changes made to the output buffers of components and reacts on the appearance of a new message awaiting delivery.

6.6 TOWARD MODEL DECOMPOSITION

Now we arrive at a centralized specification of resilient CPS. Our next goal is to derive its distributed implementation by refinement. We employ modularization facilities of Event-B to achieve this.

To model a distributed architecture of the system, we decompose by refinement the centralized model into a machine defining middleware functionality and two separate indexed modules representing the managers and workers. Moreover, we explicitly model communication between them. We assume that all managers and workers are identical in terms of communication. Since indexed module instantiation allows for the creation of an arbitrary number of copies of the same module, we define one interface for a manager and one interface for a worker. These modules will be imported to the refined machine instantiated by the finite set of managers *Init_Managers* and workers *Init_Workers*, respectively.

The generic module interfaces reflecting the described functionality for a manager component are presented in Figure 6.1. The interface contains processes that model the described behavior of a manager (see **Process** clause). Moreover, to model component communication, the callable operations for each component are defined as well (see **Operations** clause). These operations will be invoked by the middleware every time a message is delivered or received.

The generic module interface for a worker is presented in Figure 6.2. The refined machine middleware models all types of communication and invokes these operations in the bodies of its events. As a result of the decomposition, we arrive at a specification of a distributed

```
Interface Manager
  Variables ...
  Invariants ...
  Process ...
  AssignTask          // a manager sends a task to a worker
  TaskCompleted       // a manager records a task as completed and
                         broadcast to others
  TaskFailed          // a manager records a task as failed
  UpdateKnowledge     // a manager updates its local knowledge
  NewResponsibility   // reassigning workers and tasks from other
                         manager
  GiveResponsibility  // reassigning workers from other manager
  ManagerFailure      // a manager fails
  Operations
  FromWorker          // incoming message from a worker about
                         accomplished task
  FromManager         // incoming message from other manager
  WorkerFailureNotif  // a worker has failed
  ManagerFailureNotif // a manager has failed
  NewWorkers          // a manager receives new worker(s)
```

FIGURE 6.1   Base station interface component.

```
Interface Worker
  Variables ...
  Invariants ...
  Process
  ProcessTask         // a worker starts processing a task
  FinishTask          // a worker records a task as completed
  SendToManager       // a worker sends a message to its manager about
                         task completeness
  WorkerFailure       // a worker fails
  Operations
  TaskFromManager     // message from a manager about the assigned task
  NewManager          // message about new manager
```

FIGURE 6.2   Worker interface component.

system. The manager and worker components are represented by the corresponding modules. The middleware serves as a communication infrastructure.

6.7   RELATED WORK AND CONCLUSIONS

6.7.1   Related Work

Significant work has been done on goal-oriented requirement engineering approaches. The foundational work on *goal-oriented development* belongs to van Lamsweerde [3,4,22]. The proposed KAOS framework [22] introduces a goal-oriented approach for requirements modeling, specification, and analysis as well as addresses both functional and nonfunctional system requirements. Based on the KAOS framework, van Lamsweerde [23] has proposed a method for deriving the software architecture from its requirements. Specifically, according to the method, a software specification is developed from the given system requirements and then used to build the architectural system design. The design is developed by consecutive refinements, which take into account the system constraints and nonfunctional goals. The KAOS approach is supported by the GRAIL tool [22].

Over the last decade, the goal-oriented approach has also received several extensions that allow the designers to link it with formal modeling [24,25]. In particular, the work [24] presents a translation technique of KAOS operational models into event-based tabular specifications, which can then be analyzed by the SCR* toolset [26]. The technique consists of a number of transformation steps, each of which solves semantic, structural, or syntactic differences between the KAOS and SCR (Software Cost Reduction) languages.

A significant body of research has also been devoted to translating formal specifications built according to the KAOS goal-oriented method into event-based transition systems. For example, the work [27] presents an approach to use the formal analysis capabilities of LTSA (Labelled Transition System Analyser) to analyze and animate KAOS operational models. The mapping allows the designers to translate goal-oriented operational requirements into a black-box, event-based model of the software behavior, expressed in a formalism appropriate to reason about system behaviors at the architectural level.

One of the first attempts to bridge the KAOS goal-oriented framework with the B formalism was presented in Reference 28. More recently, the study to formalize KAOS requirements in Event-B was attempted in Reference 29. The paper proposes a constructive approach that allows linking of high-level system requirements expressed as linear temporal logic formulas to the corresponding Event-B elements. The notion of a triggered event is used to translate time operators that are used in KAOS models. Similarly, Matoussi et al. [30,31] present works on coupling requirements engineering methods with formal methods. In contrast, our work relies on goals to facilitate structuring of the system behavior, while connecting them with the required agent collaboration and system reconfiguration mechanisms.

The goal behind our research is to both formally model and verify systems with intricate relationships between system components and goal structures that are able to dynamically reconfigure themselves in order to tolerate various failures or changes.

The problem of formal verification and validation of CPS has been largely studied in the software engineering research community. For instance, in the ADVANCE project [32], the main focus of the proposed methodologies supports the construction of CPS and augmenting formal modeling and verification with simulation and testing. The resulting proposed simulation-based approach combines Event-B development and cosimulation with tool-independent physical components via the FMI interface [33].

The DESTECS project [34] also studies the problem of engineering resilient CPS by integration of model-based formal methods with discrete-event models. The research conducted within the project proposes an approach to the model-based design through cosimulation of discrete-event models in the Vienna Development Method (VDM) and continuous-time models in 20-sim. These models are coupled with a cosimulation tool that coordinates execution of the developed models in their respective simulators. The resulting models can also be augmented with descriptions of potential failures and fault tolerance mechanisms [35–37]. In contrast, our work focuses only on modeling the architectural aspects of CPS and does not yet consider separately the modeling of the physical processes.

The similar idea of modeling a dynamic system architecture with the integrated reconfiguration mechanisms and assessing system resilience characteristics is presented in Reference 38. Using Event-B and its refinement technique, the authors derive a complex architecture of data processing capabilities of CPS. To assess the resilience of the obtained data processing architecture, the authors rely on the statistical model checking in UPPAAL.

The problem of system reconfiguration and its connection to predictable dynamic resilience is presented in References 39 and 40. The proposed architectural framework—MetaSelf—is suitable to development of dynamically resilient systems using a specific architectural model. The main idea of the approach is to separate the functional and nonfunctional descriptions of the system components, formulate necessary resilience policies, and reason about system reconfiguration based on these policies. The key feature of the proposed solution is the need for metadata—information about system components—sufficient to enable decision-making about dynamic system reconfiguration. The metadata are used to guide the system reconfiguration with respect to the given reconfiguration policies.

In Reference 41 the authors propose a goal-oriented approach for self-reconfiguration by employing the idea of a monitoring feedback loop. The proposed framework uses goal models as software requirements models and employs SAT solvers to check the current execution records against the models to diagnose task failures. The monitoring component monitors requirements and records data, while the diagnostic component analyzes the recorded data and identifies failures of the system behavior. In our work, we focus on modeling the system behavior and dependencies between components while not distinguishing the system components to be the monitoring or diagnostic ones. However, this idea can be incorporated into our system design.

6.7.2 Conclusions

In this work we have proposed a formal goal-oriented approach to development of resilient CPS by refinement in Event-B. We considered resilience as an ability of the system to achieve its goals despite changes. We believe that a goal-oriented approach offers a suitable basis for reasoning about functional aspects of system resilience.

In our modeling, we have formalized the main concepts of goal-oriented development and formally defined connections among system goals, functional capabilities of the components, and component interactions. The derived hierarchical distributed architecture can be seen as a generic pattern for defining supervisory control of reconfigurable CPS.

We formalized the main functional properties of the functional aspect of resilience in the goal-oriented style and demonstrated how refinement can help to structure complex requirements and scale proof-based verification. The formal development has allowed us to rigorously define and verify the intricate relationships between system components and goal structures (tasks). The presented development is generic. It defines a number of specification patterns that allow the designers to explicitly define architectural reconfiguration mechanisms to build the resilient system.

In this work, we focused on reasoning about functional aspect of resilience. In future work, we are planning to address nonfunctional properties of resilience, such as reliability and performance, as well as investigate how to formally model learning—an intrinsic part of future CPS.

REFERENCES

1. E. A. Lee. Cyber physical systems: Design challenges. In *11th IEEE International Symposium on Object-Oriented Real-Time Distributed Computing (ISORC 2008)*, pp. 363–369. IEEE Computer Society, Orlando, Florida, USA, May 5–7, 2008.

2. J.-C. Laprie. From dependability to resilience. In *DSN 2008, Dependable Systems and Networks*. IEEE Computer Society, Anchorage, Alaska, USA, June 24–27, 2008.

3. A. van Lamsweerde. Goal-oriented requirements engineering: A guided tour. In *Requirements Engineering*, pp. 249–263. *5th IEEE International Symposium on Requirements Engineering (RE)*, Toronto, Canada, August 27–31, 2001.

4. A. van Lamsweerde. *Requirements Engineering: From System Goals to UML Models to Software Specifications*. Wiley, 2009.

5. I. Pereverzeva. Formal development of resilient distributed systems. PhD thesis, Aabo Akademy University, Turku, 2015.

6. J.-R. Abrial. *Modeling in Event-B*. Cambridge University Press, Cambridge, 2010.

7. Rodin. Event-B Platform. online at http://www.event-b.org/.

8. M. Y. Said, M. J. Butler, and C. F. Snook. Language and tool support for class and state machine refinement in UML-B, volume 5850 of *Lecture Notes in Computer Science*, pp. 579–595. Springer, Eindhoven, the Netherlands, November 2–6, 2009.

9. C. F. Snook and M. J. Butler. UML-B: Formal modeling and design aided by UML. *ACM Trans. Softw. Eng. Methodol.*, 15(1):92–122, 2006.

10. D. Mery and R. Monahan. Transforming Event-B models into verified C# implementations. In A. Lisitsa and A. Nemytykh, editors, *VPT 2013*, volume 16 of *EPiC Series*, pp. 57–73, EasyChair, Saint Petersburg, Russia, July 12–13, 2013.

11. D. Méry and N. K. Singh. Automatic code generation from Event-B models. In *Proceedings of the Second Symposium on Information and Communication Technology*, SoICT '11, pp. 179–188. ACM, Hanoi, Viet Nam, October 13–14, 2011.

12. A. Edmunds. Templates for Event-B code generation. In *Abstract State Machines, Alloy, B, TLA, VDM, and Z—4th International Conference, ABZ 2014*, volume 8477 of *Lecture Notes in Computer Science*, pp. 284–289. Springer, Toulouse, France, June 2–6, 2014.

13. A. Edmunds, A. Rezazadeh, and M. J. Butler. Formal modelling for Ada implementations: Tasking Event-B. In *Reliable Software Technologies–Ada-Europe 2012: 17th Ada-Europe International Conference on Reliable Software Technologies*, volume 7308 of *Lecture Notes in Computer Science*, pp. 119–132. Springer, Stockholm, Sweden, June 11–15, 2012.

14. T. S. Hoang, A. Iliasov, R.Silva, and W. Wei. A survey on Event-B decomposition. *ECEASST*, 46:1–15, 2011.

15. J.-R. Abrial and S. Hallerstede. Refinement, decomposition, and instantiation of discrete models: Application to Event-B. *Fundam. Inform.*, 77(1–2):1–28, 2007.

16. M. J. Butler. Decomposition structures for Event-B. In *Integrated Formal Methods, 7th International Conference, IFM 2009*, Düsseldorf, Germany, February 16–19, 2009. Proceedings, pp. 20–38, 2009.

17. T. S. Hoang and J.-R. Abrial. Event-b decomposition for parallel programs. In *Abstract State Machines, Alloy, B and Z, Second International Conference, ABZ 2010*, Orford, QC, Canada, February 22–25, 2010. Proceedings, pp. 319–333, 2010.

18. A. Iliasov, E. Troubitsyna, L. Laibinis, A. Romanovsky, K. Varpaaniemi, D. Ilic, and T. Latvala. Supporting reuse in Event-B development: Modularisation approach. In *Proceedings of Abstract*

State Machines, Alloy, B, and Z (ABZ 2010), volume 5977, pp. 174–188. Lecture Notes in Computer Science, Springer, Orford, QC, Canada, February 22–25, 2010.

19. RODIN Modularisation Plug-in. Documentation at http://wiki.event-b.org/index.php/ Modularisation_Plug-in.

20. R. Silva, C. Pascal, T. S. Hoang, and M. J. Butler. Decomposition tool for Event-B. *Softw., Pract. Exper.*, 41(2):199–208, 2011.

21. A. Tarasyuk, I. Pereverzeva, E. Troubitsyna, and T. Latvala. The formal derivation of mode logic for autonomous satellite flight formation. In *SAFECOMP 2015, International Conference on Computer Safety, Reliability and Security*, volume 9337 of *Lecture Notes in Computer Science*, pp. 29–43. Springer, Delft, the Netherlands, September 23–25, 2015.

22. R. Darimont, E. Delor, P. Massonet, and A. van Lamsweerde. GRAIL/KAOS: An environment for goal-driven requirements engineering. In *Proceedings of the 19th International Conference on Software Engineering*, pp. 612–613. ACM, Boston, Massachusetts, USA, May 17–23, 1997.

23. A. van Lamsweerde. From system goals to software architecture. In *Formal Methods for Software Architectures, Third International School on Formal Methods for the Design of Computer, Communication and Software Systems: Software Architectures, SFM 2003*, volume 2804 of *Lecture Notes in Computer Science*, pp. 25–43. Springer, Bertinoro, Italy, September 22–27, 2003.

24. R. De Landtsheer, E. Letier, and A. van Lamsweerde. Deriving Tabular event-based specifications from goal-oriented requirements models. *Requir. Eng.*, 9(2):104–120, 2004.

25. C. Ponsard, G. Dallons, and M. Philippe. From rigorous requirements engineering to formal system design of safety-critical systems. *ERCIM News*, 75:22–23, 2008.

26. C. L. Heitmeyer, J. Kirby, B. G. Labaw, and R. Bharadwaj. SCR*: A Toolset for specifying and analyzing software requirements. In *Computer Aided Verification, 10th International Conference, CAV '98*, volume 1427 of *Lecture Notes in Computer Science*, pages 526–531. Springer, Vancouver, BC, Canada, June 28–July 2, 1998.

27. E. Letier, J. Kramer, J. Magee, and S. Uchitel. Deriving event-based transition systems from goal-oriented requirements models. *Autom. Softw. Eng.*, 15(2):175–206, 2008.

28. C. Ponsard and E. Dieul. From requirements models to formal specifications in B. In *Proceedings of the CAISE*06 Workshop on Regulations Modelling and Their Validation and Verification ReMo2V '06*, Luxemburg, June 5–9, 2006.

29. B. Aziz, A. Arenas, J. Bicarregui, C. Ponsard, and P. Massonet. From goal-oriented requirements to Event-B specifications. In *First NASA Formal Methods Symposium—NFM 2009*, pp. 96–105. Moffett Field, California, USA, April 6–8, 2009.

30. A. Matoussi, F. Gervais, and R. Laleau. A first attempt to express KAOS refinement patterns with Event-B. In *Abstract State Machines, B and Z, First International Conference, ABZ 2008*, p. 338. London, UK, September 16–18, 2008.

31. A. Matoussi, F. Gervais, and R. Laleau. A goal-based approach to guide the design of an abstract Event-B specification. In *16th IEEE International Conference on Engineering of Complex Computer Systems, ICECCS 2011*, pp. 139–148. Las Vegas, Nevada, USA, April 27–29, 2011.

32. Advanced Design and Verification Environment for Cyber-physical System Engineering (Advance). FP7 Information and Communication Technologies (ICT) Programme. online at http://www.advance-ict.eu/.

33. V. Savicks, M. Butler, J. Colley, and J. Bendisposto. Rodin multi-simulation plug-in. In *5th Rodin User and Developer Workshop*, Toulouse, France, June 2–3, 2014.

34. Design Support and Tooling for Embedded Control Software (DESTECS). FP-7 project. online at http://www.destecs.org.

35. J. S. Fitzgerald, P. G. Larsen, K. G. Pierce, and M. Verhoef. A formal approach to collaborative modelling and co-simulation for embedded systems. *Math. Struct. Comp. Sci.*, 23(4):726–750, 2013.

36. J. S. Fitzgerald, K. G. Pierce, and P. G. Larsen. Co-modelling and co-simulation in the engineering of systems of cyber-physical systems. In *9th International Conference on System of Systems Engineering, SoSE 2014*, pp. 67–72. IEEE, Glenelg, Australia, June 9–13, 2014.

37. J. S. Fitzgerald, K. G. Pierce, and C. Gamble. A rigorous approach to the design of resilient cyber-physical systems through co-simulation. In *IEEE/IFIP International Conference on Dependable Systems and Networks Workshops, DSN 2012*, pp. 1–6. IEEE, Boston, MA, USA, June 25–28, 2012.

38. L. Laibinis, D. Klionskiy, E. Troubitsyna, A. Dorokhov, J. Lilius, and M. Kupriyanov. Modelling resilience of data processing capabilities of CPS. In *Software Engineering for Resilient Systems—6th International Workshop, SERENE 2014*, volume 8785 of *Lecture Notes in Computer Science*, pages 55–70. Springer, Budapest, Hungary, October 15–16, 2014.

39. G. Di Marzo Serugendo, J. S. Fitzgerald, and A. Romanovsky. MetaSelf: An architecture and a development method for dependable self-* systems. In *Proceedings of the 2010 ACM Symposium on Applied Computing (SAC)*, pp. 457–461. ACM, Sierre, Switzerland, March 22–26, 2010.

40. G. Di Marzo Serugendo, J. S. Fitzgerald, A. Romanovsky, and N. Guelfi. A metadata-based architectural model for dynamically resilient systems. In *Proceedings of the 2007 ACM Symposium on Applied Computing (SAC)*, pp. 566–572. ACM, Seoul, Korea, March 11–15, 2007.

41. Y. Wang, S. A. McIlraith, Y. Yu, and J. Mylopoulos. An automated approach to monitoring and diagnosing requirements. In *Proceedings of the Twenty-second IEEE/ACM International Conference on Automated Software Engineering*, ASE '07, pp. 293–302. ACM, Atlanta, Georgia, USA, November 5–9, 2007.

CHAPTER 7

Formal Reasoning about Resilient Cyber-Physical Systems

Linas Laibinis and Elena Troubitsyna

CONTENTS

CYBER-PHYSICAL SYSTEMS (CPS) are complex software intensive systems consisting of assemblies of heterogenous components. To guarantee the resilience of CPS, that is, to ensure that systems can adapt to changing operating conditions, CPS architectures should integrate mechanisms enabling adaptation. In this chapter, we employ Event-B to formally derive the reconfiguration mechanisms that allow the systems to dynamically interconnect components to cope with failures. The proposed approach formalizes relationships among system functionality, component functional capabilities, and failures. It serves as a basis for developing a system health monitoring infrastructure and facilitates rigorous construction of resilient systems.

7.1 INTRODUCTION

Cyber-Physical Systems promise great economic and societal benefits in such domains as aerospace, production automation, transportation, and healthcare. However, to unlock their full potential, rigorous and powerful methods for ensuring system trustworthiness need to be developed and employed.

CPSs operate in continuous interaction with the physical world. They need to constantly monitor external operating conditions and react to them to provide the required services in a correct and safe way. At the same time, CPSs should also continuously monitor their internal state and mitigate the effect of component failures on service provisioning. The latter task is usually handled by employing dynamic system reconfiguration.

In this chapter, we demonstrate how to reason about dynamically reconfigurable CPS. Our goal is to explicitly define a link between the services that the systems should deliver, and the functional capabilities of the components and their health. We demonstrate how to formalize the proposed approach to reconfiguration, that is, the mechanisms that allow the components to dynamically interconnect and cooperatively perform the required services despite the loss of certain functional capabilities by some of the components.

We rely on Event-B formalism [1] to formally reason about dynamic reconfiguration. Event-B is a state-based approach that promotes correct-by-construction systems development. Development starts from creating an abstract system specification that defines only the most essential system properties and behavior. In a number of correctness-preserving steps, called refinements, the initial specification is transformed into a detailed system model. Each refinement step is accompanied by proofs that guarantee that the externally observable system behavior is preserved during such model transformations. The Rodin platform [2] provides an automated support for modeling and verification in Event-B.

The top-down approach to systems development promoted by Event-B allows us to gradually unfold the system architecture and formally define the requirements that should be imposed on it to enable dynamic system reconfiguration. As a result of our development, we derive a number of generic specification and refinement patterns that represent the essential properties of reconfigurable architectures and define coordination mechanisms for cooperative service provisioning.

The proposed generic approach explicitly defines the main concepts that should be analyzed while developing a resilient CPS. It facilitates a structured derivation of a resilient system architecture by a gradual unfolding of architectural details and revealing the necessary interrelationships between service provisioning and the related reconfiguration mechanisms. The approach is illustrated by a small example from the aerospace domain.

The chapter is structured as follows. In Section 7.2, we provide an overview of the characteristics and properties of systems that we are interested in modeling and analyzing. Section 7.3 briefly describes our formal framework—Event-B. In Section 7.4, we present our main contribution—formal generic development of a resilient CPS. Section 7.5 illustrates the proposed approach with a small example of a satellite data processing unit. Finally, in Section 7.6, we present some concluding remarks.

7.2 CYBER-PHYSICAL SYSTEMS

In this section, we explicitly state our assumptions about the characteristics and properties of the systems that we are interested in modeling and analyzing in this chapter. The focus of our investigation is on service provisioning by CPS in the presence of constantly changing operating conditions and fluid system configurations as well as how possible dynamic reconfiguration mechanisms of such systems can improve their resilience.

In general, in this chapter we consider systems that exhibit the following characteristics:

- Architecturally, the systems consist of a number of different components that may communicate and collaborate with one another in order to provide a common service

- Components are heterogenous, that is, they may have different functional capabilities. A specific functional capability is provided by a so-called component *functional block*. In other words, we associate the component capabilities with a number of distinct functional blocks it is capable of executing

- Each component contains its own collection of functional blocks of different types. On the other hand, functional blocks of the same type can reside on different components. Therefore, we assume here a certain degree of system redundancy with respect to provided functionalities, which, in turn, is the main source of system resilience

- The systems accept a number of external requests to execute specific services and respond with either successful service execution or an error message

- Each request is split within the systems into the necessary functional block types that must be completed

- The execution scenario for a particular service request then may involve different components with available functional blocks of the required types

- Additional component functional blocks may provide external or internal communication or other auxiliary services

- Both the component internal state and its environment can affect the availability of a specific component functional block; in other words, the functional block in question can be enabled or disabled depending on both the internal and external component operating states

- A component functional block can fail. For our systems, this means that the functional block becomes permanently disabled

- Each component can be in one of many *roles* (sometimes called modes of operation). The component role is associated with a particular subset of functional blocks; it should be currently able to perform and/or with particular duties within the service execution scenario

- The collective roles of all the systems components can be interpreted as the current configuration of the systems, while the current role of a component and all the statuses of its functional blocks represent the configuration of this component

- We assume that there is a predefined minimal condition for each role that a component or its current configuration (i.e., collective statuses of the component functional blocks) should satisfy. If this condition is not satisfied, the component in question should change its role to a more restrictive or degraded one

- We also a distinguish a so-called *primary role* (i.e., leading or coordinating) that is responsible for orchestrating the overall service execution and also often serving as a front end for the outside world, that is, receiving service requests and sending system responses. It is assumed that the system typically has only one component in such a primary role. If the component cannot carry on in the primary role, another component satisfying the minimal condition for the primary role can take over the assigned duties

In Section 7.4, we will show how we can gradually formalize the concepts and characteristics of a system presented above. Before doing that, we briefly introduce the formal framework we rely on—Event-B.

7.3 BACKGROUND: EVENT-B

Event-B [1] is a state-based framework that promotes a correct-by-construction approach to systems development and formal verification by theorem proving. In Event-B, a system model is specified using the notion of an *abstract state machine* [1]. An abstract state machine encapsulates the model state, represented as a collection of variables, and defines operations on the state, that is, it describes the dynamic behavior of a modeled system. The variables are strongly typed by the constraining predicates that, together with other important properties of the systems, are defined in the model *invariants*. Usually, a machine has an accompanying component, called *context*, that includes user-defined sets, constants, and their properties given as a list of model axioms.

A general form for Event-B models is given in Figure 7.1. The machine is uniquely identified by its name, M. The state variables, v, are declared in the **Variables** clause and initialized in the *Init* event. The variables are strongly typed by the constraining predicates I given in the **Invariants** clause. The invariant clause might also contain other predicates defining properties (e.g., safety invariants) that should be preserved during system execution.

The dynamic behavior of the systems is defined by a set of atomic *events*. Generally, an event has the following form:

$$e \mathrel{\widehat{=}} \textbf{any } a \textbf{ where } G_e \textbf{ then } R_e \textbf{ end},$$

where e is the event's name, a is the list of local variables, and *guard* G_e is a predicate over the local variables of the event and the state variables of the system. The body of an event is defined by a *multiple* (possibly nondeterministic) assignment over the system variables. In

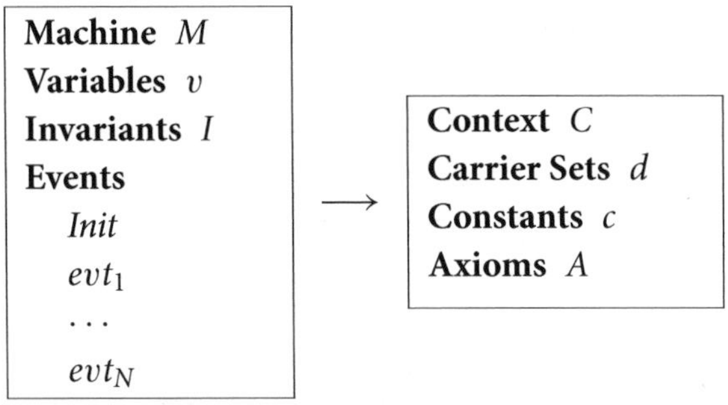

FIGURE 7.1 Event-B machine and context.

Event-B, an assignment represents a corresponding next-state relation, R_e. Later on, using the concrete syntax in our Event-B models, we will rely on two kinds of assignment statements: deterministic ones, expressed in the standard form $x := some\_expression(x, y)$, and nondeterministic ones, represented as $x :| some\_condition(x, y, x')$. In the latter case, the state variable x gets nondeterministically updated by the value, x', which may depend on the initial values of the variables, x and y.

The guard defines the conditions under which the event is *enabled*, that is, its body can be executed. If several events are enabled at the same time, any of them can be chosen for execution nondeterministically.

If an event does not have local variables, it can be described simply as

$$e \,\widehat{=}\, \textbf{when } G_e \textbf{ then } R_e \textbf{ end}.$$

If the event guard, G_e, is always true, the event syntax becomes $e \,\widehat{=}\, \textbf{begin } R_e \textbf{ end}$.

Event-B employs a top-down refinement-based approach to system development. Development starts from an abstract specification that nondeterministically models the most essential functional requirements. In a sequence of refinement steps, we gradually reduce nondeterminism and introduce detailed design decisions. In particular, we can add new events and split events as well as replace abstract variables by their concrete counterparts, that is, perform *data refinement*.

The consistency of Event-B models, that is, verification of well-formedness and invariant preservation as well as correctness of refinement steps, is demonstrated by discharging a number of verification conditions—proof obligations. For instance, to verify *invariant preservation*, we should prove the following logical formula:

$$A(d, c),\ I(d, c, v),\ G_e(d, c, x, v),\ R_e(d, c, x, v, v') \vdash I(d, c, v'), \quad \text{(INV)}$$

where A are the model axioms, I are the model invariants, d and c are the model constants and sets, respectively, x are the event's local variables, and v, v' are the variable values before and after event execution.

In turn, each refinement step generates additional proof obligations ensuring that the transformation is performed in a correctness-preserving way. In particular, to formally demonstrate refinement between the corresponding events of the abstract and concrete models, the *guard strengthening* and *event simulation* proof obligations must be discharged. The full definitions of all the proof obligations are given in Reference 1.

The Rodin platform [2] provides an automated support for formal modeling and verification in Event-B. In particular, it automatically generates the required proof obligations and attempts to discharge them. The remaining unproven conditions can be dealt with by using the provided interactive tools.

7.4 FORMAL DEVELOPMENT OF A RESILIENT CPS IN EVENT-B

In this section, we present a formal Event-B development that will gradually incorporate the essential concepts and relationships between roles for the systems described above. The development is generic, that is, the abstract sets, functions, and relations are defined via their essential properties and thus can be interpreted as generic parameters of the whole development. As such, models represent a family of suitable systems. In Section 7.5, we show how such a generic development can be instantiated for a concrete system.

We present our development as the initial model of CPS and several of its subsequent refinements. In each step, we introduce some new concepts and their properties or elaborate on the already introduced ones. We separately discuss both static and dynamic system aspects, represented in the context and machine components, respectively. The developed models are rather big, thus in many cases we only highlight the most important changes or introduced features.

7.4.1 The Initial Model

In our initial model, we abstract away from many internal details of CPS, focusing instead on handling external requests for particular services arriving from outside the system boundaries. The execution of such requests is usually split into smaller tasks, which in turn are handled by specific functional blocks of the system. For the moment, we do not care where such blocks reside physically, that is, what particular components they belong to. Instead, we concentrate on modeling the connection between the execution of an arrived request and the dynamic availability of the required functional blocks.

In the context component of the initial model (presented in Figure 7.2), we introduce necessary static data structures. The abstract set, *FBLOCK*, models all the functional blocks residing within boundaries of our system. Each functional block belongs to a specific type, an element of the defined abstract set, *FB_TYPE*. In other words, there can be multiple blocks (instances) for the same functional block type, which is the main source for designing various resilience mechanisms for such a system. The defined function, *FB_type*, associates a given functional block with its type.

The system configuration in terms of its functional capabilities (i.e., availability of specific functional blocks) is very fluid. In the context, we model this by introducing an enumerated type, *FB_STATUS,*[*] which contains three values: *Enabled*, signifying that a block is healthy

---

[*] The predefined Event-B operator *partition* is used to defined an enumerated set with all the necessary axioms for its elements.

```
CONTEXT Data0
SETS FBLOCK, FB_TYPE, FB_STATUS, REQUEST, OUTPUT
CONSTANTS FB_type, Enabled, Disabled, Failed, Req_blocks, NIL,
             Success_resp, Failure_resp
AXIOMS
   axm1: FB_type ∈ FBLOCK ⟶ FB_TYPE
   axm2: partition(FB_STATUS, {Enabled}, {Disabled}, {Failed})
   axm3: Req_blocks ∈ REQUEST → ℙ₁(FB_TYPE)
   axm4: partition(OUTPUT, {NIL}, {Success_resp}, {Failure_resp})
END
```

FIGURE 7.2   The context Data0.

and ready to accept tasks, *Disabled*, indicating that a block is temporarily unavailable due to some external or internal factors, and *Failed*, denoting that a block can be considered permanently (at least, with respect to the current request) unavailable.

All possible incoming requests are represented by the abstract set, *REQUEST*. We assume that any such request can be decomposed into a set of the required functional blocks or, more precisely, functional block types. The introduced function, *Req_blocks*, associates a given request with such a set. Finally, outgoing system responses are abstractly modeled by the enumerated set *OUTPUT*, containing three elements: *NIL, Success_resp*, and *Failure_resp*.

In the machine component (the structure of which is shown in Figure 7.3), we describe the dynamics of our abstract systems. Essentially, the machine models the arrival of new requests, splitting them into the required functional block types, executing these blocks if they are available, and finally returning a success or failure response. In parallel, the systems regularly monitor the health of its functional blocks and update their status.

```
MACHINE M0
   SEES Data0
   VARIABLES bstatus, rem_blocks, output
   INVARIANT
       bstatus ∈ FBLOCK → FB_STATUS
       rem_blocks ∈ ℙ(FB_TYPE)
       output ∈ OUTPUT
       output = Success_resp ⇒ rem_blocks = ∅
       output = Failure_resp ⇒ rem_blocks ≠ ∅
   INITIALISATION
       bstatus :| bstatus' ∈ FBLOCK → {Enabled, Disabled}
       rem_blocks, output := ∅, Success_resp
   EVENTS
       Start = ...
       Monitor = ...
       Execute = ...
       Finish_success = ...
       Finish_failure = ...
END
```

FIGURE 7.3   The machine M0.

The three machine variables—*bstatus, rem_blocks*, and *output*—represent the current status of the system blocks, the remaining block types to be executed for the current request,[*] and the outgoing system response, respectively. Initially, all the system blocks are nonfailed, which modeled as the following nondeterministic initialization statement:

$$bstatus :| \ bstatus' \in FBLOCK \rightarrow \{Enabled, Disabled\}.$$

In the model invariants, the non-NIL system response is directly associated with success or failure of request execution which, in turn, is formulated as the condition on the remaining blocks, for example:

$$output = Failure\_resp \ \Rightarrow \ rem\_blocks \neq \varnothing.$$

Next, we briefly describe some machine events. The event, Start, models an arrival of a new request and splits it into a set of functional block types to be executed. The context function *Req_blocks* does the splitting for the received request.

```
EVENT Start
    ANY new_request
    WHERE
        grd1 : new_request ∈ REQUEST
        grd2 : output ≠ NIL
    THEN
        act1 : rem_blocks := Req_blocks(new_request)
        act2 : output := NIL
    END
```

The event, Monitor, regularly monitors the health status of the system functional blocks and updates the variable *bstatus* accordingly. The event nondeterministic assignment allows pretty much arbitrary change of the block status with the only restriction that a functional block, once failed, stays failed. At this level, there are no guard conditions for the event so it can happen at any time.

```
EVENT Monitor
    BEGIN
        act1 : bstatus :| bstatus' ∈ FBLOCK → FB_STATUS ∧
                    ∀bb. bstatus(bb) = Failed ⇒ bstatus'(bb) = Failed
    END
```

The event, Execute, models successful execution of one block type belonging to the variable *rem_blocks*. The block should be enabled to perform the task. As a result of event execution, the finished block type is removed from *rem_blocks*.

---

[*] We assume here that a functional block of a particular type should be executed only once for a request.

```
EVENT Execute
  ANY block, btype
  WHERE
    grd1 : block ∈ FBLOCK
    grd2 : btype ∈ FB_TYPE
    grd3 : btype ∈ rem_blocks
    grd4 : bstatus(block) = Enabled
  THEN
    act1 : rem_blocks := rem_blocks\{btype}
  END
```

The remaining two events respectively model success and failure responses of the system. The success response is returned when *rem_blocks* becomes empty, while the failure one is returned if all the functional blocks are failed for some remaining block type, thus making successful service completion impossible. An alternative or additional way can be used to associate failure responses with some middleware timeouts, for example, indicating that some required blocks were disabled for too long.

7.4.2 The First Refinement

In the first refinement step, we introduce the system components and partition the functional blocks among them. Moreover, we explicitly model both the component internal (collective) state and the component environment (external) state. The monitored states are used to determine the status of specific functional blocks, which in turn affects their availability for execution of the remaining service tasks. We also enforce a specific cyclic behavior within a component when the functional block execution step always comes immediately after the component monitoring step.

In the context part of the refined model (presented below), we introduce three additional abstract sets—*COMPONENT, CSTATE*, and *ESTATE*—standing for all possible components, (collective) component states, and environment states, respectively. The function *Block_comp* associates each block with one of the system components. Since it is often needed to reason about the functional blocks of a particular component, we also introduce the opposite function, *Comp_blocks*, which is required to be the exact inverse of *Block_comp* in the axiom, axm3. Finally, we require that the updated status of a functional block should depend on the latest monitored internal and external component states by defining the abstract function, *Block_monitor*.

```
CONTEXT Data1
EXTENDS Data0
SETS COMPONENT, CSTATE, ESTATE
CONSTANTS Block_comp, Comp_blocks, Block_monitor
AXIOMS
  axm1: Block_comp ∈ FBLOCK ↠ COMPONENT
  axm2: Comp_blocks ∈ COMPONENT → ℙ₁(FBLOCK)
  axm3: ∀bb, cc. cc = Block_comp(bb) ⇔ bb ∈ Block_comp(cc)
  axm4: Block_monitor ∈
          FBLOCK → (ESTATE × CSTATE → FB_STATUS)
END
```

In the machine component (the structure of which is shown below), we introduce model variables, *estate* and *cstate*, for storing the current values of the internal state and external component states, respectively. The array variable, *monit_flag*, is added to enforce the specific execution order within each component, that is, monitoring followed by task execution. Finally, the functional block status is now partitioned among the system components with each component storing only statuses of the functional blocks belonging to it. This is achieved by data refinement, where the abstract variable, *bstatus*, is now replaced by the array variable, *comp_bstatus*, separately keeping the status information for every component. Since components are heterogenous and may contain different functional blocks, the resulting status information is constrained by the invariant:

$$\forall cc.\ \mathrm{dom}(comp\_bstatus(cc)) = Comp\_blocks(cc)$$

MACHINE M1
   SEES Data1
   REFINES M0
   VARIABLES *rem_blocks, output, estate, cstate, comp_bstatus, monit_flag*
   INVARIANT
      estate ∈ COMPONENT → ESTATE
      cstate ∈ COMPONENT → CSTATE
      comp_bstatus ∈ COMPONENT → (FBLOCK ⇸ FB_STATUS)
      ∀*cc*. dom(*comp_bstatus(cc)*) = *Comp_blocks(cc)*
      ∀*bb*. *bstatus(bb)* = *comp_bstatus(Block_comp(bb))(bb)*
      monit_flag ∈ COMPONENT → BOOL
   INITIALISATION
      ...
   EVENTS
      ...
END

EVENT Monitor
   ANY *cc, new_estate, new_cstate, new_bstatus*
   WHERE
      grd1 : *cc* ∈ COMPONENT
      grd2 : *new_estate* ∈ ESTATE
      grd3 : *new_cstate* ∈ CSTATE
      grd4 : *new_bstatus* ∈ FBLOCK ⇸ FB_STATUS
      grd5 : dom(*new_bstatus*) = *Comp_blocks(cc)*
      grd6 : !*bb*. *bb* ∈ dom(*new_bstatus*) ∧ *comp_bstatus(cc)(bb)* = *Failed* ⇒
            new_bstatus(bb) = *Failed*
      grd7 : !*bb*. *bb* ∈ dom(*new_bstatus*) ∧ *comp_bstatus(cc)(bb)* ≠ *Failed* ⇒
            new_bstatus(bb) = *Block_monitor(bb)(new_estate* ↦ *new_cstate)*
      grd8 : *monit_flag(cc)* = FALSE
   THEN
      act1 : *comp_bstatus(cc)* := *new_bstatus*
      act2 : *monit_flag(cc)* := *TRUE*
      act3 : *estate(cc)* := *new_estate*
      act4 : *ecstate(cc)* := *new_cstate*
   END

The introduced changes mostly affect two model events—Monitor and Execute. Both these events are now distributed; they are defined locally for each system component. The event parameter, *cc*, defines a specific for which component the event is executed.

The event, Monitor (presented above), relies on the evaluation procedure encoded in the context function, *Block_monitor*, which is applied to the observed values of the internal and external state of the component, *cc – new_cstate* and *new_estate*. The evaluation is done only for the blocks belonging to the component, defined by *Comp_blocks(cc)*. The blocks that were detected as failed earlier remain as such.

Like Monitor, the event, Execute (shown below), now executes a functional block for a particular component, defined by the first parameter, *cc*. However, the execution is allowed only after the monitoring step was executed, which is indicated by the condition, *monit_flag(cc) = TRUE*. As before, the result of event execution is removal of the completed functional block type from the variable, *rem_blocks*.

Despite introduction of the system component and distribution of functional blocks among them, the information about the progress of service execution (stored in the variable, *rem_blocks*) and overall service coordination remains global (centralized). We address this issue in the subsequent refinement steps where we model component roles and the internal communication between components.

EVENT Execute
   ANY *block, btype, cc, new_cstate*
   WHERE
      *grd*1 : *block* ∈ *FBLOCK*
      *grd*2 : *btype* ∈ *FB_TYPE*
      *grd*3 : *cc* ∈ *COMPONENT*
      *grd*4 : *btype* ∈ *rem_blocks*
      *grd*4 : *comp_bstatus(cc)(block)* = *Enabled*
      *grd*5 : *monit_flag(cc)* = *TRUE*
      *grd*5 : *new_cstate* ∈ *CSTATE*
   THEN
      *act*1 : *rem_blocks* := *rem_blocks\{btype}*
      *act*2 : *monit_flag(cc)* := *FALSE*
      *act*2 : *cstate(cc)* := *new_cstate*
   END

7.4.3 The Second Refinement

In the second refinement step, we introduce the notion of component roles. The component role is associated with a particular subset of functional blocks it should be able to perform currently and/or with particular duties within the service execution scenario. The examples of such roles are Manager, Standby, Master, Spare, Location, Load Balancer, Front End, etc. Sometimes roles can also signify the current operational mode of a component, especially when a component is forced to take a degraded role because of failures of its functional blocks.

```
CONTEXT Data2
EXTENDS Data1
SETS ROLES
CONSTANTS Role_blocks, Comp_roles, Init_config, min_role_cond
AXIOMS
    axm1: Role_blocks ∈ ROLE → ℙ₁(FB_TYPE)
    axm2: Comp_roles ∈ COMPONENT → ℙ₁(ROLE)
    axm3: Init_config ∈ COMPONENT → ROLE
    axm4: min_role_cond ∈
            ROLE → ((FBLOCK ⇸ FB_STATUS) → BOOL)
    axm5: ∀cc, bb. bb ∈ Comp_blocks(cc) ⇒
            ∃rr. rr ∈ Comp_roles(cc) ∧ FB_type(bb) ∈ Role_blocks(rr)
    axm6: ∀cc. cc ∈ COMPONENT ⇒
            Init_config(cc) ∈ Comp_roles(cc)
    axm7: ∀rr, bb. bb ∈ dom(dom(min_role_cond(rr))) ⇒
            FB_type(bb) ∈ Role_blocks(rr)
END
```

FIGURE 7.4   The context Data2.

In the context part of the refined model (see Figure 7.4), we introduce the abstract set, *ROLE*, for all possible component roles. The abstract function, *Role_blocks*, associates a given role with a number of functional block types, while the function, *Comp_roles*, relates each component with a number of the roles it can take on. These two functions are constrained together by the axiom, **axm5**, stating that, for each block of a component, there is a role that the component is able perform which is associated with the type of that functional block.

As its name suggests, the abstract function, *min_role_cond*, defines the minimal condition the component should satisfy while taking on a particular role. For a given role, the function takes the component configuration, that is, its status function of the type, *FBLOCK ⇸ FB_STATUS*, and returns a boolean value indicating whether this configuration is acceptable.

Finally, the function, *Init_config*, determines the initial system configuration, that is, all the initial roles for the system components. Obviously, the configuration should be achievable, that is, the components should be able to take these roles as formulated by the axiom, **axm6**.

In the machine component (see Figure 7.5), we add a new variable, *curr_role*, to store the current role of each component. We also restrict the status update of component functional blocks to reflect the expectation that if the component is in a particular role, it can employ only those functional blocks associated with that role. Consequently, all the other functional blocks can be either disabled or failed. This property is added as an additional invariant and enforced by the initialization and the refined version of the Monitor event.

A component may change its role. Sometimes, it is forced to do so after monitoring its internal state. In other cases, it can be asked to take a more active role in orchestrating the overall service. In both cases, the component should satisfy the minimal role condition for such a change. The role changing functionality is added as a new event, Change_role, shown below. The guard, *grd5*, of this event checks that the current component configuration (all its block statuses) is acceptable for taking *new_role*. The component block status is also updated to reflect the necessary changes.

```
EVENT Change_role
    ANY cc, new_role, new_bstatus
    WHERE
        grd1 : cc ∈ COMPONENT
        grd2 : cc ∈ ROLE
        grd3 : new_role ∈ Comp_roles(cc)
        grd4 : new_role ∉ curr_role(cc)
        grd5 : min_role_cond(new_role)(comp_bstatus(cc)) = TRUE
        grd6 : new_bstatus ∈ FBLOCK ⇸ FB_STATUS
        grd7 : dom(new_bstatus) = Comp_blocks(cc)
        grd8 : ∀bb. bb ∈ dom(new_bstatus) ∧ comp_bstatus(cc)(bb) = Failed ⇒
                    new_bstatus(bb) = Failed
        grd9 : ∀cc, bb. bb ∈ Comp_blocks(cc) ∧ FB_type(bb) ∉ Role_blocks(new_role) ⇒
                    new_bstatus(bb) ∈ {Disabled, Failed}
        grd10 : ∀cc, bb. bb ∈ Comp_blocks(cc) ∧ FB_type(bb) ∈ Role_blocks(new_role) ⇒
                    new_bstatus(bb) = comp_bstatus(cc)(bb)
    THEN
        act1 : comp_bstatus(cc) := new_bstatus
        act2 : curr_role(cc) := new_role
    END
```

```
CONTEXT Data3
EXTENDS Data2
SETS INT_REQUEST
CONSTANTS Primary_role, INT_NIL, FromPrimary,
                ToPrimary, IntReq_blocks
AXIOMS
    axm1: Primary_role ∈ ROLE
    axm2: partition(INT_REQUEST, {INT_NIL}, FromPrimary,
                ToPrimary)
    axm3: IntReq_blocks ∈ INT_REQUEST → ℙ₁(FB_TYPE)
    axm4: card({cc. cc ∈ COMPONENT ∧
                Init_config(cc) = Primary_role}) = 1
END
```

7.4.4 The Third Refinement

In the third refinement step, we elaborate on the concept of component roles. Even if the roles give us a nice abstraction to reason about different subsets of component functionalities or operational modes, we need to designate some role(s) a special status to ensure the overall orchestration and coordination of service execution. We do so by introducing the notion of a *primary* (i.e., leading or coordinating) role. This is the role responsible for orchestrating the overall service execution and also often for serving as a front end for the outside world, that is, receiving service requests and sending system responses. For simplicity, in this chapter, we assume that there is a single component in this role.

Moreover, we also need to introduce the mechanisms for internal communication when a component in the primary role delegates some remaining tasks to nonprimary (e.g., standby, spare) components as well as receives some intermediate results from them.

In the context component of the refined model (presented above), we explicitly designate one role as the primary one by introducing the constant *Primary_role*. We also add a

```
MACHINE M2
    SEES Data2
    REFINES M1
    VARIABLES rem_blocks, output, estate, cstate, comp_bstatus,
            monit_flag, curr_role
    INVARIANT
        ...
        curr_role ∈ COMPONENT → ROLE
        ∀cc. cc ∈ COMPONENT ⇒ curr_role(cc) ∈ Comp_roles(cc)
        ∀cc, bb.
            bb ∈ Comp_blocks(cc) ∧ FB_type(bb) ∉ Role_blocks(curr_role(cc))
            ⇒ comp_bstatus(cc)(bb) ∈ {Disabled, Failed}
    INITIALISATION
        ...
        comp_bstatus :| ... ∧ (FB_type(bb) ∉ Role_blocks(curr_role(cc)) ⇒
            comp_bstatus'(cc)(bb) ∈ {Disabled, Failed})
        curr_role := Init_config
    EVENTS
        ...
        Change_role = ...
END
```

FIGURE 7.5   The machine M2.

new abstract set, *INT_REQUEST*, for modeling all the internal requests. We explicitly partition this set to contain *INT_NIL*, the nil value indicating the absence of such requests, *FromPrimary*, the subset of requests coming from the primary role component, and *ToPrimary*, a subset of the requests received by the primary component from the nonprimary components.

Since the main purpose of internal communication is to facilitate coordinated execution of the requested service, we assume that an internal request contains the remaining tasks (functional block types) that the primary role delegates to other components to be executed. The returning requests to the primary role components then contain the tasks still to be completed. To extract the task information from an internal request, we introduce the abstract function *IntReq_blocks* of the type $INT\_REQUEST \rightarrow \mathbb{P}_1(FB\_TYPE)$.

In the machine component (shown on Figure 7.6), we explicitly store the component with the primary role in the variable *primary_comp* and require that there should be no more than one such component in the systems as an invariant property. Moreover, we introduce the variable *internal_bus* to model the bus for internal operations and define four new events for sending and receiving internal requests via this bus, distinguishing between the components in the primary and nonprimary roles. Finally, the variable *execution_token* is added to ensure that only one component proceeds with the service execution.

After initialization, the primary component gets the token. At some point of service execution, it can delegate the remaining task(s), via sending an internal request, to some nonprimary component which then receives the token. When such component can proceed no further, it returns the token by sending an internal request to the primary component. The service execution ends when the primary component sends an external response with either a success or failure message.

```
MACHINE M3
   SEES Data3
   REFINES M2
   VARIABLES rem_blocks, output, estate, cstate, comp_bstatus,
                monit_flag, curr_role, primary_comp, internal_bus,
                execution_token
   INVARIANT
       ...
       primary_comp ∈ COMPONENT
       curr_role(primary_comp) = Primary_role
       internal_bus ∈ INT_REQUEST
       execution_token ∈ COMPONENT → BOOL
       card({cc. cc ∈ COMPONENT ∧ curr_role(cc) = Primary_role}) ≤ 1
       card({cc. cc ∈ COMPONENT ∧ execution_token(cc) = TRUE}) ≤ 1
   INITIALISATION
       ...
       primary_comp :| Init_config(primary_comp′) = Primary_role
       internal_bus := INT_NIL
       execution_token : |
           (∀cc. execution_token′(cc) = bool(Init_config(cc) = Primary_role))
   EVENTS
       ...
       Send_internal_request_primary = ...
       Send_internal_request = ...
       Receive_internal_request_primary = ...
       Receive_internal_request = ...
END
```

FIGURE 7.6   The machine M3.

7.4.5 The Fourth Refinement

The coordination and communication scheme described in the previous subsection is a very "liberal" one in the sense that the primary component just puts out an outgoing internal request, essentially "hoping" that some other component would be able to proceed with the service execution. We assume that if such a request is not taken for processing for a predefined period of time (e.g., because none of the remaining components have the required nonfailed functional blocks), the corresponding timeout signal is issued by the system middleware consequently leading to the respective failure message by the primary component to the outside world.

In the fourth refinement, we introduce some discipline into component coordination when the primary component explicitly targets the component to which service execution task are delegated. To be able to do that, some information about the internal configuration (e.g., current statuses of the component functional blocks) of the nonprimary components should be accessible to the prime component. In this refined model, we add additional communication by which the nonprimary components must inform their primary counterpart once they detect that any of their functional components have failed. With such extra information, the primary component can decide which of the components is most suitable to handle the remaining service tasks.

In the case of failures that may force the primary component to relinquish its primary role, the collected information is transferred to a new primary component, if possible.

For brevity, we just sketch the highlights of this refinement step. As a refinement result, the previous model is elaborated by

- Defining an additional field in the existing messages for service orchestration containing the target component (in the context component)

- Adding a new type of internal requests for failure messages (in the context component)

- Constraining the existing events for receiving internal requests by checking the request target (in the machine component)

- Introducing new events for sending information about the failed blocks to the primary component (in the machine component)

- Introducing a new event for the primary component to decide on the most suitable component to which to delegate the remaining tasks (in the machine component)

- Introducing a new event for transferring the collected information to a new primary component (in the machine component)

There are many aspects of resilient CPS that are still omitted in the obtained formal models. Some of them can be introduced in the subsequent refinement steps. For instance, the actual service execution scenario is very simplistic, and many details about the order and branching of specific service tasks can be introduced. No parallel execution (either within the execution of a single service request or handling multiple service requests at the same time) is considered for the moment. We restrict ourselves to modeling a single primary component and therefore just one subordination level between the system components. We hope to address some of these issues in our future work on the subject.

7.5 EXAMPLE: SATELLITE DATA PROCESSING UNIT

7.5.1 Description

Our work is partially inspired by the actual solution to circumvent double failure that occurred in a currently operational onboard satellite system. The architecture of this system is similar to that of the Data Processing Unit (DPU)—a subsystem of European Space Agency mission BepiColombo [3] that is currently in development. The main goal of the mission is to carry various scientific measures to explore the planet Mercury. DPU—an important part of the Mercury Planetary Orbiter—consists of four independent components (computers) responsible for receiving and processing data from four sensor units: SIXS-X (x-ray spectrometer), SIXS-P (particle spectrometer), MIXS-T (telescope), and MIXS-C (collimator).

The behavior of DPU is managed by Telecommands (TCs) received from the spacecraft. Processing of each TC results in producing a telemetry (TM) message. As a result of decoding TC, DPU might produce a housekeeping report, switch to some mode, or initiate/continue production of *scientific data*. Correspondingly, TM would contain either housekeeping data, an acknowledgement of the requested mode transition, or scientific data. Each component acquires fresh scientific data from the corresponding sensor unit (SIXS-X, SIXS-P, MIXS-T, and MIXS-C), preprocesses and then makes it available to DPU which eventually forms the entire TM package.

To cope with errors that may occur during the satellite mission, the redundant DPUs are usually used, that is, in case of failure of any component of the active DPU, the entire TM package cannot be formed and the spare DPU is activated. Let us consider a duplicated system that consists of two identical DPUs—DPU_A and DPU_B. As was explained above, each DPU contains four components responsible for controlling the corresponding sensors.

Traditionally, the satellite systems are designed to implement the following simple redundancy scheme. Initially, DPU_A is active while DPU_B is a cold spare. DPU_A allocates tasks on its components to achieve the system goal—processing of a TC and producing a TM. When some lower-level component of DPU_A fails, DPU_B is activated to achieve the goal. Failure of DPU_B results in failure of the overall system. However, let us observe that even though none of the DPUs can accomplish the overall goal on its own, it might be the case that the components that remained operational can perform the entire set of tasks required to reach the goal. This observation allows us to define the following dynamic reconfiguration strategy.

Initially, DPU_A is active and is assigned to reach the goal. If any of its components fail to execute one of four scientific tasks (let it be $task_j$), the spare DPU_B is activated and DPU_A is deactivated. DPU_B performs $task_j$ and the consecutive tasks required to reach the goal. It becomes responsible for achieving the overall goal until some of its component fail. To remain operational, the system then performs *dynamic reconfiguration*. Namely, it reactivates DPU_A and tries to assign the failed task to the corresponding component of DPU_A. If such a component is operational, then DPU_A continues to execute the consequential tasks until it encounters a failed component. Then, the control is passed to DPU_B again. Obviously, the overall system stays operational until two identical components of both DPUs have failed.

7.5.2 Instantiation of Generic Models

It is rather easy to demonstrate that the satellite system described above is a special instance of a generic CPS we modeled in this chapter. To do so, we need to show a valid instantiation of our generic parameters with concrete values pertaining to the described systems.

We start by instantiating the abstract set *COMPONENTS* with two system components, DPU_A and DPU_B:

$$COMPONENTS = \{DPU_A, DPU_B\}.$$

We can distinguish four essential functional blocks—SIXS-X, SIXS-P, MIXS-T, and MIXS-C. In addition, we need the functional blocks responsible for external and internal communication—EXT-IO and INT-IO. As a result the abstract set *FB_TYPE* is instantiated as

$$FB\_TYPE = \{SIXSX, SIXSP, MIXST, MIXSC, EXT\_IO,$$
$$INT\_IO\}.$$

The two components are functionally identical, thus we define the data structures *FBLOCKS*, *FB_type*, and *Block_comp*, as follows:

$$FBLOCKS = \{SIXSX_A, SIXSP_A, MIXST_A, MIXSC_A,$$

$$EXT\_IO_A, INT\_IO_A, SIXSX_B, SIXSP_B,$$

$$MIXST_B, MIXSC_B, EXT\_IO_B, INT\_IO_B\},$$

$$FB\_type = \{SIXSX_A \mapsto SIXSX, SIXSP_A \mapsto SIXSP,$$

$$MIXST_A \mapsto MIXST, ...\},$$

$$Block\_comp = \{SIXSX_A \mapsto DPU_A, SIXSX_B \mapsto DPU_B,$$

$$MIXST_A \mapsto DPU_A, ...\}.$$

We can distinguish three different component roles—*Master*, *Cold_spare*, and *Hot_spare*. The initial system configuration is as follows:

$$Init\_config = \{DPU_A \mapsto Master, DPU_B \mapsto Cold\_spare\}.$$

The defined roles are associated with the following functional blocks:

$$Role\_blocks =$$

$$\{Master \mapsto$$

$$\{SIXSX, SIXSP, MIXST, MIXSC, EXT\_IO, INT\_IO\},$$

$$Cold\_spare \mapsto \{INT\_IO\},$$

$$Hot\_spare \mapsto$$

$$\{SIXSX, SIXSP, MIXST, MIXSC, EXT\_IO, INT\_IO\}\}.$$

While the *Master* and *Hot_spare* roles have identical functionalities, *Cold_spare* has only the internal communication block, which may be used to activate it and change its current role into either the *Master* or *Hot_spare* role.

The dynamic reconfiguration mechanisms described in the previous section can be easily shown as a special case of the formalized ones. The minimal role condition for the *Master* role is defined by requiring that its *EXT_IO* block not fail. If this condition is not satisfied, some other component should take over by switching to the *Master* role if possible.

7.6 CONCLUSIONS

Our work presented in this chapter combines formal modeling of CPS with enhancing the resilience of such systems by integrating dynamic system reconfiguration mechanisms. The main focus is on gradually describing the involved concepts and their interrelationships as well as unfolding the system architecture with the integrated dynamic reconfiguration mechanisms to enhance system resilience.

There is very active ongoing research in both these areas. The DESTECS project [4] studies the problem of engineering resilient CPS. The research conducted within the project proposes an approach to model-based design through cosimulation of discrete event models in the Vienna Development Method (VDM) and continuous-time models in 20-sim. These models are coupled by a cosimulation tool that coordinates execution of the developed models in their respective simulators. The resulting models can be also augmented with descriptions of potential failures and fault tolerance mechanisms [5–7].

The problem of formal verification and validation of CPS has also been addressed in the ADVANCE project [8]. The main focus of the proposed methodologies supports the construction of CPS and augmenting formal modeling and verification with simulation and testing. The resulting proposed simulation-based approach combines Event-B development and cosimulation with tool-independent physical components via the FMI interface [9]. The ongoing EU H2020 project Into-CPS [10] aims at further advancing these results by exploring the idea of creating a multiobjective formal modeling framework and an integrated tool chain for model-based design of CPS. In contrast, in our work, we focused only on modeling the architectural aspects of CPS and did not consider separately the modeling of the physical processes. Essentially, we assume that the systems will be able to monitor the changes in both the environment and internal states and focus on designing more flexible and resilient mechanisms to react to these changes.

In Reference 11, Inverardi et al. investigate system adaptation based on the assume–guarantee concept. In particular, they propose a framework that allows the developers to efficiently define under which conditions adaptation can be performed by preserving the desired system-invariant properties. The framework also allows the designers to split the systems into parts that can be substituted. In order to guarantee the correctness of adaptation, the special conditions are formulated and must be proven at run-time. In our approach, the reconfiguration strategies are already defined at development phase and are incorporated into the system architecture. In the case of failures or changes, the system is able to reconfigure by changing interdependencies among components as well as between the components and the system goals (e.g., providing the required system services).

An extensive body of research investigates the quality of service characteristics of dynamically reconfigurable service-oriented systems. Among the most prominent works in this area is the approach proposed by Calinescu [12]. It aims at defining the optimal configuration with respect to quality of service by assessing the quality of service attributes of various service components that are available at run-time.

The idea of achieving system dependability via reconfiguration is described in Reference 13. The authors present a method for constructing systems where general properties of reconfiguration can be ensured via formal proofs. The idea of the proposed approach is to introduce a formal definition of reconfiguration as well as a set of high-level properties. Then, a system architecture is introduced which guarantees those reconfiguration properties. In our research, we follow the same idea to enable the systems to be reconfigurable already at a high-level system specification.

The problem of system reconfiguration and its connection to predictable dynamic resilience is presented in References 14 and 15. The proposed architectural framework—MetaSelf—is suitable for development of dynamically resilient systems using a specific architectural model. The main idea of the approach is to separate the functional and non-functional descriptions of the system components, formulate necessary resilience policies, and reason about system reconfiguration based on these policies. The key feature of the proposed solution is the need for metadata—information about system components—sufficient to enable decision-making about dynamic system reconfiguration. The metadata are used to guide the system reconfiguration with respect to the given reconfiguration policies.

In general, in our work, we see reconfigurability as an ability of components to redistribute their responsibilities (roles) to ensure continued system operation and service provision. Therefore, the proposed reconfiguration mechanisms are built by constantly monitoring both internal and external component states and then reacting by changing links and associations between components.

REFERENCES

1. J.-R. Abrial. *Modeling in Event-B*. Cambridge University Press, New York, 2010.
2. Rodin. Event-B Platform. Online at www.event-b.org.
3. BepiColombo. ESA Media Center, Space Science. Online at www.esa.int/esaSC/ SEM-NEM3MDAF_0_spk.html.
4. Design Support and Tooling for Embedded Control Software (DESTECS). FP-7 project. Online at www.destecs.org.
5. J. S. Fitzgerald, P. G. Larsen, K. G. Pierce, and M. Verhoef. A formal approach to collaborative modelling and co-simulation for embedded systems. *Mathematical Structures in Computer Science*, 23(4):726–750, 2013.
6. J. S. Fitzgerald, K. G. Pierce, and C. Gamble. A rigorous approach to the design of resilient cyber-physical systems through co-simulation. In *IEEE/IFIP International Conference on Dependable Systems and Networks Workshops, DSN 2012*, pp. 1–6. IEEE, Boston, USA, 2012.
7. J. S. Fitzgerald, K. G. Pierce, and P. G. Larsen. Co-modelling and co-simulation in the engineering of systems of cyber-physical systems. In *9th International Conference on System of Systems Engineering, SoSE 2014*, pp. 67–72. IEEE, Adelaide, Australia, 2014.
8. Advanced Design and Verification Environment for Cyber-physical System Engineering (Advance). FP7 Information and Communication Technologies (ICT) Programme. Online at www.advance-ict.eu.
9. V. Savicks, M. Butler, J. Colley, and J. Bendisposto. Rodin multi-simulation plug-in. In *5th Rodin User and Developer Workshop*, Southampton, UK, 2014.
10. Into-CPS—Integrated Tool Chain for Model-based Design of Cyber-Physical Systems. EU H2020 project. Online at into-cps.au.dk.
11. P. Inverardi, P. Pelliccione, and M. Tivoli. Towards an assume-guarantee theory for adaptable systems. In *ICSE Workshop on Software Engineering for Adaptive and Self-Managing Systems, SEAMS 2009*, pp. 106–115. IEEE, Vancouver, Canada, 2009.
12. R. Calinescu, L. Grunske, M. Kwiatkowska, R. Mirandola, and G. Tamburrelli. Dynamic QoS management and optimization in service-based systems. *IEEE Computer Society*, 37:387–409, 2011.
13. E. A. Strunk and J.C. Knight. Dependability through assured reconfiguration in embedded system software. *IEEE Transactions on Dependable and Secure Computing*, 3(3):172–187, 2006.

14. G. Di Marzo Serugendo, J. S. Fitzgerald, and A. Romanovsky. Metaself: An architecture and a development method for dependable self-* systems. In *Proceedings of the 2010 ACM Symposium on Applied Computing (SAC)*, pp. 457–461. ACM, Sierre, Switzerland, 2010.
15. G. Di Marzo Serugendo, J. S. Fitzgerald, A. Romanovsky, and N. Guelfi. A metadata-based architectural model for dynamically resilient systems. In *Proceedings of the 2007 ACM Symposium on Applied Computing (SAC)*, pp. 566–572. ACM, Seoul, Korea, 2007.

CHAPTER 8

Collaborative Modeling and Simulation for Cyber-Physical Systems

Peter Gorm Larsen, John Fitzgerald, Jim Woodcock, and
Thierry Lecomte

CONTENTS

THE ENGINEERING OF DEPENDABLE CYBER-PHYSICAL SYSTEMS (CPS) demands model-based methods and tools that support—at a semantic level—the diversity of notations necessary to describe both the computational and physical elements of the systems of interest. This extends to the need for integrated tool chains that support the collaborative construction of models, the exploration of alternative designs, and the validation of key system properties. We discuss research toward an open framework for the collaborative construction and simulation of models of CPS, including foundations that support semantic diversity and method guidelines that assist the activities from requirements to realizations. The approach under examination integrates existing industry-strength tools based around Functional Mock-up Interface-compatible co-simulation. Integrating the well-founded semantics of different simulators should allow us to deliver collaborative simulation (co-simulation) of multiple models (co-models). We demonstrate the intended use of this technology with an industrial case study from the railway domain.

8.1 INTRODUCTION

A *system* is a combination of interacting elements organized to achieve a stated purpose [1], and a *dependable* system is one on which reliance may justifiably be placed [2]. Dependable systems include *safety-critical systems* whose failure or malfunction may result in one (or more) of the following outcomes: (1) death or serious injury to people; (2) loss or severe damage to equipment/property; or (3) environmental harm. CPS are systems in which some elements are *computational* and some *physical* [3].

In this chapter we discuss foundations, methods, and tools to support the necessarily collaborative and multidisciplinary model-based design of dependable CPS. CPS present major business and societal opportunities in a variety of application areas—if they can be developed economically [4], and provided they merit the reliance placed upon them.

Model-based design (MBD) has the potential to enhance the development of CPS, increasing the competitiveness of the industry by shortening time to market and reducing development costs. However, the nature of CPS design raises various challenges not currently met in the "state of the art" [5]:

- The design space for CPS is large, with many design decisions and trade-offs among physical (typical mechanical) components, hardware, and software still to be made. Given the cost of producing and evaluating full-sized prototypes, it is essential to be able to define test cases and explore designs early in the development process. Flexibility is also essential: solutions should allow designs to be altered and rapidly reevaluated under changing requirements [6].

- Collaboration between multiple disciplines is paramount for successful CPS design, yet these disciplines have distinct cultures and formalisms. Systems engineers work with notations such as the *Systems Modeling Language* (SysML) [7], control engineers use continuous-time (CT) formalisms, and software engineers use discrete-event formalisms. Enhancing the dialogue between these disciplines is essential [8].

- Support is needed to help maintain the complex collections of artifacts produced in CPS development. Traceability between all artifacts must be enabled, allowing the provenance of all elements to be recorded and the final system to be linked to the requirements.

- Configuration management is needed to ensure that specific versions of models are used in the production of certain results; this is particularly important when providing evidence for certification such that the CPS can be *trusted*.

We propose support for the holistic modeling of CPS in the project INTO-CPS.[*] This will allow system models to be built and analyzed that may otherwise not be possible using standalone tools. This will be achieved by integrating existing industry-strength tools-based centrally around functional mock-up interface (FMI)-compatible co-simulation.[†] The tool chain will be underpinned by well-founded semantic foundations that ensure the analysis results are reliable.

We envisage a tool chain able to support powerful analysis techniques for CPS, including connection to SysML, generation and static checking of FMI interfaces, model checking, and Hardware-in-the-Loop (HiL) and Software-in-the-Loop (SiL) simulation, all supported by code generation. The tool chain will allow for both test automation (TA) and design space exploration (DSE) of CPS. The technologies will be accompanied by a comprehensive set of method guidelines that describe how to adopt the suggested approach, lowering entry barriers for CPS development.

The main message of this chapter is how the INTO-CPS technology can be used to model and analyze complex CPS and produce results in a heterogeneous virtual setting that will be applicable as evidence for the safety of the corresponding realization given that the methods and tools are well-founded.

In the remainder of this chapter, we first discuss related work, which we take as input for the proposed approach (Section 8.2). We present an example of the type of CPS engineering challenge that has motivated our work in Section 8.3, drawing on experience in the railway domain. In order to achieve an integrated tool chain, we make use of a collection of existing baseline tools, which we integrate together in the INTO-CPS project in Section 8.4. Section 8.5 presents the foundations for the development of trusted CPS that we propose. Afterward, Section 8.6 discusses the methods that we aim to develop on these foundations, with the ultimate aim of delivering trustworthy CPS. We discuss the architecture and content of a tool platform to support these methods in Section 8.7, with a focus on collaborative simulation (co-simulation). Finally, Section 8.8 concludes the chapter and points to the envisaged future work.

8.2 BACKGROUND

Many authors have called for better modeling notations to support the development of CPS and their associated challenges, including Broy [9–11], Lee [3,12], Wan [13], Derler [14,15], and Horvath [16].

A language dedicated to modeling CPS should fulfill the following requirements [17]: (a) it must model the interaction between discrete controllers and their continuous environments; (b) it must have precise operational semantics to support faithful simulation; and

---

[*] Integrated Tool Chain for Model-Based Design of Cyber-Physical Systems (INTO-CPS), see http://into-cps.au.dk/.
[†] In Section 8.7.1 FMI is going to be explained in more detail.

(c) it must have denotational or algebraic semantics that support automated analysis [18]. Furthermore, there is a need for a contract-based tool chain for CPS [19].

Typically, CPS cut across different domains involving diverse components, including algorithms, control systems, communication networks, and physical systems, and they are subject to expensive deployment costs and complex network interactions. They require comprehensive modeling, simulation, and verification to make sure that they function as required before they are deployed. It is not sufficient to apply individual tools to individual domains; rather, tools must be integrated into comprehensive tool chains, with sound links bridging the semantic gaps. The benefit of an integrated tool chain would be the creation of a collaborative design process that could be carried out within a product ecosystem in a well-integrated "virtual company" [20]. It would also permit an efficient DSE by providing a framework where different architectures could be quickly assembled and evaluated according to the design task being considered in the chain [21].

There is usually inadequate support for design chains that span different models, with most designers using a collection of tools that are not linked together. Implementation is then carried out with informal techniques that involve a great deal of ineffective human interaction, thus creating unnecessary and unwanted iterations among groups of designers, possibly in different companies, that share little understanding of their respective knowledge domains. Tools are linked by manual or empirical translation of intermediate formats without any guarantees that the design semantics is preserved; this is a source of errors that are difficult to identify and debug. One such tool is HybridSim [22], which supports importing existing system components from multidomains into SysML blocks, where Functional Mock-up Units (FMUs) and configuration scripts can be generated. FMUs can then be co-simulated, using the FMI standard to synchronize their corresponding simulators and exchange information between them [23,24]. However, unfortunately it seems that this initiative has been discontinued.

The Vanderbilt model-based prototyping tool chain [25] provides another integrated framework for embedded control-system development, providing multiple views, including Simulink/Stateflow models, CyPhyML models, software architecture, hardware modeling, and deployment. Each view requires a different metamodel. The tool chain uses a time-triggered language, ESMoL, which results in the Simulink blocks being restricted to periodic execution. There is no support for verification models or for physical descriptions, nor is there a notion of consistency. In addition, this tool chain lacks recording of traceability between the different artifacts produced in the development of CPS.

Techniques for modeling CPS are mostly based on those for hybrid systems, which have included hybrid statecharts [26,27] and hybrid automatons [28], where finite control graphs describe discrete behaviors and differential equations describe continuous behaviors. More recently, several languages have been proposed to describe hybrid systems and to support simulation and verification. Carloni [29] surveys various languages and tools, such as Simulink/Stateflow, Modelica [30], HyVisual [31], Charon [32], Masaccio [33], Shift [34], Hysdel [35], and SAL [36]. All of these techniques are only able to solve a part of the needs for generic CPS developments with natural ways of expressing the interesting properties.

8.3 INDUSTRIAL CASE STUDY: DISTRIBUTED INTERLOCKING

In railway signaling, an *interlocking* is an arrangement of signal apparatuses that prevents conflicting movements through an arrangement of tracks, such as junctions or crossings. Based on the status of the railway system as seen from sensors and on its short-term history, the interlocking computes the status of the actuators (switches, signals). This computation is determined by signaling safety rules that depend on different countries, but also by various optimization issues.

To create a route from A to B for a train in Figure 8.1, track circuits T2, T3, T4, T5, T6, and T7 have to be free. Once they are allocated to the route, switches have to be positioned correctly: switch SW1 has to be in direct position. SW1 and SW2 remain in a fixed position as long as their respective track circuit is allocated to the route. Signals S1 and S3 turn green to allow the train to enter the route. Then S1 turns red as soon as the train enters T3 in order to prevent another train coming from T1 to collide at the rear. However, other conditions have to be set up in order to prevent collision with trains entering T8 or T17 and going through T13, and trains entering T16, T24, or T26 and going through T11. In the first case, signals S7 and S12 have to be turned red and track circuits leading to T13 have to be free. In the second case, signals S9, S14, and S16 have to be turned red and the track circuits leading to T11 have to be free.

There are also degraded modes that are required to ensure an exploitation of the overall design space in case trains, sensors, or actuators fail. A route is allocated only for a given period of time: after a given delay (for example 2 mins), an allocated route is freed and its constituting track circuits are likely to be used to set up other routes. In Figure 8.1, a train stuck on track circuit T20 would block trains going straight from track circuit T24 to T17. An alternative route could be T24, T23, T22, T21, T12, T11, T10, T19, T18, and T17 but would have to allocate some track circuits that could conflict with other routes like the one from T16 to T8. Rerouting trains is decided by supervision systems or human operators, but interlocking is in charge of ensuring that incompatible routes are not allocated at the same

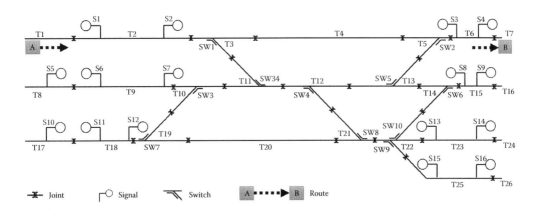

FIGURE 8.1   Partial scheme plan of a train line as seen from the sky, including track circuits, switches, and signals.

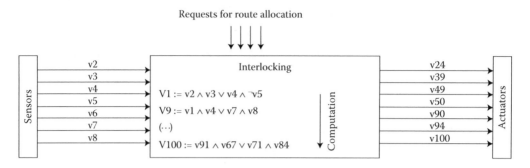

FIGURE 8.2   Interlocking transfer function, using sensor inputs and requests for route allocation to command actuators.

time. Signaling rules are implemented as so-called "binary equations"* that encode safety principles, exploitation rules, and industrial know-how. Usually interlocking is in charge of a complete line, dealing with hundreds of track circuits, signals, and switches. A typical metro interlocking is in charge of managing 180,000 equations that have to be computed several times per second. These equations compute the commands to be issued to track-side devices based on the perception of the status of the railways system and requests for route allocation to trains (see Figure 8.2).

A central interlocking is able to deal with a complete line; all decisions are made globally. However, the distance between devices distributed along the tracks and the interlocking system may lead to a significant delay in updating device status. Moreover, this architecture, well dimensioned for metro lines, is often overkill for simpler infrastructures like tramway lines. So there is room for an alternative solution: a distributed interlocking. A train or metro line is then divided into overlapping interlocked zones, each zone being controlled by an interlocking. Such interlockings would be smaller as fewer local devices have to be taken into account—a local decision could be made in a shorter time and would result in potentially quicker train transfers.

However, overlapping zones have to be carefully designed (a train cannot appear by magic in a zone without prior notice), and some variable states have to be exchanged between interlocking systems as the previously mentioned Boolean equations have to be distributed accordingly over the interlocking systems (see Figure 8.3).

This distribution implies several engineering challenges. An "optimal distribution," that is, the decomposition of the line into overlapping areas so as to minimize delays and costs, requires a smart exploration of the design space; decomposition is directly linked with railways signaling rules but also depends on the skills and habits of the signaling engineer. It also implies defining the information that has to be exchanged between interlocking computers and how many equations have to run on any of them (20,000 equations maximum, for

---

* Equations are used to bind values to variables. Left hand variable is assigned with right hand Boolean expression that needs to be computed: all related variables are valued, hence no solving is required. Equations are processed one after the other, always in the same order. While a variable may be computed and modified several times during this traversal, only its final value is considered.

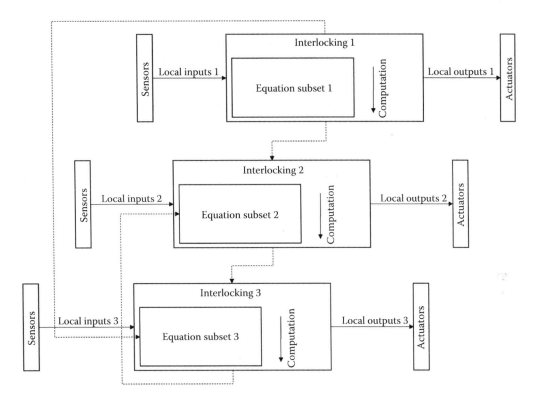

FIGURE 8.3   Distributed interlocking computers. Boolean equations are divided into subsets. Missing information (inputs or internal variables) has to be transmitted from one interlocking computer to another (dotted line).

example). This would help to evaluate different solutions in terms of communication bandwidth, computing power, amount of memory, etc. in order to improve architecture costs and efficiency.

Figure 8.1 shows a simplified view of the real system:

- Real distances are hidden (distances between stations are several orders of magnitude larger than station dimensions).

- Track slopes are hidden (gradients have a direct impact on security as they increase breaking distances).

- Trains are seen as occupying one or several track circuits without much precision.

In order to assess the final railway system, the discrete interlocking logic model and the physical models of the trains and the tracks have to be combined. The physical model will include fine-grain train movement, accurate 3D track topology (curves, slopes), and an ideal (fault-free) driver. Our goal is the ability to accurately model specific situations, for example, where trains are at different altitudes and where train movement could result in oscillations after braking is complete. The main property to check is the absence of collision between trains during the co-simulations. In order to check that distances are sufficient to ensure safety (distance between signal and track circuit entrance, track circuit lengths, etc.), several scenarios have to be considered, including maximum descending slope, maximum train weight, minimum braking capability, maximum acceleration, and speed. When checking if a configuration is compatible with the requirements, scenarios combining several calculations have to be assessed against diverse configuration regarding the time to complete all routes.

The objective of this case study is to obtain an integrated modeling environment capable of performing dimensioning engineering activities with a sufficient level of precision and handling an industry-strength system (a recent French tramway line). Safety-critical properties considered include, for example, physical ability to comply with signal protection mechanisms (a train has to stop before crossing a red signal). Such strong assumptions will be validated with the help of fine-grained simulations for a wide range of scenarios. Target code running on interlocking (binary equations) is expected to be generated from such models, aimed at several hardware platforms such as microcontrollers or PLCs.

8.4 TOWARD INTEGRATED TOOLCHAINS: THE INTO-CPS PROJECT

The overall objectives of the INTO-CPS project can be listed as

1. *Build an open, well-founded tool chain for a multidisciplinary model-based design of CPS that covers the full development life cycle of CPSs.* The tool chain will support multiple modeling paradigms and will cover multiple development activities, including requirements modeling, analysis, simulation, validation, verification, and traceability of artifacts throughout all development activities across disciplinary boundaries.

2. *Provide a sound semantic basis for the tool chain.* We will produce mathematical foundations to support CPS co-modeling and to underpin the tool chain. This will include semantics for FMI co-simulation, as well as SysML, discrete-event, and CT paradigms.

3. *Provide practical methods in the form of guidelines and patterns that support the tool chain.* The INTO-CPS methodology will be developed to ensure that adoption of the tool chain is cost-effective, providing industrial users with pragmatic guidance to help them determine the best modeling technologies and patterns to meet their needs.

4. *Demonstrate, in an industrial setting, the effectiveness of the methods and tools in a variety of application domains.* Four complementary industry case studies have been selected from four distinct domains that currently experience pressure to develop reliable CPS (automotive, agricultural, railways, and building automation). The case studies will be used to drive the production of the tools and methods and evaluate them.

5. *Form an INTO-CPS association to ensure that project results extend beyond the life of the project.* Membership of the association will allow future case study owners access to information, training, and competitively priced licenses at various levels of support. Tool vendors will be offered services to help integrate their products into the tool chain.

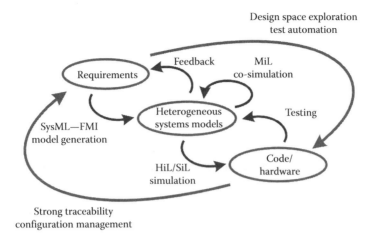

FIGURE 8.4 Connections in the INTO-CPS tool chain.

The overall workflow and services offered by the tool chain are illustrated in Figure 8.4.

At the top level, the tool chain will allow requirements to be formalized using SysML, supported by guidelines for capturing the requirements on CPS. A SysML profile will be developed that allows the architecture of CPS to be described, including both software, physical, and networking elements. From the architectural model, an FMI interface can be generated, along with initial draft models, to reduce effort in producing initial models. We envisage that we will export model descriptions for each of the constituent models that can then be imported by different simulation tools, indicating the interfaces that are needed for the corresponding FMUs inspired by the work from HybridSim.

Heterogeneous system models can then be built around this FMI interface, using the initial models as a starting point. A number of industry-strength tools will be connected here, permitting these heterogeneous "co-models" to contain discrete-event models of software, CT models of physical elements, and the networks between them. The tool chain will permit static analysis of these co-models, including model checking and static analysis of the FMI interfaces. The element models can either be in the form of discrete-event (DE) models or in the form of CT models combined in different ways.

A Co-simulation Orchestration Engine (COE) is created by combining existing co-simulation solutions and scaling them to the CPS level, allowing these CPS co-models to be evaluated through co-simulation. The COE will also allow real software and physical elements to participate in co-simulation alongside models, enabling both HiL and SiL simulation. Code generation from some of the baseline tools (see Section 8.4) will help support automated HiL simulation.

The COE will also allow multiple co-simulations to be defined and executed, and the results collated and presented automatically. The tool chain will allow these multiple co-simulations to be defined via DSE or through TA based on test cases generated from the SysML requirement diagrams.

Developing CPS will produce a large number of artifacts, including requirements, models, analysis results, and generated code. The tool chain will allow these artifacts to be stored, organized, and easily retrieved at a later date. It will allow the provenance of all artifacts to be recorded and traced back to the requirements. This data can be used at a later stage as evidence in documenting the adequacy of a design to meet the requirements. This results in a complete engineering approach to manage, track, and monitor model artifacts used in collaborative heterogeneous modeling.

The following list describes the existing baseline tools that will be linked in the INTO-CPS tool chain:

Modelio[*] is an open-source modeling environment supporting industry standards like UML and SysML. INTO-CPS will make use of Modelio for high-level system architecture modeling using the SysML language and proposed extensions for CPS modeling.

Overture[†] is another open-source tool that supports modeling and analysis in the design of discrete computer-based systems using the VDM-RT notation. This tool was used in the DESTECS[‡] project for modeling and simulation of DE controllers.

20-sim[§] was used in DESTECS as the main tool for modeling and simulation of CT systems. INTO-CPS will expand this use by incorporating results of systems engineering. The code generation and deployment capabilities of 20-sim will be used for HiL testing.

OpenModelica[¶] is an open-source Modelica-based modeling and simulation environment. Modelica is an object-oriented, equation-based language to model complex CPS. A large number of Modelica model libraries are available.

Crescendo[**] is the co-simulation tool developed in the DESTECS project. This tool enables the collaborative simulation of a DE controller modeled from the Overture tool, and a CT model of the physical plant from the 20-sim tool. The custom-built co-simulation interface was expanded to support DE models from Matlab/Simulink as well.

TWT co-sim engine[††] is a framework for configuring and running co-simulations. The individual simulations each run in their own native tool or are supplied as FMUs. The simulations are connected via definition of signals to be exchanged. These signals are passed between the simulations using the co-sim router. Among the currently supported tools are Matlab/Simulink, Modelica (both OpenModelica and Dymola[‡‡]), StarCCM+,[§§] and Qucs.[¶¶]

---

[*] http://www.modelio.org/
[†] http://overturetool.org/
[‡] http://destecs.org/
[§] http://www.20sim.com/
[¶] https://www.openmodelica.org/
[**] http://crescendotool.org/
[††] http://www.twt-gmbh.de/produkte/co-simulationen/co-simulation-framework.html/
[‡‡] http://www.3ds.com/products-services/catia/capabilities/systems-engineering/modelica-systems-simulation/dymola
[§§] http://www.cd-adapco.com/products/star-ccm
[¶¶] http://qucs.sourceforge.net/

RT-*Tester*[*] is a TA tool for automatic test generation, test execution, and real-time test evaluation. The RT-Tester Model-Based Test Case and Test Data Generator supports model-based testing, that is, automated generation of test cases, test data, and test procedures from UML/SysML models.

8.5 CPS FOUNDATIONS

CPS are, in general, inherently complex; much of this complexity is due to the necessary integration of the cyber and the physical world. Trustworthy CPS engineering requires hybrid and heterogeneous models that deal compositionally with the modeling and analysis of networked CPS, including the four dimensions of computational, physical, human, and regulatory requirements. Within each dimension there is a rich texture of different modeling issues, and cutting across these four dimensions are conceptual concerns that include distribution, concurrency, and time. For example, all four dimensions involve concurrency: computational models can be synchronous or asynchronous; the physical world can be divided between co-existing physical dynamics in a time continuum; human agents can have competing objectives and motives; and the system may have to conform to several different regulatory requirements. Reconciling these divergent concerns, and ensuring interoperability and communication between their components, is a central challenge in the development of CPS.

CPS are intrinsically heterogeneous. Our approach to handling the wide variety of semantics needed for modeling CPS is to describe each semantic concept separately and then to link them together in a coherent way. We use Galois connections to specify the interfaces between heterogeneous modeling concepts. These concepts can then be assembled to form the semantics for different modeling languages, and additional Galois connections allow heterogeneous modeling using different languages. This compositional approach also leads to compositional analysis techniques.

Our chosen meta-modeling notation for giving semantics to different semantic concepts is Unifying Theories of Programming (UTP) [37]. Our technique is to isolate important language features and give them denotational semantics; algebraic, axiomatic, and operational semantics can then be proved sound against this model. This allows different languages and paradigms to be linked together.

The semantic model is an alphabetized version of Tarski's relational calculus, presented in a predicative style that is reminiscent of the schema calculus in the Z notation [38]. Each programming construct is formalized as a relation between an initial and an intermediate or final observation. The collection of these relations forms a *theory* of the paradigm being studied, and it contains three essential parts: an alphabet, a signature, and healthiness conditions. *The alphabet* is a set of variable names that gives the vocabulary for the theory being studied. Names are chosen for any relevant external observations of behavior. For instance, a program with variables x, y, and z would contain these names in its alphabet. Theories for particular programming paradigms require the observation of extra

---

[*] http://www.verified.de/products/rt-tester/

information. Some examples are a flag that says whether the program has started (*ok*); the current time (*clock*); the number of available resources (*res*); a trace of the events in the life of the program (*tr*); a set of refused events (*ref*); and a flag that says whether the program is waiting for interaction with its environment (*wait*). *The signature* gives the syntax rules for denoting objects of the theory. For instance, in a theory of imperative programming, this would include operators like sequential composition, assignment, if-then-else, and iteration. *Healthiness conditions* identify properties that characterize the predicates of the theory. Each healthiness condition embodies an important fact about the computational model for the programs being studied.

Relations are used as a semantic model for unified languages of specification and programming. Specifications are distinguished from programs only by the fact that the latter use a restricted signature. As a consequence of this restriction, programs satisfy a richer set of healthiness conditions.

An important application of UTP is to build a foundational tool chain where unifying theories link tools to ensure the suggested approach is well founded. Tool chains can be either longitudinal or transverse; a longitudinal tool chain involves a series of tools where the output from one tool is used as the input to another. An example is Isabelle/HOL [39] and its sledgehammer tool, which invokes a number of other proof tools [40]. The outputs from these tools are proofs that must be interpreted as Isabelle proofs. A transverse tool chain provides a collection of tools for a particular language. This is illustrated in Figure 8.5.

For a particular language, we might like to provide a compiler, an interpreter, a model checker, a refinement calculator, and a theorem prover. The diagram shows the UTP approach to building this collection. Modern languages have heterogeneous semantics, so at the base of the diagram there is the mathematical semantics: the alphabetized relational calculus. The UTP theories for each semantic paradigm are built on top of this and linked together to form the gold standard for the language definition—its denotational semantics. Next, the operational semantics and axiomatic semantics for the language are derived from the denotational semantics so that they are consistent and complementary. The operational semantics can then be used as the basis for tools such as a compiler (code generator), an interpreter (simulation engine), and a model checker. The axiomatic semantics form the basis for verification tools and refinement calculators.

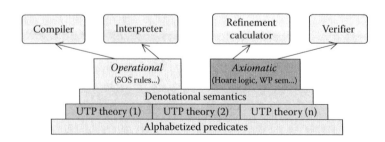

FIGURE 8.5 UTP semantic foundations.

8.5.1 Example Illustrating Heterogeneous Semantics with Railway Safety

In the next example, we describe a simple specification of one aspect of railway trains to illustrate the use of heterogeneous semantics (the example is inspired by Reference 41).

Example 8.1 (Railway Safety). As an example, consider the emergency brake for a train involved in the industrial case study in Section 8.3. The train is moving at an acceleration a until the train reaches an *Emergency Brake Intervention* speed. We specify this as follows:

$$\langle (\dot{s} = v, \dot{v} = a) \wedge (v < v_{ebi}) \rangle .$$

The formulas contained in angle brackets form a continuous statement defining the evolution of a continuous component using a differential equation. If $y = s(t)$ represents the position function for the train at time t, then $v = \dot{s}(t)$ represents its velocity and $a = \dot{v}(t) = \ddot{s}(t)$ represents its acceleration. The second part of this statement, $v < v_{ebi}$, defines a domain for v such that if $v < v_{ebi}$ is violated, then the statement terminates; otherwise it goes forward. The value v_{ebi} is the emergency brake intervention velocity.

Now, we know that the previous statement terminates if the velocity goes above v_{ebi}; so what happens next? The train must now decelerate until it reaches a safe velocity, specified by v_s. Again, we can specify this using a continuous statement:

$$\langle (\dot{s} = v, \dot{v} = sb) \wedge (v > v_s) \rangle .$$

The differential equations specify that the acceleration will be the safe braking value $sb \; \text{ms}^{-2}$; this is actually deceleration, since sb is negative. The two behaviors are composed in sequence

$$\langle (\dot{s} = v, \dot{v} = a) \wedge (v < v_{ebi}) \rangle \; ; \langle (\dot{s} = v, \dot{v} = sb) \wedge (v > v_s) \rangle .$$

This continuous behavior can be interrupted by a signal from the controller to cause an emergency stop. This signal is the event eb (for "emergency brake"). Call the continuous behavior we have described above P and call the continuous behavior after the event Q, then the syntax for describing the interruption is $P \triangle eb \rightarrow Q$. The description of Q is straightforward now, the train must decelerate safely until it stops

$$\langle (\dot{s} = v, \dot{v} = sb) \wedge (v > 0) \rangle .$$

Putting all this together, we have

$$TrainEB \;\; \hat{=} \;\; ((\langle (\dot{s} = v, \dot{v} = a) \wedge (v < v_{ebi}) \rangle \; ; \langle (\dot{s} = v, \dot{v} = sb) \wedge (v > v_s) \rangle))$$
$$\triangle$$
$$eb \rightarrow \langle (\dot{s} = v, \dot{v} = sb) \wedge (v > 0) \rangle .$$

Here $P \triangle eb \rightarrow Q$ behaves like P until interrupted by the event eb and then behaves like Q. We can now model aspects of the controller. For example, the controller may wait d seconds and then signal an emergency halt to the train

$$ControllerEB \; \widehat{=} \; \textbf{wait } d \; ; eb \rightarrow SKIP.$$

Our train emergency brake intervention system is made up of the parallel composition of the two pieces that we have specified:

$$TEBIS \; \widehat{=} \; TrainEB \parallel ControllerEB.$$

In this example, we have seen the need for constructs to deal with differential equations, communications between train and controller, real-time aspects, communications, sequence, parallelism, and process interruptions.

8.6 METHODS FOR MODEL-BASED CPS ENGINEERING

We have discussed the need for well-founded tool chains to support the successful integration of diverse models involved in the design of CPS. The provision of such tool chains requires semantic integration between equally diverse modeling and analysis tools, and we have outlined a framework based on UTP that could underpin this semantic integration. In this section, we outline methods that need to be supported for successful co-model-based CPS engineering but which also present particular challenges. In Section 8.7 we outline the architecture of tools to support these activities.

8.6.1 Co-Models

We advocate a model-based approach to collaborative design of dependable CPS. Models are abstract representations of systems of interest. The specific abstractions that are made in a given model depend on why the model is constructed and analyzed. As a consequence, models developed to represent the architecture of CPS might make quite different abstractions from those that focus on the computational elements and those that focus on the physical aspects. For example, in the railway interlocking example introduced in Section 8.3 we noted some details of track topography omitted from the logical representation shown in Figure 8.1.

CPS are inherently mutidisciplinary, integrating the computational, physical, human, and regulatory aspects identified in Section 8.5. CPS models are therefore likely to reflect this, being composed of complementary constituent models that use very different abstractions for handling these diverse dimensions. Further, since CPS may often be the result of integrating separately owned and managed preexisting systems, these models of similar aspects might also be represented in diverse notations (e.g., Simulink and 20-sim models of physical aspects of integrated constituent systems). In our approach, we use the term *co-model* to describe a collection of semantically diverse constituent models describing complementary aspects of CPS of interest.

The goal of our work is to enable engineers to use co-modeling to allow early stage analysis of system-level behaviors in CPS, so that risks to dependability and performance can be identified and handled appropriately in the design process. Co-model-based methods must therefore integrate with design workflows, permit the analysis of a range of design alternatives at strategic points during development, and help provide the evidence that is needed to justify the dependance placed on CPS once created.

In the rail example introduced in Section 8.3, there is a need to model both the logic of the interlocking transfer functions in the distributed solution and the physics of trains on the track, given the track topography. Co-modeling linking discrete event models of the former with CT models of the latter are appropriate here. It is even conceivable that analysis of trade-offs between design alternatives could influence the design of both computing and physical elements.

8.6.2 Workflows

Within the application domains represented in INTO-CPS, we see mainly iterative and V life cycles, sometimes involving the development of an entire system *ab initio*, but often integrating existing components contributed by multiple engineers. Workflows are managed by a variety of tools ranging from spreadsheets or manual documents in word processors to special-purpose systems such as Rational DOORS. There is considerable diversity in development practices, and it is not our aim to mandate a specific process to replace established approaches wholesale. Indeed, systems and software process standards such as IEEE 15288 [42] and 12207 [43] do not specify particular life cycles but do identify development activities that go to make up a development process. Typically, these include requirements definition and analysis, architectural design, detailed design, implementation and integration, and operation and maintenance. Within each of those activities, we may expect to see transverse tool chains, as alternative tools are required to manage the construction, simulation, and analysis of diverse constituent models. Between activities, we might expect to see longitudinal tool chains.

The process of co-model construction may depend on a range of factors, including the development team capabilities available at any time and the use of legacy models. In the case of embedded systems, guidelines have been developed for co-model construction in a range of such scenarios [44].

Figure 8.6 illustrates a possible flow for the development of CPS, supported by integrated tools. In addition to Model-in-the-Loop (MiL) co-simulation, the figure also shows SiL and HiL simulations, enabling a gradual transition of either DE or CT elements from models to their corresponding realizations. In order for this to work appropriately, it is not enough just to have competent co-models; one also needs to consider their fidelity in terms of the accuracy with which they predict the behavior of both cyber and physical elements of the CPS. Refinement theories enabled by the semantic approaches outlined in Section 8.5 allow the relationship between abstract models and successively more concrete counterparts to be addressed. Considering our baseline technologies, we might see architecture models represented in SysML in the Modelio tool, with detailed modeling of cyber elements using VDM in Overture, and models of physical elements and the environment in 20-sim and/or OpenModelica as introduced in Section 8.4.

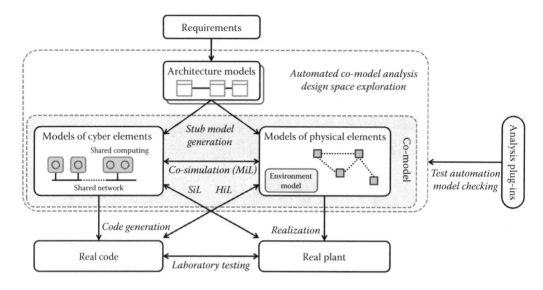

FIGURE 8.6 Design workflow features in a co-modeling environment [45].

CPS can be large-scale distributed integrations of diverse elements that include existing systems, devices, and infrastructures that are independently owned and managed. Reliance becomes placed on the behaviors that emerge from the interactions between these diverse elements, and so CPS have characteristics in common with *systems of systems* [46]. Among these, independence of ownership and management means that the development of co-models might take place as a collaboration between distributed and possibly independent development teams. Within model construction, there is, therefore, a need to support distributed specification of interfaces and to manage the results of co-simulation [47].

8.6.3 Design Space Exploration

A significant goal of our research is to support the exploration of alternative system architectures at the modeling level, allocating and reallocating responsibilities to cyber and physical elements in an effort to deliver the functional and extra-functional behaviors required at the CPS level. We use the term *DSE* to refer to the construction and evaluation of a range of designs of alternative co-models, with an assessment of the outcomes to inform the selection of a preferred alternative. These types of activities are normally undertaken in an informal or ad hoc manner, but the use of well-founded co-simulation technology enables scripted simulations of co-models with a range of values for design parameters, with (optionally, weighted) evaluations of the performance of the co-model on each run.

A number of candidate design parameters can be considered. Within the DE elements, one might wish to consider different control loop frequencies, especially if there is interaction with the physical environment. On the CT element side, parameters might include physical configurations with positioning of sensors or actuators as well as alternative equipment, for example. In our rail example, DSE could be of value in arriving at optimal and safe allocations of responsibilities between the distributed controllers (Figure 8.3). Variations might be considered in terms of the number and positioning of distributed controllers.

Exploration of large design spaces caused by large numbers and ranges of design parameters presents challenges. However, candidate techniques such as Taguchi tables, simulated annealing, etc., have the potential to help manage this. Tool performance is always going to be under pressure, however.

8.6.4 Model Management, Traceability, and Provenance

There are several reasons for taking a disciplined approach to model management during the development of dependable CPS. Co-models are composed of a numbers of diverse constituent models, each of which may have been prepared using different versions of different tools. Dynamic or static analyses conducted on the co-models may have been performed by various tools using specific input data or parameters.

Models produced by individual development engineers or teams will evolve continuously, and requirements will change during development. It is therefore essential to keep track of the increasingly complex set of models and their constituent parts as the overall CPS design set changes, in order to avoid costly redesign. Perhaps most importantly, dependability requires that reliance can *justifiably* be placed on the functioning of CPS. It is not enough to simply engineer a dependable CPS; it is imperative to record and maintain the rationale and evidence required to justify that claim as part of safety and certification criteria.

Traceability refers to the existence of evidence of an association between elements of the design set. In a document-centric approach to CPS development, traceability matrices are maintained, providing bidirectional links mapping sources to code via levels of design and software requirements to test cases [48]. Maintaining these matrices can be labor-intensive, but no tools maintain this automatically [49], limiting the support available for regression testing and evidence generation in the complex space of CPS models. Some techniques are promising and merit investigation in INTO-CPS. In particular, the W3C provenance notation (PROV-N) model [50] permits the maintenance of graph representations of relations between artifacts. From a tool support perspective, graph queries and abstractions are necessary to reduce potentially large provenance graphs to smaller but semantically correct versions [51].

8.7 CPS CO-SIMULATION TOOL SUPPORT

8.7.1 The FMI Standard

The starting point for enabling the production of the tool chain vision presented previously is the FMI standard version 2.0 for co-simulation.[*] However, there is a working group discussing enhancements of the standard, and here the INTO-CPS project is also represented in order to influence the future improvements that will be made to the standard. We (and others [52]) have already identified that a number of the parameters in the FMI standard that are made optional need to be present for our COE to function efficiently, inspired by Broman [53,54]. Here we wish to use the formalization of the core COE semantics from Section 8.5 as a kind of oracle to test the realization of the COE. This will enable us to

---

[*] See https://www.fmi-standard.org/downloads for more information.

explore optimizations for example using concurrency and a multiprocessor platform in a semantically sound fashion.

8.7.2 Tool Usage for the Distributed Interlocking Case Study

Considering the railway case study introduced in Section 8.3, it is envisaged that the tool will be able to support this all the way from requirements expressed in SysML using Modelio via heterogeneous models to a final realization of the system, with full traceability of all the artifacts produced in the process. The intent is to model the centralized interlocking both at the cyber side (the interlocking logic) as well as at the physical side (the trains and the tracks) using the different baseline tools introduced in Section 8.4. Here the timing delays on both sides will be central to ensuring a sufficiently high level of fidelity of the model for it to be trustworthy. The initial way of assessing the models produced will be using MiL simulations. High-level consistency requirements for avoiding collisions between trains will be formulated, and the simulators will be monitoring this continuously. Because of the complexity of the full scale, system scalability and speed of the collaborative simulation will be essential parameters for this case study. Thus, it is also envisaged that using code generation for the discrete event models can be a way to speed up the overall simulation time.

In order to certify a system like this interlocking system, different safety standards needs to be followed. In the INTO-CPS project we will investigate whether it is possible to produce results from the MBD analysis that will be good enough to serve as elements in the safety case needed for the application. This also means that fault modeling and investigating how to make the overall system tolerant against the most important potential faults will be taken into account. Here the accuracy of the model predictions in relation to the real physical behavior will be of paramount importance for the success of creating such a safety case.

8.8 CONCLUDING REMARKS

In this chapter we have presented what we believe will be the technology to take collaborative modeling and analysis of CPS to a new level. We will build an open, well-founded tool chain for a multidisciplinary, model-based design of CPS that covers the full development life cycle of CPS. The tool chain will support multiple modeling paradigms and will cover multiple development activities, including requirements modeling, analysis, simulation, validation, verification, and traceability of artifacts throughout all development activities across disciplinary boundaries.

This tool chain will be based on mathematical foundations formulated in UTP to support CPS co-modeling. This will include semantics for FMI co-simulation, as well as SysML, discrete-event, and CT paradigms.

The INTO-CPS methodology will be developed to ensure that adoption of the tool chain is cost-effective, providing industrial users with pragmatic guidance in determining the best modeling technologies and patterns to meet their needs.

In the INTO-CPS project we will supply four complementary industry case studies, which have been selected from four distinct domains that currently experience pressure to develop reliable CPS (automotive, agricultural, railways, and building automation). In this chapter we have focused mainly on the railway sector. However, in general the case studies will be used to drive the production of the tools and methods as well as evaluate them. Finally, we expect to establish an association that will allow future case study owners access to information, training, and competitively priced licenses at various levels of support. Tool vendors will be offered services to help integrate their products into the tool chain.

In the future we expect that many more baseline tools will be incorporated into the INTO-CPS tool chain suggested in this chapter. We expect that the future directions of the work here will be driven by the needs both of the end-users of the INTO-CPS project as well as the members of an Industrial Follower Group (IFG).

ACKNOWLEDGMENTS

The work presented here is partially supported by the INTO-CPS project funded by the European Commission's Horizon 2020 program under grant agreement number 664047. We would also like to thank all of the participants in the INTO-CPS project as well as the support from the members of the IFG. Finally, we would like to thank Nick Battle and Stefan Hallerstede for valuable review comments on earlier drafts of this chapter.

REFERENCES

1. INCOSE. *Systems Engineering Handbook. A Guide for System Life Cycle Processes and Activities, Version 3.2.2.* Technical Report INCOSE-TP-2003-002-03.2.2, International Council on Systems Engineering (INCOSE), October 2011.
2. A. Avizienis, J.-C. Laprie, B. Randell, and C. Landwehr. Basic concepts and taxonomy of dependable and secure computing. *IEEE Transactions on Dependable and Secure Computing*, 1:11–33, 2004.
3. E.A. Lee. CPS foundations. In *Proceedings of the 47th Design Automation Conference*, DAC '10, pp. 737–742, New York, NY, USA, 2010. ACM.
4. M.V. Cengarle, S. Bensalem, J. McDermid, R. Passerone, A. Sangiovanni-Vincentelli, and M. Törngren. Characteristics, capabilities, potential applications of cyber-physical systems: A preliminary analysis. Project Deliverable D2.1, EU Framework 7 Project: Cyber-Physical European Roadmap & Strategy (CyPhERS), November 2013.
5. M. Törngren, S. Bensalem, M.V. Cengarle, D.-J. Chen, J. McDermid, R. Passerone, A. Sangiovanni-Vincentelli, and T. Runkler. CPS: State of the Art. Project Deliverable D5.1, EU Framework 7 Project: Cyber-Physical European Roadmap & Strategy (CyPhERS), March 2014.
6. B. Penzenstadler and J. Eckhardt. A requirements engineering content model for cyber-physical systems. *Requirements Engineering for Systems, Services and Systems-of-Systems (RES4)*, pp. 20–29, 2012 IEEE Second Workshop on, Chicago, IL, 2012, doi: 10.1109/RES4.2012.6347692.
7. OMG Systems Modeling Language (OMG SysML^{TM}). Technical Report Version 1.3, SysML Modelling team, June 2012. http://www.omg.org/spec/SysML/1.3/.
8. Thomas A. Henzinger and Joseph Sifakis. The Embedded Systems Design Challenge. In *FM 2006: Formal Methods, 14th International Symposium on Formal Methods, Hamilton, Canada, August 21-27, 2006, Proceedings*, pages 1–15, 2006.
9. Acatech. *Cyber-Physical Systems—Driving Force for Innovation in Mobility, Health, Energy and Production (Acatech Position Paper)*. Technical Report, Acatech, 2011.
10. M. Broy. Engineering cyber-physical systems: Challenges and foundations. In M. Aiguier, Y. Caseau, D. Krob, and A. Rauzy, editors, *Complex Systems Design & Management*, pp. 1–13. Springer, Berlin Heidelberg, 2013.

11. M. Broy, M.V. Cengarle, and E. Geisberger. Cyber-physical systems: Imminent challenges. In R. Calinescu and D. Garlan, editors, *Large-Scale Complex IT Systems. Development, Operation and Management*, volume 7539 of *Lecture Notes in Computer Science*, pp. 1–28. Springer, Berlin Heidelberg, 2012.

12. E.A. Lee. *Cyber Physical Systems: Design Challenges.* Technical Report UCB/EECS-2008-8, EECS Department, University of California, Berkeley, Jan 2008.

13. K. Wan, D. Hughes, K.L. Man, and T. Krilavicius. Composition challenges and approaches for cyber physical systems. In *Networked Embedded Systems for Enterprise Applications (NESEA), 2010 IEEE International Conference on*, pp. 1–7, Suzhou, China, 2010.

14. P. Derler, E.A. Lee, and A. Sangiovanni-Vincentelli. Modeling cyber-physical systems. *Proceedings of the IEEE (special issue on CPS)*, 100(1):13–28, January 2012.

15. P. Derler, E.A. Lee, and A.L. Sangiovanni-Vincentelli. *Addressing Modeling Challenges in Cyber-Physical Systems.* Technical Report UCB/EECS-2011-17, EECS Department, University of California, Berkeley, Mar 2011.

16. I. Horvath and B.H.M. Gerritsen. Outlining nine major design challenges of open, decentralized, adaptive cyber-physical systems. In *Proceedings of the ASME 2013 International Design Engineering Technical Conferences and Computers and Information in Engineering Conference IDETC/CIE 2013*, Portland, Oregon, USA, August 2013.

17. K. Bauer. A new modelling language for cyber-physical systems. PhD thesis, Kaiserslautern University, 2012.

18. X. Zheng, C. Julien, M. Kim, and S. Khurshid. *On the State of the Art in Verification and Validation in Cyber Physical Systems.* Technical Report TR-ARiSE-2014-001, The University of Texas at Austin, The Center for Advanced Research in Software Engineering, 2014.

19. A. Sangiovanni-Vincentelli, W. Damm, and R. Passerone. Taming Dr. Frankenstein: Contract-based design for cyber-physical systems. *European Journal of Control*, 18(3):217–238, 2012.

20. CyPhERS: Cyber-Physical European Roadmap & Strategy. *Deliverable D5.1, CPS: State of the Art.* Technical Report, Project co-funded by the European Unions Seventh Framework Programme (FP/2007-2013), 2014.

21. J. El-Khoury, F. Asplund, M. Biehl, F. Loiret, and M. Törngren. A roadmap towards integrated CPS development environments. In *First Open EIT ICT Labs Workshop on Cyber-Physical Systems Engineering*, Trento, Italy, 2013.

22. B. Wang and J.S. Baras. HybridSim: A modeling and co-simulation toolchain for cyber-physical systems. In *17th IEEE/ACM International Symposium on Distributed Simulation and Real Time Applications, DS-RT 2013*, Delft, The Netherlands, October 30–November 1, 2013, pp. 33–40. IEEE Computer Society, 2013.

23. T. Blochwitz. Functional mock-up interface for model exchange and co-simulation. https://www.fmi-standard.org/downloads, July 2014. Torsten Blochwitz Editor.

24. D. Broman, C. Brooks, L. Greenberg, E.A. Lee, M. Masin, S. Tripakis, and M. Wetter. Determinate composition of fmus for co-simulation. In *Embedded Software (EMSOFT), 2013 Proceedings of the International Conference on*, pp. 1–12, Montreal, Canada, September 2013.

25. P.J. Mosterman, J. Sztipanovits, and S. Engell. Computer-automated multiparadigm modeling in control systems technology. *IEEE Transactions on Control Systems Technology*, 12(2):223–234, 2004.

26. Y. Kesten and A. Pnueli. Timed and hybrid statecharts and their textual representation. In J. Vytopil, editor, *Formal Techniques in Real-Time and Fault-Tolerant Systems, Second International Symposium*, Nijmegen, The Netherlands, January 8–10, 1992, *Proceedings*, volume 571 of *Lecture Notes in Computer Science*, pp. 591–620. Springer, 1992.

27. O. Maler, Z. Manna, and A. Pnueli. From timed to hybrid systems. In J.W. de Bakker, C. Huizing, W.P. de Roever, and G. Rozenberg, editors, *Real-Time: Theory in Practice, REX Workshop*, Mook,

The Netherlands, June 3–7, 1991, *Proceedings*, volume 600 of *Lecture Notes in Computer Science*, pp. 447–484. Springer, 1992.

28. R. Alur, C. Courcoubetis, N. Halbwachs, T. A. Henzinger, P.-H. Ho, X. Nicollin, A. Olivero, J. Sifakis, and S. Yovine. The algorithmic analysis of hybrid systems. *Theoretical Computer Science*, 138:3–34, 1995.

29. L.P. Carloni, R. Passerone, A. Pinto, and A.L. Sangiovanni-Vincentelli. Languages and tools for hybrid systems design. *Foundations and Trends in Electronic Design Automation*, 1(1/2):1–193, 2006.

30. P. Fritzson and V. Engelson. Modelica—A unified object-oriented language for system modelling and simulation. In *ECCOP '98: Proceedings of the 12th European Conference on Object-Oriented Programming*, pp. 67–90. Springer-Verlag, Brussels, Belgium, 1998.

31. E.A. Lee and H. Zheng. Operational semantics of hybrid systems. In *Hybrid Systems: Computation and Control (HSCC), volume LNCS 3414*, pp. 25–53. Springer-Verlag, Zurich, Switzerland, 2005.

32. R. Alur, R. Grosu, Y. Hur, V. Kumar, and I. Lee. Modular specification of hybrid systems in charon. In N.A. Lynch and B.H. Krogh, editors, *Hybrid Systems: Computation and Control, Third International Workshop, HSCC 2000*, Pittsburgh, PA, USA, March 23–25, 2000, *Proceedings*, volume 1790 of *Lecture Notes in Computer Science*, pp. 6–19. Springer, 2000.

33. T.A. Henzinger. Masaccio: A formal model for embedded components. In J. van Leeuwen, O. Watanabe, M. Hagiya, P.D. Mosses, and T. Ito, editors, *Theoretical Computer Science, Exploring New Frontiers of Theoretical Informatics, International Conference IFIP TCS 2000*, Sendai, Japan, August 17–19, 2000, *Proceedings*, volume 1872 of *Lecture Notes in Computer Science*, pp. 549–563. Springer, 2000.

34. A. Deshpande, A. Göllü, and P. Varaiya. Shift: A formalism and a programming language for dynamic networks of hybrid automata. In P.J. Antsaklis, W. Kohn, A. Nerode, and S. Sastry, editors, *Hybrid Systems IV*, volume 1273 of *Lecture Notes in Computer Science*, pp. 113–133. Springer, Berlin Heidelberg, 1997.

35. F.D. Torrisi and A. Bemporad. HYSDEL—A tool for generating computational hybrid models for analysis and synthesis problems. *IEEE Transactions on Control Systems Technology*, 12(2):235–249, 2004.

36. Leonardo Mendonça de Moura, S. Owre, H. Rueß, J.M. Rushby, N. Shankar, M. Sorea, and A. Tiwari. Sal 2. In R. Alur and D. Peled, editors, *Computer Aided Verification, 16th International Conference, CAV 2004*, Boston, MA, USA, July 13–17, 2004, *Proceedings*, volume 3114 of *Lecture Notes in Computer Science*, pp. 496–500. Springer, 2004.

37. T. Hoare and H. Jifeng. *Unifying Theories of Programming*. Prentice Hall, London, April 1998.

38. J. Woodcock and J. Davies. *Using Z—Specification, Refinement, and Proof*. Prentice Hall International Series in Computer Science, Upper Saddle River, NJ, USA, 1996.

39. T. Nipkow, M. Wenzel, and L.C. Paulson. *Isabelle/HOL: A Proof Assistant for Higher-Order Logic*. Springer-Verlag, Berlin, Heidelberg, 2002.

40. L.C. Paulson. Three years of experience with sledgehammer, a practical link between automatic and interactive theorem provers. In R.A. Schmidt, S. Schulz, and B. Konev, editors, *Proceedings of the 2nd Workshop on Practical Aspects of Automated Reasoning, PAAR-2010*, Edinburgh, Scotland, UK, July 14, 2010, volume 9 of *EPiC Series*, pp. 1–10, 2010.

41. J. Liu, J. Lv, Z. Quan, N. Zhan, H. Zhao, C. Zhou, and L. Zou. A calculus for hybrid CSP. In K. Ueda, editor, *Programming Languages and Systems—8th Asian Symposium, APLAS 2010*, volume 6461 of *LNCS*, pages 1–15. Springer, Shanghai, China, 2010.

42. IEEE. *International Standard ISO/IEC/IEEE 15288:2015(E), Systems and Software Engineering—System Life Cycle Processes*. ISO/IEC and IEEE Computer Society, 2015.

43. IEEE. *International Standard ISO/IEC 12207:2008(E), IEEE Std 12207-2008 (Revision of IEEE/EIA 12207.0-1996) Systems and Software Engineering—Software Life Cycle Processes.* ISO/IEC and IEEE Computer Society, 2008.

44. J. Fitzgerald, Peter Gorm Larsen, K. Pierce, and M. Verhoef. A formal approach to collaborative modelling and co-simulation for embedded systems. *Mathematical Structures in Computer Science*, 23(4):726–750, 2013.

45. J. Fitzgerald, Peter Gorm Larsen, and M. Verhoef. From embedded to cyber-physical systems: Challenges and future directions. In J. Fitzgerald, Peter Gorm Larsen, and M. Verhoef, editors, *Collaborative Design for Embedded Systems—Co-Modelling and Co-Simulation*, pp. 289–298. Springer, Berlin Heidelberg, 2014.

46. C.B. Nielsen, Peter Gorm Larsen, J. Fitzgerald, J. Woodcock, and J. Peleska. Model-based engineering of systems of systems. *ACM Computing Surveys*, 48(2):18:1–18:41, September 2015.

47. C.B. Nielsen, K. Lausdahl, and Peter Gorm Larsen. Distributed simulation of formal models in system of systems engineering. In *4th IEEE track on Collaborative Modelling and Simulation in IEEE WETICE 2014*, pp. 211–216, Parma, Italy, June 2014.

48. O.C.Z. Gotel and A.C.W. Finkelstein. An analysis of the requirements traceability problem. In *Proceedings of the First International Conference on Requirements Engineering*, pp. 94–101, Colorado Springs, CO, USA, April 1994.

49. P. Mäder. *Rule-Based Maintenance of Post-Requirements Traceability*. MV Wissenschaft. MV-Verlag, 2010.

50. L. Moreau and P. Missier. *PROV-DM: The PROV Data Model*. Technical Report, World Wide Web Consortium, 2012.

51. P. Missier, J. Bryans, C. Gamble, V. Curcin, and R. Danger. ProvAbs: Model, policy, and tooling for abstracting PROV graphs. In *Procs. IPAW 2014 (Provenance and Annotations)*. Springer, Cologne, Germany, 2014.

52. B. van Acker, J. Denil, H. Vangheluwe, and P. De Meulenaere. Generation of an optimised master algorithm for FMI co-simulation. In *DEVS '15 Proceedings of the Symposium on Theory of Modeling & Simulation*, Alexandria, VA, USA, January 2015.

53. D. Broman, C. Brooks, L. Greenberg, E.A. Lee, M. Masin, S. Tripakis, and M. Wetter. Determinate composition of fmus for co-simulation. In *13th International Conference on Embedded Software (EMSOFT)*, pp. 1–12. Montreal, September 2013.

54. D. Broman, C. Brooks, L. Greenberg, E.A. Lee, M. Masin, S. Tripakis, and M. Wetter. *Determinate Composition of Fmus for Co-Simulation*. Technical Report UCB/EECS-2013-153, EECS Department, University of California, Berkeley, August 2013.

CHAPTER 9

Verifying Trustworthy Cyber-Physical Systems Using Closed-Loop Modeling

Neeraj Kumar Singh, Mark Lawford, Thomas S. E. Maibaum, and Alan Wassyng

CONTENTS

T RUSTWORTHY-CYBER PHYSICAL SYSTEMS (TCPS) are safety-critical systems, in which
  failures can lead to injuries and loss of life. These systems demand strong integration
and coordination among the computing sciences, network communication, and physical
modeling. Analyzing systems requirements is a major challenge in the area of safety-critical
software, where requirements quality is also an important issue in building a dependable
cyber-physical system. Most projects fail due to a lack of understanding of user needs, inadequate knowledge of the system's environment, and inconsistent system specifications. This
combination typically results in poor systems requirements. Since software plays such an
important role in critical systems embedded in a physical environment, it is essential that
we trace unidentified and hidden requirements by validating and checking the consistency
of the systems requirements. To this end, formal methods that model the closed-loop system are invaluable. In this chapter, we present an incremental proof-based development of a
closed-loop model of the Cardiac Resynchronization Therapy (CRT) and heart. We analyze
the prime benefits of a closed-loop modeling approach in requirements engineering to validate the appropriateness and correctness of system behaviors in the early stage of systems
development, including new research directions.

9.1 INTRODUCTION

Trustworthy cyber-physical systems (TCPS) are dependable critical systems that refer to the
tight integration of and coordination between computational and physical resources [1]. The
TCPS innovate several industrial sectors related to avionic, transportation, medical, space,
and automotive domains, in which the main research goal is to improve our own ability to
understand and exploit interfaces between the cyber and physical worlds and to analyze new
behaviors and capabilities from their seamless integration. An increasing demand for new
technology forces the rapid adoption of commercial firmware and software for TCPS. The
rapid adoption of TCPS increases vulnerabilities that could lead to devastating systems failures. A failure in these systems could result in loss of life, as well as loss of reputation and
economic damage. In fact, any failure in the medical domain is a serious public health problem and poses a threat to patient safety. For example, the US Food and Drug Administration
(FDA) has reported several recalls for cardiac pacemakers and implantable cardioverter-
defibrillators (ICDs). These recalled devices are responsible for a large number of serious
illnesses and deaths. During 1900–2002, 17,323 devices (8,834 pacemakers and 8,489 ICDs)
were explanted and 61 deaths (30 Pacemaker patients, 31 ICD patients) were reported due
to erroneous behavior according to the FDA report. The FDA found that the deaths and
adverse events associated with the cardiac pacemakers and ICDs were caused by product
design and engineering flaws, including firmware problems [2].

Software play a vital role in developing and controlling the critical embedded systems.
Over the past 40 years, formal techniques have shown some promising results in several

domains, including healthcare, automotive, avionic, and nuclear by identifying possible errors through formal reasoning. The formal reasoning has great impact in developing systems requirements or checking the correctness of functional requirements. In current industrial practices, formal methods have been used to meet the standard requirements or certification requirements. For example, ISO 26262 [3] standard has adopted formal methods to design a passenger vehicle, particularly to meet the safety requirements of Automotive Safety Integrity Level (ASIL) D. Validation of requirements specification is an integral and essential part of requirements engineering. Validation is a process of checking, together with stakeholders, whether the requirements specification meets its stakeholders' intentions and expectations [4].

Ideally, emergent behaviors of a given physical environment (e.g., the heart in the case of the cardiac pacemaker), missing requirements, or inconsistencies can be identified early on, in order to develop safe and dependable systems. We can then look beyond the system itself and into its working environment, including human interaction, to specify and verify the given systems requirements. In this chapter, we demonstrate the results of our new work on the formalization of a closed-loop model of the *Cardiac Resynchronization Therapy (CRT)* pacemaker and heart. The closed-loop model is an integration of a systems model (CRT pacemaker) and an environment model (heart), in which both the system and environment models are formalized using formal techniques. For developing the closed-loop model, we use the Event-B modeling language that supports stepwise refinement. This stepwise refinement is used to introduce safety properties at each layer of refinement to guarantee the safe behavior of the CRT pacemaker and to help in the certification of the CRT pacemaker. The closed-loop modeling approach also helps to establish confidence in the early stage of systems development by providing required safety properties in the virtual environment. Our main objectives and contributions are given below. We suggest consideration of all these objectives for developing any other TCPS that will use the closed-loop modeling approach for verifying systems requirements using formal techniques.

1. Closed-loop modeling in the early stage of TCPS development

2. Identifying gaps or inconsistencies in the requirements of TCPS

3. Verifying and validating the behavior requirements of TCPS

4. Strengthening the given TCPS requirements

5. Supporting "what-if" analysis during the formal reasoning of TCPS

6. Traceability of missing behaviors that leave a TCPS in undesirable states

7. Automatic identification of emergent behaviors

8. Validation of the TCPS assumptions

9. Demonstrating how we can help to meet the FDA requirements for certifying the medical TCPS using the closed-loop formal modeling

The structure of the chapter is as follows. In Section 9.2, we review preliminary material: requirement engineering and CRT pacemaker. Section 9.3 presents an environment modeling of the heart, and the CRT pacemaker control requirements are presented in Section 9.4. Section 9.5 explores an incremental proof-based formal development of a closed-loop system of the CRT pacemaker and heart. A brief discussion is provided in Section 9.6. Section 9.7 presents related work and, finally, in Section 9.8, we conclude the chapter.

9.2 PRELIMINARIES

9.2.1 Requirements Engineering

The Institute of Electrical and Electronics Engineers (IEEE) defines a requirement as a condition or capability that must be met or possessed by a system or system components to satisfy the contract, standard, specification, or other formally imposed document [5]. Requirements engineering is a branch of software engineering that allows the use of systematic techniques to analyze systems requirements for checking the required properties of completeness and consistency of a given system [6]. Requirements engineering is a complex process that contains several small steps, such as elicitation, specification, validation, analysis, and management, for developing a system. In these steps, requirements elicitation is a process for identifying, reviewing, checking, and documenting the stakeholder requirements; requirements specification is used to precisely document stakeholder needs and constraints using formal or semiformal techniques; requirements analysis checks the stakeholder requirements and system constraints using formal and informal techniques; requirements verification ensures the completeness, correctness, comprehensibility, and consistency of system behavior according to stakeholders; and, finally, requirements management is used for managing, coordinating, and documenting the system development life cycle [7,8].

Requirements analysis plays an important role in determining the required properties of the systems and software development process. In addition, it also supports feedback mechanisms to improve the systems requirements by incorporating useful information [9]. There are several techniques that help improve the quality of requirements for both the ordinary and dependable systems. In our work, we present a conceptual treatment for analyzing the systems requirements by developing a closed-loop model of TCPS and a virtual operating environment for identifying the emergent properties and peculiar requirements, which eventually provide us with a theoretical lens with which to examine this adoption in a systematic manner [10]. This approach is useful for analyzing the systems requirements as long as its adoption decision is present, preferably during the early stages of systems development, when we need to understand how a decision on analyzing requirements is made and which factors influence adoption of the requirements engineering.

9.2.2 CRT Pacemaker

The cardiac pacemaker is a complex electronic device that is designed to maintain an adequate heart rate in cases of *bradycardia*. This device is equipped with a microprocessor that controls heart rhythm intelligently by observing the actual behavior of the heart. The pacemaker generally serves two main roles, known as *pacing* and *sensing*. A sensor is used to sense the heart's intrinsic activity, and an actuator is used to deliver a short intense electrical pulse to the heart. Several sensors and actuators work together to sense and actuate into multiple heart chambers, and all these sensors and actuators are controlled by a microprocessor [11].

A Cardiac Resynchronization Therapy (CRT) or multisite pacing device is one of the advanced pacemakers that is designed to maintain heart rate by treating a specific form of heart failure—poor synchronization of the two lower heart chambers. This type of device has three electrodes, equipped with sensors and actuators, for the right atrium, right ventricle, and left ventricle. The sensors of this device sense intrinsic activities from the chambers and the actuators deliver a short intense electrical pulse to the chambers of the heart when needed to help them beat together synchronously. The basic elements of a CRT pacemaker are as follows:

1. *Leads:* A set of insulated flexible wires for transmitting electrical impulses between microprocessor and heart to fulfill the requirements of pacing and sensing.

2. *The CRT Generator:* A metal case that contains the microprocessor and battery. The microprocessor is also called the brain of the CRT pacemaker, because it controls all system functionalities.

3. *Device Controller-Monitor (DCM):* This is an external device that communicates with an implanted CRT pacemaker through a wireless connection, and it helps set new parameters, changing configurations and monitoring the actual behavior of the heart and implanted CRT pacemaker.

4. *Accelerometer:* This is a specific sensor that is used to measure body motion or dynamic activities to allow modulated pacing and sensing to control the heart rhythm according to a patient's physical needs.

9.2.3 Definition of the NBG Code

The NASPE/BPEG generic (NBG) code is summarized in Table 9.1. There are five columns in the table, where each column represents a sequence of letters for presenting the different types of operating modes. The first letter of the code indicates the chamber(s) in which pacing occurs, the second letter indicates the chamber(s) in which sensing occurs, and the third letter indicates each instance of sensing on the triggering or inhibition of subsequent pacing stimuli. The fourth letter is optional, and indicates the presence (R) or absence (O) of an adaptive rate modulation. The last letter indicates whether multisite pacing is present in (O) none of the cardiac chambers, (A) one or both atria, (V) one or both ventricles, and (D) any combination of atria and ventricles.

9.2.4 Event-B

For our work, we chose the Event-B modeling language [12,13] that enables us to formalize a system using a *correct by construction* approach. The *correct by construction* approach allows us to design a complex system incrementally by adding concrete details in each new refinement level. The incremental development gradually builds a concrete system by

TABLE 9.1 The NASPE/BPEG Generic Code for Antibradycardia Pacing

Chambers Paced	Chambers Sensed	Response to Sensing	Rate Modulation	Multisite Pacing
O-None	**O**-None	**O**-None	**O**-None	**O**-None
A-Atrium	**A**-Atrium	**T**-Triggered	**R**-Rate Modulation	**A**-Atrium
V-Ventricle	**V**-Ventricle	**I**-Inhibited		**V**-Ventricle
D-Dual(A+V)	**D**-Dual(A+V)	**D**-Dual(T+I)		**D**-Dual(A+V)

introducing new safety properties and checking the correctness of required behavior at each refinement layer. The basic system modeling components are *context* and *machine*. The main elements of the *context* are *carrier set*, *constant*, *axiom*, and *theorem* that, taken together, describe the static properties of a system. The main elements of another component *machine* are *variable*, *invariant*, *event*, and *theorem*, that specify the dynamic properties of a system. An *event* is composed of the *guard* and *action* that model a changing state of a system. At each refinement step, we can either refine abstract events or introduce a new set of variables, invariants, and events. In addition, we can also introduce new safety properties and theorems for developing a safe system. The abstract events can be refined by (1) keeping the event as it is by strengthening the abstract guards; (2) splitting an event into several events; or (3) refining it into several other events to specify a change of the state variables. At each new refinement level, the developing system always preserves its abstract functional behavior and required safety properties.

Rodin [13] is an open source Eclipse-based Integrated Development Environment (IDE) for developing the Event-B models. This is a collection of plug-ins that support model management, model development, refinement based modeling, model composition/decomposition, proof obligation generation, discharging the generated proof obligations using automated theorem provers, and code generation. Due to page limitation, we will not discuss the Event-B modeling language in detail. There are several publications and books [12,13] available for fundamental and refinement strategies for gaining experience and knowledge in Event-B.

9.3 ENVIRONMENT MODELING (THE HEART)

The heart, a biological muscular organ, pumps blood to circulate throughout the entire body. It consists of four chambers. The left and right atrium chambers collect blood and pump it into the lower ventricle chambers to pump blood out to the lungs and other parts of the body. The heart requires an electrical stimulus to contract and relax periodically. This stimulus is generated by the small mass of specialized tissue called the *sinus node*, and travels down through the conduction network. The flow of an electrical impulse varies, and it is time-dependent to synchronize the heart chambers: atria and ventricles. For example, the atria contract before the ventricles, so that the blood pumps out from the atria and into the ventricles. The basic components of the heart are depicted in Figure 9.1a. To model the heart behavior abstractly, we consider a set of landmark nodes (A, B, C, D, E, F, G, and H) on the conduction network (see Figure 9.1b). All these landmarks are identified through a survey of the literature [14–17] and extensive discussions with cardiologists and physiologists.

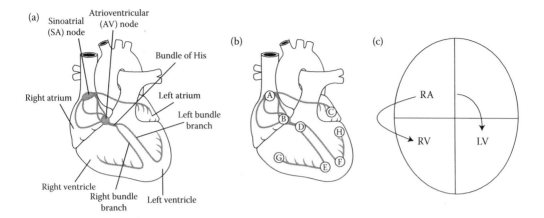

FIGURE 9.1 The electrical conduction and landmarks of the heart [18,19]. (a) Basic electrical conduction system. (b) Landmarks in networks. (c) Biventricular pacing.

In this section, we present a formal definition of the heart, and the required properties related to impulse propagation time and impulse propagation speed. Moreover, we also discuss heart blocks and cellular automata that are used for correctly specifying heart behavior. This brief introduction allows users to understand the modeling concepts of the heart and the approach used to develop a closed-loop model together with the CRT pacemaker. A detailed description of the heart and formalization steps are available in References 18 through 20. In the following discussion we introduce only the elements necessary to formally define the heart. The required definitions and figures are taken from the literature [18,19].

Definition 9.1 (The Heart). *Given a set of nodes N, a transition (conduction) t is a pair (i, j), with i, j ∈ N. A transition is denoted by i ⤳ j. The heart system is a tuple HSys = (N, T, N₀, TW$_{time}$, CW$_{speed}$) where:*

- $N = \{A, B, C, D, E, F, G, H\}$ *is a finite set of landmark nodes in the conduction pathways of the heart.*

- $T \subseteq N \times N = \{A \leadsto B, A \leadsto C, B \leadsto D, D \leadsto E, D \leadsto F, E \leadsto G, F \leadsto H\}$ *is a set of transitions to represent electrical impulse propagation between two landmark nodes.*

- $N_0 = A$ *is the initial landmark node (SA node).*

- $TW_{time} \in N \rightarrow TIME$ *is a weight function as time delay of each node, where TIME is time delay in range.*

- $CW_{speed} \in T \rightarrow SPEED$ *is a weight function as impulse propagation speed of each transition, where SPEED is propagation speed in range.*

Property 9.1 (Impulse Propagation Time). *In the biological heart, an electrical impulse originates from the SA node (node A) and then travels through the conduction network* *and terminates in the atrial muscle fibers (node C) and at the end of the Purkinje fibers into both sides of the ventricular chambers (node G and node H). The impulse propagation time delay differs for each landmark node (N). The impulse propagation time is represented as a total function $TW_{time} \in N \to \mathbb{P}(0..230)$. The impulse propagation time delay for each node is represented as $TW_{time}(A) = 0..10$, $TW_{time}(B) = 50..70$, $TW_{time}(C) = 70..90$, $TW_{time}(D) = 125..160$, $TW_{time}(E) = 145..180$, $TW_{time}(F) = 145..180$, $TW_{time}(G) = 150..210$, and $TW_{time}(H) = 150..230$.*

Property 9.2 (Impulse Propagation Speed). *Similar to the impulse propagation time, the impulse propagation speed also differs for each transition ($i \rightsquigarrow j$, where $i, j \in N$). The impulse propagation speed is represented as a total function $CW_{speed} \in T \to \mathbb{P}(5..400)$. The impulse propagation speed for each transition is represented as $CW_{speed}(A \rightsquigarrow B) = 30..50$, $CW_{speed}(A \rightsquigarrow C) = 30..50$, $CW_{speed}(B \rightsquigarrow D) = 100..200$, $CW_{speed}(D \rightsquigarrow E) = 100..200$, $CW_{speed}(E \rightsquigarrow G) = 300..400$, and $CW_{speed}(F \rightsquigarrow H) = 300..400$.*

The sinoatrial (SA) node spontaneously emits some electrical current that spreads through the walls of the atria, causing them to contract. This SA node, known as the physiological pacemaker of the heart and is represented by node A in Figure 9.2a is responsible for maintaining the heart rhythm. From the SA node, an electrical impulse propagates through the atria chambers and reaches nodes B and C (see Figure 9.2b) at the end of the muscle fibers without crossing the boundary between atria and ventricles.

An electrical impulse generated from the SA node only enters through the atrioventricular (AV) node. The AV node, labeled node B (see Figure 9.1b), is located at the boundary between atria and ventricles. There is a small delay at this node to synchronize the atria and ventricles so that blood will flow effectively. The distal portion of the AV node is made of the bundle of His denoted as the landmark node D (see Figure 9.1b). The bundle of His splits into two branches in the interventricular septum, the left bundle branch and the right bundle branch. Electrical impulses enter at the base of the ventricle at the Bundle of His (node D) and then follow the left and right bundle branches along the interventricular septum (see Figure 9.2c).

Two separate, left and right, bundle branches propagate together on each side of the septum. Two landmark nodes on the lower area of the heart near the left and right bundle branches are denoted as E and F (see Figure 9.1b). The specialized fibers of left and right bundle branches conduct an impulse rapidly; the left bundle branch activates the left ventricle and the right bundle branch activates the right ventricle (see Figure 9.2d).

The bundle branches are divided into an extensive system of Purkinje fibers that conduct the impulses at high velocity (see Table 9.2) throughout the ventricles. The Purkinje fibers stimulate contraction of individual groups of myocardial cells. Two landmark nodes, G and H (see Figure 9.1b), are denoted at the end of the Purkinje fibers in the ventricles (see Figure 9.2e). At the end of the Purkinje fibers, the electrical impulse is transmitted through the ventricular muscles [15,17].

The heart's electrical behavior plays an important role in the synchronization of atria and ventricles and it helps to optimize the haemodynamic. Minor changes in conduction time

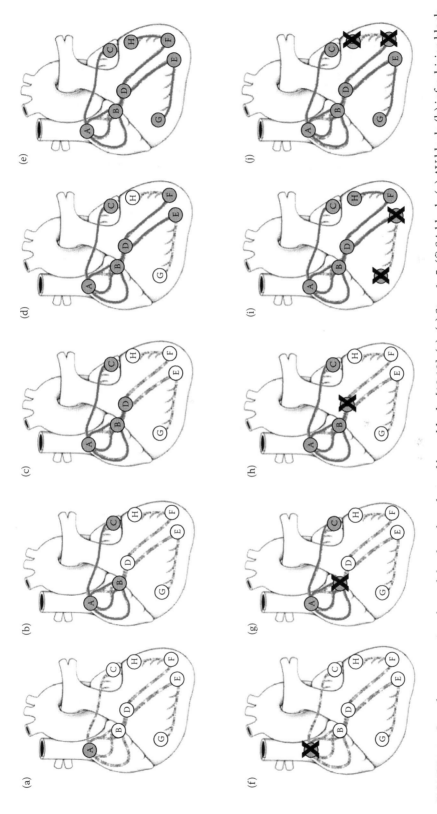

FIGURE 9.2 Impulse propagation through landmark nodes and heart blocks [18,19]. (a)–(e) Step 1–5, (f) SA block, (g) AV block, (h) infra-hisian block, (i) right bundle branch block, and (j) left bundle branch block.

TABLE 9.2   Cardiac Activation Time and Cardiac Velocity

Location in the Heart	Cardiac Activation Time (ms)	Location in the Heart	Conduction Velocity (cm/s)
SA node (A)	0..10	A ⤳ B	30..50
Left atrium muscle fibers (C)	70..90	A ⤳ C	30..50
AV node (B)	50..70	B ⤳ D	100..200
Bundle of His (D)	125..160	D ↦ E	100..200
Right bundle branch (E)	145..180	D ⤳ F	100..200
Left bundle branch (F)	145..180	E ⤳ G	300..400
Right Purkinje fibers (G)	150..210	F ⤳ H	300..400
Left Purkinje fibers (H)	150..230		

Source: Adapted from R. Plonsey and J. Malmivuo. *Bioelectromagnetism.* Oxford University Press, 1995. ISBN 0-19-505823-2.

or conduction speed between landmark nodes can cause different types of abnormalities known as *arrhythmias*. These arrhythmias can be categorized as bradycardia (slow heart rate) or tachycardia (rapid heart rate). All possible range of values for conduction speed and conduction time are given in Table 9.2.

9.3.1 Heart Block

Heart block is a disorder of impulse conduction that stimulates heart muscle contraction. The normal cardiac impulse emits from the SA node that spreads throughout the atria and ventricles. Disturbances of conduction may present as slow conduction, intermittent condition failure, or complete conduction failure. These types of conduction failures are also known as 1st, 2nd, and 3rd degree blocks. Figure 9.2 depicts different kinds of heart blocks throughout the conduction network using a set of landmark nodes.

9.3.1.1 SA Block

This type of block occurs within the SA node (A) and is known as a sinoatrial (SA) nodal block or sick sinus syndrome. In this block, the SA node fails to originate an impulse and the heart misses one or two beats at regular or irregular intervals (see Figure 9.2f).

9.3.1.2 AV Block

The AV block occurs due to conduction defects between the atria and ventricles. This block may originate in the AV node (B), bundle of His (D), or both nodes B and D (see Figure 9.2g).

9.3.1.3 Infra-Hisian Block

This type of block occurs due to a defect after the AV node (B) known as an Infra-Hisian block (see Figure 9.2h).

9.3.1.4 Left Bundle Branch Block

A left bundle branch block occurs when conduction into the left branch of the bundle of His is interrupted. A block that occurs within the fascicles of the left bundle branch is known as a hemiblock (see Figure 9.2i).

9.3.1.5 Right Bundle Branch Block

A right bundle branch block occurs when conduction into the right branch of the bundle of His is interrupted (see Figure 9.2j).

9.3.2 Cellular Automata Model

To correctly describe a biological structure at the cellular level is a challenging task. Several existing models fail to describe the correct biological structure and the required properties at the cellular level. In fact, cellular automata (CA) is considered to provide biological structure at both the cellular and subcellular levels [21].

A CA model is a set of spatially distributed cells that contains uniform connection patterns among the neighboring cells and local computation laws. In 1940, Ulam and von Neumann [22] proposed the CA for investigating the behavior of complex and distributed systems. CA is a discrete dynamic system corresponding to space and time that provides uniform properties for state transitions and interconnection patterns. A CA model can have an infinite number of cells in any dimension. In our work, we only consider a finite number of cells in two dimensions as shown in Figure 9.3a. A two-dimensional CA model is defined below.

Definition 9.2. *The Cellular Automata Model: Cellular Automata (CA)* $= \langle S, N, T \rangle$ *: Discrete Time System*

   S *: a set of states*

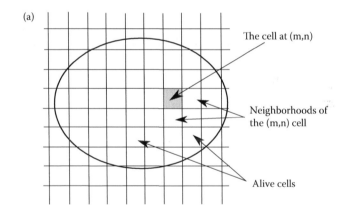

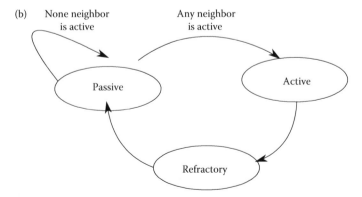

FIGURE 9.3   Two-dimensional cellular automata (CA) and state transition model [18,19]. (a) A two-dimensional cellular automata model. (b) Simple state transition.

N: *a set of neighboring patterns*

T: *a transition function*

subject to the following constraints:

$N \subseteq I^D$

$T : S^{|N|} \rightarrow S$

A typical case of the CA is realized in a D-dimensional grid, N consists of D-tuples of indices from a coordinate set. In the case of a 2D cellular model, the constraints become $N \subseteq I^2$ and $T : S^{|N|} \rightarrow S$.

To consider the automaton specified by the CA, let

$\Sigma = \{\tau | \tau : I^2 \rightarrow S\}$ *be the global state space*

$\alpha : \Sigma \rightarrow \Sigma$ *is the global transition function*

For a given global state λ (i.e., $\lambda : \Sigma$), the global transition function is

$\alpha(\lambda) = \{(i,j)|(i,j) \mapsto T(\lambda_{|N+(i,j)|})\}$

where $T(\lambda_{|N+(i,j)|})$ is application of the transition function T to a set of neighboring patterns at (i,j) of the global state λ.

The Greenberg–Hastings CA [23] is a discrete dynamical system that can be used to describe the physiological behavior of the heart in the form of electrical conduction. The cells are interconnected to each other to form a spatial space in 2D. The phase activity of cells are coded into discrete states. These discrete states are *Active, Refractory,* and *Passive,* that represent different transition states of the cells. Definition 9.3 is proposed to mimic the dynamic of a myocyte.

Definition 9.3. *State Transition of a Cell. The heart muscle is composed of heterogeneous cells, the cellular automata model of the muscle, CAM_{CA}, is characterized with no dependencies on the type of cells. CAM_{CA} is defined as follows:*

$CAM_{CA} = \langle S, N, T \rangle$

$S = \{Active, Passive, Refractory\}$

$N_{m,n} = \{(m,n), (m+1,n), (m-1,n), (m,n+1), (m,n-1)\}$

$T : S^{|N|} \rightarrow S$

$$
s_{t+1}(m,n) = \begin{cases} Refractory & \text{if } s_t(m,n) = Active \\ Passive & \text{if } s_t(m,n) = Refractory \\ Active & \text{if } s_t(m,n) = Passive \text{ and any neighbor is in an Active state} \\ Passive & \text{if } s_t(m,n) = Passive \text{ and no neighbor is in an Active state} \end{cases}
$$

where, $s_t(m,n)$ denotes the state of the cell located at (m,n).

In our formal model, we use this definition as the basis for modeling the CA. The cardiac heart muscle cells have three main states: *Active, Passive,* and *Refractory.* Initially, all the cells are in the *Passive* state. In the *Passive* state, the cells are electrical discharged and they do not affect any neighboring cells. When an electrical impulse passes through the cell is then charged and eventually activated, and the current state of the cell switches to the *Active* state. An active cell can transmit an electrical impulse to its neighboring cells, and then the active cell can switch into the *Refractory* state, in which the cell cannot be reactivated instantly. After delaying, the *Refractory* state cells can switch into the *Passive* state to await the next impulse (see Figure 9.3b).

9.4 CRT PACEMAKER CONTROL REQUIREMENTS

The CRT pacemaker is an advanced electronic device that controls the heart rate by means of sensing and pacing in various heart chambers. In this section, we describe the systems requirements for biventricular sensing with biventricular pacing (BiSP), considering other complex operating modes. BiSP allows pacing and sensing in the right atrium, left ventricle, and right ventricle (see Figure 9.1c).

Biventricular pacing coordinates the left ventricle (LV) and right ventricle (RV), and intraventricular regional wall contractions, by synchronizing with the sinus rhythm. There are various intrinsic activities related to pacing and sensing events that can reset escape intervals, such as atrioventricular interval (AVI) and ventriculoatrial interval (VAI). Biventricular pacing controls the heart rate using various combinations of timing from events in either the LV or RV. For example, the first ventricular sense either from the left or right ventricular chamber can reset the VAI, and the heart rate depends on intervals between the first ventricular events in each cycle. However, heart rate intervals can vary due to stimulation in the opposite chambers.

Minor delays between RV and LV pacing introduce complications in biventricular timings. These timings allow several definitions of escape intervals, such as AVI and VAI. The pacing rate is the sum of the VAI and AVI for dual chamber timing. The definition allows for biventricular timing of the VAI and AVI for pacing either the RV for RV-based timing (RVI), or the LV for LV-based timing (LVI). The pacing delay can be represented by the RVI–LVI interval. It can be negative, positive, or zero, as per the occurrence order of stimulation in both the left and right ventricles. Possible scenarios for biventricular sensing and pacing are depicted in Figure 9.4 that are described below assume normal sensing and pacing activities in the right atrium.

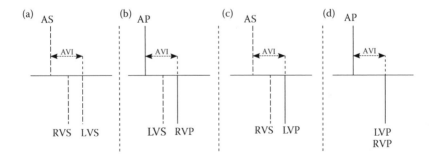

FIGURE 9.4   Possible scenarios for biventricular sensing and pacing. (a) No Pacing in RV and LV. (b) Pacing in RV. (c) Pacing in LV. (d) Pacing in RV and LV. AS = atrial sensed; AP = atrial paced; LVS = left ventricular sensed; LVP = left ventricular paced; RVS = right ventricular sensed; and RVP = right ventricular paced.

1. *Scenario A* shows the case where pacing activities are inhibited in both left and right ventricles due to the sensing of intrinsic activities in both left and right ventricles.

2. *Scenario B* shows a situation in which the CRT pacemaker paces in the right ventricle only after an AVI, while LV pacing is inhibited due to the sensing of an intrinsic activity in the left ventricle.

3. *Scenario C* shows a situation in which the CRT pacemaker paces in the left ventricle only after an AVI, while RV pacing is inhibited due to the sensing of an intrinsic activity in the right ventricle.

4. *Scenario D* shows a situation in which the CRT pacemaker paces in both left and right ventricles after an AVI in which no intrinsic heart activity is detected.

In the following section, we provide a detailed description of given scenarios for biventricular sensing and pacing in order to capture possible behavioral requirements.

- We consider biventricular sensing for RV-based timing with the positive RVI–LVI interval in the absence of intrinsic conduction (see Figure 9.5a). Following the AVI, an LVP event follows an RVP event.

- An event sense related to the right ventricle resets all the pacing intervals for both the right and left ventricles, so pacing is not allowed in the right ventricle or in the left ventricle following an RVS event. Figure 9.5b shows that an RVS event resets the timing cycle and starts a new VAI.

- An event sense related to the right ventricle may reset the VAI so the right ventricle is not paced, but pacing is allowed in the left ventricle if any intrinsic activity is not detected in that ventricle. Figure 9.5c shows that the RVS event starts a new VAI but is also followed by an LVP event (*) unless an LVS event occurs first (**). Alternatively, the VAI may be initiated from the right ventricle pace at the end of an AVI. Figure 9.5d shows that a delivery of RVP starts a new VAI but also is followed by an LVP event (*) unless an LVS event occurs first (**).

- An event sense related to the right ventricle may allow an immediate trigger to pace in the left ventricle without delaying to synchronize the left ventricle and right ventricle contractions. Figure 9.5e shows that the RVS event results in an immediate LVP event.

- Sometimes, pacing in opposite chambers can be advantageous after some delay rather than immediately. Figure 9.5f shows that an RVS event is followed by an LVP event after the modified RVI–LVI interval. This interval may be different from the RVP–LVP interval.

- An event sense related to the left ventricle in an AV or RVI–LVI interval before pacing in the left ventricle may inhibit the left ventricle pace, and the right ventricle pace would be unaffected unless the right ventricle sense occurs before the AVI. Figure 9.5g

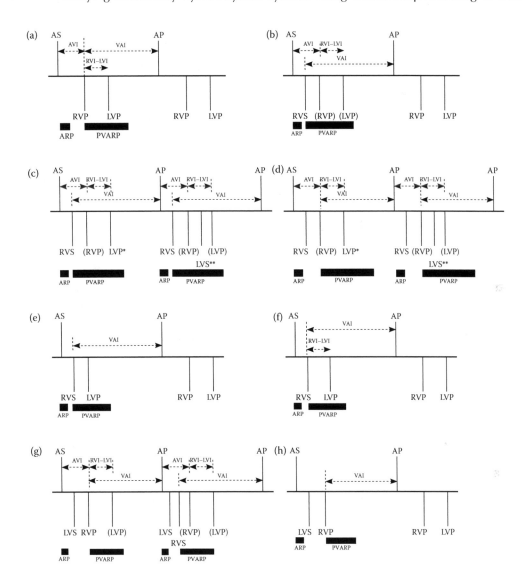

FIGURE 9.5   Biventricular sensing and pacing (BiSP) requirements. (a) BiSP in the absence of AV conduction. (b) BiSP with an RVS event. (c) The RVS event starts a new VAI in BiSP. (d) The theoretical point of delivery of RVP starts a new VAI in BiSP. (e) BiSP with triggering. (f) An RVS event is followed by an LVP event after a modified RVI–LVI interval in BiSP. (g) BiSP with RV-based timing. (h) BiSP with right ventricular-based timing and trigger.

shows that the LVS event does not inhibit the RVP event. However, if left-ventricle-to-right-ventricle conduction occurs quickly enough, the RVS event starts a new timing cycle.

- During an AVI, an event sense related to the left ventricle may trigger an immediate pace in the right ventricle, which can reset the pacing intervals. Figure 9.5h shows that the left ventricle sense event does not reset the timing cycle but initiates an immediate RVP event.

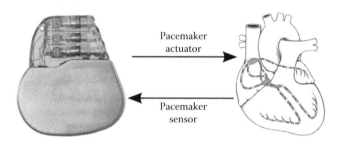

FIGURE 9.6   The closed-loop model.*

9.5   CLOSED-LOOP MODEL OF THE CRT PACEMAKER AND HEART

In this chapter, we present a closed-loop formal model of the CRT pacemaker and heart, in which the formal model of the heart is used as a virtual environment, and the formal model of the CRT pacemaker is used as a TCPS that guarantees to respond according to intrinsic activities of the heart (see Figure 9.6). The main objective of this closed-loop model is to verify and validate the complex properties of the CRT pacemaker under the virtual environment, identifying new emergent behaviors and strengthening the given systems requirements. As far as we know, this is the first closed-loop formal model of the CRT pacemaker and heart to analyze the functional behavior of the CRT pacemaker under the virtual environment by satisfying the required safety properties. To develop the closed-loop model, we use the previously developed and verified formal models of the CRT pacemaker [24] and heart [18,19]. In fact, we use our previous works as the basis for developing a closed-loop model of the CRT pacemaker and heart using stepwise refinement from scratch. To check the correctness of the closed-loop system, we introduce several safety properties and discharge all the generated proof obligations at each refinement level. An abstract model and a series of gradually refined models are described below.

9.5.1   Abstract Model

To model the functional behavior of the heart abstractly, we formalize the characteristics of electrical impulse propagation using conduction networks in various chambers. An impulse propagation controls the temporal activities of the heart, and it allows heart muscle to contract and relax periodically. We identify a set of landmark nodes (see Figure 9.1b) on the conduction network. These nodes are connected to form a path for impulse propagation from the SA node to the end of the Purkinje fibers. To specify the static properties of the heart, we declare the enumerated set *ConductionNode* and three constants, *ConductionTime*, *ConductionPath*, and *ConductionSpeed*. The enumerated set *ConductionNode* is a collection of landmarks nodes, and the constants *ConductionTime*, *ConductionPath*, and *Conduction-Speed* are impulse propagation time for each landmark node, the impulse propagation

---

* The image of a CRT pacemaker is adapted from: http://www.amayeza.co.za/files/content/images/img331.jpg.

network of the heart, and impulse propagation speed for each path, respectively. These are defined using axioms ($axm1$–$axm4$), that are extracted from the heart definition and the given properties (see Section 9.3).

An abstract model of the CRT pacemaker formalizes only sensing and pacing behaviors for each chamber (RA, RV, and LV) without considering any temporal requirements. To model the functional behavior of the CRT pacemaker abstractly, we define the enumerated set *Status* in $axm5$ that shows the *ON* and *OFF* states of the actuators and sensors.

$axm1 : partition(ConductionNode, \{A\}, \{B\}, \{C\}, \{D\}, \{E\}, \{F\}, \{G\}, \{H\})$
$axm2 : ConductionTime \in ConductionNode \rightarrow \mathbb{P}(0 .. 230)$
$axm3 : ConductionPath \subseteq ConductionNode \times ConductionNode$
$axm4 : ConductionSpeed \in ConductionPath \rightarrow \mathbb{P}(5 .. 400)$
$axm5 : partition(Status, \{ON\}, \{OFF\})$

To model an abstract dynamic behavior of the closed-loop system, we had to first develop a formal model of the CRT pacemaker and heart together. In this closed-loop model, the CRT pacemaker behaves appropriately by observing the normal and abnormal behaviors of the heart. A virtual environment model of the heart simulates according to impulse propagation in the conduction network using conduction nodes and the defined properties. An abstract model declares a list of variables to model the closed-loop system. For modeling the heart, we define four variables ($inv1$–$inv4$): *ConductionNodeState*—to show the Boolean states of the landmark nodes to distinguish the visited (*TRUE*) and unvisited (*FALSE*) landmark nodes; *CConductionTime*—to present the current impulse propagation time in the conduction network; *CConductionSpeed*—to present the current impulse propagation speed in the conduction network; and *HeartState*—to show the Boolean states of the heart to represent normal (*TRUE*) or abnormal (*FALSE*) conditions. The CRT pacemaker contains three electrodes, equipped with sensors and actuators, that delivers pacing stimulus in the heart chambers (RA, RV, and LV) as per the physiological needs through sensing intrinsic activities of a patient's heart. All three electrodes synchronize to sense and to pace appropriately in different heart chambers whenever required. A list of variables is declared to define different actuators and sensors for each chamber. The actuators are defined as *PM_Actuator_A*, *PM_Actuator_LV*, and *PM_Actuator_RV*, and the sensors are defined as *PM_Sensor_A*, *PM_Sensor_LV*, and *PM_Sensor_RV*. These actuators and sensors are defined as the type of *Status* using invariants ($inv5$–$inv10$).

$inv1 : ConductionNodeState \in ConductionNode \rightarrow BOOL$
$inv2 : CConductionTime \in ConductionNode \rightarrow 0 .. 300$
$inv3 : CConductionSpeed \in ConductionPath \rightarrow 0 .. 500$
$inv4 : HeartState \in BOOL$
$inv5 : PM\_Actuator\_A \in Status$
$inv6 : PM\_Actuator\_RV \in Status$
$inv7 : PM\_Actuator\_LV \in Status$
$inv8 : PM\_Sensor\_A \in Status$
$inv9 : PM\_Sensor\_RV \in Status$
$inv10 : PM\_Sensor\_LV \in Status$

The abstract model of the closed-loop system contains a set of events to show changing states of the CRT pacemaker and heart. For modeling the heart, we define three events, namely *HeartOK* to show a normal state of the heart; *HeartKO* to express an abnormal state of the heart; and *HeartConduction* to trace the current updated values of each landmark node in the conduction network. The event *HeartOK* models the required behavior of the heart when the heart is in its normal state. The guards of this event state that all the landmark nodes of the conduction network must be visited during the impulse propagation, the current impulse propagation time for every landmark node must belong to the given range of the conduction time, and the impulse propagation speed for every path must belong to the given range of the conduction speed (see Properties 9.1 and 9.2). The action of this event shows that the heart is in its normal state when all the guards of this event are satisfied.

```
EVENT HeartOK
  WHEN
    grd1 : ∀i·i ∈ ConductionNode ⇒ ConductionNodeState(i) = TRUE
    grd2 : ∀i·i ∈ ConductionNode ⇒ CConductionTime(i) ∈ ConductionTime(i)
    grd3 : ∀i, j·i ↦ j ∈ ConductionPath⇒
              CConductionSpeed(i ↦ j) ∈ ConductionSpeed(i ↦ j)
    THEN
    act1 : HeartState := TRUE
  END
```

The event *HeartKO* models the required behavior of an abnormal condition of the heart. The guards of this event state that if any landmark node is not visited during the impulse propagation, the current impulse propagation time of any landmark node does not belong to the given range of the conduction time, or the impulse propagation speed of any path does not belong to the given range of the conduction speed (see Properties 9.1 and 9.2). The action of this event shows that the heart is in the abnormal state as *FALSE* when the given guard (*grd1*) of this event is satisfied.

```
EVENT HeartKO
  WHEN
    grd1 : ∃i·i ∈ ConductionNode ∧ ConductionNodeState(i) = FALSE)
            ∨
            (∃j·j ∈ ConductionNode ∧ CConductionTime(j) ∉ ConductionTime(j))
            ∨
            (∃m, n·m ↦ n ∈ ConductionPath ∧ CConductionSpeed(m ↦ n)
            ∉ ConductionSpeed(m ↦ n))
    THEN
    act1 : HeartState := FALSE
  END
```

The event *HeartConduction* models the dynamic behavior of heart conduction abstractly. This event allows the setting of new updated values to the visited state of the landmark nodes, impulse propagation time, impulse propagation velocity, and state of the heart. This event is described to show updated values only in one image. This event is further refined to model the concrete behaviors of the heart.

```
EVENT HeartConduction
  BEGIN
    act1 : ConductionNodeState :∈ ConductionNode → BOOL
    act2 : CConductionTime :∈ ConductionNode → 0 .. 300
    act3 : CConductionSpeed :∈ ConductionPath → 0 .. 500
    act4 : HeartState :∈ BOOL
  END
```

In the abstract model of the closed-loop system, the CRT pacemaker model describes only discrete functional behaviors for modeling sensors and actuators without considering any temporal requirements. For modeling the CRT pacemaker, we define 12 new events to formalize the sensing and pacing activities in the form of changing states for actuators and sensors for each chamber (RA, RV, and LV). All these events are very simple, and these events contain only a single guard to specify the current state of the actuators/sensors, and then the action of these events allows a change to the current state of the actuators/sensors. As examples, we present two events, *PM_Pacing_On_RV* and *PM_Sensing_On_RV*. The event *PM_Pacing_On_RV* is used to set *ON* for the right ventricle actuator, when the right ventricle actuator is *OFF*, and the event *PM_Sensing_On_RV* is used to set *ON* for the right ventricle sensor, when the right ventricle sensor is *OFF*. Other events are formalized in a similar way.

```
EVENT PM_Pacing_On_RV
  WHEN
    grd1 : PM_Actuator_RV = OFF
  THEN
    act1 : PM_Actuator_RV := ON
  END
```

```
EVENT PM_Sensing_On_RV
  WHEN
    grd1 : PM_Sensor_RV = OFF
  THEN
    act1 : PM_Sensor_RV := ON
  END
```

9.5.2 First Refinement: Impulse Propagation and Timing Requirements

This refinement step allows us to add more detail in the abstract model of the closed-loop system by refining the CRT pacemaker and heart models together. The heart model is refined by adding conduction behavior that introduces the functional behavior of the impulse propagation in the conduction network. The SA node originates an impulse that passes through the conduction network by visiting all the landmark nodes to reach the Purkinje fibers of the ventricles. For example, an impulse generates from node *A* and finally sinks to the terminal nodes (*C*, *G*, and *H*). The conduction model uses a clock counter to specify the required temporal properties for impulse propagation.

For modeling the CRT pacemaker, we define a list of constants as static properties for describing the timing requirements for controlling the pacing and sensing events, and to simulate the desired heart behavior. We define four constants, the atrioventricular interval (*AVI*); the ventriculoatrial interval (*VAI*); the left ventricular interval (*LVI*); and the right ventricular interval (*RVI*). All these constants are defined in axioms (*axm1–axm4*). An extra axiom (*axm5*) is defined as a constraint that specifies that the *RVI* is greater than or equal to the *LVI*. In this study, we consider all times to be in milliseconds.

$$
\begin{aligned}
&axm1 : AVI \in 50 .. 350 \\
&axm2 : VAI \in 350 .. 1200 \\
&axm3 : LVI \in 0 .. 50 \\
&axm4 : RVI \in 0 .. 50 \\
&axm5 : RVI \geq LVI
\end{aligned}
$$

In this refinement, we introduce a logical clock for modeling the timing requirements for the CRT pacemaker and heart. To model the clock behavior, we declare a variable *now* to specify the current clock counter in *inv*1. This clock counter always progresses by 1 ms in every clock tick. In the CRT pacemaker model, we define a variable *PSRecord* to store a time whenever any pacing or intrinsic activity related to sensing occurs within the chambers in *inv*2. The stored time can be used to control the future activities related to pacing and sensing events. The CRT pacemaker model contains a list of variables to synchronize the sensing and pacing events by observing the various states of sensors and actuators for all three chambers (RA, RV, and LV) in order to specify the required behavior. In the heart model, we declare only a variable *CCSpeed_CCTime_Flag* to synchronize and preserve the desired behavior of the heart for capturing the current values of the impulse propagation time and impulse propagation speed in the conduction network.

$$
\begin{aligned}
&inv1 : now \in \mathbb{N} \\
&inv2 : PSRecord \in \mathbb{N} \\
&inv3 : CCSpeed\_CCTime\_Flag \in BOOL \\
&inv4 : HeartState = TRUE \Rightarrow (\forall node \cdot node \in ConductionNode \Rightarrow \\
&\qquad CConductionTime(node) \in ConductionTime(node)) \\
&inv5 : HeartState = TRUE \Rightarrow (\forall m, n \cdot m \mapsto n \in ConductionPath \Rightarrow \\
&\qquad CConductionSpeed(m \mapsto n) \in ConductionSpeed(m \mapsto n)) \\
&inv6 : now = 0 \Rightarrow PM\_Sensor\_RV = OFF \wedge PM\_Actuator\_RV = OFF \\
&inv7 : now = 0 \Rightarrow PM\_Sensor\_LV = OFF \wedge PM\_Actuator\_LV = OFF \\
&inv8 : now = 0 \Rightarrow PM\_Sensor\_A = OFF \wedge PM\_Actuator\_A = OFF \\
&inv9 : PM\_Actuator\_RV = ON \Rightarrow now \geq AVI \vee Immd\_Pace\_RV = 1 \\
&inv10 : PM\_Actuator\_LV = ON \Rightarrow \\
&\qquad now \geq AVI + (RVI - LVI) \vee Immd\_Pace\_LV = 1 \vee Delay\_Pace\_LV = 1 \\
&inv11 : PM\_Actuator\_A = ON \Rightarrow \\
&\qquad now \geq PSRecord + VAI \vee now \geq AVI + VAI
\end{aligned}
$$

We define a list of safety properties to model the correct functionalities of the closed-loop system. Most of the safety properties are defined independently for both the CRT pacemaker and heart. A set of invariants (*inv4–inv5*) defines safety properties for the heart model that state that if the heart is in its normal state then the current impulse propagation speed and current impulse propagation time are always within the given range for each landmark node and the defined path of the conduction network. Another list of safety properties is introduced for the CRT pacemaker. Invariants (*inv6*, *inv7*, and *inv8*) state that when the current clock counter is zero then the sensors and actuators are *OFF* of the right ventricle, left ventricle, and right atrium. The next safety property, *inv9*, states that if the current clock counter elapses then an AVI or an immediate pacing is required in the right ventricle then the actuator of the right ventricle must pace. Similarly, the next safety property, $inv10$, states that if the current clock counter elapses the total duration of the atrioventricular interval and pacing delay, an immediate pacing is required in the left ventricle, or a delay pacing is detected in the left ventricle, then the actuator of the left ventricle must pace. The last safety property, $inv11$, states that if the current clock counter elapses the total duration of VAI and PSRecord, or the current clock counter elapses the total duration of the AVI and VAI, then the actuator of the right atrium must pace.

In this refinement, we introduce several events to specify the desired behavior of the CRT pacemaker and heart according to the given timing requirements and impulse propagation. A few events are refinements of the abstract events, and other new added events at this level are the refinement of the *skip*. In the heart modeling, we introduce the event *SinusNodeFire* that refines the abstract event *HeartConduction*. This event models the functional behavior of the SA node that initially generates an electrical impulse for propagating in the conduction network. The guards of this event state that all the landmark nodes are unvisited, the current impulse propagation time of each node is 0 ms., and the impulse propagation speed for every path is 0 cm/sec. The actions of this event show that the SA node is visited, and the current impulse propagation time and speed flag becomes *TRUE*.

EVENT SinusNodeFire Refines HeartConduction
  WHEN
    grd1 : $\forall n \cdot n \in ConductionNode \Rightarrow ConductionNodeState(n) = FALSE$
    grd2 : $\forall n \cdot n \in ConductionNode \Rightarrow CConductionTime(n) = 0$
    grd3 : $\forall n, m \cdot n \in ConductionNode \wedge m \in ConductionNode \wedge$
        $n \mapsto m \in ConductionPath \Rightarrow CConductionSpeed(n \mapsto m) = 0$
  THEN
    act1 : $ConductionNodeState(A) := TRUE$
    act2 : $CCSpeed\_CCTime\_Flag := TRUE$
  END

We also introduce a list of events to model the stepwise progression of an electrical impulse from an SA node to the Purkinje fibers. The introduced events synchronize all the heart chambers by progressing an impulse in the conduction network. This event is the refinement of the abstract event *HeartConduction*. An event *HeartConduction_NODES* is formalized to show the basic formalization steps of the heart conduction. The guards of this event state that there are two nodes and the nodes are directly connected; the first node p is visited, the second node q is not visited, the current conduction time of the second node q belongs to the given range of the conduction time, the conduction speed from the first node p to node q belongs to the given range of the conduction speed, and the current impulse propagation time and speed flag is *FALSE*. The actions of this event show that the first node q is visited, and the current impulse propagation time and speed flag becomes *TRUE*.

```
EVENT HeartConduction_NODES Refines HeartConduction
  ANY  p, q
  WHERE
    grd1 : p ∈ ConductionNode ∧ q ∈ ConductionNode ∧ p ↦ q ∈ ConductionPath
    grd2 : ConductionNodeState(p) = TRUE
    grd3 : ConductionNodeState(q) = FALSE
    grd4 : CConductionTime(q) ∈ ConductionTime(q)
    grd5 : CConductionSpeed(p ↦ q) ∈ ConductionSpeed(p ↦ q)
    grd6 : CCSpeed_CCTime_Flag = FALSE
  THEN
    act1 : ConductionNodeState(q) := TRUE
    act2 : CCSpeed_CCTime_Flag := TRUE
  END
```

A new event, *Update_CCSpeed_CCtime*, is introduced that is the refinement of the event *HeartConduction*. The main functionality of this event is to update the current impulse propagation time and impulse propagation speed according to the progressive conduction flow using landmark nodes in the conduction network. The guards of this event are used to select a pair of nodes that belongs to the defined conduction paths, and to select the current conduction speed and current conduction time from the given range at the present time. The actions of this event update the current conduction time and current conduction speed according to the selected nodes and path.

```
EVENT Update_CCSpeed_CCtime Refines HeartConduction
  ANY  i, j, CSpeed, CTime
  WHERE
    grd1 : i ∈ ConductionNode
    grd2 : j ∈ ConductionNode
    grd3 : i ↦ j ∈ ConductionPath
    grd4 : CSpeed ∈ 0 .. 500
    grd5 : CTime ∈ 0 .. 300
    grd6 : ConductionNodeState(i) = TRUE
    grd7 : CCSpeed_CCTime_Flag = TRUE
    grd8 : HeartState = FALSE
    grd9 : tic = CTime
  THEN
    act1 : CConductionTime(j) := CTime
    act2 : CConductionSpeed(i ↦ j) := CSpeed
    act3 : CCSpeed_CCTime_Flag := FALSE
  END
```

In the CRT pacemaker modeling, we introduce a total of 18 events, in which 17 events are refinements of the abstract events. For example, the event *PM_Pacing_On_RV* refines the abstract event *PM_Pacing_On_RV* by adding new guards and new actions. The guards of this event state that the actuator of the right ventricle is *OFF*; the current clock counter is equivalent to the atrioventricular interval (AVI), no immediate pacing is required, there is no pace in the left or right ventricles, and the right ventricular sensor is *OFF*, or an immediate pacing is required in the right ventricle; the current clock counter is not 0; there is no pacing in the right ventricle; and the actuator and sensor of the right atrium are *OFF*. The actions of this event state that the actuator of the right ventricle is *ON* and it is paced, and the present time is stored for controlling future pacing or sensing activity. In the other refined events, we have added the temporal requirements for specifying the desired pacing and sensing behaviors considering synchronization in the heart chambers (RA, RV, and LV) by observing the heart behavior.

```
EVENT PM_Pacing_On_RV Refines PM_Pacing_On_RV
  WHEN
    grd1 : PM_Actuator_RV = OFF
    grd2 : ((now = AVI ∧ Immd_Pace_RV = 0 ∧ No_Pace_LV_RV = 0∧
            PM_Sensor_RV = OFF)
            ∨
            Immd_Pace_RV = 1)
    grd3 : ¬now = 0
    grd4 : Pace_RV = 0
    grd4 : PM_Actuator_A = OFF ∧ PM_Sensor_A = OFF
  THEN
    act1 : PM_Actuator_RV := ON
    act2 : Pace_RV := 1
    act3 : PSRecord := now
  END
```

A logical clock plays an important role for modeling the temporal requirements of the CRT pacemaker and the functional behavior of the heart. In order to design a logical clock, we introduce a new event, *tic*, that allows us to increase the current clock counter by 1 ms. progressively. This event *tic* does not have a guard, but in further refinements, we introduce a guard to control different functionalities of the CRT pacemaker and heart within the restricted time intervals.

```
EVENT tic
  WHEN
  THEN
    act1 : now := now + 1
  END
```

9.5.3 Second Refinement: Threshold and Heart Blocks

This refinement introduces the functional properties of abnormal behaviors of the heart, such as heart blocks, and *threshold* is used by the CRT pacemaker to detect an appropriate intrinsic activity for each chamber. The heart block introduces perturbation in the heart conduction network that generates some difficulties in the electrical impulse propagation, and it influences the normal behavior of the heart. We use the landmark nodes to show the different types of heart blocks in Figure 9.2. For modeling the heart blocks, we define an enumerated set *HeartBlockSets* to present different types of blocks of the heart.

```
axm1 : partition(HeartBlockSets, {SA_nodal_blocks}, {AV_nodal_blocks},
        {Infra_Hisian_blocks}, {LBBB_blocks}, {RBBB_blocks}, {None})
```

The sensors play an important role in sensing an intrinsic activity of the heart. The CRT pacemaker actuator delivers stimulation for a short period of time by observing the sensed values and required safety margins. Each chamber of the heart has different range of threshold values that can be specified by the physiologist by monitoring the detected intrinsic activities. We define three constants, *STA_THR_A*, *STA_THR_RV*, and *STA_THR_LV* to hold the standard threshold values for the right atrium, right ventricle, and left ventricle, respectively in *axm1–axm3*. We define a new constraint using an axiom (*axm4*) to show that the threshold of the atrium chamber is less than the threshold of the left and right ventricles.

$$
\begin{aligned}
&axm1 : STA\_THR\_A \in \mathbb{N}1 \\
&axm2 : STA\_THR\_RV \in \mathbb{N}1 \\
&axm3 : STA\_THR\_LV \in \mathbb{N}1 \\
&axm4 : STA\_THR\_A < STA\_THR\_LV \wedge STA\_THR\_A < STA\_THR\_RV
\end{aligned}
$$

For modeling the dynamic properties of the closed-loop system in this refinement, we declare some new variables. These variables are: *HeartBlocks*—to show the different types of heart blocks; *C_Thr*—to hold the current sensing value that is produced by the heart, and it can be used by the CRT pacemaker for detecting an intrinsic activity of the heart; *Thr_State_A*—a Boolean state of the right atrium; *Thr_State_RV*—a Boolean state of the right ventricle; and *Thr_State_LV*—a Boolean state of the left ventricle. Three state variables are defined for each chamber (RA, RV, and LV) to synchronize and maintain the order of sensing activities. We introduce three safety properties using invariants (*inv4*, *inv5*, and *inv6*) that state that the sensor of each chamber is *OFF* when the detected sensor value is greater than or equal to the standard threshold value and the Boolean state of the chamber is *TRUE*.

$$
\begin{aligned}
&inv1 : HeartBlocks \in HeartBlockSets \\
&inv2 : C\_Thr \in \mathbb{N} \\
&inv3 : Thr\_State\_A \in BOOL \wedge Thr\_State\_LV \in BOOL \wedge Thr\_State\_RV \in BOOL \\
&inv4 : \forall i \cdot i \in \mathbb{N}_1 \wedge i \geq STA\_THR\_A \wedge Thr\_State\_A = TRUE \\
&\qquad \Rightarrow PM\_Sensor\_A = OFF \\
&inv5 : \forall i \cdot i \in \mathbb{N}_1 \wedge i \geq STA\_THR\_LV \wedge Thr\_State\_LV = TRUE \\
&\qquad \Rightarrow PM\_Sensor\_LV = OFF \\
&inv6 : \forall i \cdot i \in \mathbb{N}_1 \wedge i \geq STA\_THR\_RV \wedge Thr\_State\_RV = TRUE \\
&\qquad \Rightarrow PM\_Sensor\_RV = OFF
\end{aligned}
$$

We introduce a set of events in this refinement to simulate the desired behavior of the heart block and detect the intrinsic activities of the heart chambers using the CRT pacemaker's sensors. A sensor can detect an intrinsic activity when the threshold value of the detected signal is greater than or equal to the standard threshold constant. A set of events is introduced to model the possible behaviors of the heart blocks. A conduction disturbance in the heart is generated when an impulse produced from the sinus node (A) is blocked or delayed from depolarizing the atria known as the SA block [15,17]. To model the SA block, we introduce a new event, *HeartConduction_Block_A_B_C*, that is the refinement of the abstract event *HeartKO*. The guard of this event states that the landmark node (*A* or *C*) is not visited, the current impulse propagation time of the node (*B* or *C*) does not belong to the given range of the impulse propagation time, or the current impulse propagation speed of the path ($A \mapsto B$ or $A \mapsto C$) does not belong to the given range of impulse propagation speeds. When the guard provided is satisfied, the actions show that the heart is in an abnormal state and has an SA nodal block. Other heart block events are also part of the refinement of the abstract event *HeartKO*, and are modeled similarly.

EVENT HeartConduction_Block_A_B_C Refines HeartKO
  WHEN
    grd1 : $(ConductionNodeState(A) = FALSE) \vee$
        $(ConductionNodeState(C) = FALSE) \vee$
        $(CConductionTime(B) \notin ConductionTime(B)) \vee$
        $(CConductionTime(C) \notin ConductionTime(C)) \vee$
        $(CConductionSpeed(A \mapsto B) \notin ConductionSpeed(A \mapsto B)) \vee$
        $(CConductionSpeed(A \mapsto C) \notin ConductionSpeed(A \mapsto C))$
  THEN
    act1 : $HeartState := FALSE$
    act2 : $HeartBlocks := SA\_nodal\_blocks$
  END

For modeling the threshold for detecting the appropriate intrinsic activities of the heart using the CRT pacemaker sensors, we do not introduce any new event in this refinement step. We introduce the functional behavior of threshold by strengthening the guards of the abstract events. The newly added guards are used to detect an intrinsic activity of the heart by comparing the value of a sensed signal with the standard threshold value for the right atrium, right ventricle, and left ventricle. For example, the event *PM_Sensing_Off_A* is the refinement of the abstract event *PM_Sensing_Off_A*. In the refined event, we introduce new variable *Thr_A* and some new guards (*grd8* and *grd9*). The guard *grd8* declares the type of local variable (*Thr_A*), and *grd9* compares the sensed value with the selected standard threshold value of the atria chamber and the sensed value is equivalent to the current sensing value that is produced by the heart model.

EVENT PM_Sensing_Off_A Refines
        PM_Sensing_Off_A
  ANY *Thr_A*
  WHERE
    grd1 : $PM\_Sensor\_A = ON$
    grd2 : $PM\_Sensor\_RV = OFF$
    grd3 : $PM\_Actuator\_A = OFF$
    grd4 : $PM\_Actuator\_RV = OFF$
    grd5 : $PM\_Sensor\_LV = OFF$
    grd6 : $PM\_Actuator\_LV = OFF$
    grd7 : $(now < AVI + VAI \wedge No\_Pace\_LV\_RV = 0 \wedge$
        $RV\_Delay\_AVI = 0 \wedge Delay\_Pace\_LV = 0 \wedge$
        $Immd\_Pace\_RV = 0 \wedge Immd\_Pace\_LV = 0) \vee$
        $(now < PSRecord + VAI)$
    grd8 : $Thr\_A \in \mathbb{N}$
    grd9 : $Thr\_A \geq STA\_THR\_A \wedge Thr\_A = C\_Thr$

```
THEN
   act1 : PM_Sensor_A := OFF
   act2 : PSRecord := 0
   act3 : now := 0
   act4 : Immd_Pace_LV := 0
   act5 : Delay_Pace_LV := 0
   act6 : No_Pace_LV_RV := 0
   act7 : Immd_Pace_RV := 0
   act8 : Pace_LV := 0
   act9 : Pace_RV := 0
   act10 : Pace_A := 0
   act11 : RV_Delay_AVI := 0
   act12 : Thr_State_LV := FALSE
   act13 : Thr_State_RV := FALSE
END
```

The variable *C_Thr* plays an important role in synchronizing the behavior of the CRT pacemaker and heart in the closed-loop model and monitoring the activity of the right atrium by comparing the sensed value with the selected threshold value. In this event, we introduce two extra actions (*act*12 and *act*13) to set a *FALSE* state for both the left and right ventricles. The rest of the guards and actions of this event are similar to the abstract event. We also modify other events related to the left and right ventricles to model the desired behavior of the sensors for synchronizing between the CRT pacemaker and heart models.

9.5.4 Third Refinement: Refractory and Blanking Periods and a Cellular Model

This is the last refinement of the closed-loop system that introduces CA for modeling the concrete behavior of the heart, and the refractory and blanking periods for modeling the concrete behavior of the CRT pacemaker. This final refinement of the heart model provides a simulation model that includes an impulse propagation at the cellular level in the heart chambers.

```
axm1 : partition(CellStates, {PASSIVE}, {ACTIVE}, {REFRACTORY})
axm2 : NeighbouringCells = (λx ↦ y·x ∈ ℤ ∧ y ∈ ℤ|
         {(x ↦ y), ((x + 1) ↦ y), ((x − 1) ↦ y), (x ↦ (y + 1)), (x ↦ (y − 1))})
axm3 : NEXT ∈ ran(NeighbouringCells) → CellStates
axm4 : CellS ∈ ran(NeighbouringCells) → CellStates
axm5 : (∀m, n·{m ↦ n} ∈ ran(NeighbouringCells)∧
         CellS({m ↦ n}) = ACTIVE
         ⇒
         NEXT({m ↦ n}) = REFRACTORY
axm6 : ∀m, n·{m ↦ n} ∈ ran(NeighbouringCells)∧
         CellS({m ↦ n}) = REFRACTORY
         ⇒
         NEXT({m ↦ n}) = PASSIVE
axm7 : ∀m, n·{m ↦ n} ∈ ran(NeighbouringCells)∧
         {m + 1 ↦ n} ∈ ran(NeighbouringCells)∧
         {m − 1 ↦ n} ∈ ran(NeighbouringCells)∧
         {m ↦ n + 1} ∈ ran(NeighbouringCells)∧
         {m ↦ n − 1} ∈ ran(NeighbouringCells)∧
         CellS({m ↦ n}) = PASSIVE∧
```

$$(CellS(\{m + 1 \mapsto n\}) = ACTIVE\vee$$
$$CellS(\{m - 1 \mapsto n\}) = ACTIVE\vee$$
$$CellS(\{m \mapsto n + 1\}) = ACTIVE\vee$$
$$CellS(\{m \mapsto n - 1\}) = ACTIVE)$$
$$\Rightarrow$$
$$NEXT(\{m \mapsto n\}) = ACTIVE$$
$$axm8 : \forall m, n \cdot \{m \mapsto n\} \in ran(NeighbouringCells)\wedge$$
$$\{m + 1 \mapsto n\} \in ran(NeighbouringCells)\wedge$$
$$\{m - 1 \mapsto n\} \in ran(NeighbouringCells)\wedge$$
$$\{m \mapsto n + 1\} \in ran(NeighbouringCells)\wedge$$
$$\{m \mapsto n - 1\} \in ran(NeighbouringCells)\wedge$$
$$CellS(\{m \mapsto n\}) = PASSIVE\wedge$$
$$CellS(\{m + 1 \mapsto n\}) \neq ACTIVE\wedge$$
$$CellS(\{m - 1 \mapsto n\}) \neq ACTIVE\wedge$$
$$CellS(\{m \mapsto n + 1\}) \neq ACTIVE\wedge$$
$$CellS(\{m \mapsto n - 1\}) \neq ACTIVE$$
$$\Rightarrow$$
$$NEXT(\{m \mapsto n\}) = PASSIVE$$

We define a set of constants and mathematical properties to formalize the desired behavior of the heart using CA (see Figure 9.3). Each biological cell can have one of the following states: *Active*, *Passive*, or *Refractory*. To define the possible cell states, we declare the enumerated set *CellStates* in *axm*1. To simplify the modeling of CA, we consider only the two-dimensional structure of the connected cells. In order to define the two-dimensional structure, we declare a function *NeighbouringCells* to specify a set of coordinated positions of neighboring cells (*axm*2). The two functions *NEXT* and *CellS* are declared to define the state of the neighboring cells in *axm*3 and *axm*4. A set of properties (*axm*5–*axm*8) is introduced to model the desired functionalities of the CA in two dimensions according to the Greenberg–Hastings CA. The first property states that every cells, belong to neighboring cells in 2D, change to *Refractory* state if theses cells are in *Active* state. The second property states that every cells, belong to neighboring cells in 2D, change to *Passive* state if theses cells are in *Refractory* state. The third property states that every cells, belong to neighboring cells in 2D, change to *Active* state if theses cells are in *Passive* state and any neighbor is in *Active* state. The last property states that every cells, belong to neighboring cells in 2D, change to *Passive* state if theses cells are in *Passive* state and none neighbor is in *Active* state (see Definitions 9.2 and 9.3).

For modeling the concrete behavior of actuators and sensors of the CRT pacemaker, we introduce the refractory and blanking periods [25] for the right atrium, right ventricle, and left ventricle. These refractory and blanking periods play an important role in the suppression of device-generated artifacts and unwanted signal artifacts generated from intrinsic activities of the heart. Moreover, these periods also help to identify appropriate sensing events, and to prevent over-sensing events in other chambers. To define the static behavior of the CRT pacemaker, we declare eight constants: Atrial Refractory Period (ARP); Right Ventricular Refractory Period (RVRP); Left Ventricular Refractory Period (LVRP); Post-Ventricular Atrial Refractory Period (PVARP); Right Ventricular Blanking Period (RVBP); Left Ventricular Blanking Period (LVBP); A-Blank after Right Ventricular Activity (ABaRV); and A-Blank after Left Ventricular Activity (ABaLV), using axioms (*axm*1–*axm*8).

$$
\begin{aligned}
&axm1 : ARP \in 30 .. 500 \\
&axm2 : RVRP \in 20 .. 500 \\
&axm3 : LVRP \in 20 .. 500 \\
&axm4 : PVARP \in 50 .. 500 \\
&axm5 : RVBP \in 5 .. 60 \\
&axm6 : LVBP \in 5 .. 60 \\
&axm7 : ABaRV \in 5 .. 150 \\
&axm8 : ABaLV \in 5 .. 150
\end{aligned}
$$

The dynamic behavior of CA is defined by introducing four new variables. These four variables are m, n, *Transition*, and *NextCellState*. The first two variables m and n are declared to model the coordinate positions in two dimensions of an active cell during an impulse propagation. The next variable, *Transition*, is defined as a Boolean type to enable transition between different states of the cells to model the behavior of a tissue. The last variable, *NextCellState*, is used to store the state of a cell after each transition. In this refinement, we introduce six new safety properties. The first two safety properties state that the actuator and sensor of the right atrium become *ON* when the current clock counter elapses the refractory and blanking periods. To check the refractory period after pacing or sensing activity, we need to store the time of pacing or sensing activity of the ventricular chambers. In these safety properties, we use the variable *PSRecord* to record the time of previous occurrence of a pacing or sensing event for initiating the refractory periods. The last four safety properties show that the actuators and sensors of the right and left ventricles become *ON* when the current clock counter elapses the refractory and blanking periods. This means that the sensing and pacing events always occur after elapsing the refractory and blanking periods.

$$
\begin{aligned}
&inv1 : m \in \mathbb{Z} \wedge n \in \mathbb{Z} \\
&inv2 : Transition \in BOOL \\
&inv3 : NextCellState \in CellStates \\
&inv4 : PM\_Actuator\_A = ON \Rightarrow now \geq PSRecord + PVARP \wedge \\
&\qquad\qquad now \geq PSRecord + RVRP \wedge now \geq PSRecord + LVRP \wedge \\
&\qquad\qquad now \geq PSRecord + ABaLV \wedge now \geq PSRecord + ABaRV \\
&inv5 : PM\_Sensor\_A = ON \Rightarrow now \geq PSRecord + PVARP \wedge \\
&\qquad\qquad now \geq PSRecord + RVRP \wedge now \geq PSRecord + LVRP \\
&\qquad\qquad \wedge now \geq PSRecord + ABaLV \wedge now \geq PSRecord + ABaRV \\
&inv6 : PM\_Actuator\_RV = ON \Rightarrow now \geq ARP \wedge now \geq RVBP \\
&inv7 : PM\_Actuator\_LV = ON \Rightarrow now \geq ARP \wedge now \geq LVBP \\
&inv8 : PM\_Sensor\_RV = ON \Rightarrow now \geq ARP \wedge now \geq RVBP \\
&inv9 : PM\_Sensor\_LV = ON \Rightarrow now \geq ARP \wedge now \geq LVBP
\end{aligned}
$$

For modeling an impulse propagation at the cellular level in the heart for two-dimensional structures, we define the two events *HeartConduction_Cellular* and *HeartConduction_Next_UpdateCell*. It should be noted that we define these events abstractly so that they can be further refined to formalize the concrete behavior of the CA at the tissue level corresponding to various states of the cell. The event *HeartConduction_Cellular* enables transition for a switching state of the neighboring cells of the conduction network by propagating electrical current. This event allows the setting of a Boolean state as *TRUE* of the *Transition*. The guards of this event state that there is a valid path between two landmark nodes that belong to a set of pairs of the conduction network; the current impulse propagation speed and time flag is *TRUE* and therefore permits synchronization and preservation of the desired behavior of the heart for capturing the current values of the impulse propagation time and impulse propagation speed for each node; a cell (s, t) is selected that belongs to neighboring cells at (m, n), and the selected cell is in any state that is equivalent to *NextCellState*, which stores the next state of the previous transition; and the current transition state is *FALSE*.

```
EVENT HeartConduction_Cellular
  ANY   p, q, s, t
  WHERE
    grd1 : p ↦ q ∈ ConductionPath
    grd2 : CCSpeed_CCTime_Flag = TRUE
    grd3 : m ↦ n ∈ dom(NeighbouringCells) ∧ s ↦ t ∈ NeighbouringCells(m ↦ n)
    grd4 : {s ↦ t} ∈ dom(CellS) ∧ CellS({s ↦ t}) ∈ {PASSIVE, ACTIVE, REFRACTORY}
    grd5 : NextCellState = CellS({s ↦ t})
    grd6 : Transition = FALSE
  THEN
    act1 : Transition := TRUE
  END
```

The event *HeartConduction_Next_UpdateCell* is a new event for calculating the state of neighboring cells and to update the position of the current cell (m, n). The guards of this event state that a cell (s, t) is selected that belongs to neighboring cells at (m, n); the selected cell is domain of the function *NEXT*; and the current transition state is *TRUE*. The actions of this event calculates the state of selected cell (s, t) that is assigned to *NextCellState*, sets the current transition state as *FALSE*, and updates the current cell (m, n) nondeterministically to propagate an impulse in the conduction network.

```
EVENT HeartConduction_Next_UpdateCell
  ANY   s, t
  WHERE
    grd1 : m ↦ n ∈ dom(NeighbouringCells) ∧ s ↦ t ∈ NeighbouringCells(m ↦ n)
    grd2 : s ↦ t ∈ dom(NEXT)
    grd3 : Transition = TRUE
  THEN
    act1 : NextCellState := NEXT({s ↦ t})
    act2 : Transition := FALSE
    act3 : m :∈ {m − 1, m, m + 1}
    act4 : n :∈ {n − 1, n, n + 1}
  END
```

For modeling the concrete temporal behavior of the CRT pacemaker considering the refractory and blanking periods, we introduce many guards in several events. For example, we introduce an event *PM_Pacing_On_A* that refines the abstract event *PM_Pacing_On_A* by strengthening the guards. The guards of this event state that the actuator of the right atrium is *OFF*; the current clock counter elapses the AVI + VAI or PSRecord + VAI interval considering the possible scenarios related to delay pacing, no pacing, and immediate pacing for either one chamber (LV or RV) or both chambers (LV and RV); the pacing state of the atrium chamber is 0; and the current clock counter elapses the refractory and blanking periods after detecting an activity of pacing or sensing in the heart. The actions of this event state that the actuator of the right atrium generates a small electrical current to pace, and to set the pacing state of the atrium chamber for synchronizing the pacing and sensing events of the other heart chambers.

```
EVENT PM_Pacing_On_A Refines PM_Pacing_On_A
  WHEN
    grd1 : PM_Actuator_A = OFF
    grd2 : (now = AVI + VAI ∧ No_Pace_LV_RV = 0 ∧ Delay_Pace_LV = 0)
              ∨
           (now = PSRecord + VAI ∧ (Delay_Pace_LV = 2 ∨ Delay_Pace_LV = 1
              ∨No_Pace_LV_RV = 1 ∨ RV_Delay_AVI = 1 ∨ Immd_Pace_RV = 1
              ∨Immd_Pace_LV = 1))
    grd3 : Pace_A = 0
    grd4 : now ≥ PSRecord + PVARP
    grd5 : now ≥ PSRecord + RVRP ∧ now ≥ PSRecord + LVRP
    grd6 : now ≥ PSRecord + ABaRV ∧ now ≥ PSRecord + ABaLV
  THEN
    act1 : PM_Actuator_A := ON
    act2 : Pace_A := 1
  END
```

We also introduce a new guard in the event *tic* to describe the correct temporal behavior of the sensing and pacing events. The provided guard synchronizes the pacing and sensing activities of the CRT pacemaker according to the biventricular sensing and pacing (BiSP) requirements by monitoring the heart functionalities. The *tic* event contains a very large guard to represent all the possible timing requirements (see Section 9.4). We present below only a slice of the complete formalized timing requirements to show the progress of the event *tic* in the guard conditions according to Figure 9.5b.

```
EVENT tic
  WHEN
    grd1: (now < AVI ∧ PM_Sensor_LV = OFF ∧ PM_Sensor_RV = OFF∧
          No_Pace_LV_RV = 1)
            ∨
          (now ≥ AVI ∧ now < PSRecord + PVARP ∧ PM_Sensor_LV = OFF∧
          PM_Sensor_RV = OFF ∧ No_Pace_LV_RV = 1∧
          (Pace_RV = 2 ∨ Thr_State_RV = TRUE))
            ∨
          (now ≥ PSRecord + PVARP ∧ now < PSRecord + VAI∧
          PM_Sensor_LV = OFF ∧ PM_Sensor_RV = OFF ∧ No_Pace_LV_RV = 1
          ∧PM_Sensor_A = ON)
            ∨
          . . .
          . . .
  THEN
    act1 : now := now + 1
  END
```

9.5.5 Model Validation and Analysis

This section presents the proof statistics of the developed closed-loop model, and the validity of functional behavior of the CRT pacemaker and heart by simulating the proven formal model in ProB [26]. The Event-B language supports *consistency checking* and *model analysis*, in which the *consistency checking* shows that the defined events always preserve the given invariants and the refinement checking ensures that the concrete machine is a valid refinement of an abstract machine, and the *model analysis* is used to animate the formal specifications to check the required functional behavior of the system. Model validation plays an important role for gaining confidence in the developed formal model in order to check the consistency with the systems requirements. In addition, the ProB tool also allows *automated consistency checking* and *constraint-based checking*. This tool helps to identify possible deadlocks and hidden properties that may be exposed by the generated proof obligations. In our work, the ProB tool is used to animate the closed-loop model of the CRT pacemaker and heart at each refinement level to check the required functional behavior by considering numerous scenarios for validating the developed formal models. This tool effectively assists us in finding potential problems and improving the guard conditions in each layer of the refinement. To use the ProB model checker at each refinement level, we were able to animate all the abstract and a series of refined models to prove the absence of errors (no counterexample). It should be noted that the ProB tool automatically imports static and dynamic properties, including safety properties, of the closed-loop system that are used for consistency checking and model checking to discover violations of the given safety properties against the formalized system behavior.

Table 9.3 shows the proof statistics of the closed-loop model developed in the RODIN tool [13]. This formal development generates 275(100%) proof obligations (POs), in which 246(90%) POs are proven automatically with the help of inbuilt RODIN provers, and the remaining 29(10%) POs are proven interactively by simplifying the predicates using the Rodin provers. It should be noted that the simplifying predicates are quite simple. An integration of the heart model and CRT pacemaker model generates some extra POs related to the joint behavior of the closed-loop system and by the sharing of some common variables in both the heart and CRT models. For example, the current clock counter variable (*now*) is shared, and has been used in the events of the CRT pacemaker and heart models. The CRT pacemaker shows functional properties of pacing and sensing modes under the virtual biological environment of the heart. The heart model represents normal and abnormal states of the heart, that are estimated by the physiological analysis. A list of safety properties

TABLE 9.3 Proof Statistics

Model	Total Number of POs	Automatic Proof	Interactive Proof
Abstract model	29	25(86%)	4(14%)
First refinement	138	126(91%)	12(9%)
Second refinement	36	27(75%)	9(25%)
Third refinement	72	68(94%)	4(6%)
Total	**275**	**246(90%)**	**29(10%)**

is introduced in the incremental refinements to guarantee the correctness of the functional requirements of the closed-loop model of the heart and CRT pacemaker.

9.6 DISCUSSION

This chapter presents an approach for modeling closed-loop trustworthy cyber physical systems (TCPS). The main goal of this work is to provide a novel modeling technique that helps system engineering community to develop a closed-loop model of the trustworthy cyber physical system (TCPS) and virtual environment to analyze the TCPS requirements precisely, and to address the enumerated objectives (see Section 9.1). If any fault presents in the system then the formalized system does not behave appropriately. In fact, we need precise knowledge of the error states for finding any effective heuristics. All the relevant states that lead to error/hazard that can be exploited to discover the desired correct functionalities or improve existing systems requirements.

In this chapter, we have formalized a closed-loop system of the CRT pacemaker and heart using stepwise incremental refinement. The stepwise refinement approach is a well known technique for developing dependable systems, and was championed by Harlan Mills in his work on box structures [27]. In developing the closed-loop system, we have used the formal development of the CRT pacemaker [24] and the formal development of the heart [18] as the foundation for this work. It should be noted that our formalized heart model is also validated by medical experts and cardiologists. The developed case study demonstrates that this approach is applicable in the early stage of the systems development for developing the closed-loop model of TCPS and its environment using informal requirements to specify the desired behavior for checking the functional correctness. The formalized environment shows environmental conditions that present conditional properties that help to design a system whether or not the system provides an appropriate action or solution as feedback according to the environmental situation. In our case study, the required behavior is checked according to the formalized heart model to guarantee the correctness of sensing and pacing functionalities work in all three electrodes. In fact, in our preliminary version of the model, we have found some undesired pacing and sensing activities that have been corrected with the help of the heart model. The closed-loop model of the CRT pacemaker and heart allows us to evaluate whether the CRT pacemaker provides an appropriate therapy for any arrhythmias. Additionally, an environment model also helps to validate the TCPS requirements and strengthens the given system requirements. For example, the heart model validates all the CRT requirements and helps to strengthen the CRT behaviors. In our case, we have found that all the given requirements are correct with respect to the heart model, but we have found some weaknesses in the existing requirements, such as a lack of synchronization between pacing and sensing activities in the heart chambers. The closed-loop model is used to improve the guards to synchronize the pacing and sensing activities of the electrodes with respect to time; these guards were missing in our informal requirements.

The closed-loop model is well suited for *what-if* analysis during the formal reasoning of TCPS. For instance, we have selected different ranges of constants and variables for checking the correctness of desired behavior under the given heart model. We have found a range of values for constants that can allow implementation of a patient-specific CRT pacemaker.

Moreover, this technique also allows tuning of the parameters, and alteration of several CRT functions to analyze the possible effects on the global system behavior, and to check the optional behavior of the CRT pacemaker. A simple system without an environment may satisfy the specified systems requirements, but when we use the TCPS and environment model together, it helps in traceability of missing behaviors and detection of new emergent behavior with the help of newly generated POs. The combined closed-loop model of the TCPS and virtual environment generates several new POs that do not appear in either the TCPS model or environment model, independently. All the newly generated POs of the closed-loop system are produced according to the required behavior of TCPS under the given environment. Moreover, the generated POs allow for strengthening the guards to make systems deterministic to meet stakeholder needs by describing the concrete behavior of TCPS and removing flaws in the developing model. In addition, this closed-loop modeling approach validates TCPS assumptions under the specified environment. We have provided a list of safety properties in each refinement to verify the correctness of the defined system behavior. The given safety properties guarantee that all possible executions of the closed-loop system are safe, if the generated POs are successfully discharged.

This can be an effective approach to guarantee the correctness of the functional behavior and requirements of TCPS. In fact, the developed closed-loop model is not limited only to analyzing systems requirements, but can also be used for other purposes during the systems development, such as automated code generation, automated test case generation, use as an evidence for safety assurance cases, and to assist in evaluating the product for certification purpose. This approach also provides proof artifacts, including formal models and safety properties that can help to obtain the certification requirements. For instance, the generated formal model of CRT and heart, including safety properties, may help the FDA when certifying the CRT pacemakers. The final results of this experiment show that the closed-loop model has potential to verify the TCPS requirements, to identify newly emergent behavior, to check requirements consistency, and to strengthen the given systems requirements. Although the TCPS modeling and environment modeling can be formalized in many ways by using various techniques, the modeling methodologies for developing the closed-loop system and analyzing the systems requirements described in this chapter would remain the same.

9.7 RELATED WORK

Requirements analysis is an important and challenging phase in the software development lifecycle for eliciting, analyzing, and recording systems requirements according to stakeholder needs. The recorded requirements must be precisely defined, unambiguous, formally verified, documented, and traceable by covering the possible requirements [9]. Most common errors related to software requirements are listed in Reference 28 that presents a detailed survey on the types of errors and associated root causes for failure of critical systems. These root causes include be human error, process flaws, and program faults. There are several popular techniques for requirements analysis, such as simulation and prototyping. A prototype is an early model of a product that contains partial features of the system when the systems requirements are unclear or indefinite [29]. An approach is proposed in Reference 30 for constructing a prototype using an algebraic specification language and for executing the developed specification. A run-time technique for monitoring the systems requirements is presented in Reference 31 that allows the developer to monitor violating properties of the system behavior and to adapt the new dynamic behaviors by satisfying the higher-level system goals.

Ian et al. [32] proposed an approach to specify the requirements and environment of the system, and then capture the assumptions on the physical components by recording rely-conditions, and later derive a specification of the computational part of the control system. Moreover, the proposed approach does not claim that the developed system can be perfectly safe, but it claims that the proposed approach will help to identify the assumptions for physical components of the system and ensure that the requirements are formally documented. Butler et al. [33] proposed an approach for modeling an action system using refinements that contains both the system and its environment. This approach allows to abstract away from the communication mechanism between system and physical environment, and clearly specifies the system assumptions in order to meet a changing environment. The model is developed as a unit and the elements are then separated. This approach is evaluated through formal reasoning of the Steam Boiler. Kishi et al. [34] proposed an environment modeling approach for designing embedded systems and rectifying possible bugs. A new language is proposed for modeling an environment and specifying the dynamic behavior for simulating the virtual environment in Reference 35. Several other papers [36,37] also report work on environment modeling and simulation using different techniques.

An environment modeling is an essential approach that is not limited to simulation only, but can be used for checking the desired systems requirements and during systems testing. Auguston et al. [38] propose an environmental behavior model based on Attributed Event Grammar for testing the embedded systems. Heisel et al. [39] discuss the testing approach by using the specified requirements of the system and environment models in the UML state machines. Based on environmental constraints, a testing approach is presented in Reference 40 for describing the behavior of synchronous reactive software using temporal logic.

Méry et al. [18] proposed the first heart model considering all the required normal and abnormal behaviors in the Event-B modeling language. Formal techniques based a closed-loop model of the cardiac pacemaker, for one- and two-electrodes and the heart is presented in References 19 and 41. This approach is based on formal modeling and verification of the cardiac pacemaker. We have adopted this approach for modeling the closed-loop system of the TCPS and environment, and we have used a case study, the CRT pacemaker and heart, as an example to demonstrate the results and benefits.

9.8 CONCLUSION

The trustworthy cyber-physical systems (TCPS) are dependable critical systems that play a major role in several industrial sectors related to medicine, avionics, and transportation. In response to increasing demand for new technology and growing interest in developing safe and dependable TCPS, we have discussed an approach for requirements analysis using closed-loop modeling. We have developed the closed-loop model, an integration of TCPS and environment, to identify new emergent behavior, missing system requirements, validating assumptions, identifying undesired states of the system, strengthening the given requirements, *what-if* analysis, and to check inconsistencies in the given TCPS. In our work, for modeling both the TCPS and environment models by supporting the *correct-by-construction* approach, we use the Event-B modeling language and Rodin tools [12,13] for managing, developing, verifying, and simulating the desired requirements under the given safety constraints.

To demonstrate the effectiveness of the closed-loop modeling approach, our goal is to illustrate by integrating the formal models of the CRT pacemakers and heart to model the closed-loop system for verifying the desired behavior under relevant safety properties, and to be able to guarantee the correctness of the CRT pacemaker's functional behavior. Our experiment involves formalizing and reasoning about aspects of the behavior of a CRT pacemaker, and allows actuators to pace appropriately whenever required, under normal and abnormal heart conditions. A set of general and patient condition-specific temporal requirements is specified that is used to formalize an interactive and physiologically relevant closed-loop model for verifying the basic and complex operations of the CRT pacemaker. With the use of model checkers, we demonstrate that the proposed system is capable of testing both common and complex heart conditions across a variety of CRT pacemaker modes. This system is a step toward a modeling approach for medical cyber-physical systems with the patient in the loop.

Applying the closed-loop approach for developing the TCPS has many benefits: the exposure of errors which might have not been detected without the environment model; validation of the given assumptions; increased confidence and decreased risk of failure; and promotion of the use of the closed-loop modeling approach for identifying emergent behavior and improving systems requirements for developing quality TCPS. Moreover, this approach also allows us to consider the feedback from domain experts by simulating the desired behavior. There are scientific and legal applications for using the closed-loop modeling approach for better understanding, identifying the desired functional behavior, improving the systems requirements, and meeting certification requirements for developing a dependable TCPS.

REFERENCES

1. M.C. Bujorianu and H. Barringer. An integrated specification logic for cyber-physical systems. In *Engineering of Complex Computer Systems, 2009 14th IEEE International Conference on*, Potsdam, Germany, pp. 291–300, June 2009.
2. W.H. Maisel, M. Moynahan, B.D. Zuckerman, T.P. Gross, O.H. Tovar, D.-B. Tillman, and D.B. Schultz. Pacemaker and ICD generator malfunctions: Analysis of food and drug administration annual reports. *JAMA*, 295(16):1901–1906, 2006.
3. International Standard Organization. ISO 26262: Road Vehicles—Functional Safety, 2011.
4. J. McDermid. *Software Engineer's Reference Book*. CRC Press, Inc., Boca Raton, FL, USA, 1991.
5. IEEE Standard Glossary of Software Engineering Terminology. *IEEE Std 610.12–1990*, pp. 1–84, Dec 1990.
6. I. Sommerville and P. Sawyer. *Requirements Engineering: A Good Practice Guide*. John Wiley & Sons, Inc., New York, NY, USA, 1st edition, 1997.

7. Bubenko, J.A., Jr. Challenges in requirements engineering. In *Requirements Engineering, Proceedings of the Second IEEE International Symposium on*, Washington, DC, pp. 160–162, March 1995.

8. D. Firesmith. Common requirements problems, their negative consequences, and the industry best practices to help solve them. *Journal of Object Technology*, 6(1):17–33, 2007.

9. A. van Lamsweerde. Goal-oriented requirements engineering: A guided tour. In *Requirements Engineering, 2001. Proceedings. Fifth IEEE International Symposium on*, Toronto, ON, Canada, pp. 249–262, 2001.

10. D. Méry and Neeraj Kumar Singh. Analyzing requirements using environment modelling. In *Digital Human Modeling—Applications in Health, Safety, Ergonomics and Risk Management: Ergonomics and Health—6th International Conference, DHM 2015, Held as Part of HCI International*, pp. 345–357, Los Angeles, CA, USA, August 2–7, 2015.

11. S. Serge Barold, R.X. Stroobandt, and A.F. Sinnaeve. *Cardiac Pacemakers Step by Step*. Futura Publishing, Oxford, United Kingdom, 2004. ISBN 1-4051-1647-1.

12. J.-R. Abrial. *Modeling in Event-B: System and Software Engineering*. Cambridge University Press, New York, NY, USA, 1st edition, 2010.

13. Project RODIN. Rigorous open development environment for complex systems. http://rodin-b-sharp.sourceforge.net/, 2004.

14. V.N. Bayes de Luna, A. Batcharov, and M. Malik. In A. John Camm, Thomas F. Lüscher, and Patrick W. Serruys, editors. *The Morphology of the Electrocardiogram in The ESC Textbook of Cardiovascular Medicine*. Blackwell Publishing Ltd., Oxford, UK, 2006.

15. R. Plonsey and J. Malmivuo. *Bioelectromagnetism*. Oxford University Press, New York, 1995. ISBN 0-19-505823-2.

16. Societe franaise cardiologie Jean-Yves Artigou, Jean-Jacques Monsuez. *Cardiologie et maladies vasculaires (Heart and Vascular Diseases)*. Elsevier Masson, Paris, 2006.

17. M. Gabriel Khan. *Rapid ECG Interpretation*. Humana Press, Totowa, NJ, 2008.

18. D. Méry and Neeraj Kumar Singh. Formalization of heart models based on the conduction of electrical impulses and cellular automata. In Z. Liu and A. Wassyng, editors, *Foundations of Health Informatics Engineering and Systems*, volume 7151 of *Lecture Notes in Computer Science*, pp. 140–159. Springer, Berlin Heidelberg, 2012.

19. Neeraj Kumar Singh. *Using Event-B for Critical Device Software Systems*. Springer-Verlag GmbH, 2013.

20. D. Méry and Neeraj Kumar Singh. *Technical Report on Formalisation of the Heart Using Analysis of Conduction Time and Velocity of the Electrocardiography and Cellular-Automata*. Technical Report, http://hal.inria.fr/inria-00600339/en/, 2011.

21. A.R.A. Anderson. A hybrid mathematical model of solid tumour invasion: The importance of cell adhesion. *Mathematical Medicine and Biology*, 22(2):163–186, 2005.

22. J. von Neumann. *Theory of Self-Reproducing Automata*. University of Illinois Press, Champaign, IL, 1966 Edited by A.W. Burks.

23. J.M. Greenberg and S.P. Hastings. Spatial patterns for discrete models of diffusion in excitable media. *SIAM Journal on Applied Mathematics*, 34(3):515–523, 1978.

24. Neeraj Kumar Singh, M. Lawford, T.S.E. Maibaum, and A. Wassyng. Formalizing the cardiac pacemaker resynchronization therapy. In Vincent G. Duffy, editor, *Digital Human Modeling. Applications in Health, Safety, Ergonomics and Risk Management, HCII 2015*, LNCS. Springer International Publishing, Los Angeles, CA, 2015.

25. https://www.bostonscientific.com/content/dam/bostonscientific/quality/education-resources/english/acl_cross-chamber_blanking_20081219.pdf.

26. M. Leuschel and M. Butler. *ProB: A Model Checker for B*, pp. 855–874. LNCS. Springer, Berlin Heidelberg, 2003.

27. H.D. Mills. Stepwise refinement and verification in box-structured systems. *IEEE Computer*, 21(6):23–36, 1988.

28. R.R. Lutz. Analyzing software requirements errors in safety-critical, embedded systems. In *Requirements Engineering, 1993, Proceedings of IEEE International Symposium on*, San Diego, CA, pp. 126–133, Jan 1993.

29. A.M. Davis. Operational prototyping: A new development approach. *IEEE Software*, 9:70–78, September 1992.

30. J. Goguen and J. Meseguer. Rapid prototyping: in the obj executable specification language. *SIGSOFT Software Engineering Notes*, 7:75–84, April 1982.

31. S. Fickas and M. S. Feather. Requirements monitoring in dynamic environments. In *Proceedings of the Second IEEE International Symposium on Requirements Engineering*, RE '95, pp. 140–147, Washington, DC, USA, 1995. IEEE Computer Society.

32. I.J. Hayes, M.A. Jackson, and C.B. Jones. Determining the specification of a control system from that of its environment. In K. Araki, S. Gnesi, and D. Mandrioli, editors, *FME 2003: Formal Methods*, volume 2805 of *Lecture Notes in Computer Science*, pp. 154–169. Springer, Berlin Heidelberg, 2003.

33. M.J. Butler, E. Sekerinski, and K. Sere. An action system approach to the steam boiler problem. In *Formal Methods for Industrial Applications, Specifying and Programming the Steam Boiler Control (the book grew, out of a Dagstuhl Seminar, June 1995)*, pp. 129–148, London, UK, 1996. Springer-Verlag.

34. N. Noda and T. Kishi. Aspect-oriented modeling for embedded software design. In *Software Engineering Conference, 2007. APSEC 2007. 14th Asia-Pacific*, pp. 342–349, Dec 2007.

35. G. Karsai, S. Neema, and D. Sharp. Model-driven architecture for embedded software: A synopsis and an example. *Science of Computer Programming*, 73(1):26–38, September 2008.

36. K. Choi, S. Jung, H. Kim, and Doo hwan Bae. Uml-based modeling and simulation method for mission-critical real-time embedded. In *System Development, IASTED Conference on Software Engineering 2006*, Innsbruck, Austria, 160–165. Mittal, Zeigler and De la Cruz, 2006.

37. C. Kreiner, C. Steger, and R. Weiss. Improvement of control software for automatic logistic systems using executable environment models. In *Euromicro Conference, 1998. Proceedings. 24th*, volume 2, pp. 919–923, Aug 1998.

38. M. Auguston, James Bret Michael, and M.-T. Shing. Environment behavior models for automation of testing and assessment of system safety. *Information and Software Technology*, 48(10):971–980, 2006.

39. D. Hatebur, M. Heisel, and Thomas Santen Dirk Seifert. Testing against requirements using uml environment models. In *Fachgruppentreffen Requirements Engineering und Test, Analyse & Verifikation*, pp. 28–31, 2008.

40. L. du Bousquet, F. Ouabdesselam, J.-L. Richier, and N. Zuanon. Lutess: A specification-driven testing environment for synchronous software. In *Proceedings of the 21st International Conference on Software Engineering*, ICSE '99, pp. 267–276, New York, NY, USA, 1999. ACM.

41. D. Méry and Neeraj Kumar Singh. Closed-loop modeling of cardiac pacemaker and heart. In J. Weber and I. Perseil, editors, *Foundations of Health Information Engineering and Systems*, volume 7789 of *Lecture Notes in Computer Science*, pp. 151–166. Springer, Berlin Heidelberg, 2013.

CHAPTER 10

Stop-and-Go Adaptive Cruise Control

A Case Study of Automotive Cyber-Physical Systems

Sasan Vakili, Neeraj Kumar Singh, Mark Lawford,
Alan Wassyng, and Ben Breimer

CONTENTS

Automotive cyber-physical systems (Auto-CPS) are safety critical systems in which failure can lead to grave consequences, such as financial loss, severe injuries, or even loss of life. Auto-CPS consists of tightly coupled interactions of computational units and physical systems that involve interplay among embedded systems, control theory, real-time systems, and software engineering. Increases in software-related recalls of vehicles are driving demand for methods to cost-effectively produce safe, secure, and reliable cars with more than 100 electronic control units (ECUs) communicating over vehicle networks. Since formal methods have played a significant role in developing safe and dependable systems, the role of formal methods in developing Auto-CPS should be taken into account. In this work, we will demonstrate the benefits and limitations of formal methods for Auto-CPS by considering a relatively new driver assistance feature.

Stop and Go Adaptive Cruise Control (ACC+) is an extension of Adaptive Cruise Control (ACC) that assists drivers by regulating the speed of the driver's vehicle relative to the vehicle it is following for speeds from 0 to a drive set maximum cruising speed. In this chapter, we present the formal verification of a practical, robust ACC+ design using differential dynamic logic ($d\mathcal{L}$) to formally state environmental assumptions and prove safety goals, including collision freedom. The verification is done in two stages. First, we identify the invariant required to ensure the safe operation of the system and we formally verify that the invariant preserves the safety property of any system with similar dynamics. This procedure provides a high-level abstraction of a class of safe solutions for ACC+ system designs. Second, we show that our ACC+ system design is a safety preserving refinement of the abstract model. The safety of the closed-loop ACC+ system is proven by verifying bounds on the system variables using the KeYmaera verification tool for hybrid systems. The work provides a method that could be used to verify more complicated ACC+ controller designs that are optimized for fuel economy, passenger comfort, etc. We use our formal specification to create a Matlab/Simulink implementation to validate the performance of the proposed designs and compare our design with other formally verified ACC designs from the literature.

10.1 INTRODUCTION

A promising and challenging application area for formal methods in the automotive domain is specifying and designing software for embedded systems since such systems are exhibiting exponential growth in complexity [1]. Software has become a decisive factor in the automotive industry; increasing demands for high quality, safety requirements, and the shortcomings of the informal techniques applied in traditional development have motivated the examination of semiformal or formal methods to facilitate a higher degree of automation and tool support for verification and validation activities to ensure safety.

Every year, car crashes result in the loss of thousands of lives and permanent disabilities, resulting in annual costs of billions of dollars in the United States alone [2]. Although the majority of these car crashes are due to human error, failure in hardware or software components can lead to accidents and unduly risk human life [3]. While hardware failures are typically some kind of *random failure* that is caused by different wear effects such as corrosion, thermal stressing, etc., software failures are *systematic failures* that may be introduced by human error during the system development, and these failures always appear in the same circumstances until the error is removed. However, it is difficult to predict the occurrence of systematic software-related failures or detect all of them by classical means such as testing and inspection [4].

Formal verification is an effective approach to help ensure the safety and reliability of complex hybrid systems that can provide an additional level of confidence. This work contributes to the formal verification of automotive hybrid systems by using *dynamic logic* ($d\mathcal{L}$) to analyze the correctness of a high-level ACC+ design. The verification method used assures safety of the system is robust with respect to variations in the plant models. Formal development and verification of hybrid systems using dynamic logic provides satisfaction of safety and performance requirements if the models used correctly represent the system. Using parametric constraints, we find a region of safe operation for a continuous controller when we have an upper bound limit on response time in the presence of disturbances and uncertainties. In addition to the advantages of formal verification given above, making system descriptions more precise can serve to expose problematic parts in the requirements. Based on the formal model of the system, we want the analysis techniques to establish the system correctness to be consistent with requirements. The main contributions of this chapter are

1. A new high-level design for ACC+

2. Formalization of the ACC+ requirements using $d\mathcal{L}$

3. Formal verification of the new design's safety properties using the KeYmaera tool [5]

The outline of the remainder of the chapter is as follows. Section 10.2 presents related work. A brief outline of ACC+ and the tools and techniques used in the verification of the system are described in Section 10.3. Section 10.4 presents a high-level design of an ACC+ controller, and Section 10.5 describes the formal verification of the ACC+ design. Section 10.6 concludes the chapter and discusses future work.

10.2 RELATED WORKS

Automotive control is a wide and interesting area that has been studied by academic and industrial researchers in an effort to minimize the risk and to improve the safety of driving. Since these kinds of systems deal with the safety of humans, even a small error or mistake in the design of these systems can lead to irreparable harm. Therefore, sufficient assurance is necessary before deployment of any system.

Guven et al. [6] presented a low-cost, real-time driver-in-the-loop vehicle simulator that was used to analyze ACC behavior for highway traffic. Several papers [7–10] reported work on the simulation of ACC. However, these simulations are not enough to guarantee that the tested system is safe and collision-free in all traffic conditions.

Our ACC+ design is a hybrid system. Hybrid systems integrate both continuous and discrete dynamics, bringing together several research fields to address challenging problems in order to demonstrate safe operations. Logic plays a significant role in formal verification of hybrid systems from reachability analysis to undecidability in theory and practice.

Several approaches have been proposed in the literature to verify safety properties of hybrid systems. An inductive method proposed in Reference 11 is based on the PVS theorem prover to verify the required safety properties of the parallel hybrid systems. SMT solvers have been used to verify safety properties and time-bounded reachability of a class of "reasonable" parametric linear hybrid automata [12]. Stursberg et al. [13] presented a counterexample-guided verification approach based on a model checker to identify unwanted behaviors in the verification of a cruise control system, where a sequence of abstractions was used to reduce the computational cost. Jairam et al. [14] present verification of a MEMS-based ACC system using simulation and semiformal approaches, where they have used Matlab Simulink to develop the case study and validate their system using a transformation-based approach. An interesting piece of work is presented in Reference 15, where a theoretically ACC system is described by process algebra (*timed distributed pi-calculus*) to analyze the informal requirements of the system, and some properties such as deadlocks are then verified using the Mobility Workbench model checker.

Platzer et al. proposed dynamic logic and proof calculus for verifying hybrid systems [16,17]. Dynamic logic considers continuous evolutions between discrete behaviors via transitions between the states. In the past few years Loos et al. have published several papers applying KeYmaera to ACC [18,19]. All these papers address fundamental principles of ACC, including important safety properties in various scenarios. However, none of them provide a feasible solution for implementation purposes. For instance, Loos et al. [18] discussed formal verification of ACC considering the required following distance for avoiding collision when there are arbitrarily many cars driving on a highway, including the case in which another car enters the lane. Another paper, Loos et al. [19] proposed ACC modeling based on different acceleration choices for different modes of operation using various conditions, however, it has overlap in the modes that causes thrashing due to improper guard conditions defined in the paper. The mode thrashing has the potential to result in changes of acceleration that would be unacceptable in terms of driver comfort and fuel efficiency. The second and fourth controller mode in Reference 19 is unreachable regardless of the plant model. Moreover, Loos et al. [19] proposed an acceleration formula for the third controller mode using a square root term involving variables such as communication time (τ). However, the optimal τ assumed as the maximum communication time (3.2 s) is unrealistic for a real-time application. These two proposed solutions for ACC have not taken any desired set point velocity or distance into account, which is also required to consider when formalizing a complete system that could serve as a potential ACC system. In another paper, Aréchiga et al. [20] proposed a PID controller based on the previous paper [18] to maintain a desired distance between two successive vehicles when the host vehicle follows a lead vehicle, describing acceleration in terms of position and velocity of vehicles. However, a large desired following distance is attained in their controller design, which makes the system unrealistic. The system gets into an unsafe region if a small desired following distance is considered. The problem with this system is that the specified operating regime for the controller is in the unsafe boundary for small set point values and the controller will not take any action in some unsafe scenarios. Therefore, only unrealistically large following distance set point values can be used in order to satisfy the safety condition.

Our motivation is to provide a complete solution for a practical, generic ACC+ system design that guarantees the safety properties outlined in References 18 through 20 while incorporating practical ACC+ design requirements such as headway reference tracking and respecting the user-specified maximum velocity constraint, both of which are missing from the works of Loos et al. Our proposed solution allows the host vehicle to maintain a desired velocity when there is no slower lead vehicle or obstacle, and to safely approach a slower lead vehicle to a desired safe following distance. Moreover, we investigate the system's safety critical behavior as a separate mode of operation used to guarantee safety and collision freedom properties.

10.3 PRELIMINARIES

10.3.1 ACC+

In order to understand an ACC+ system, one should first understand prior iterations such as cruise control (CC) and adaptive cruise control (ACC) systems. The main function of ACC is to maintain a desired speed that is set by the driver unless a slower "lead vehicle" is encountered ahead of the driver's vehicle—the "host vehicle" for the ACC system. For example, if the current speed of a lead vehicle is slower than the speed of the host vehicle, then the ACC system starts to control the current speed of the host vehicle to maintain a desired safe distance between lead vehicle and host vehicle. Automatic adjustment of the host vehicle's acceleration allows the host vehicle to adjust its speed according to traffic conditions without driver intervention. Any required braking action carried out by an ACC system will typically not exceed 30% of the host vehicle's maximum deceleration. When stronger deceleration is needed, the driver is warned by an auditory signal and a warning message is displayed on a driver-information screen. The driver can override the ACC system at any time to take back control of the vehicle. The ACC system must guarantee that it will always behave correctly and safely while respecting rules regarding passenger comfort (i.e., trying to avoid excessive changes in acceleration). Consideration of traffic conditions can be included in an ACC system design to help decrease traffic congestion by trying to provide a smooth flow of traffic. An ACC system has a minimum threshold speed (e.g., 30 km/h) below which it stops operating, hence an ACC system does not deal with stop and go traffic [21,22].

Stop and Go Adaptive Cruise Control, also known as Adaptive Cruise Control Plus (ACC+), is a system that operates at all velocities greater than or equal to 0 km/h. It is an extension of the ACC system that is basically a superset of the features found in ACC. In particular, ACC+ is designed to provide controllability in very low-speed driving scenarios. In Reference 23 the development of an ACC+ system is discussed, including some of the challenges that arise from low-speed driving such as smaller intervehicular spacing and more frequent changes in velocity. An ACC+ system must obtain information about its environment, such as the speed of the lead vehicle, the user's requested speed, what constitutes a safe distance between the host vehicle and the lead vehicle, etc., in order to meet its requirements. This can be achieved by using a set of sensors that monitor the environment at sufficiently high sampling rates to capture the continuous behavior of the hybrid system in sufficient detail. For example, many ACC+ systems use a frequency modulated continuous wave (FMCW) Doppler radar sensor mounted on the front of the vehicle to measure the distance to the lead vehicle and its relative velocity [24]. Since the system must both accelerate and decelerate the host vehicle based on the information obtained from the environment, an ACC+ system must be able to adjust the throttle and/or brake using appropriate control signals.

Uncertainty is another fact that cannot be ignored in the design of this system. No mathematical model can represent the exact physical system [25]. Therefore, different types of uncertainty should be taken into account during the process of controller design. Considering all of these aspects makes the system complex and increases the difficulty in assuring safety and correctness. The goal of this work is to design a robust ACC+ system that provides sufficient assurances of safety for typical operating conditions.

10.3.2 Differential Dynamic Logic (d$\mathcal{L}$)

Differential dynamic logic (d$\mathcal{L}$) is a first-order dynamic logic for specification and verification of hybrid systems. A program notation is used to describe hybrid systems as hybrid programs, with symbolic parameters being used during the verification process in dynamic logic. A free variable sequential composition proof calculus with real arithmetic and quantifier elimination allows deductive verification of hybrid programs [16,17,26].

Symbolic parameters are represented by a set of logical variables in first-order logic, while dynamic logic describes the continuous behavior of a system. This logic can be used to verify the operation of a system with discrete and continuous state transitions by introducing hybrid programs with discrete assignments and differential actions and then applying a deductive method rather than using abstractions and exhaustive state-space exploration as is typically done in model checking approaches [27]. The limited knowledge of d$\mathcal{L}$ needed to understand this chapter is summarized below. More detailed descriptions are available in References 16 and 26.

Dynamic logic (d$\mathcal{L}$) consists of nonlinear real arithmetic, real-valued quantifiers, and modal operators, such as $\langle \alpha \rangle$ or $[\alpha]$ for expressing reachable state conditions during system execution, where α presents the continuous evolution of a system. A set of logical variables V, a signature Σ, and a set of real-valued function and predicate symbols are used to define the well-formed terms and formulas that are given as follows:

$$\theta ::= x \mid f(\theta_1, \ldots, \theta_n)$$

where $\theta_1, \ldots, \theta_n$ are terms, f is a function symbol of arity n, and x is a real-valued constant symbol.

$$\phi, \psi ::= p(\theta_1, \ldots, \theta_n) \mid \neg\phi \mid \phi \wedge \psi \mid \phi \vee \psi \mid \phi \rightarrow \psi \mid \forall x \phi \mid \exists x \phi$$

where ϕ and ψ are first-order formulas, θ_i are terms, p is a predicate symbol of arity n, and $x \in V$ is a logical variable.

Hybrid programs consist of discrete jump sets, systems of differential equations, and a control structure. The discrete transitions assign values to the state variables, and the differential equations are used to express a continuous dynamic evolution of the system, which may change from one discrete state to another. The control structure plays an important role in combining the discrete and continuous transitions using regular expression operators, such as ($\cup$, *, ;). The grammar for designing the hybrid programs is given as follows:

$$\alpha, \beta ::= x_1 := \theta_1, \dots, x_n := \theta_n \mid x_1' = \theta_1, \dots, x_n' = \theta_n \& \chi \mid ?\chi \mid \alpha \cup \beta \mid \alpha; \beta \mid \alpha^*$$

where α and β are hybrid programs, θ_i are terms, $x_i \in \Sigma$ are state variables, and χ is a formula of first-order logic. $x_1 := \theta_1, \dots, x_n := \theta_n$ shows a discrete jump, in which θ_i assigns to state variables x_i. $x_1' = \theta_1, \dots, x_n' = \theta_n \& \chi$ presents a list of differential equations for describing dynamic behavior with additional first-order constraints χ. $?\chi$ and $\alpha \cup \beta$ are used to test the state variables and represent nondeterministic choice, respectively. $\alpha; \beta$ and α^* present sequential composition and nondeterministic repetition, respectively. Dynamic logic (d$\mathcal{L}$) can be used to design other structures by combining the control structure operators ($\cup$, *, ;) with $?\chi$, such as in conditional statements such as **if** χ **then** α **else** β, **while** χ **do** α. Formulas of dynamic logic (d$\mathcal{L}$) based on first-order logic together with some modal operators ($\langle \alpha \rangle$ or $[\alpha]$) are defined as follows:

$$\phi, \psi ::= p(\theta_1, \dots, \theta_n) \mid \neg\phi \mid \phi \wedge \psi \mid \phi \vee \psi \mid \phi \rightarrow \psi \mid \forall x \phi \mid \exists x \phi \mid [\alpha]\phi \mid \langle \alpha \rangle \phi$$

where ϕ, ψ are dynamic logic (d$\mathcal{L}$) formulas, θ_i are terms, p is a predicate symbol of arity n, $x \in V$ is a logical variable, and α is a hybrid program. The syntax of dynamic logic (d$\mathcal{L}$) allows real arithmetic predicate expressions, negation, conjunction, disjunction, implication, universal and existential quantification, and modalities to express the validity of formula ϕ for any terminating execution of hybrid program α ($[\alpha]\phi$) or at least one terminating execution of hybrid program α ($\langle \alpha \rangle \phi$).

10.3.3 Verification Tool: KeYmaera

KeYmaera [5] is a hybrid verification tool integrated with an automated and an interactive theorem prover to formalize and verify hybrid systems. It supports dynamic logic (d$\mathcal{L}$) and combines different methods, such as deductive logic, real algebraic, and computer algebraic rules. Moreover, KeYmaera also supports nonlinear discrete jumps, nonlinear differential equations, differential-algebraic equations, differential inequalities, and nondeterministic discrete or continuous input for hybrid systems to express the functional behaviors. KeYmaera allows decomposition of hybrid system specifications into symbolic form and into subsystems to simplify the proof strategy. However, a bottom-up approach employing compositional verification allows KeYmaera to verify large, complex systems by proving the required properties of the subsystems and then the main system.

10.4 A HIGH-LEVEL SAFETY CONCEPT ABSTRACTION FOR ACC+ SYSTEMS

ACC+ systems can be formalized using *differential dynamic logic* (d$\mathcal{L}$) to state and prove safety properties and performance requirements by capturing the system constraints together with the desired behaviors and controller designs. A high-level conceptual safety design of ACC+ is proposed in this section, where we consider that the host vehicle is equipped with ACC+, and the host vehicle follows a lead vehicle in the same lane. We use a high-level conceptual model of the ACC+ system to formalize an abstraction of the systems requirements to satisfy the desired safety properties. According to References 18 and 20, collision freedom for these kinds of systems can be achieved if and only if there is always a safe distance between two successive vehicle. This distance, which we will denote by sc_{gap}, can be derived from Newton's formula of motion as in Equation 10.1, where B is the absolute value of maximum deceleration achieved by maximum brake force, and v_l and v_h are lead and host vehicles' velocities, respectively.

$$sc_{gap}(v_l, v_h) = \frac{v_h^2 - v_l^2}{2 \times B} \tag{10.1}$$

The length of $sc_{gap}(v_l, v_h)$ should be such that the host vehicle can fully stop at the rear end of the lead vehicle or end of $sc_{gap}(v_l, v_h)$ in the worst case scenario when the lead vehicle may itself be suddenly using the same maximum brake force to come to a full stop. In the case in which the relative distance between the two vehicles is less than or equal to this safe distance, the host vehicle has no choice but to use its maximum braking power to exit the critical zone in order to make the system collision free. This fact is critical to the safe, collision-free operation of any ACC or ACC+ design.

In addition, the system uses sensors to provide the required values for the control system; however, because there is some lag associated with acquiring sensor readings, the controller needs some time to react to any new sensor values, and the actuators take some time to react. Therefore, a safety margin should be taken into account related to the maximum delays in the system. This extra padding distance can be determined by the following formula 10.2 according to Reference 18.

$$margin_{sc_{gap}}(v_h) = \left(\frac{A_{max}}{B} + 1 \right) \left(\frac{A_{max}}{2} \times \epsilon^2 + \epsilon \times v_h \right) \tag{10.2}$$

Here, A_{max} is the maximum acceleration of the host vehicle and ϵ is the worst case delay time, which is close to zero. Equation 10.2 is considered as the worst case scenario where the host vehicle is traveling with maximum acceleration (A_{max}) when the ACC+ system requests the maximum negative acceleration B. The host vehicle will continue to accelerate at A_{max}, increasing its velocity v_h for ϵ seconds before it starts to decelerate at $-B$. Therefore, the extra distance given in Equation 10.2 is required for acceleration $-B$ to return the host vehicle to what was its initial velocity, v_h, when the negative deceleration was first requested. Consequently, $margin_{sc_{gap}}(v_h)$ is the total of these two distances that the host vehicle travels during the ϵ delay. Thus, the ACC+ system can react safely if the relative distance between host and lead vehicle ($d_{gap} = x_l - x_h$) is always greater than the sum of $sc_{gap}(v_l, v_h)$ and $margin_{sc_{gap}}(v_h)$ as in Equation 10.3.

$$d_{gap} > sc_{gap}(v_l, v_h) + margin_{sc_{gap}}(v_h) \tag{10.3}$$

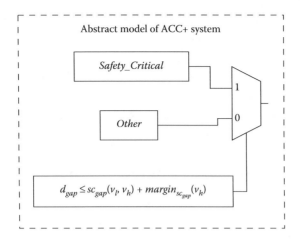

FIGURE 10.1 High-level abstract conceptual design block diagram.

TABLE 10.1   Decision-Making Structure of Abstract ACC+

	a_h	Mode
$d_{gap} \leq sc_{gap}(v_l, v_h) + margin_{sc_{gap}}(v_h)$	$-B$	Safety_Critical
$d_{gap} > sc_{gap}(v_l, v_h) + margin_{sc_{gap}}(v_h)$	$[-B, A_{\max}]$	Other

Therefore, the High Level Safety Concept Abstraction for ACC+ can be specified as shown in Figure 10.1. This figure depicts an independent safety system that intervenes only when necessary. This system monitors the relative distance to the lead vehicle (d_{gap}) and sets the host vehicle acceleration a_h to $-B$ whenever the relative distance is less than or equal to safety distance ($d_{gap} \leq sc_{gap}(v_l, v_h) + margin_{sc_{gap}}(v_h)$). Effectively it activates a *Safety_Critical* mode that applies the maximum brake force. Four components have been considered for this purpose: a *guard condition* block, a switching block, *safety_Critical*, and *Other*. The *guard condition* block checks the validity of the safety condition and the switching block changes the active mode from *Other* (normal ACC+) functionalities to *Safety_Critical* in critical cases when $d_{gap} \leq sc_{gap}(v_l, v_h) + margin_{sc_{gap}}(v_h)$ becomes valid. A tabular representation of this decision-making structure is shown in Table 10.1, where *Other* is used to consider all other behaviors ACC+ could have in different scenarios and *Safety_Critical* is used to apply maximum brake.

10.4.1 Verification

Since some of the system parameters come from the environment or are yet to be determined by a more detailed design, we need to define symbolic constraints for some parameters such as acceleration of the vehicles. Before doing this we first define the state variables of the host and lead vehicles that will be used to model their continuous behaviors:

$$host = (x_h, v_h, a_h) \tag{10.4}$$

$$leader = (x_l, v_l, a_l) \tag{10.5}$$

where x_h is position, v_h is velocity, a_h is acceleration of the host vehicle, and x_l is position, v_l is velocity, and a_l is acceleration of the lead vehicle. These variables from the tuples can then be used to specify the dynamics of the real-time system, where the relationships between position, velocity, and acceleration are $x'_h = v_h$ and $v'_h = a_h$ for the host vehicle, and $x'_l = v_l$ and $v'_l = a_l$ for the lead vehicle.

The velocity of the host (lead) vehicle changes continuously according to the current acceleration of the host (lead) vehicle. We assume the maximum acceleration for both the host and lead vehicles is $A_{max} > 0$, and similarly the maximum deceleration due to braking with the maximum braking force is $-B$ where $B > 0$. Therefore,

$$-B \leq a_h \leq A_{max} \quad \& \quad -B \leq a_l \leq A_{max} \tag{10.6}$$

The complete formalization of our abstract ACC+ is presented in Model 10.1. The model contains both discrete and continuous dynamic behaviors. Model 10.1 can be derived in a similar fashion to Reference 18, where Loos et al. also define an abstract model for an autonomous vehicle. However, Model 10.1 presented here is simpler and more abstract than the model in Reference 18. The *Local Lane Control* of the ACC system in Reference 18 always sets the acceleration of the host vehicle to zero in the case that its velocity is zero. Thus once stopped, the ACC system in Reference 18 remains stopped regardless of the behavior of the lead vehicle. Also, *Local Lane Control* in Reference 18 chooses a nondeterministic brake value within a particular range for the safety critical situation, which makes the system more complicated than a safety concept abstraction. Despite its complexity, *Local Lane Control* [18] is more realistic than a basic safety concept abstraction since it might not always be possible to achieve $-B$, for example when a road is wet.

The host and lead vehicles can repeatedly choose an acceleration from the range $[-B, A_{max}]$ in Model 10.1. This behavior is specified by the nondeterministic repetition $*$ in line (1). The host and lead vehicles operate in parallel as defined in (2). The lead vehicle is free to use brake or acceleration at any time so a_l is assigned nondeterministically in (3), and the model continues if a_l is within its accepted range $[-B, A_{max}]$.

The host vehicle's movement depends on the distance between the host vehicle and the lead vehicle. The most crucial functionality of ACC+ is formalized as successive actions to capture the decision on entering the safety critical mode as the last action in (4) before the system's continuous state is updated. The *safety critical following distance* ($sc_{gap}(v_l, v_h)$) and the *extra safety margin for delays* ($margin_{sc_{gap}}(v_h)$) are calculated in (5). The last line in (5) assigns the relative distance to d_{gap}. The host vehicle can choose any arbitrary acceleration value in the valid range $-B$ to A_{max} for the *Other* mode in (6) to capture all dynamic behaviors of possible ACC+ system designs. The safety requirement that the system applies maximum brake force when the host vehicle is within the safe following distance is formalized as the overriding action of the *Safety_Critical* mode in line (7). The continuous state of the system then evolves over time that is measured by a clock variable t. The sampling time of the system has been considered as the delay of the system $t \leq \epsilon$ where slope is considered as $t' = 1$. Therefore, the system is piecewise continuous and the physical laws for movement as formalized by simplified versions of Newton's formula are contained in line (8).

$$
\begin{aligned}
\text{ACC+} &\equiv \text{(Vehicle; Drive)}^* &(1)\\
\text{Vehicle} &\equiv \text{host} \mathbin{||} \text{leader}; &(2)\\
\text{leader} &\equiv a_l := {*}; ?(-B \le a_l \le A_{\max}) &(3)\\
\text{host} &\equiv \text{Calc\_sc}_{gap}; \text{Other}; \text{Safety\_Critical}; &(4)\\
\text{Calc\_sc}_{gap} &\equiv sc_{gap}(v_l, v_h) := \tfrac{v_h^2 - v_l^2}{2 \times B}; \\
&\quad margin_{sc_{gap}}(v_h) := (\tfrac{A_{\max}}{B} + 1)(\tfrac{A_{\max}}{2} \times \epsilon^2 + \epsilon \times v_h); \\
&\quad d_{gap} := x_l - x_h; &(5)\\
\text{Other} &\equiv a_h := {*}; ?(-B \le a_h \le A_{\max}); &(6)\\
\text{Safety\_Critical} &\equiv \text{if} \left(d_{gap} \le sc_{gap}(v_l, v_h) + margin_{sc_{gap}}(v_h) \right) \text{ then} \\
&\quad a_h := -B \\
&\quad \text{fi}; &(7)\\
\text{Drive} &\equiv t := 0; (x_h' = v_h \wedge v_h' = a_h \wedge x_l' = v_l \wedge \\
&\quad v_l' = a_l \wedge t' = 1 \wedge v_h \ge 0 \wedge v_l \ge 0 \wedge t \le \epsilon) &(8)
\end{aligned}
$$

Model 10.1 Formalization of an abstract model for ACC+ systems.

With the system dynamics specified, we can now use the KeYmaera [5] tool to verify the required collision-freedom safety property.

Property 10.1. *If the host vehicle is following at a safe distance behind the lead vehicle, then the vehicles will never collide in any operation when the host vehicle controllers follow the defined dynamics given by the safety constraints.*

In KeYmaera this property will take the form:

$$
\text{Controllability Condition} \rightarrow [\text{Abstract ACC+}] \ x_h < x_l \tag{10.7}
$$

The controllability condition will be given below in Equation 10.8. We now explain how we will arrive at the appropriate precondition for the safety property. To complete (10.7), we must establish a precondition that says that the host vehicle is behind the lead vehicle and both vehicles are moving in a forward direction. The relation (10.7) indicates that for all iterations of the hybrid program in Model 10.1 the position of the host vehicle is always less than the lead's vehicle's position ($x_h < x_l$) if the given controllability condition is satisfied. In other words, the relative distance between the vehicles is always greater than zero ($d_{gap} > 0$) if the precondition holds. One of the most important conditions is the safe distance formula, which is an invariant during the proof of this hybrid program. This condition can be considered as a controllability property and must always be satisfied by every operation of the ACC+ system.

10.4.2 Controllability

The controllability formula states that for every possible evolution of the ACC+ system, it can satisfy the safety property by applying maximum brake before it has passed the

Safety_Critical distance. The vehicle is controllable if there is enough distance in order to fully stop the car by the rear end of the lead vehicle or it exits the critical zone. The assumption is that both vehicles only move forward (i.e., their velocity is greater than or equal to zero). Therefore, the ACC+ system will be safe if it can satisfy condition (10.8), which is an invariant for the defined systems dynamics of Model 10.1. This controllability property in condition (10.8) is a safety concept invariant not only for ACC+ systems, but also for any kind of system with similar continuous motion dynamics.

$$x_l > x_h \wedge v_h^2 - v_l^2 < 2 \times B \times d_{gap} \wedge v_l \geq 0 \wedge v_h \geq 0 \tag{10.8}$$

An important fact in this verification is that there must be required distance to make it physically possible to stop the host vehicle by the rear end of any obstacle appearing in front of the host vehicle. This fact has been mathematically indicated in (10.8) as $v_h^2 - v_l^2 < 2 \times B \times d_{gap}$. The system checks whether it can satisfy the safety property in case of detecting any obstacle and once it enters the critical zone, it uses maximum brake until the safety property holds again.

This model has been written in the KeYmaera theorem prover [5] and the required safety property (10.7) has been successfully proven. In this abstract model of an ACC+ system, we considered a viable range of accelerations for the host vehicle that admits a variety of desired behaviors for a concrete ACC+ system in different scenarios. The focus here is the desired behavior of any of these concrete models in the safety critical case required to guarantee the safety requirement of collision freedom. In the next section, we will refine this system with respect to other requirements to create a more realistic concrete ACC+ design, the safety of which has already been proven if we can show the new design refines this abstract ACC+ safety concept.

10.5 REFINEMENT OF THE SAFETY CONCEPT INTO A PRACTICAL ACC+ MODEL

An ACC+ design requires information about the host vehicle's continuous state (velocity, acceleration, etc.), as well as information about the presence and behavior of the lead vehicle. While the most important requirement for ACC+ systems is to safely adjust the host vehicle's speed in the presence of a lead vehicle, some additional functional requirements and assumptions have to be considered in the design of ACC+ systems. Assumptions can help to make the design more reliable and practical. Also, understanding additional functional requirements can allow us to scope the design and verification effort.

Assumptions:

1. The ACC+ system will never be operating when the vehicle is moving backward (velocity < 0).

2. The driver is responsible for steering the host vehicle in a safe manner.

3. It is assumed that the maximum range of the sensors for detecting objects in front of the host vehicle (d_{range}) is always greater than the safety gap obtained in the previous

section (Section 10.4):

$$d_{range} > sc_{gap}(v_l, v_h) + margin_{sc_{gap}}(v_h).$$

4. Errors will be detected by a separate subsystem, a Fault Detection System, that will alert the driver to intervene in the case of a fault.

Given Requirements:

1. The user has the ability to override the ACC+ system settings such as desired velocity v_{set} and desired headway h_{set}, at any point in the system's operation except in safety critical cases.

2. The accessible parameters of the ACC+ system, such as desired velocity v_{set} and desired headway h_{set}, should be restricted to an acceptable range in order to meet the assumptions and limitations of the design.

3. The ACC+ system must regulate the velocity of the host vehicle to maintain the user's expected velocity in the absence of a slower lead vehicle.

4. The ACC+ system must slow down the host vehicle's velocity and maintain the desired headway when approaching a slower lead vehicle.

5. The acceleration of the system must be restricted to a comfortable range. Therefore, rapid deceleration should not be applied during the normal operation of the ACC+ system.

6. The ACC+ system should return the operation of the vehicle to the user in the presence of any failure in the system or when the throttle/brake is touched by the user.

Among all the requirements, we consider the implementation of the first to fifth one in our design. The third and fourth requirements, which are not typically discussed in other related works such as [18,19], play a major role in our ACC+ design. The restriction on v_{set}, as described by the second requirement, is derived in Sections 10.5.1 and 10.5.2. The required restriction on h_{set} can be derived in a similar fashion to v_{set}. The sixth requirement is not directly addressed in our work. It can be designed in a separate block by using fault diagnosis techniques as in Reference 28. The ACC+ system controls the speed of the host vehicle according to the different scenarios that are considered during the high-level design. Figure 10.2 depicts a high-level design of the ACC+ that contains four components: a *low-level (continuous) controller*, an extended *finite state machine* (FSM), a *sensor*, and the *host vehicle*. We consider the fifth component, the *lead vehicle*, as being external to the ACC+ system. All the components of the ACC+ system are connected by arrows that represent the system data flow. Thus this block diagram shows the flow of information that is required to design the ACC+ system, providing the relationship between the ACC+ subsystems and the lead vehicle. *Mode* is the value of the current state of the FSM that is used by the low-level controller to select a particular continuous controller. The value of *Mode* belongs to the set

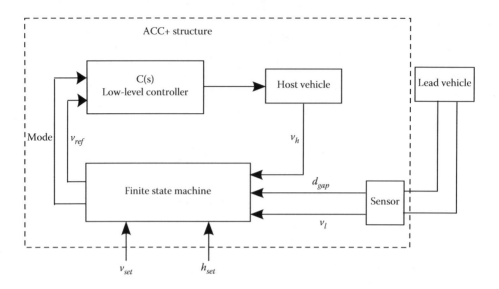

FIGURE 10.2   High-level conceptual design block diagram.

TABLE 10.2   Terms Used in ACC+ Specification and Controller Design

Term	Definition
	Specification Terms
v_h	Velocity of the host vehicle
v_l	Velocity of the lead vehicle
a_h	Acceleration of host vehicle
a_l	Acceleration of lead vehicle
d_{gap}	Relative distance between the host vehicle and lead vehicle
	Controller Terms
v_{set}	Desired velocity of the host vehicle
B	Absolute value of deceleration achieved by maximum brake force, which depends upon the current vehicle weight and road conditions
h_{set}	Desired following time gap between two successive vehicles (headway)
ϵ	Maximum response delay from any actuators (i.e., engine, brake, etc.)
$f_{gap}(v_l, v_h, a_h)$	The distance it takes for the host vehicle to match the lead vehicle's velocity and be following at the desired headway h_{set} using acceleration a_h
$sc_{gap}(v_l, v_h)$	The distance at which the ACC+ system switches into safety critical mode

{*Cruise, Follow, Safety_Critical*}. Signal v_{ref} is a reference signal for the target velocity for the continuous controller selected inside the low-level controller. A list of the other symbols for describing vehicle behavior is given in Table 10.2.

10.5.1   Controller Modes

There are three main operational modes of our ACC+ design (see Figure 10.4). These modes are as follows:

Cruise, which implements a standard cruise control system (CC) when no lead vehicle is detected or the lead vehicle exceeds the desired maximum velocity of the host vehicle (v_{set}) and is outside of the safety critical region.

Follow, which tries to match the lead vehicle's velocity at distance $h_{set} \times v_l$.

Safety_Critical, where the vehicle has to apply maximum braking force to avoid a collision as discussed in the previous section (Section 10.4).

Figure 10.3a and b shows the headway diagrams describing the possible scenarios. The first mode is similar to a conventional cruise control system (CC) that regulates the speed of the host vehicle to the desired set point (v_{set}) within acceleration limits based on requirements such as comfort and fuel efficiency. If the host vehicle detects a leader or other object, the

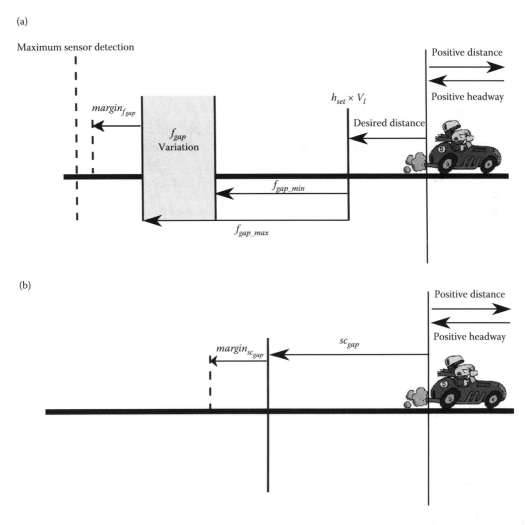

FIGURE 10.3 Required distance in *Follow* and *Safety Critical* mode. (a) Required distance for approaching the lead vehicle in *Follow* mode. (b) Minimum required distance in *Safety_Critical* mode.

system determines whether or not the sensed object is going faster than v_{set}. If the lead vehicle is traveling faster than v_{set} and is outside of the safety critical zone ($d_{gap} > sc_{gap}(v_l, v_h)$) then the ACC+ system will not change its operating mode.

The second mode, *Follow*, becomes active when the host vehicle follows a slower lead vehicle outside the safety critical zone. In this situation, the objective is to maintain the desired headway gap $h_{set} \times v_l$ while diverse aspects such as driver comfort, fuel economy, etc., are considered. When a slower lead vehicle is present, the goal is to reduce the host vehicle's velocity as it approaches the lead vehicle, matching the lead vehicle's velocity when the gap closes to the desired headway $h_{set} \times v_l$. To achieve this behavior the system picks a negative acceleration for the host vehicle ($a_h < 0$). An additional restriction for the host vehicle's acceleration a_h is the maximum available deceleration B available by applying full brake force, that is, $a_h \geq -B$. For a chosen value of a_h in this range, the distance required to reduce the host vehicle's velocity to match the lead vehicle's velocity at the desired following distance $h_{set} \times v_l$ is $f_{gap}(v_l, v_h, a_h)$. The size of $f_{gap}(v_l, v_h, a_h)$ is derived from Newton's formula of motion as follows:

$$f_{gap}(v_l, v_h, a_h) = \frac{v_h^2 - v_l^2}{-2 \times a_h}, \quad (-B \leq a_h < 0) \tag{10.9}$$

Equation 10.9 describes the distance required for the host vehicle to achieve v_l as its new velocity, where $-B \leq a_h < 0$ is the deceleration of the host vehicle. According to Equation 10.9, if the host vehicle wants to use a negative, constant acceleration a_h to achieve the leader's velocity by the time it reaches following distance $h_{set} \times v_l$, it has to start decelerating at distance $f_{gap}(v_l, h_v, a_h)$. Note that once the host vehicle achieves the leader's velocity the size of $f_{gap}(v_l, v_h, a_h)$ will become zero (Equation 10.9).

Based on the constraints on acceleration and the maximum range of the distance sensor, there are constraints on possible values for f_{gap} to a value between f_{gap_min} and f_{gap_max}. As shown in Figure 10.3a, the system bounds on $f_{gap}(v_l, v_h, a_h)$ by restricting the values of a_h that the ACC+ system will use. The upper bound f_{gap_max} and lower bound f_{gap_min} will be defined based upon the upper and lower bound of $a_h < 0$. An upper bound of a_h is the minimum deceleration that the ACC+ system will use by, for example, easing up on the throttle at the current vehicle velocity, while a lower bound is achieved by the maximum braking deceleration B that the vehicle can generate based on the current vehicle weight and road conditions (i.e., $a_h \geq -B$).

To make the system more realistic, another safety margin has been taken into account related to the system delay ϵ that is required to respond to messages from the ACC+ system to the engine and brake controllers and the time they require to activate their respective actuators and have them respond. This margin can be determined by the following formula:

$$margin_{f_{gap}}(v_h, a_h) = \left(\frac{A_{max}}{-a_h} + 1 \right) \left(\frac{A_{max}}{2} \times \epsilon^2 + \epsilon \times v_h \right) \tag{10.10}$$

The size of this margin for the response delay (Equation 10.10) can be derived in similar fashion to the derivation of $margin_{sc_{gap}}(v_h)$ (Equation 10.2) by replacing the maximum braking deceleration B with the deceleration a_h. The value of a_h is considered to be negative in all of the formulas given when approaching the lead vehicle, assuming a slower lead vehicle. Once the host vehicle reaches the leader's velocity, it will attempt to track the lead vehicle's velocity and those formulas are no longer required. Finally, the *Follow* mode will be activated if the relative distance between vehicles, d_{gap}, is less than or equal to $f_{gap}(v_l, v_h, a_h) + margin_{f_{gap}}(v_h, a_h) + (h_{set} \times v_l)$ but greater than the safety critical distance. Note that the value of $f_{gap}(v_l, v_h, a_h)$ becomes negative in the case when the leader's velocity is greater than the host vehicle's velocity ($v_l > v_h$). Therefore, the system always chooses the $\max(f_{gap}(v_l, v_h, a_h), 0)$ for system safety. Although $h_{set} \times v_l$ converges to zero as v_l goes to zero, $margin_{f_{gap}}(v_h, a_h) > 0$ ensures a minimum following distance.

$$d_{gap} \leq \max(f_{gap}(v_l, v_h, a_h), 0) + margin_{f_{gap}}(v_h, a_h) + (h_{set} \times v_l) \qquad (10.11)$$

The velocity of the host vehicle should be in a range such that the right side of Equation 10.11 is within the maximum range of the *Sensor*, as shown in Figure 10.3a. Assume the *Sensor* component of the ACC+ system shown in Figure 10.2, which measures the velocity and position of the lead vehicle relative to the host vehicle, has a maximum range of d_{range} meters. Then the maximum v_{set} that can be employed by the system assuming a realistically comfortable deceleration a_h as 30% of maximum deceleration B ($a_h = -0.3B$), time delay ϵ in the response of the system, and a worst case zero velocity of lead vehicle ($v_l = 0$) results in the following equation:

$$v_{set} \leq \sqrt{2 \times 0.3B \times (d_{range} - margin_{f_{gap}}(v_{set}, -0.3B))} \qquad (10.12)$$

Note that Equation 10.12 represents an approximation of v_{set} because $margin_{f_{gap}}(v_h, a_h)$ has been considered to be a fixed value.

The third mode is the *Safety_Critical* mode that activates when a vehicle suddenly cuts in the lane or an obstacle appears in front of the vehicle, and the relative distance is less than or equal to the minimum stopping distance for the vehicle when full braking power is applied. In this case, the host vehicle has no choice but to use its maximum braking power to exit the critical zone where $d_{gap} \leq sc_{gap}(v_l, v_h)$ (see Figure 10.3b). Although this situation should not normally occur when the host vehicle is in the *Follow* mode, it may happen in critical situations such as a cut-in scenario. Figure 10.3b illustrates $sc_{gap}(v_l, v_h)$. The size of this zone and the margin for the response delay have been derived in Equations 10.1 and 10.2. According to the discussion in Section 10.4, this mode implies safety and collision-freedom for any ACC+ system. Note that in this scenario, we assume the same maximum braking deceleration for both the host and lead vehicles.

10.5.2 Mode Switching

It may, in fact, be the case that the lead and host vehicles have different values for the maximum brake deceleration; hence, the equation for the distance $sc_{gap}(v_l, v_h)$ (Equation 10.1) may be changed accordingly. Let us assume that the maximum negative acceleration due to maximum brake force for host and lead vehicles are B and b, respectively, then the value of $sc_{gap}(v_l, v_h)$ becomes

$$sc_{gap}(v_l, v_h) = \frac{v_h^2}{2 \times B} - \frac{v_l^2}{2 \times b} \qquad (10.13)$$

The procedure for deriving this version of sc_{gap} is trivial. The ACC+ system switches to *Safety_Critical* mode when the relative distance becomes less than or equal to the value of $sc_{gap}(v_l, v_h)$ as formalized in Equation 10.14.

$$d_{gap} \leq sc_{gap}(v_l, v_h) + margin_{sc_{gap}}(v_h) \qquad (10.14)$$

According to Equation 10.1, the safety critical gap shown in Figure 10.3b converges to zero when the host vehicle attains the same velocity as the lead vehicle (i.e., $sc_{gap}(v_l, v_h) \rightarrow 0$). This is under the assumption that both vehicles have the same maximum braking deceleration. In the case in which the maximum braking deceleration differs (Equation 10.13), $sc_{gap}(v_l, v_h) > 0$ when $v_h = v_l$ and $B < b$, providing the extra required braking margin due to the lesser maximum deceleration of the host vehicle. In the case when $B > b$, $sc_{gap}(v_l, v_h) < 0$ when $v_h = v_l$ so we take the maximum of $sc_{gap}(v_l, v_h)$ and 0.

Figure 10.4 and Table 10.3 are alternative representations of the high-level design of our ACC+ system. Figure 10.4 shows the finite state machine (FSM) with the three major modes as separate states. Guard conditions are attached to transitions in this figure. The tabular representation of transition from one state to another is given in Table 10.3. Here the notation $Mode_{-1}$ denotes the previous mode of the FSM.

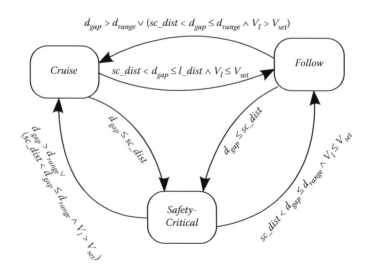

FIGURE 10.4 Finite state machine.

TABLE 10.3   Decision-Making Structure of ACC+

						Mode
$d_{gap} \leq d_{range}$		$d_{gap} \leq sc\_dist$				Safety_Critical
	$d_{gap} > sc\_dist$	$v_l > v_{set}$				Cruise
		$v_l \leq v_{set}$	$d_{gap} \leq l\_dist$			Follow
			$d_{gap} > l\_dist$	$Mode_{-1} \neq Cruise$		Follow
				$Mode_{-1} = Cruise$		Cruise
$d_{gap} > d_{range}$						Cruise

Let

$$l\_dist := \max(f_{gap}(v_l, v_h, a_h), 0) + margin_{f_{gap}}(v_h, a_h) + (h_{set} \times v_l)$$

$$sc\_dist := \max(sc_{gap}(v_l, v_h), 0) + margin_{sc_{gap}}(v_h) \text{ in Table 10.3.}$$

The actual value of deceleration applied in the *Follow* state could be chosen to be comfortable for the user while achieving a high level of fuel economy, traffic flow, and safety. This deceleration can be described as minimizing a_h. The point at which the vehicle switches to the *Follow* mode can be determined by an optimization process selecting a_h and then the desired velocity reference will be computed using Equation 10.15 and sent from the FSM to the low-level controller.

$$v_{ref} = \sqrt{v_l^2 - 2 \times a_h \times (d_{gap} - v_l \times h_{set})} \tag{10.15}$$

Equation 10.15 can be derived based on Equation 10.11 for the normal following action where $-B \leq a_h < 0$. This velocity reference signal is defined for the case in which the host vehicle detects a slower lead vehicle and the ACC+ system needs to decrease the velocity such that the host vehicle can achieve the leader's velocity by the desired headway. The term under the square root in Equation 10.15 will not be negative as long as the velocities are greater than or equal to zero. In Equation 10.15 $d_{gap} - v_l \times h_{set}$ represents the distance it takes for the host vehicle to achieve the leader's velocity. If this term is less than zero $(v_l^2 - 2 \times a_h \times (d_{gap} - v_l \times h_{set}) < 0)$, it means that the host vehicle should move backward, which is in contradiction with the first assumption of the ACC+ system. Therefore, the system always picks a maximum value between this term $(v_l^2 - 2 \times a_h \times (d_{gap} - v_l \times h_{set}))$ and zero (i.e., $max(v_l^2 - 2 \times a_h \times (d_{gap} - v_l \times h_{set}), 0)$).

In a cut-in scenario when the velocity of the lead vehicle (v_l) and relative distance (d_{gap}) change abruptly from one set of specific values to another, there is no continuous trajectory for the parameters of the system. It is possible that the *Mode* of the system was *Follow* before the change and remains in *Follow* after updating the sensor values for the new leader. For example, the host vehicle is decreasing the velocity of the host in the *Follow* mode to achieve a leader's velocity by the desired headway when suddenly another lead vehicle l_{new} with velocity less than v_{set}, that is, $v_{l_{new}} < v_{set}$, cuts into the lane. However, the host vehicle's velocity is less than this new leader's velocity ($v_h < v_{l_{new}}$) and the relative distance between the host vehicle and new leader is not less than safety critical distance (i.e., $d_{gap_{new}} > \max(sc_{gap}(v_l, v_h), 0) + margin_{sc_{gap}}(v_h))$. In this example, the ACC+ system will not change the mode of operation and will remain in *Follow*. Therefore, the ACC+ system may accelerate to match the new leader's velocity by the desired headway. After updating different parameters such as v_l and d_{gap}, the range of a_h for this purpose can be between 0 and A_{max}. Finally, Equation 10.15 can be used as v_{ref} with $0 \le a_h \le A_{max}$. The optimization process should find a valid value for a_h from this range based on the relative distance, the distance it takes the host vehicle to achieve the leader's velocity, and the desired headway. However, it should be bounded so that the system will not thrash between *Follow* and *Safety_Critical*. This is a subset of the behavior we are formally verifying. Although test cases do not reveal any thrashing in this scenario, further analysis and formal verification are necessary to ensure that the system is free of thrashing. This formal analysis is left to future work that could be done using techniques such as those of Reference 29.

In the *Safety_Critical* state the desired velocity v_{ref} is set to zero by the FSM. The goal is to continuously apply the maximum brake force to get the host vehicle out of this critical zone. Therefore, the maximum brake command is passed to the low-level controller. In this case, the reference signal will be zero velocity with the *Safety_Critical* mode signal from the FSM to the Low-Level Controller being interpreted as "apply the maximum brake power and close the throttle."

Typically it is important in hybrid systems design to avoid rapid mode switching. In the ACC+ system as designed, mode switching between states could cause rapid changes in acceleration which is not comfortable for passengers. Thus in Table 10.3 we avoid rapid mode switching by using hysteresis. When $v_l \le v_{set}$ and $d_{gap} > l\_dist$, the table checks the previous value of the FSM state, denoted *Mode*$_{-1}$. The system remains in *Cruise* if the previous value of *Mode* is *Cruise* (*Mode*$_{-1}$ = *Cruise*), otherwise (*Mode*$_{-1}$ ≠ *Cruise*) it remains in *Follow* or switches from *Safety_Critical* to *Follow*. This behavior is similarly defined in the finite state machine Figure 10.4. Once the current state becomes *Follow* or *Safety_Critical*, the FSM will switch to *Cruise* only in the case that the leader is traveling faster than v_{set} or in the absence of a lead vehicle or object ($d_{gap} > d_{range} \vee (sc\_dist < d_{gap} \le d_{range} \wedge v_l > v_{set})$). Consequently, the FSM changes state from *Cruise* to *Follow* only in the case that $d_{gap} \le l\_dist$. The only other potential source of mode thrashing is between *Follow* and *Safety_Critical*. According to Table 10.4, the reference velocity signal in *Follow* mode is defined as in Equation 10.15. This v_{ref} ensures that in a typical following of a slower leader, the FSM does not switch back and forth between *Follow* and *Safety_Critical*. In the case that a lead vehicle cuts in the lane, once the host vehicle exits the critical distance, the ACC+ system switches from *Safety_Critical* to *Follow* through the guard condition $sc\_dist < d_{gap} \le d_{range} \wedge v_l \le v_{set}$ (Figure 10.4) and the system does not fall back in to *Safety_Critical* due to the definition

TABLE 10.4 Velocity Reference Signal with Respect to the State

Mode	v_{ref}
Cruise	v_{set}
Follow	$\sqrt{\max(V_l^2 - 2 \times a_h \times (d_{gap} - v_l \times h_{set}), 0)}$
Safety_Critical	0

of v_{ref} in *Follow* mode (Equation 10.15). Finally, we can provide v_{ref} for the continuous controller based upon the *Mode* of operation (Figure 10.2). Table 10.4 defines the value of v_{ref} for each *Mode*.

The desired objective in this mode switching is to avoid the sudden application of full brake with the resulting severe jerk in noncritical scenarios. The system should not switch to *Safety_Critical* mode unless a lead vehicle cuts into the lane and there is not enough distance between the host and lead vehicles. There is a particular circumstance under which the mentioned desired objective may be violated during the "normal" functioning of our ACC+ system. This scenario happens when the host vehicle is traveling with reference velocity v_{set} and the ACC+ system's current *Mode* is *Cruise*. If there is a slower lead vehicle in the lane (i.e., $v_l < v_{set}$), but the ACC+ system has not yet changed its *Mode* to *Follow*, we expect the system to switch from *Cruise* to *Follow* when $d_{gap} \leq l\_dist$. However, if the host vehicle's driver decides to change the value of v_{set} to a new value that is less than v_l (i.e., $v_{set_{new}} < v_l$), then according to Figure 10.4 and Table 10.3, the ACC+ system will not switch the *Mode* from *Cruise* to *Follow* after $d_{gap} \leq l\_dist$. Therefore, the ACC+ system will try to maintain the new desired velocity $v_{set_{new}}$ in *Cruise* mode. In this situation, for a fixed acceleration $a_h < 0$, the required distance for the host vehicle to slow to $v_{set_{new}}$ can be obtained from the following formula:

$$dist_v = \frac{v_h^2 - v_{set_{new}}^2}{-2 \times a_h}, \quad (-B \leq a_h < 0) \tag{10.16}$$

If this required distance is greater than or equal to the difference between d_{gap} and the safety critical distance $sc_{gap}(v_l, v_h)$, the system will eventually transition directly from *Mode Cruise* to *Safety_Critical*. Therefore, some additional functionality should be defined in *Cruise* to avoid this undesired behavior. The system should restrict the driver to choosing the set point velocity from a range of values that do not lead to a full brake in *Safety_Critical* mode. This range of values can be derived from the above explanation, and is formulated in Equation 10.17.

$$d_{gap} - sc_{gap}(v_l, v_h) > dist_v \tag{10.17}$$

Note that in Equation 10.17, $margin_{sc_{gap}}(v_h)$ is not considered for the sake of simplicity. A lower bound for $v_{set_{new}}$ can be calculated by replacing $sc_{gap}(v_l, v_h)$ and $dist_v$ in Equation 10.17 with their formulas. Equation 10.18 demonstrates the lower bound of v_{set} in *Cruise* mode when the velocity of the lead vehicle (v_l) is lower than the initial set point velocity v_{set} and the driver decides to change v_{set} to a value lower than v_l.

$$v_{set_{new}} > \sqrt{v_h^2 \left(\frac{B - a_h}{B}\right) + 2a_h \left(d_{gap} + \frac{v_l^2}{2B}\right)} \tag{10.18}$$

This lower bound for v_{set} in Equation 10.18 is only for avoiding *Safety_Critical* mode in normal operation of the ACC+ system. Although the continuous controller in *Cruise* mode manipulates the throttle to decrease or increase the velocity, some percentage of brake can be added to the control action in *Cruise* mode. Therefore, a_h can be picked, for instance, as 10% of the maximum deceleration achieved by full brake B ($a_h = -0.1B$). Equation 10.19 provides the lower bound for v_{set} by considering 10% of B.

$$v_{set_{new}} > \sqrt{1.1v_h^2 - 0.1v_l^2 - 0.2 \times B \times d_{gap}} \qquad (10.19)$$

Note that, if the term under the square root in Equations 10.18 or 10.19 becomes negative, it means that $v_{set_{new}}$ can be any value greater than zero ($v_{set_{new}} \geq 0$). Therefore, the lower bound, derived in Equation 10.19, can be defined by the maximum function in Equation 10.20.

$$v_{set_{new}} > \sqrt{\max(1.1v_h^2 - 0.1v_l^2 - 0.2 \times B \times d_{gap}, 0)} \qquad (10.20)$$

As a conclusion, *Cruise* mode operation should be refined based on the following conditions:

No leader/Faster leader: v_{set} can be defined in the interval from zero to the upper limit in Equation 10.12.

Slower leader: v_{set} can be defined in the interval from the lower limit in Equation 10.20 to the upper limit in Equation 10.12.

The implementation of this conditioning operation in *Cruise* mode has been left for future work.

10.5.3 Continuous Controller Design

Other than the case in which we are in *Safety_Critical* mode, the Low-Level (continuous) controller can implement a standard continuous feedback controller designed to meet tracking and disturbance rejection performance requirements. In mode *Cruise* we can use a simple Single-Input Single-Output (SISO) controller to try to have v_h track v_{ref}. In the case when we are in mode *Follow*, we have to use a Multiple-Input Multiple-Output (MIMO) controller in order to have v_h track the v_l at a distance of d_{gap}. A typical performance specification of a control system is "good tracking" of the reference signal(s). This is usually interpreted as asymptotic tracking of a single step or ramp reference signal and is commonly met with a standard design such as a PID controller. However, PID controllers typically do not maintain reasonable performance in the presence of uncertainties in the plant model and set of reference signals. Therefore, robust controller design techniques have been developed to achieve performance in terms of a weighted norm bound that result in strong performance in the presence of plant uncertainties such as weight of the loaded vehicle, friction of the road, etc. Among all possible structured and unstructured uncertainties, we choose simple disk-like multiplicative uncertainty to simplify our analysis. We then design our Low-Level Controller based on the Loopshaping analysis technique of Reference 25. As a result, our ACC+ system attains reliable performance in the presence of plant uncertainty for a variety of reference signals (Table 10.4).

10.5.4 Simulation Results

In this section, a test case is presented in Figure 10.5 to evaluate the behavior of the proposed ACC+ system design on a scale model vehicle. Although all the possible scenarios cannot be captured with one test case, we try to capture the most significant behaviors to examine the performance of our ACC+ system.

Figure 10.5 illustrates the behavior of the host vehicle controlled by our ACC+ system in various conditions such as when a lead vehicle is present at varying velocities or absent. The first plot in Figure 10.5 depicts the velocity behavior of the lead vehicle (v_l), measured in centimetre per second (cm/s). According to this plot, the leader starts from an initial velocity of zero ($v_l = 0$) and changes its velocity as shown. The host vehicle starts at a certain desired headway at the beginning of the simulation. Therefore, the host vehicle tracks the leader's velocity in *Follow* mode until the leader travels faster than the host vehicle's set point velocity

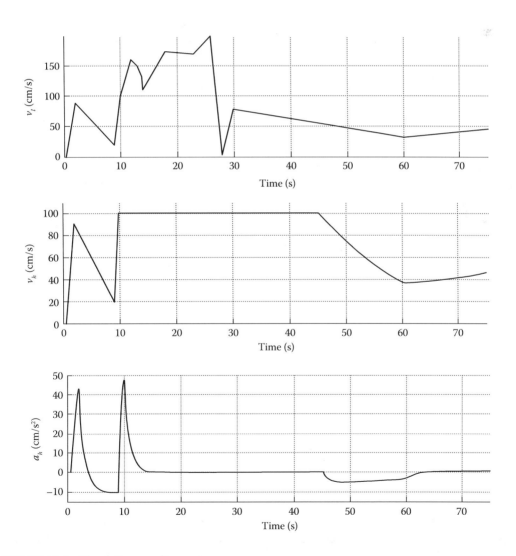

FIGURE 10.5   Simulation results.

(v_{set}) at time $t = 10$. This behavior can be seen in the second plot of Figure 10.5, where the host vehicle's velocity is shown. The host vehicle's set point velocity is defined as 100 cm/s ($v_{set} = 100$). Therefore, the second plot of Figure 10.5 shows the host vehicle tracking the leader's velocity until it exceeds $v_{set} = 100$ cm/s and the ACC+ mode of operation is changed from *Follow* to *Cruise*. As shown in the first plot of Figure 10.5, the lead vehicle decreases its velocity to lower values than the host's v_{set} at approximately $t = 27$. However, the ACC+ system does not immediately change its mode of operation and the host vehicle continues with $v_h = 100$ cm/s in *Cruise* mode as long as the relative distance between the two vehicles is greater than the required distance for following the leader. The ACC+ system changes the mode from *Cruise* to *Follow* only when d_{gap} becomes less than or equal to l_dist. This occurs around $t = 45$ when the host vehicle decreases its velocity in order to attain the leader's velocity by the desired headway. Consequently, the host vehicle matches its velocity with that of the lead vehicle while maintaining the desired headway.

The third plot of Figure 10.5 shows the host vehicle's acceleration. According to this plot, the acceleration increases when the host vehicle is initially accelerating its velocity from zero to approximately 40 cm/s^2, while the host vehicle's velocity increases from zero to 90 cm/s. The acceleration does not become negative right after the host vehicle starts decreasing its velocity from 90 to 20 cm/s. The reason for the host vehicle still having a positive acceleration, even when the lead vehicle is decelerating, is the presence of an integral term in the continuous controller. When initially developing the safety verification of the refined ACC+ controller that appears in the following section, the specification stated that if $v_l < v_h$ in *Follow* mode, then the acceleration of the host vehicle chosen by the controller must satisfy $a_h < 0$. Clearly, a reasonable linear control system design, such as the one simulated in Figure 10.5, does not satisfy this property. The verification in the following subsection was modified to allow positive accelerations even in the case when $v_l < v_h$ precisely for this reason. The lesson here is that one has to be careful to make sure that the formal model that is verified faithfully models the actual system.

The host vehicle changes its acceleration in order to maintain a required velocity. However, due to the nature of the continuous controller, the acceleration control it generates does not change instantaneously due to the continuous dynamic of the system. This ACC+ system attains the leader's velocity by a desired headway if the leader travels more slowly than the host vehicle's set point velocity. In addition, this system will continue to track the leader' velocity after the desired headway is achieved. In the absence of a slower leader, the system's objective is to track a desired set point velocity. As shown in Figure 10.5, our ACC+ system behaves safely and will not switch to *Safety_Critical* mode in this particular normal operation scenario. This control structure is not conservative because the required safety constraints are considered as a separate mode of operation and do not affect the operation of *Cruise* and/or *Follow* modes. Therefore, the required safety constraints and the desired performance could be obtained simultaneously by this design in this scenario.

10.5.5 Verification

In this section, we provide a formalization for the refined mode switching of the ACC+ system described in Section 10.5.2 using *differential dynamic logic* (d$\mathcal{L}$). This formalization is presented in Model 10.2. We use the dynamic operations of the host and lead vehicles as defined in Section 10.4 (Equations 10.4 and 10.5). The lead vehicle behavior is the same as that in Model 10.1 in Section 10.4, where acceleration can be chosen from the valid range (Equation 10.6) (Line(3)). The nondeterministic repetition * and parallel operation of the host and lead vehicles has already been defined in Model 10.1 (Line (1-2)). The *Other* functionality of the ACC+ system in Model 10.1 is now formalized as successive actions to capture the other driving modes. The host vehicle controller takes action in a more restricted manner. Safety constraints must be satisfied related to relative distance, velocities, and the selected velocity for cruise mode. The host vehicle has three different operating modes that are represented sequentially in (4).

Three operating modes, *Cruise*, *Follow*, and *Safety_Critical*, always use the current value of $sc_{gap}(v_l, v_h)$, $margin_{sc_{gap}}(v_h)$, and d_{gap} in Line (5) to randomly choose the desired acceleration nondeterministically within the valid range to control the speed of the host vehicle under given safety margins.

The *Cruise* operating mode states that if d_{gap} is not less than or equal to $sc_{gap}(v_l, v_h) + margin_{sc_{gap}}(v_h)$ and the speed of the lead vehicle is greater than v_{set} then either the acceleration of the host vehicle a_h can be assigned nondeterministically from $-B \leq a_h \leq A_{max}$ when the speed of the host vehicle v_h is less than or equal to v_{set}, or a_h can be assigned nondeterministically from $-B \leq a_h \leq A_{max}$ when the speed of the host vehicle v_h is greater than v_{set}. This operating mode is formalized in (6), where the speed of the host vehicle always maintains according to the selected driver speed considering the speed of the lead vehicle and safety margins. The range of deceleration is formalized based on the band limit of the continuous controller. For instance, if the continuous controller has an integral term, the acceleration might continue for some time with a value greater than zero even when $v_h > v_{set}$; hence, when performing the verification the range of a_h cannot be restricted to a value between $-B$ and 0 in the case when $v_h > v_{set}$. By a similar reasoning, the range of a_h cannot be restricted to a value between 0 and A_{max} in the case that $v_h \leq v_{set}$. Thus, we consider the range of $-B \leq a_h \leq A_{max}$ for both mentioned cases. In the work of [19], the behavior of the continuous controller is not taken into account. The verified ACC controller in Loos et al. [19] may not apply to certain controllers, such as a PID controller, since the band limit of the continuous controller is not included in the verification. This concept was explained in detail in Section 10.5.4. Although the host vehicle's acceleration can be formalized based on the continuous controller (i.e., for a PID controller $a_h := k_1 + k_2 \times \int v_h(t).dt + k_3 \times (d/dt)v_h(t)$), we use the possible physical range $(-B \leq a_h \leq A_{max})$ to capture a class of possible continuous controllers. Accordingly, the continuous controller can be an arbitrary design without any concern about the safety properties of the ACC+ system.

The *Follow* operating mode is applicable only when the speed of the lead vehicle is less than or equal to the driver-selected desired speed (i.e., $v_l \leq v_{set}$), and the vehicle separation is not in the safety critical zone. The *Follow* operating mode is specified in statement (7) of Model 10.2 that specifies that if d_{gap} is not less than or equal to $sc_{gap}(v_l, v_h) + margin_{sc_{gap}}(v_h)$ and the speed of the lead vehicle is less than or equal to v_{set}, then the acceleration of the host vehicle a_h can be assigned nondeterministically based on the current status of the host and lead vehicles' velocities (v_h and v_l). If the host's velocity is greater than the leader's velocity ($v_h > v_l$) then a_h should be negative ($-B \leq a_h < 0$). This case occurs normally when the ACC+ system detects a slower lead vehicle and should decrease its current velocity gradually in order to maintain v_l. Although a_h should be negative in this case, we considered its value between $-B$ and $A_{\max}$ ($-B \leq a_h \leq A_{\max}$) in the formalization. This demonstrates that the continuous controller may work with a positive acceleration for a short time interval until obtaining a negative value. Therefore, $-B \leq a_h \leq A_{\max}$ is not in contradiction with the expected behavior in *Follow* mode by the reasoning provided in Section 10.5.4. After maintaining v_l, the ACC+ system should track the leader's behavior. Therefore, if the leader accelerates, and $v_h \leq v_l$, then of the possible values in $-B \leq a_h \leq A_{\max}$ one would expect the controller to trend toward values of $a_h \geq 0$. As mentioned earlier, a_h cannot be switched from a negative to a positive value instantly due to the continuous behavior of the system. Therefore, a_h cannot be formalized between 0 and $A_{\max}$ in the last case when $v_h \leq v_l$, otherwise significant jerk may occur in the system. As a result, $-B \leq a_h \leq A_{\max}$ is a reasonable range to be considered for any arbitrarily continuous controller.

The ACC+ system can make either of these two choices according to the situation. Furthermore, the current values of $f_{gap}(v_l, v_h, a_h)$ and $margin_{f_{gap}}(v_h, a_h)$ are calculated sequentially, where the system must satisfy $d_{gap} - (h_{set} \times v_l) \leq f_{gap}(v_l, v_h, a_h) + margin_{f_{gap}}(v_h, a_h)$. The test checks that the host vehicle is within $f_{gap}(v_l, v_h, a_h)$ to make sure that the transition to *Follow* is done properly and there is enough distance to achieve v_l as the new host velocity. If the test condition does not hold ($d_{gap} - (h_{set} \times v_l) > f_{gap}(v_l, v_h, a_h) + margin_{f_{gap}}(v_h, a_h)$), then execution will fail. Therefore, this assertion forces the system operation to maintain enough distance for taking an appropriate action in the *Follow* mode. Although this test does not have any impact on the proof of safety and collision-freedom of Model 10.2, we have defined this assertion to allow the conformity of this formalization to the actual mode switching system of Section 10.5.2.

In the case that this test cannot be satisfied, there are two possible outcomes. First, there is the normal behavior in the presence of a slower leader when the sensor detects a slower leader, but the host vehicle is still able to travel at $v_h = v_{set}$ until it comes within $f_{gap}(v_l, v_h, a_h)$ of the lead vehicle. The second case for this violation is the opposite problem where there is not enough of a gap to reduce the host vehicle to v_l by the time the host vehicle is within the desired headway h_{set}. However, the $f_{gap}(v_l, v_h, a_h)$ distance is always greater than the minimum stopping distance $sc_{gap}(v_l, v_h)$ in the presence of a slower lead vehicle. Further, the maximum delay for the ACC+ system to react to a change, ϵ, has also been taken into consideration during the calculation to estimate the additional safe distance margin for $f_{gap}(v_l, v_h, a_h)$ in order to provide sufficient time for the controllers to react. Line (7) formalizes the behavior of the *Follow* mode in the case when the ACC+ system starts to decrease the host vehicle's velocity to match a slower leader's velocity by the time the host vehicle reaches distance $h_{set} \times v_l$, and then tracks the leader's velocity at an appropriate distance as long as the leader does not travel faster than v_{set}. This formalization also captures the behavior of the *Follow* mode after the host vehicle starts to track the leader's velocity. Line (8) formalizes the *Safety_Critical* mode as defined previously in Model 10.1.

The sampling time and dynamic evolution of the system are defined in Line (9), similar to Model 10.1.

The main purpose of the refined ACC+ formalization given in Model 10.2 is to investigate the safety of the ACC+ system in the presence of a leader in front of the host vehicle. Additionally, we also want to ensure that the vehicle behaves safely when switching between the different modes of operation. The host vehicle's behavior when the lead vehicle is out of range of the sensor is the same as conventional cruise control systems. The *Cruise* mode controls the speed of the host vehicle on behalf of the driver. In the current formal model of the refined ACC+ system, d_{range} has not been defined (i.e., we assume that the sensor range is effectively infinite). It is left as future work to prove the correctness of the system with a limited range sensor under the conditions outlined in (Equation 10.12).

For now, we consider that there is a lead vehicle in the same lane as the host vehicle in Model 10.2. The system checks whether it can satisfy the safety property in the case when an obstacle or lead vehicle is detected. Once the path is cleared of any obstacle or there is no longer a lead vehicle, then it can switch back to the *Cruise* mode to maintain the desired speed (v_{set}).

The proposed ACC+ design has three operating modes, where the system is switching from one mode to another according to a desired situation considering safety constraints. The safe distance formula is the most important invariant that must always be satisfied by the ACC+ system in all operating modes as stated by the Controllability property (Equation 10.8) in Section 10.4. We wrote Model 10.2 in the KeYmaera theorem prover's input language to further demonstrate that this ACC+ system design is safe and collision-free as long as the safety critical distance condition (Equation 10.8) has not been violated.

$$\text{Controllability Condition (10.8)} \rightarrow [\text{Refined ACC+}] \; x_h < x_l \qquad (10.21)$$

The precondition for the formula (10.21) is similar to that of formula (10.7). It indicates that for all iterations of the Refined ACC+ (Model 10.2), the system is collision-free ($x_h < x_l$) if the controllability condition (10.8) is satisfied. This fact confirms that the ACC+ system in Model 10.2 is a refinement of Model 10.1. Therefore any system, such as Model 10.2, will be safe as long as the controllability condition (10.8) is maintained. We will further investigate the refinement and refactoring relations between Models 10.1 and 10.2 in the next subsection. The safety of a complex model, such as Model 10.2, can be proved based on an abstract model, such as Model 10.1. The refactoring relation makes the proof procedure easier than the procedure we have done for proving relation (10.21). Additionally, the refinement relation allows designers to add new requirements to a system and/or change some parts of the system without violating the required safety properties. Consequently, it can be shown that Model 10.2 is derived from the abstract model of Section 10.4 (Model 10.1) by adding some new states and refining the system's behavior while preserving the required safety properties.

10.5.6 Safe Refactoring

Direct proof of safety and other properties of a complex CPS is often difficult if not impossible due to the complex interaction between software and hardware models. Different

$$ACC+ \equiv (Vehicle; Drive)^* \tag{1}$$

$$Vehicle \equiv host \parallel leader; \tag{2}$$

$$leader \equiv a_l = *; ?(-B \leq a_l \leq A_{\max}) \tag{3}$$

$$host \equiv Calc\_sc_{gap}; Cruise; Follow; Safety\_Critical; \tag{4}$$

$$Calc\_sc_{gap} \equiv sc_{gap}(v_l, v_h) := \frac{v_h^2 - v_l^2}{2 \times B};$$
$$margin_{sc_{gap}}(v_h) := (\frac{A_{\max}}{B} + 1)(\frac{A_{\max}}{2} \times \epsilon^2 + \epsilon \times v_h);$$
$$d_{gap} := x_l - x_h; \tag{5}$$

$$Cruise \equiv \text{if} \left(\neg(d_{gap} \leq sc_{gap}(v_l, v_h) + margin_{sc_{gap}}(v_h)) \wedge v_l > v_{set}\right) \text{then}$$
$$\left(?(v_h \leq v_{set}); a_h := *; ?(-B \leq a_h \leq A_{\max})\right) \bigcup$$
$$\left(?(v_h > v_{set}); a_h := *; ?(-B \leq a_h \leq A_{\max})\right)$$
$$\text{fi}; \tag{6}$$

$$Follow \equiv \text{if} \left(\neg(d_{gap} \leq sc_{gap}(v_l, v_h) + margin_{sc_{gap}}(v_h)) \wedge v_l \leq v_{set}\right) \text{then}$$
$$\left(?(v_h > v_l); a_h := *; ?(-B \leq a_h \leq A_{\max})\right) \bigcup$$
$$\left(?(v_h \leq v_l); a_h := *; ?(-B \leq a_h \leq A_{\max})\right)$$
$$f_{gap}(v_l, v_h, a_h) := \frac{v_h^2 - v_l^2}{-2 \times a_h};$$
$$margin_{f_{gap}}(v_h, a_h) := (\frac{A_{\max}}{-a_h} + 1)(\frac{A_{\max}}{2} \times \epsilon^2 + \epsilon \times v_h);$$
$$?(d_{gap} - (h_{set} \times v_l) \leq f_{gap}(v_l, v_h, a_h) + margin_{f_{gap}}(v_h, a_h))$$
$$\text{fi}; \tag{7}$$

$$Safety\_Critical \equiv \text{if} \left(d_{gap} \leq sc_{gap}(v_l, v_h) + margin_{sc_{gap}}(v_h)\right) \text{then}$$
$$a_h := -B$$
$$\text{fi}; \tag{8}$$

$$Drive \equiv t := 0; (x_h' = v_h \wedge v_h' = a_h \wedge x_l' = v_l \wedge$$
$$v_l' = a_l \wedge t' = 1 \wedge v_h \geq 0 \wedge v_l \geq 0 \wedge t \leq \epsilon) \tag{9}$$

Model 10.2   Formalization of the refined ACC+ system.

approaches have been investigated to overcome this fact such as over-approximating the reachable set of states and defining an abstract model in order to reduce the complexity [30]. The abstract model can then be verified for safety purposes. However, an important part of this method, which is typically disregarded in this area, is to prove that the original, complex system model is a property preserving refinement of the proposed abstraction. After verifying safety of an abstract model of a CPS, any update in any part of that model requires reverification of the whole new system. Refinement reasoning makes the reverification process easier by assuring that the new additional part of the system does not violate the safety of the whole system. Platzer and his coworkers recently proposed a refinement relation for systems described in differential dynamic logic (d$\mathcal{L}$) in [31]. Mitsch et al. [31] introduced two notions of refinement: "*Projective Relational Refinement*" and "*Partial Projective Relational Refinement.*" According to Reference 31

Projective Relational Refinement: Let $V \subseteq \Sigma$ be a set of variables. Let $|_V$ denote the projection of relations or states to the variables in V. We say that hybrid program α refines hybrid program γ w.r.t the variables in V ($\alpha \sqsubseteq^V \gamma$) iff $\rho(\alpha)|_V \subseteq \rho(\gamma)|_V$.

where ρ is the transition relation used to specify reachable states.

Although we used KeYmaera [5] to prove the safety properties (Equation 10.21) of Model 10.2 in Section 10.5.5, we want to further investigate refinement reasoning. We defined a safe abstract model of any ACC or ACC+ system in Section 10.4 and proved the collision-freedom property of that model. In this section, we want to show that Model 10.2 refines Model 10.1. The Projective Relational Refinement definition holds for Model 10.2 with respect to Model 10.1 since the reachable states of Model 10.2 are a subset of the reachable states of Model 10.1. We can thus conclude that Model 10.2 refines Model 10.1 with respect to the variables of these models (Model 10.2 $\sqsubseteq^V$ Model 10.1). Therefore, Model 10.2 inherits the collision-freedom safety property from Model 10.1. We will use refactoring methods from [31] to demonstrate the validity of this claim.

Mitsch et al. [31] developed "proof-aware refactoring" and proposed some rules with associated proof obligations to define a refinement relation in terms of refactoring. Two refactorings, *Structural* and *Behavioral*, are defined in Reference 31. "*Structural refactoring changes the structure of a hybrid program without changing its reachable states*"; whereas Behavioral refactoring partially changes the reachable states. Therefore, some auxiliary proof obligations are necessary to demonstrate inheritance of safety or correctness properties in behavioral refactoring. We use the "safety relational refinement" and the "auxiliary safety proof" from Reference 31 for refinement reasoning.

Safety relational refinement. Prove that all reachable states from the refactored model α are already reachable in the original model γ.

Auxiliary safety proof. Prove that a refactored model α satisfies some safety properties under the assumption of an existing proof about the original model γ. The auxiliary safety proof patches this proof w.r.t. the changes made by the refactoring. Let $\forall^\gamma$ quantify universally over all variables that are changed in γ. The intuition is that, assuming $\models \forall^\gamma(\phi \rightarrow [\gamma]\phi)$ (ϕ is an inductive invariant of γ), we can close the identical parts in the proof from the assumption by axiom and only need to show correctness for the remaining, new parts of the refactored model. For auxiliary safety use an invariant of $\mathcal{I}(\phi) \equiv (\phi \wedge \forall^\gamma(\phi \rightarrow [\gamma]\phi))$ for the refactored program α to prove $(F \wedge \mathcal{I}(\phi)) \rightarrow [\alpha^*]\psi$.

where F is some formula based on the definition of partial projective relational refinement. A hybrid program α is a partial refinement of γ with respect to some variables in the set of variables V and some formula F ($\alpha \sqsubseteq_F^V \gamma$) if and only if $(?F; \alpha) \sqsubseteq_F^V (?F; \gamma)$. In the case that $F \equiv true$, this partial refinement relation becomes a total refinement relation ($\alpha \sqsubseteq^V \gamma$ iff $\alpha \sqsubseteq_{true}^V \gamma$). Therefore, F in $(F \wedge \mathcal{I}(\phi)) \rightarrow [\alpha^*]\psi$ is an additional condition for partial refinement cases.

According to the above definitions from Reference 31, we want to show that if an abstract model, Model 10.1, guarantees a safety property, that is, collision-freedom, then this safety property can be proven for a refactored model, Model 10.2. We translate our problem using the auxiliary safety proof method as shown below:

- α is "Refined ACC+" (Model 10.2)

- γ is "Abstract ACC+" (Model 10.1)

- ϕ is "Condition (10.8)"

- ψ is $x_h < x_l$

We already proved that: $(\phi \rightarrow [\text{Abstract ACC+}] \ x_h < x_l)$ as Equation 10.7. We want to show the same collision-freedom $(x_h < x_l)$ is valid for refactored model, in this case the refined ACC+ $(\phi \rightarrow [\text{refined ACC+}] \ x_h < x_l)$. Therefore, we should strengthen the inductive invariant of Model 10.1, Controllability Condition (10.8), with the safety approved assumption for the abstract model.

$$\mathcal{I}(\phi) \equiv (\phi \wedge \forall x \forall v (\phi \rightarrow [\text{Abstract ACC+}] \ \phi))$$

We want to formally prove that: $\mathcal{I}(\phi) \rightarrow [\text{Refined ACC+}] \ x_h < x_l$

The *event- to time-triggered architecture* refactoring changes a hybrid program from event-triggered to time-triggered. This refactoring process separates the continuous evolution of the system from control choices. Figure 10.6a shows the time-triggered architecture of Model 10.1 (Abstract ACC+) as a state transition system. The procedure of deriving this architecture can be found in Reference 31. Figure 10.6b demonstrates removing one branch (*Other*) from the original model (Figure 10.6a) while the safety property is still preserved and then Figure 10.6c introduces two new branches (*Cruise* and *Follow*) to Figure 10.6b without changing the *Safety_Critical* branch. Both figures (Figure 10.6b and c) depict the "*Introduce Control Path*" refactoring, which is defined under the category of "*Behavioral Refactorings*" in [31]. Finally, Figure 10.6c shows the time-triggered architecture of Model 10.2 (Refined ACC+). The safety proof procedure of this refactored model can then be constructed from Figure 10.6b and c. The details of this proof are left for future work. Although this procedure provides easier steps in the safety proof of the refined system, we want to further demonstrate that one transition is split into two transitions. The two new transitions cover the same guard condition, while each transition has a subset of the old transition's behavior. Therefore, we want to further prove the case splitting the *Other* (old transition), which means that *Cruise* and *Follow*, as new transitions, result in a subset of the behavior of *Other*.

Another use for the refinement of these two models (i.e., Models 10.1 and 10.2) is to demonstrate that the abstract model can be improved and adapted to a more complex model in order to meet new requirements. We want to use a refactoring from Reference 31 to prove a safety property of a refined model based on the abstract one. However, proof-aware refactoring in Reference 31 does not propose any path-split (case splitting) refactoring. In other

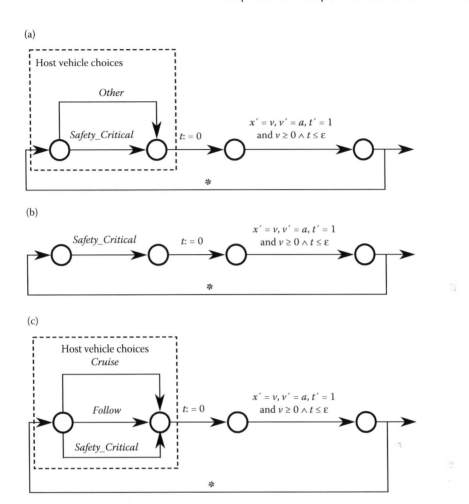

FIGURE 10.6 Time-triggered architecture of ACC+. (a) Time-triggered architecture of abstract ACC+. (b) Removed branch model. (c) Time-triggered architecture of refined ACC+.

words, we need to add an additional refactoring proof in *differential dynamic logic* (d$\mathcal{L}$) to show that a transition in the abstract model can be split into two or more new branches. In this case, the *Other* transition is split without touching the *Safety_Critical* case in order to preserve safety of the whole new system.

We want to show that the *Other* mode in Model 10.1, Figure 10.1, is split into two new modes, *Cruise* and *Follow*, in Model 10.2, Figure 10.4, without affecting the *Safety_Critical* mode. This fact can be established by proving that new branches apply in the same situations as the old branches and each will not violate the acceptable range for parameters of the old branch. In our example, a_h in *Cruise* and *Follow* will not be out of the acceptable range, which has been already defined in *Other* ($-B \leq a_h \leq A_{max}$). Therefore, another notion of refactoring and refinement (path-split) can be introduced in the proof refactoring of *differential dynamic logic* (d$\mathcal{L}$) that can be shown to preserve safety properties.

Consequently, the whole refactoring procedure with path-split refinement can be done more easily than the steps that we have done based on "*proof-aware refactoring.*" Using this technique we can deduce Figure 10.6c from Figure 10.6a directly, without the intermediate step shown in Figure 10.6b. Providing the formal syntax and semantics of path-split refinement is again left as future work.

10.6 CONCLUSION

CC and ACC systems have been used by several car companies to regulate the speed of the car in restricted traffic situations with a required minimum speed. These systems are not suitable for a very low speed, traffic jam environment. ACC+ extends CC and ACC features to provide automatic regulation of the speed of the car in these types of traffic environment. In this work we have verifed an abstract ACC+ system behavior that guarantees that the system is collision-free and safe under all possible scenarios other than when a car cuts in front of the host vehicle inside the safety critical stopping distance. ACC and CC systems have already been formalized to verify their correct behavior and safe operation, however, formal methods have not previously been applied to ACC+ systems to verify the systems requirements and desired system behavior. We have presented formal verification of a more realistic hybrid control system for ACC+ systems using differential dynamic logic (d$\mathcal{L}$) and proposed a new path-split refinement in the process. Future work includes the implementation of the ACC+ system on a hardware platform to validate that formal models faithfully capture the system behavior, requirements, and safety properties.

REFERENCES

1. J. Bowen and V. Stavridou. Safety-critical systems, formal methods and standards. *Software Engineering Journal*, 8(4):189–209, Jul 1993.
2. E. Zaloshnja, T. Miller, F. Council, and B. Persaud. Comprehensive and human capital crash costs by maximum police-reported injury severity within selected crash types. *Annual Proceedings/Association for the Advancement of Automotive Medicine*, 48:251–263, September 2004.
3. G.A. Peters and B.J. Peters. *Automotive Vehicle Safety*. CRC Press, Boca Raton, FL, 2003.
4. D.L. Parnas, A.J. van Schouwen, and Shu Po Kwan. Evaluation of safety-critical software. *Communications of the ACM*, 33(6):636–648, 1990.
5. A. Platzer and J.-D. Quesel. KeYmaera: A hybrid theorem prover for hybrid systems (system description). In A. Armando, P. Baumgartner, and G. Dowek, editors. *Automated Reasoning*, pp. 171–178. IJCAR 2008, LNCS 5195, Springer-Verlag, Berlin Heidelberg, 2008.
6. B.A. Guvenc and E. Kural. Adaptive cruise control simulator: A low-cost, multiple-driver-in-the-loop simulator. *Control Systems, IEEE*, 26(3):42–55, 2006.
7. H. Arioui, S. Hima, and L. Nehaoua. 2 DOF low cost platform for driving simulator: Modeling and control. In *Advanced Intelligent Mechatronics, 2009. AIM 2009. IEEE/ASME International Conference on*, SUNTEC City Convention Center Singapore, Singapore, pp. 1206–1211, 2009.
8. O.J. Gietelink, J. Ploeg, B. De Schutter, and M. Verhaegen. Development of a driver information and warning system with vehicle hardware-in-the-loop simulations. *Mechatronics*, 19(7):1091–1104, 2009. Special Issue on Hardware-in-the-Loop Simulation.
9. L. Nehaoua, H. Mohellebi, A. Amouri, H. Arioui, S. Espie, and A. Kheddar. Design and control of a small-clearance driving simulator. *Vehicular Technology, IEEE Transactions on*, 57(2):736–746, 2008.

10. D.J. Verburg, A.C.M. Van der Knaap, and J. Ploeg. Vehil: Developing and testing intelligent vehicles. In *Intelligent Vehicle Symposium, 2002. IEEE*, Oakland, CA, volume 2, pp. 537–544 vol. 2, 2002.

11. E. Abrahám-Mumm, U. Hannemann, and M. Steffen. Verification of hybrid systems: Formalization and proof rules in PVS. In *Engineering of Complex Computer Systems, 2001. Proceedings. Seventh IEEE International Conference on*, Skovde, Sweden, pp. 48–57. IEEE, 2001.

12. W. Damm, C. Ihlemann, and V. Sofronie-Stokkermans. Decidability and complexity for the verification of safety properties of reasonable linear hybrid automata. In *Proceedings of the 14th International Conference on Hybrid Systems: Computation and Control*, Chicago, pp. 73–82. ACM, 2011.

13. O. Stursberg, A. Fehnker, Z. Han, and B.H. Krogh. Verification of a cruise control system using counterexample-guided search. *Control Engineering Practice*, 12(10):1269–1278, 2004.

14. S. Jairam, K. Lata, S.K. Roy, and N. Bhat. Verificaton of a MEMS based adaptive cruise control system using simulation and semi-formal approaches. In *Electronics, Circuits and Systems, 2008. ICECS 2008. 15th IEEE International Conference on*, St. Julien's, Malta, pp. 910–913, 2008.

15. G. Ciobanu and S. Rusu. *Verifying Adaptive Cruise Control by π-Calculus and Mobility Workbench*. Technical Report FML-08-01, Institute of Computer Science Iaşi, December 2008.

16. A. Platzer. Differential dynamic logic for hybrid systems. *Journal of Automated Reasoning*, 41(2):143–189, 2008.

17. A. Platzer. The complete proof theory of hybrid systems. In *Logic in Computer Science (LICS), 2012 27th Annual IEEE Symposium on*, pp. 541–550. IEEE, Dubrovnik, Croatia, June 25–28, 2012.

18. S.M. Loos, A. Platzer, and L. Nistor. Adaptive cruise control: Hybrid, distributed, and now formally verified. In *Proceedings of the 17th International Conference on Formal Methods*, FM'11, pp. 42–56, Springer-Verlag, Berlin, Heidelberg, 2011.

19. S.M. Loos, D. Witmer, P. Steenkiste, and A. Platzer. Efficiency analysis of formally verified adaptive cruise controllers. In *Intelligent Transportation Systems—(ITSC), 2013 16th International IEEE Conference on*, The Hague, Netherlands, pp. 1565–1570, Oct 2013.

20. N. Aréchiga, S.M. Loos, A. Platzer, and B.H. Krogh. Using theorem provers to guarantee closed-loop system properties. In *American Control Conference (ACC), 2012*, Montreal, Canada, pp. 3573–3580. IEEE, 2012.

21. J.E. Naranjo, C. González, R. García, and T. De Pedro. ACC+ Stop & Go maneuvers with throttle and brake fuzzy control. *Intelligent Transportation Systems, IEEE Transactions on*, 7(2):213–225, 2006.

22. P. Shakouri and A. Ordys. Application of the state-dependent nonlinear model predictive control in adaptive cruise control system. In *Intelligent Transportation Systems (ITSC), 2011 14th International IEEE Conference on*, Washington, DC, pp. 686 –691, Oct. 2011.

23. Y. Yamamura, M. Tabe, M. Kanehira, and T. Murakami. *Development of an Adaptive Cruise Control System with Stop-and-Go Capability*. Technical Report 2001-01-0798, SAE, Warrendale, PA, 2001.

24. M.E. Russell, A. Crain, A. Curran, R.A. Campbell, C.A. Drubin, and W.F. Miccioli. Millimeter-wave radar sensor for automotive intelligent cruise control (icc). *Microwave Theory and Techniques, IEEE Transactions on*, 45(12):2444–2453, Dec 1997.

25. J.C. Doyle, B.A. Francis, and A. Tannenbaum. *Feedback Control Theory*, volume 1. Macmillan Publishing Company, New York, 1992.

26. A. Platzer. *Logical Analysis of Hybrid Systems—Proving Theorems for Complex Dynamics*. Springer, Berlin Heidelberg, 2010.

27. A. Platzer. Differential dynamic logic for verifying parametric hybrid systems. In N. Olivetti, editor. *Automated Reasoning with Analytic Tableaux and Related Methods (TABLEAUX 2007)*, pp. 216–232. Springer-Verlag, Berlin Heidelberg, 2007.

28. R. Mohammadi. Fault diagnosis of hybrid systems with applications to gas turbine engines. PhD thesis, Concordia University, 2009.

29. P. Tabuada. *Verification and Control of Hybrid Systems: A Symbolic Approach*. Springer, New York, 2009.

30. T.T. Johnson, J. Green, S. Mitra, R. Dudley, and R.S. Erwin. Satellite rendezvous and conjunction avoidance: Case studies in verification of nonlinear hybrid systems. In *FM 2012: Formal Methods*, pp. 252–266. Springer, Berlin Heidelberg, 2012.

31. S. Mitsch, J.-D. Quesel, and A. Platzer. Refactoring, refinement, and reasoning. In *FM 2014: Formal Methods*, pp. 481–496. Springer, Berlin Heidelberg, 2014.

CHAPTER 11

Model-Based Analysis of Energy Consumption Behavior

Shin Nakajima

CONTENTS

E NERGY CONSUMPTION is one of the primary nonfunctional concerns in systems equipped with batteries. The capacity of batteries is limited and reducing the consumption of battery power is mandatory for the systems to be long-lived. The hardware components consume the battery power directly. However, application programs are responsible for energy consumption because they control the usage of the hardware components. If the programs contain some hidden bugs, unexpectedly large amounts of energy may be consumed. These energy bugs must be eliminated in the early stages of the development of application programs. In this chapter we study a model-based analysis method of energy consumption behavior. Using a variant of linear hybrid automata as a rigorous model, we reduce the problem of detecting anomalies in the energy consumption behavior to logic model checking.

11.1 INTRODUCTION

Cyber-Physical Systems (CPS) have brought forward scientific challenges to construct dependable software-intensive systems. The systems constitute social infrastructures to support our daily lives, and have strong connections with their outside environment as *embedded systems* do (e.g., [1]).

These systems are different in their shapes and capabilities. Some of them have a common nonfunctional property regarding their energy consumption behavior, because they are dependent on batteries. The capacity of a battery is limited and, unless charged, decreases monotonically to eventual depletion.

Furthermore, the systems are software-intensive so that they implement *smart* services, in which application programs are indirectly responsible for consuming the battery power. Running a buggy program may result in a large amount of unexpected energy use. This creates a new technical challenge of eliminating energy bugs (e-bugs) [2] during the development of software-intensive systems.

Model-based analysis methods are finding potential e-bugs in application software designs. Such methods use a formal *model* to account for the behavior of both application programs and functional hardware components. Although hardware components are the direct consumers of battery power, programs are responsible for energy consumption indirectly, because they control the hardware usage. An analysis model must represent discrete behaviors of programs and use real-valued variables to account for execution time or consumed energy. Thus, model-based analysis methods are studied with approaches such as the *coexistence of Booleans and reals*. These analytical methods are among the primary characteristics of CPS [3]. Establishing model-based methods of analyzing energy consumption behavior is mandatory for developing *trustworthy cyber-physical systems* (TCPS).

This chapter reports state-of-the-art methods for model-based analyses of energy consumption behavior in Android-based systems [4], in which a variant of a linear hybrid automata (LHA) plays a key role. The LHA [5] was proposed as a formal model to represent systems in which Booleans and reals coexist, and is often used in modeling various CPSs. Our key contribution is introducing a *power consumption automaton* (PCA), which is a variant of LHAs. Although a PCA is less expressive than LHAs, it is powerful enough to be applied to finding e-bugs in designs of application software.

Section 11.2 explains typical e-bug phenomena in the Android framework and the disadvantages of existing profiler-based methods for the detection of e-bugs. This discussion presents the motivation for model-based analysis methods. Section 11.3 introduces a new energy consumption model for use in the proposed model-based analysis method. Section 11.4 focuses on properties to be checked with regard to the energy consumption behavior. These properties are captured by *duration-bounded cost constraints*, where cost refers to numerical constraints on the consumed energy. In addition, checking the correctness of the properties is formalized as model checking problems. Then, two sections are devoted to automated analysis of this duration-bounded cost-constraint problem. Section 11.5 presents how the methods are encoded as time-bounded search analyses using Real-Time Maude [6,7]. Tools are effective for showing the existence of e-bugs in the designs. Section 11.6 introduces an automatic fault localization method, by which root causes of the e-bugs are identified. The method makes use of Boolean encoding of the energy consumption behavior, and relies on a maximum satisfiability method. We specifically use Yices-1 [8], a satisfiability modulo theories (SMT) solver. Finally, Section 11.7 concludes this chapter.

11.2 ENERGY BUGS OF ANDROID APPLICATIONS

Removing e-bugs is one of the primary concerns in developing Android-based systems [2]. This section illustrates typical e-bug phenomena in the Android, and discusses some existing methods for detecting e-bugs.

11.2.1 Android Framework

An Android smartphone is powered by batteries. Direct consumers of the battery power are the hardware components that constitute the smartphone. Figure 11.1 shows the multilayered architecture of Android-based systems [4].

11.2.1.1 Multilayered Architecture

Components such as the CPU and memory are used at all times. Auxiliary functional components, such as Wi-Fi or GPS, are in operation only when a particular application program requests them via APIs supported by the underlying software infrastructure. The Android framework together with the Linux kernel constitute the infrastructure that is often called the Android operating system (Android O/S) as a whole. The Android O/S encapsulates the basic computing resources in a multilayered architecture and provides appropriate abstractions to application programs. Although these abstractions help a programmer construct a nontrivial application program in a short period, they make it difficult to grasp the whole

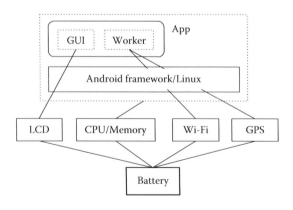

FIGURE 11.1 Multilayered architecture.

program behavior precisely. Particularly, finding the root causes of e-bugs is extremely difficult, because the Android O/S internally hides the relationship between the program and the hardware components.

11.2.1.2 Power Management

We first study a typical phenomenon that causes the behavior of application programs to have such significant impact on energy consumption. The Android O/S adapts an aggressive power-saving strategy so that the battery consumption is reduced; its power control subsystem automatically forces a smartphone to sleep when users do not touch the screen for a period of time.

However, this power-saving strategy is undesirable for some application programs such as video streaming services. If the strategy is enabled, the smartphone enters the sleep mode while the user is watching a video and not touching the screen. In this case, the sleep mode is obviously not desirable.

The Android framework provides a set of power management API objects, such as wake locks. An application program can use wake locks to request CPU resources, and then the CPU is kept awake as long as a lock remains active. Other hardware components are also supplemented with wake locks. For example, Wi-Fi lock objects provide means to control the power management aspect of Wi-Fi subsystems.

In a default setting, the wake lock is maintained by a reference-counting method. An acquired wake lock call must have a matching release call; otherwise, the wake lock is kept alive even after the caller program is terminated or even destroyed. If a wake lock remains active, the power-saving strategy is disabled. The result is undesirable consumption of battery power, known as a kind of e-bug. Because their root causes are improper use of the wake lock methods, these e-bugs are design faults and should be eliminated in the early stages of program development.

11.2.1.3 Activity Life Cycle

Next, we look at some of the internal details of the Android framework. Its life cycle management of an application is another source of potential e-bugs.

An Android application program, often called an *app*, has interactions with users via the GUI. An application is a multithreaded Java program, and the framework provides appropriate abstractions for making programming tasks easy. An application program has a main thread, which is responsible for the interactions with users. It also creates worker threads; for example, the program communicates with external network entities via Wi-Fi. The Android framework has a set of utility objects to encapsulate the technical details of manipulating concurrency primitives and to provide runtime support of event-driven style programming.

An app consists of an instance of an activity whose life cycle behavior is defined using seven callback methods. The behavior is represented concisely as a discrete transition system (see Figure 11.2). Callback methods notify an activity of changes in its life cycle. On completion of executing a callback method, an activity is in a stable life cycle stage.

The diagram in Figure 11.2 has six regular states and two special ones. An initial state marked as a black circle is special in that no activity object exists, and a final state is shown

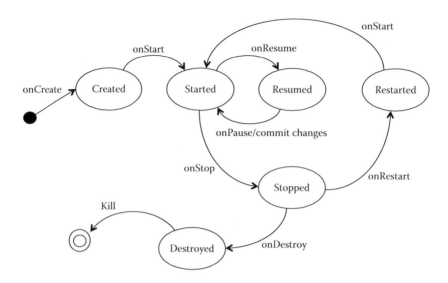

FIGURE 11.2   Life cycle behavior of activity.

as a double circle in the diagram. Each of the six regular states defines a stable state after executing a particular callback method. For example, an activity goes to the *created* state after completion of the onCreate() method.

User-visible actions consist of a sequence of callback methods. For example, starting a new activity generates a sequence of onCreate(), onStart(), and onResume().

Pressing the back button of a smartphone becomes a sequence of onPause(), onStop(), and onDestroy(), and the running activity goes to the *destroyed* state.

The memory management subsystem in the Android O/S does not release the memory space of application programs immediately. Even when an application program or an app is *destroyed*, its global states are still kept in the main memory. This design decision is based on assumption that the same app will be invoked again by users in the near future. If the app image is in the memory, restarting it can be fast because loading files is not necessary. The method is effective for quick restarts, but has a downside as well. If an app terminates while a wake lock is not completely released due to a bug, the wake lock remains active in the memory space of the *destroyed* app. The memory space is reclaimed by the low-memory killer driver of the Android O/S only when a shortage of the main memory occurs. An e-bug phenomenon appears asynchronously with the app execution, which makes finding root causes even more difficult.

11.2.2   Detecting Energy Bugs

Eliminating e-bugs is one of the primary concerns in developing Android-based systems and embedded systems as well. The energy consumption is a physical phenomenon that electric currents induce. We may use ammeters to measure the currents. An integrated development environment (IDE) must monitor running application programs to collect execution logs. Then, the measured current profile is checked against the logs to see how an unexpected amount of current is linked to the execution of a particular program fragment. This requires that the IDE must be equipped with ammeters. Although such an IDE is often used in developing embedded systems, we cannot expect every Android application programmer to have such expensive IDEs.

The IDE-based approach, in principle, makes it possible to detect e-bugs, but we must consider the fact that e-bug phenomena are asynchronous to program executions. We assume here that the recorded logs of an ammeter show an anomaly in electric currents at a time point t_d, and that a root cause of this e-bug is in an application program that uses wake locks. Since the program execution proceeds and may not be executing at t_d, establishing a link between the anomaly and the root cause in the program is not easy. Furthermore, because the Android O/S is a complex multilayered software infrastructure, finding root causes of e-bugs requires tracing the flow of information through the application program and the internal Android O/S.

Tests to detect e-bugs are usually conducted against running application programs with runtime profilers. In an early work, Pathak et al. developed Eprof [9], which was an energy profiler to monitor program executions at runtime to detect potential e-bugs. Eprof specifically uses a technique of call tracing to trace the execution flows so that asynchronous e-bugs can be detected. In another study, ADEL adopts a taint analysis technique to trace the direct and indirect use of data through the application program and the Android O/S [10]. Both methods require specific modifications to the Android O/S program codes and thus are not portable across different Android O/S versions. Note that new versions are released successively, which makes it extremely difficult to incorporate the profiler capabilities in new versions in a timely manner.

Although an energy profiler can detect some e-bugs, it is essentially a runtime monitor used for checking the energy consumption of running programs. The method has the same disadvantages that program testing methods have. Specifically, the profiler can be used only after all the programs are completed, and the coverage is limited by the supplied test cases or test environment set-ups. Moreover, because application programs run within the execution environment that the Android O/S provides, checking a program in isolation is not possible. Profilers check the target program together with many other codes in the Android framework, which introduces further complexities to the process of identifying root causes.

Another point to mention here is that the profiler approach is basically a method of measuring consumed energy. In general, these results often fluctuate because of errors in the measured values. Further, they are also affected by platform-dependent conditions that are difficult to control. Smartphones use power-saving processors such as those equipped with a dynamic-voltage frequency scaling (DVFS) mechanism (e.g., [11]). When the CPU load is small, processors run at low frequencies, or only a small number of processor cores are enabled if the CPU is multi-core. As the CPU load increases, frequencies are raised and result in increased energy consumption. Intuitively, an application program takes more physical time to complete when processors operate at lower frequencies. However, the amount of consumed energy is dependent on the duration in which the hardware components are used. In a low-frequency mode, hardware components controlled by the program are enabled longer in physical time and thus consume more battery power. This leads to an interesting observation; although the processor consumes less energy at low frequencies, the total energy consumption is larger particularly when the running program has e-bugs such as using wake locks improperly.

In order to obtain insight into the measurement fluctuations, we conducted experiments [12] to see how CPU loads with the DVFS mechanism affect the execution time of a particular application program. A governor driver of the Android O/S adjusts the voltage and frequencies so that the execution time is almost constant when CPU loads are modest.* It does not allow processors to run at their maximum performance in order to avoid using high power. The measurement results, however, show that the execution times of a given application program are mostly constant under modest CPU loads. Furthermore, measuring physical quantities is always threatened by measurement errors, which are found to be about 30% in this series of experiments. We see that e-bug phenomena may be hidden within the fluctuations in measurements if the extra energy consumption due to e-bugs is less than 30%. These results also show the disadvantages of profiler-based methods.

A model-based energy consumption analysis method is necessary in order to counter the disadvantage of runtime profiler methods. A *model* describes energy consumption behavior of a complete system, including application programs, hardware components, and the Android framework internals† at an appropriate abstract level. The abstract model faithfully represents how those elements are related. Thus, the energy consumption behavior of Android-based systems can be represented and considered early in the development.

11.3 ENERGY CONSUMPTION BEHAVIOR

This section introduces a formal model to account for energy consumption behavior. The model, PCA, is a variant of linear hybrid automata (LHA). Although a PCA is less expressive than an LHA, it is powerful enough to be applied when seeking e-bugs in designs.

11.3.1 Energy Consumption Model

We first review our energy consumption model. The presentation is based on References 13 and 14 using an example of a Wi-Fi subsystem, illustrated as a state-transition diagram shown in Figure 11.3. The diagram is introduced to explain the measured energy profile [15] of a Wi-Fi station that communicates with an external Wi-Fi access point and the specifications of Wi-Fi stations in the standard recommendation [16]. The diagram consists of four power states and several transition edges between the states.

A Wi-Fi station (Wi-Fi STA or simply STA) of smartphones operates in a passive scan mode [16]. A Wi-Fi access point (AP) periodically sends `beacon` signals to notify the STA to start a data transfer. The STA is initially in the `DeepSleep` state and enters the `HighPower` state to send or receive data frames. The STA remains in the `IdleListen` state in order to determine whether additional frames will be received. Transitions between the `IdleListen` and `HighPower` states are repeated as the data transfers continue. When the STA recognizes a `no-more-data` flag in a transferred data frame, it enters the

---

* The execution time becomes longer as the loads become greater, and real-time responses are poor.
† A library representing the frameworks abstractly is needed.

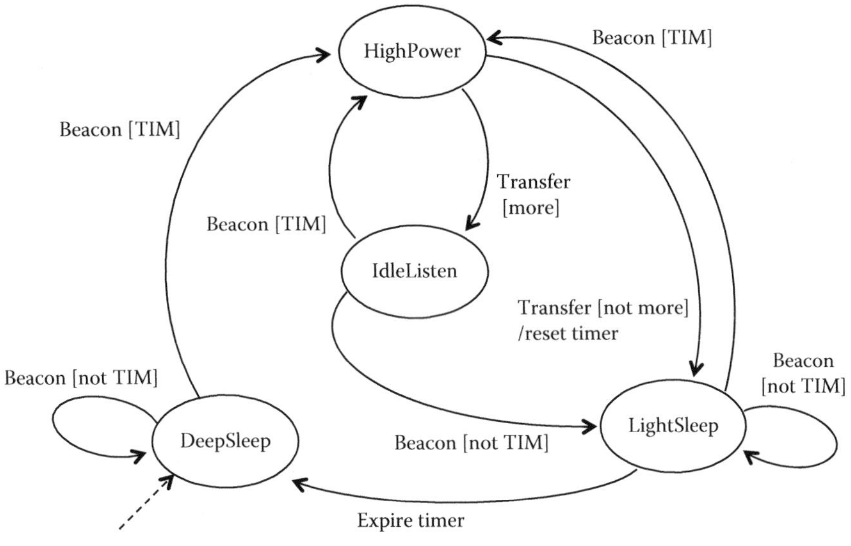

FIGURE 11.3   Energy consumption behavior of Wi-Fi station.

LightSleep state. The STA remains in this state to be ready for a quick restart when an additional data transfer is initiated. The STA does not reenter the DeepSleep state immediately. An inactivity timer is used to generate a time-out event, which forces the STA into the DeepSleep state.

In the Android, the Wi-Fi subsystem has wake locks for power management. WifiManager provides a method to control WifiLock [4]. It allows an application program to keep the Wi-Fi subsystem awake by forcing the STA into the LightSleep state, and disables transitions to the DeepSleep state.

The amount of consumed energy is different in each power state. For example, the HighPower state consumes a lot of energy in order to decode the transferred frames. In the DeepSleep state, only the power sufficient to activate a portion of the electric circuits is necessary, and thus the energy consumption rate is small. If we let $F^j(t)$ be a function of time that represents the rate of energy consumed at a state j, the total energy consumed in a time interval a to b is

$$E^j = \int_a^b F^j(t)dt.$$

Since power states are visited many times as state transitions continue, the total energy consumption is calculated as a summation over E^js,

$$E = \sum_{j=0}^{n} E^j = \sum_{j=0}^{n} \int_{a^j}^{b^j} F^j(t)dt.$$

If we introduce a linear approximation such that

$$dE^j/dt = F^j(t) = M^j$$

for a constant M^j in each power state, then

$$E = \sum_{j=0}^{n} M^j \times (b^j - a^j).$$

We assume here that at each power state ℓ, a constant M^ℓ to represent an average energy consumption rate is given as a specification of the hardware component (Wi-Fi in the above example). M^j, in the above formula, refers to a particular constant M_j^ℓ if the STA visits the power state ℓ as its jth state in a transition sequence.

For the example in Figure 11.2, ℓ refers to either `DeepSleep`, `LightSleep`, `IdleListen`, or `HighPower`. Imagine that an STA remains in the `DeepSleep` state for an amount of time t_1 and then enters the `HighPower` state. If it repeats the transitions four times between `HighPower` (t_2) and `IdleListen` (t_3), then E becomes

$$M^{DeepSleep} \times t_1 + (M^{HighPower} \times t_2 + M^{IdleListen} \times t_3) \times 4.$$

The total consumed energy E is increased as the STA transitions between the power states. Values of the variable E are used in a formula to check the correctness criteria, for example, $E \leq C_{max}$ for a given constant C_{max}.

Note that we focus on finding e-bugs by means of the PCA model. If we are interested in reproducing the precise amount of consumed energy, the linear approximation approach does not work and we must faithfully consider the real-valued time function $F^j(t)$.

11.3.2 Linear Hybrid Automata

A state-transition system captures the essential aspects of energy consumption behavior. The state-transition system has discrete transitions such as a `beacon` signal or `transfer`, and a time-dependent transition of an expiring timer. The system also carries real-valued consumed energy E. We will search for a suitable formal model, a PCA [13], to account for all these aspects, and thus start studying characteristics of various LHAs [5], that are formal models for representing real-time or hybrid systems that involve both discrete and continuous behavior. LHAs subsume several important automata. For simplicity, we study, below, these automata from the perspective of continuous dynamics. Their complete formal definitions are found in the literature [5].

Informally, an LHA is a state-transition system in which the states have continuous dynamics. If we let X be a continuous variable of time, its dynamics are defined in terms of its derivative with respect to the Newtonian time dX/dt. Various automata are classified according to the flow dynamics. An LHA allows its dynamics to be linear inequalities, $K_1 \leq dX/dt \leq K_2$ for integer constants K_1 and K_2. A timed automaton (TA) has $dX/dt = 1$, where X is called a *clock variable*. A skewed clock can change its rate K, that is $dX/dt = K$ for an integer constant $K \in \mathcal{Z}$. A multi-rate timed system (MTS) extends the TA whose clocks are skewed. An n-rate timed system (nRTS) is an MTS whose skewed clocks proceed at n different rates. A stopwatch automaton (SWA) is a TA where $dX/dt \in \{0, 1\}$. An SWA proceeds at states with a rate of one and stops at zero-valued states.

The PCA falls within a subclass of LHAs because the flow dynamics of the energy consumption variable E takes the form $dE/dt = M^\ell$, and the inactivity timer G is a clock variable such that $dG/dt = 1$ [13]. There may be a few possibilities for using a particular subclass of LHAs as PCAs. If we consider E to be a skewed clock variable, the PCA would be an nRTS in which E proceeds at ℓ different rates. The rate M^ℓ in our energy consumption model is real-valued, while the skewed values in an nRTS are integers. If we restrict the rate to be rational, then the M^ℓ values can be integers by choosing an appropriate time-scale factor [17]. Another simple model may use SWA, as in the case of MoVES [18]. A clock, which is introduced to represent energy consumption, proceeds at some states, but stops at other states. The clock value represents essentially how long it stays at the state in which energy is consumed. The total energy is calculated conceptually by multiplying a common constant energy consumption rate with the clock value. The SWA approach used in MoVES does not allow different M^ℓ values for different power states ℓ.

A PCA is, indeed, more restricted than an LHA in that E is an observer and does not have any effect on state-transition behavior, whereas G controls the timings of some state-transitions. A PCA is thus modeled as an extension of TA that has a continuous real-valued variable E. Since E represents the total energy consumption up to a particular point, it is time-dependent, $E(t)$. If we encode $E(t)$ in a weight, we can use a weighted timed automaton (WTA) [19] as a formal model of the PCA [14].

A WTA is a TA in which a weight W keeps some values as transitions proceed and its value can be retrieved when necessary. We let W store the accumulated consumed energy, $\sum_{j=0}^{n} E^j$, from an initial state to nth state.

11.3.3 Power Consumption Automata

11.3.3.1 Formal Model of a PCA

PCA $\mathcal{A}$ is defined as a variant of WTA.

WTA [19], an extension of TA [20], is a state-transition system that has a finite number of nonnegative real-valued clocks and weight variables. A PCA [14] is different from a WTA in that the PCA does not have weights on its transition edges and the weights are nonnegative and real-valued.

$$\langle Loc, X, W, \Sigma \cup \{\epsilon\}, Edg, Flow, Inv \rangle.$$

1. Loc is a finite set of locations. Each location corresponds to a power state.

2. X is a finite set of clock variables and W is a set of weight variables. X and W are disjoint ($X \cap W = \emptyset$). For a clock variable x ($\in X$), a constant n of natural number ($\in \mathcal{N}$), and an operator $\bowtie \in \{<, \leq, =, \geq, >\}$, constraints of the form $x \bowtie n$ and $x_1 - x_2 \bowtie n$ constitute a set of primitive clock constraints. We let $Z(X)$ be a set of formulas constructed from such primitive constraints using logical connectives.

3. Σ is an alphabet that is a finite set of input symbols, and ϵ is an empty symbol.

4. *Edg* is a finite set of transitions, $Loc \times Z(X) \times \Sigma \times 2^X \times Loc$. The element of *Edg*, (l_1, g, a, r, l_2), is written as $l_1 \xrightarrow{g,a,r} l_2$, where g is a guard condition in $Z(X)$, a is an input symbol $(\in \Sigma \cup \{\epsilon\})$, and r refers to a set of clock variables $(\in 2^X)$ to reset.

   Furthermore, a PCA is deterministic if the destination location l_2 is determined uniquely at a location l_1 with given a and g.

5. *Flow* represents the dynamics to account for the change rate of the weight variables. For nonnegative real $\mathcal{R}_+$ and $\mathcal{R}_+^W$ being $W \to \mathcal{R}_+$, *Flow* is $Loc \to (\mathcal{R}_+^W \to \mathcal{R}_+^W)$.

   For a particular weight variable p to keep track of energy consumption and valuation $w \in \mathcal{R}_+^W$, $Flow(\ell)(w)(p) = dp/dt = M^\ell$.

6. *Inv* is a mapping to clock constraints, $Inv : Loc \to Z(X)$.

The equation above uses valuations v for clock variables or weight variables, $v \in \mathcal{R}_+^{X \cup W}$. The equations below are the definitions of reset, delay, and multiplication, respectively, where d and e are constants $(d, e \in \mathcal{R}_+)$.

$$v[r](x) = \begin{cases} 0 & \text{if } x \in r \\ v(x) & otherwise, \end{cases}$$

$$(v + d)(x) = v(x) + d,$$

$$(v \times e)(x) = v(x) \times e.$$

We represent a complex system as a parallel composition of PCAs, where two PCAs are synchronized for the same event.

Parallel composition is defined for two given PCAs $\mathcal{A}^{(1)}$ and $\mathcal{A}^{(2)}$, where $\Sigma^{(1)} \cap \Sigma^{(2)} \neq \emptyset$ and $W^{(1)} \cap W^{(2)} = \emptyset$. The locations of the composed automata are pairs of locations, $\langle l^{(1)}, l^{(2)} \rangle \in Loc^{(1)} \times Loc^{(2)}$. An invariant at each location is a conjunction, $Inv^{(1)}(l^{(1)}) \wedge Inv^{(2)}(l^{(2)})$. Symbols common to both alphabets $(a \in \Sigma^{(1)} \cap \Sigma^{(2)})$ synchronize two automata. For such a common alphabet, $\mathcal{A}^{(1)}$ and $\mathcal{A}^{(2)}$ take transitions simultaneously.

11.3.3.2 Operational Semantics

The semantics of PCA $\mathcal{A}$ are given by a labeled transition system (LTS), $\langle S, T \rangle$. The state space S is a set of tuples consisting of a location l, a clock valuation v, and a weight valuation w. The clock invariants at the location l are satisfied.

$$S = \{ (l, v, w) \in Loc \times \mathcal{R}_+^X \times \mathcal{R}_+^W \mid v \models Inv(l) \}.$$

The transitions T consist of regular and stutter transitions.

$$T = \{ \xrightarrow{d,e} \} \cup \{ \xrightarrow{d,\epsilon} \}.$$

The details are explained below.

1. Event-trigger discrete transitions $(l_1, v_1, w) \xrightarrow{e} (l_2, v_2, w)$.

$$\exists (l_1 \xrightarrow{g,a,r} l_2) \in Edg \wedge v_1 \models g \wedge v_2 = v_1[r] \wedge v_2 \models Inv(l_2).$$

2. Delayed transitions $(l, v, w_1) \xrightarrow{d} (l, v + d, w_2)$

$$d \in \mathcal{R}_+ \wedge w_1 = f(0) \wedge w_2 = f(d) \wedge$$
$$\forall t \in \,]0, d[\, | \, v + t \models Inv(l) \wedge df/dt = Flow(l).$$

The time advances by the amount of the delay (d) to become $v + d$, and the weight is updated. In the above, $f(t)$ refers to a continuous function differentiable in the open time interval $]0, d[$. For an energy consumption variable p, $Flow(\ell) = dp/dt = M^\ell$ at a location ℓ, and thus $w_2(p) = M^\ell \times d + w_1(p)$.

3. Null transitions $(l_1, v_1, w) \xrightarrow{\epsilon} (l_2, v_2, w)$

$$\exists (l_1 \xrightarrow{g,\epsilon,r} l_2) \in Edg \wedge v_1 \models g \wedge v_2 = v_1[r] \wedge v_2 \models Inv(l_2).$$

The input symbol is empty (ϵ) and the weights are not changed.

4. Regular transitions $(l_1, v_1, w_1) \xrightarrow{d,e} (l_2, v_2, w_2)$
   A delayed transition $(l_1, v_1, w_1) \xrightarrow{d} (l_1, v, w_2)$ followed by an event-trigger discrete transition $(l_1, v, w_2) \xrightarrow{e} (l_2, v_2, w_2)$.

5. Stutter transitions $(l, v_1, w_1) \xrightarrow{d,\epsilon} (l, v_2, w_2)$
   A delayed transition $(l, v_1, w_1) \xrightarrow{d} (l, v, w_2)$ followed by a self-loop null transition at the location l, $(l, v, w_2) \xrightarrow{\epsilon} (l, v_2, w_2)$.

A transition sequence is a finite number of $\xrightarrow{d,e}$, or such sequences followed by an infinite number of $\xrightarrow{d,\epsilon}$. In the latter, the location l, at which the stutter transition is defined, is a final state. We assume that transitions are *non-Zeno*.

11.3.3.3 Timed Words

We introduce a set of timed sequences $L(\mathcal{A})$ generated by PCA $\mathcal{A}$ to be $\rho \in L(\mathcal{A})$, where ρ represents a sequence of timed states such that $\rho = \rho^0 \rho^1 \cdots$. A timed state ρ^j is defined such that $\rho^j = (\sigma^j, \tau^j)$ where $\sigma^j \in S$ is a state and $\tau^j \in \mathcal{R}_+$ is a time point. The timed points are defined such that $\tau^0 = 0$ and $\tau^{j+1} = \tau^j + d$ for a delayed transition $\xrightarrow{d}$.

11.3.3.4 Mealy Diagrams

PCAs can be extended to include *internal* variables other than clocks or weights. We introduce a finite set of variables V, whose values are elements in some finite discrete domains, Booleans, or naturals. We use D to collectively denote these discrete domains. The internal variables are updated along a transition with functions *Update*. *Edg* is now $Loc \times Z(X) \times (\Sigma \cup \{\epsilon\}) \times 2^X \times Update \times Loc$, and its element (l_1, g, a, r, u, l_2) is written as $l_1 \xrightarrow{g,a,r,u} l_2$. The formal model of PCAs is $\langle Loc, \ell_0, X, W, V, \Sigma \cup \{\epsilon\}, Edg, Flow, Update, Inv \rangle$.

We then consider a restriction of PCAs that ignores time-dependent behavior, both clocks and weights. The restriction is a Mealy diagram to represent a discrete transition system, $\langle Loc, \ell_0, V, \Sigma \cup \{\epsilon\}, Edg, Guard, Update \rangle$. *Edg* is now $Loc \times (\Sigma \cup \{\epsilon\}) \times Guard \times Update \times Loc$, and its element (l_1, a, h, u, l_2) is written as $l_1 \xrightarrow{a,h,u} l_2$. Its operational semantics is given by an LTS $\langle S, T \rangle$. The state space S is defined as a set of tuples consisting of a location $l \in Loc$ and a valuation $x \in D^V$. Transition T essentially consists of event-trigger transitions, $(l_1, x_1) \xrightarrow{e} (l_2, x_2)$:

$$\exists (l_1, a, h, u, l_2) \in Edge \land h^e(a, x_1) \land x_2 = u^e(a, x_1).$$

A transition e is fired when its guard condition h^e ($\in Guard$) is satisfied and updates x_2 by using its update function u^e ($\in Update$). Both h^e and u^e may access an input symbol a as well as x_1.

Mealy diagrams [21] are often used to represent discrete behavioral aspects of software designs such as the activity life cycle in Figure 11.2. When we take into account some elements, *Guard* and *Update*, which are related to the internal variables, PCAs essentially subsume Mealy diagrams. We hereafter consider that a PCA is a basic common model to represent the discrete behavior of programs and the energy consumption behavior as well.

11.4 PROPERTY SPECIFICATIONS FOR DETECTING ENERGY BUGS

This section studies how properties of the energy consumption behavior modeled by PCAs are represented and checks their satisfiability. We introduce a new variant of linear temporal logic (LTL) to represent the properties.

11.4.1 Duration-Bounded Cost Constraints

From a naive viewpoint, properties that must be checked appear simple enough to state that the total amount of consumed energy must be less than a specified maximum value. This relation is expressed in a linear temporal logic (LTL) formula $\Box(E \leq C_{max})$. As discussed in Section 11.3.1, a total energy consumption E is dependent on time, and the formula is $\forall t \in R_+ \mid (E(t) \leq C_{max})$. The relation is eventually violated because $E(t)$ increases monotonically. Checking the properties must be limited with respect to time intervals, $\forall t \in [t_s, t_e] \mid (E(t) \leq C_{max})$. However, usually we do not know the exact duration $[t_s, t_e]$ in advance. The interval is determined by some *events* that cause state changes in PCAs. Now, we consider a sequence of states $\{ \sigma_j \}$. If we let σ_s and σ_e be states in the sequence that determine the interval, the property is

$$\forall \sigma_i \in [\sigma_s, \sigma_e] \mid (E(\sigma_i) \leq C_{max}).$$

The interval corresponds to a *transaction*, which starts at state σ_s and ends at σ_e. If we let P_s and P_e be unique propositions that are satisfied exclusively at the states σ_s and σ_e, respectively, an LTL formula $\Diamond(P_s \wedge \Diamond P_e)$ specifies a set of transition sequences that contains the intervals. The formula actually describes the following temporal behavior; *once a state to satisfy P_s is reached, another state to satisfy P_e is eventually reached*. If we let φ_G refer to this LTL formula, then the property to be checked becomes $\varphi_G \Rightarrow \Box(E \leq C_{max})$.

Now, we redefine E to be the difference between two values such as $E(\sigma_e) - E(\sigma_s)$. More specifically, we assume here that a weight p is a variable to refer to the accumulated energy consumption. Furthermore, if we let ℓ_s be the location in σ_s. Then, $E(\sigma_s)$ is $w_s(p)$. Since $E(\sigma_e)$ is defined similarly, $E = w_e(p) - w_s(p)$. If we extract these values from the PCAs, we can check the above formula by using a bounded reachability method where the bound is specified by the intervals. For simplicity, we assume that $E(\sigma_s) = 0$. The search starts a state to satisfy P_s and ends at another state with P_e, during which $(E \leq C_{max})$ is checked for all visited states. Because E is monotonically increasing, once the property is violated $((E > C_{max}))$, then it is violated afterward, too. If we find an intermediate state σ_i such that $(E(\sigma_i) > C_{max})$ becomes true for the first time, then we can conclude that $(E(\sigma_e) > C_{max})$ is also true. Therefore, the analysis problem can be viewed as a bounded reachability search to find the violation.

This problem of energy consumption in a transaction is captured as a special instance of the more general problem of *duration-bounded cost constraints*. The cost, in general, refers to a certain numerical constraint on the amount of the consumed energy. Then, we conduct a bounded search in a specified duration to see if the constraint is violated. As can be seen in the discussions above, we must find appropriate ways to extract weight values, from a given PCA, such as $w_s(p)$ or $w_e(p)$ at states σ_s or σ_e. These states are in turn related to propositions P_s or P_e that are used to specify the intervals by a formula $\Diamond(P_s \wedge \Diamond P_e)$. We will consider an expressive and flexible property language to specify precisely and compactly both the duration and the constraints.

11.4.2 Temporal Logic with Freeze Quantifier

11.4.2.1 Temporal Logic-Based Approach

We consider the problem of duration-bounded cost constraints from the viewpoint of using LTL.

Starting with a simple LTL formula, $\Diamond(P_s \wedge \Diamond P_e)$, we extend this formula to refer to both real-time timings and real-valued consumed energy. To represent the timing, Metric Temporal Logic (MTL) [22] and TPTL [23], for example, have been proposed. However, MTL and TPTL are not applicable to real-valued energy consumption variables or weights.

We introduce freeze quantifiers that extend the original proposal in TPTL so that quantified variables can refer to either arbitrary clock or weight values as well as real-time timings. A quantified formula $\exists^D x.\phi^x$ extends TPTL so that a quantified variable x can refer to clocks or weights as well as time points. A simple relation to check the cost constraint $C^{\langle s,e \rangle}(u, v)$ in a duration is compactly written as

$$\Diamond \exists^D u. (P_s \wedge \Diamond \exists^D v. (P_e \wedge C^{\langle s,e \rangle}(u, v))),$$

where p is assumed to stand for a weight variable to keep track of accumulated consumed energy. Quantified variables u and v store the consumed energy at specified states, and that $C^{\langle s,e \rangle}(u, v) = (v - u) \leq C_{max}$.

Additionally, as for the original TPTL, our extension allows quantified variables x and y to refer to time points (τ) as well.

$$\Diamond \mathcal{F}^\tau x. \mathcal{F}^p u. (P_s \wedge \Diamond \mathcal{F}^\tau y. \mathcal{F}^p v. (P_e \wedge C^{\langle s,e \rangle}(x, y; u, v))).$$

Here, x and u are *frozen* at the state when the system is in the P_s state, and y and v are taken from the P_e state. Because we use frozen time points as well, $C^{\langle s,e \rangle}(x, y; u, v)$ can be a conjunction. For example, $C^{\langle s,e \rangle}(x, y; u, v) = ((y - x) \leq D_{max}) \wedge ((v - u) \leq C_{max})$.

The extension of LTL with freeze quantifiers is called fWLTL in our work. When we have such an extension, the model-based energy consumption analysis is formalized as a model checking problem. As seen in the simple example above, fWLTL enables the expression of durations flexibly as temporal properties. Freeze quantifiers provide the means to access weight values and time points, by which numerical constraints $C^{\langle s,e \rangle}$ regarding the consumed energy or timings are constructed. Therefore, we now have an expressive and flexible property language to specify precisely and compactly both the duration and the constraints. However, constraint expressions must further be specified, and so they are within a theory of linear real arithmetic (LRA). Below, we formally define this new extension of LTL and then investigate the model checking problem in detail in the next section.

11.4.2.2 Formal Definitions of fWLTL

We introduce fWLTL [14], a variant of LTL with freeze quantifiers. The syntax of fWLTL is shown below.

$$
\begin{aligned}
\phi := \quad & C && \text{Cost constraints} \\
| \quad & p && \text{Atomic proposition} \in Prop \\
| \quad & \neg \phi && \text{Logical negation} \\
| \quad & \phi_1 \wedge \phi_2 && \text{Conjunction} \\
| \quad & \phi_1 \, U \, \phi_2 && \text{Until operator} \\
| \quad & \mathcal{F}^m x. \phi^x && \text{Freeze quantifier over clock (X) or weight (W) variables} \\
| \quad & \mathcal{F}^\tau x. \phi^x && \text{Freeze quantifier over time points } \{\tau^j\}
\end{aligned}
$$

The variable $x \in Var$, where Var is a countable set of variables, may appear free in an fWLTL formula ϕ^x. A closed formula ϕ does not have any free variables. Cost constraints may constitute a set $Z(X \cup W)$, where X is a finite set of clock variables and W is a finite set of weight variables, energy consumption variables. Z denotes a set of formulas consisting of linear constraints. Furthermore, standard abbreviations are used. For example, $\Diamond \phi \equiv true \, U \, \phi$ is *Eventually*, and $\Box \phi \equiv \neg(\Diamond \neg \phi)$ is *Globally*.

We adapt pointwise semantics for fWLTL. We use some symbols in addition to ρ^j, which was introduced in Section 11.3.3, $\rho^j = (l_j, v_j, w_j, \tau^j)$.

1. An environment $\Gamma : Var \to \mathcal{R}_+$.

   $\Gamma[x := e]$ assigns x to a value e ($e \in \mathcal{R}_+$) in the environment Γ.

2. A labeling function Lab : $Loc \to 2^{Prop}$ where Loc is a finite set of locations and $Prop$ is a finite set of atomic propositions.

The following satisfiability relations $\models$, which are defined inductively, define when $\langle \rho, \Gamma \rangle$ satisfies an fWLTL formula ϕ, $\langle \rho, \Gamma \rangle \models \phi$.

$$
\begin{array}{ll}
\langle \rho^j, \Gamma \rangle \models C & \text{iff } \Gamma \models C \\
\langle \rho^j, \Gamma \rangle \models p & \text{iff } p \in Lab(l_j) \\
\langle \rho^j, \Gamma \rangle \models \neg\phi & \text{iff } \langle \rho^j, \Gamma \rangle \not\models \phi \\
\langle \rho^j, \Gamma \rangle \models \phi_1 \wedge \phi_2 & \text{iff } \langle \rho^j, \Gamma \rangle \models \phi_1 \text{ and } \langle \rho^j, \Gamma \rangle \models \phi_2 \\
\langle \rho^j, \Gamma \rangle \models \phi_1 \, U \, \phi_2 & \text{iff } \langle \rho^k, \Gamma \rangle \models \phi_2 \text{ for some } k \geq j \\
 & \quad \text{and } \langle \rho^i, \Gamma \rangle \models \phi_1 \text{ for all } i \, (j \leq i < k) \\
\langle \rho^j, \Gamma \rangle \models \mathcal{I}^m x . \phi^x & \text{iff } \langle \rho^j, \Gamma[x := (v_j \cup w_j)(m)] \rangle \models \phi^x \\
\langle \rho^j, \Gamma \rangle \models \mathcal{I}^\tau x . \phi^x & \text{iff } \langle \rho^j, \Gamma[x := \tau^j] \rangle \models \phi^x
\end{array}
$$

In $\mathcal{I}^m x$, m refers to a variable name, and thus the quantified variable x is bound to $(v_j \cup w_j)(m)$.

Now, we show a simple example of an fWLTL formula that represents the cost-constraint problem discussed in Section 11.4.1; $(E(\sigma_e) \leq C_{max})$. The predicate $P_e(\sigma_e)$ is satisfied.

$$(I \wedge \Diamond \mathcal{I}^p v . (P_e \wedge (v \leq C_{max}))),$$

where the weight variable p refers to the total amount of consumed energy, which is set to zero at the initial state.

11.4.3 Model Checking Problem

A problem of duration-bounded cost constraints is now reduced to model checking an fWLTL formula with respect to a given PCA. Let $L(\mathcal{A})$ be the set of timed sequences generated by PCA $\mathcal{A}$. Given a closed fWLTL formula ϕ, the model checking problem, $\mathcal{A}, \Gamma \models \phi$, is defined to ensure $\langle \rho^0, \Gamma_0 \rangle \models \phi$ with an initial empty environment Γ_0 and for all timed sequences ρ generated by $\mathcal{A}$ ($\rho \in L(\mathcal{A})$).

11.4.3.1 Decidability and Undecidability

We now study decidability issues of this model checking problem by investigating the existing results on formalisms related to the PCA and the fWLTL. First, the PCA is a kind of WTA [19], and WTAs are an extension of TAs. For a TA, model checking of MTL is undecidable. It is decidable only for a fragment of MTL with respect to a TA [22]. Furthermore, although restricted to this subset of the MTL, model checking of a WTA is undecidable. Over- and under-approximation techniques are applied to this model checking problem [24].

TPTL [23] is propositional temporal logic with freeze quantifiers, the satisfiability relation of which is defined by timed words generated by a TA. Freeze quantifiers refer to the time points of binding states (*now*). TPTL subsumes MTL as a proper subset, and thus model checking of TPTL with respect to a TA is undecidable, in general. Freeze quantifiers in Constraint LTL (cLTL) [25] can refer to variables other than time points. Model checking of cLTL is decidable for a discrete time and a restricted class of constraints only.

Based on these existing studies, model checking of an fWLTL formula with respect to a PCA is undecidable, in general. The proposed logic fWLTL is more expressive than TPTL or cLTL because fWLTL can *freeze* weight variables as well as time points. A quantified formula $x.\phi^x$ in TPTL is expressed as $\exists^\tau x.\phi^x$ in fWLTL. A formula $\downarrow_{x=m}\phi^x$ in cLTL is expressed as $\exists^m x.\phi^x$.

11.4.3.2 Approximation Methods

We now introduce some notion of approximations to enable automatic analyses [26]. The complexities are originated from various reasons.

1. There are no restrictions in a formula on the positions of freeze quantifiers.

2. There are expressive cost constraints.

3. An infinite state space is generated by continuous time.

For a practical solution to the third issue, we adapt a method of time-bounded model checking with a sampling abstraction in continuous time. The method was originally proposed in Real-Time Maude [6,7]. In addition to this, we will introduce further restrictions and consider two formula patterns.

For the first issue, we impose restrictions on the properties to be checked. We restrict the positions in a formula at which freeze quantifiers appear. Let ϕ_S be an fWLTL formula that has at least one state proposition p^g. Namely, ϕ_S is restricted such that $\phi_S = p^g \wedge \phi^x$ where $p^g \in Prop$ is called a *guard state proposition*. Now, D of $\exists^D$ refers to either τ for time points or m for clocks or weights. Furthermore, b^j is τ_j for $\exists^\tau$, and b^j is $(v_j \cup w_j)(m)$ for $\exists^m$.

$$\langle \rho^j, \Gamma[x := b^j] \rangle \models \exists^D x.\phi_S \text{ iff } \langle \rho^j, \Gamma \rangle \models p^g \text{ and } \langle \rho^j, \Gamma[x := b^j] \rangle \models \phi^x.$$

A freeze quantifier operates on states that satisfy the guard state proposition.

11.4.3.3 Monotonic Constraints

A constraint $\mathcal{C}$, a condition on consumed energy, refers to both time points and energy consumption. It is a linear inequality defined over nonnegative real numbers R^+. We further assume that $\mathcal{C}$ is monotonic in the sense discussed below.

First, we assume, for simplicity, that a constraint $\mathcal{C}$ is decomposed into a conjunction of constituent constraints, each of which refers to either time or weight. If we let $\mathcal{C}(t_1, t_2; u_1, u_2)$

be a constraint depending on the time points t_1 and t_2 and the weight variables u_1 and u_2,

$$\mathcal{C}(t_1, t_2; u_1, u_2) = \mathcal{C}^{(1)}(t_1, t_2) \wedge \mathcal{C}^{(2)}(u_1, u_2).$$

For the $C(s, e)$ introduced previously, $C^{(1)}$ and $C^{(2)}(u_1, u_2) = (u_2 - u_1 \leq C_{max})$. Because both $C^{(1)}$ and $C^{(2)}$ have similar forms, we hereafter consider a form of $C^{(1)}(t_1, t_2)$ with $t_1 \leq t_2$.

Second, when we fix t_1 to be a particular value, the constraint $C^{(1)}(t_1, t_2)$ is dependent on t_2. In general, weight as well as time increases monotonically because the weight refers to the total energy consumed up to a certain point. We assume that the constraints have a similar property of the monotonic increase. Formally, if a constraint is monotonic, then some threshold $\bar{t}$ exists such that $C^{(1)}(t_1, t)$ is satisfied for $\forall t$ with $t < \bar{t}$ and is violated for $\forall t$ with $\bar{t} \leq t$. Similarly, when we fix t_2, a certain threshold $\bar{t}$ exists such that $C^{(1)}(t, t_2)$ is violated for $\forall t$ with $t < \bar{t}$ and is satisfied for $\forall t$ with $\bar{t} \leq t$. The constraint $(t_2 - t_1 \leq D_{max})$ has this property. The discussion above is directed to $C^{(1)}(t_1, t_2)$, but the monotonicity property can be generalized to a constraint of the form $\mathcal{C}(t_1, t_2; u_1, u_2)$.

11.4.3.4 Property Patterns

Although fWLTL allows us to express intervals flexibly, we consider two practically important classes of properties. Below, we introduce *stylized* symbols for representing these properties. $F^{\langle \tau; P \rangle}$ (or $G^{\langle \tau; P \rangle}$) is a combination of an eventually (or a globally) operator with two freeze quantifiers. $F_{\mathcal{C}}^{\langle \tau; P \rangle}$ is $F^{\langle \tau; P \rangle}$ parameterized with a cost constraint $\mathcal{C}$. The constraint is evaluated within the quantified scope of $F^{\langle \tau; P \rangle}$. Q_1 and Q_2 are atomic guard state propositions.

1. Reachability

$$\begin{aligned} \phi^{ReA} &= F^{\langle \tau; P \rangle}(Q_1 \wedge F_{\mathcal{C}}^{\langle \tau; P \rangle} Q_2) \\ &= \Diamond \mathcal{I}^{\tau} x. \mathcal{I}^{P} u.(Q_1 \wedge \Diamond \mathcal{I}^{\tau} y. \mathcal{I}^{P} v.(Q_2 \wedge \mathcal{C}(x, y; u, v))). \end{aligned}$$

2. Response

$$\begin{aligned} \phi^{ReS} &= G^{\langle \tau; P \rangle}(Q_1 \Rightarrow F_{\mathcal{C}}^{\langle \tau; P \rangle} Q_2) \\ &= \Box \mathcal{I}^{\tau} x. \mathcal{I}^{P} u.(Q_1 \Rightarrow \Diamond \mathcal{I}^{\tau} y. \mathcal{I}^{P} v.(Q_2 \wedge \mathcal{C}(x, y; u, v))). \end{aligned}$$

The intended meanings are given by the corresponding fWLTL formulas as above. Precisely, the satisfiability of the formula ϕ^{ReA} is given below.

$$\begin{aligned} \langle \rho^0, \Gamma_0 \rangle &\models \phi^{ReA} \\ \text{iff } \exists i, j \mid (0 \leq i \leq j) \text{ and } \langle \rho^i, \Gamma[x := \tau^i; u := (v_i \cup w_i)(P)] \rangle &\models Q_1 \\ \text{and } \langle \rho^j, \Gamma[y := \tau^j; v := (v_j \cup w_j)(P)] \rangle &\models Q_2 \\ \text{and } \langle \rho^j, \Gamma \rangle &\models \mathcal{C}. \end{aligned}$$

The satisfiability of ϕ^{ReS} is similarly defined.

We then introduce an over-approximation method for enabling an automated analysis. First, we assume that $\langle \rho^0, \Gamma_0 \rangle \models \Diamond(Q_1 \wedge \Diamond Q_2)$ and an index j is the first occurrence such that $\langle \rho^j, \Gamma \rangle \models Q_2$. Let k be $min^{(j)}(i \mid \langle \rho^i, \Gamma \rangle \models Q_1$ and $(i \leq j))$. Because of the monotonicity, if $C(t_k, t_j)$ is satisfied, then $C(t_i, t_j)$ is satisfied for $\forall i$ with $k \leq i$. In other words, if $C(t_i, t_j)$ is violated, then $C(t_k, t_j)$ is violated. Therefore, if generating a counterexample, checking $C(t_k, t_j)$ does not miss any violation of the constraint. It is an over-approximation method that can be used for finding potential faults.

11.5 ANALYSIS USING REAL-TIME MAUDE

This section presents automatic analysis methods using Real-Time Maude. A time-bounded reachability analysis method is applicable to simple properties. For the property patterns introduced in Section 11.4.3, we discuss how the fWLTL formulas are translated into the usual LTL formulas so that we can make use of Real-Time Maude's LTL model checker.

11.5.1 A Brief Introduction to Real-Time Maude

Real-Time Maude [6,7] is an extension of Maude [27] for supporting the formal specification and analysis of real-time or hybrid systems.

11.5.1.1 Real-Time Theory

A timed module in Real-Time Maude specifies a real-time theory $\mathcal{R}$, which is written as *(Sig, Eq, IR, TR)*. Particularly, *(Sig, Eq)* is a membership equational logic theory [27], where *Sig* is a signature that constitutes sort and operator declarations. *Eq* is a set of confluent and terminating conditional equations. *(Sig, Eq)* specifies the state space of the system as an algebraic data type. *IR* is a set of labeled instantaneous rewrite rules and *TR* is a set of tick rewrite rules.

11.5.1.2 Instantaneous Rewrite Rules

Instantaneous rules (IR) are inherited from Maude, and they rewrite terms in a concurrent manner without any delay, that is, instantaneously. An instantaneous rewrite rule specifies a one-step discrete transition. The rules are applied modulo equations *Eq*. Below, A_j represents a term to play a role of argument of another term T_i.

$r : T_1(A_1) \Longrightarrow T_2(A_2)$ **if** φ.

The term $T_1(A_1)$ on the left-hand side becomes a new term $T_2(A_2)$ if the side condition φ is satisfied. Hereafter, arguments are omitted for simplicity.

11.5.1.3 Tick Rewrite Rules

Tick rules (TR), introduced in Real-Time Maude, are responsible for the passage of time. Since time is global and proceeds uniformly, tick rules manipulate the state of the entire system. Let T_1 be a term for the state of the entire system. T_1 represents a snapshot of the system, which is a term of sort System predefined in Real-Time Maude. A tick rule is introduced for a term of sort GlobalSystem, that syntactically takes the form, $\{T_1\}$.

{ _ } : System $\rightarrow$ GlobalSystem
$l : \{T_1\} \Longrightarrow \{T_2\}$ **in time** τ_l **if** φ.

The rule states that the amount of time τ_l passes when rewriting T_1 to T_2. The formula φ may refer to a condition on the time variable τ_l. Such a tick rule advances the time nondeterministically. The amount of time is not chosen exactly, but can be any value that satisfies the condition φ.

11.5.1.4 Sampling Abstraction Method

Real-Time Maude adapts *sampling abstractions* for time-nondeterministic systems in which the maximum time elapsed (mte) plays an important role. Each term T_1 is accompanied by two functions δ and *mte*. Function δ returns a new term T_2, which is a modification of T_1 after the passage of time τ_l. Function *mte* returns the maximum elapsed time, during which the term T_1 is assumed to be unchanged.

δ : System Time $\rightarrow$ System
mte : System $\rightarrow$ TimeInf

A tick rule takes into account these two functions. A new term is calculated with δ, and the condition φ refers to *mte*.

$l : \{T\} \Longrightarrow \{\delta(T, \tau_l)\}$ **in time** τ_l **if** $\tau_l \leq mte(T)$.

The $mte(T)$ is the upper limit of the time advancement. The transition is fired at least once in the time interval specified by the function $mte(T)$. Usually, system term T is decomposed into a set of constituent terms T^j. Function $mte(T^j)$ must be defined for each T^j. In order for all $mte(T^j)$ to be satisfied, their minimum value $(\min(mte(T^j)))$ is chosen. Therefore, this control strategy may result in an over-sampling for some of the components T^j, but does not miss any sampling points.

11.5.2 Time-Bounded Reachability Analysis

The bounded reachability analysis method using Real-Time Maude can be applied to duration-bounded cost-constraint problems if the properties to be checked are simple. In particular, we consider a case of $(E(\sigma_e) \leq C_{max})$ presented in Section 11.4.1. Remember that this property is written in fWLTL as $(I \wedge \Diamond_d^{P} v. (P_e \wedge (v \leq C_{max})))$.

We consider first the translation of a given PCA [26,28], which is basically encoding its LTS in Real-Time Maude.

11.5.2.1 System State

We first define a term $S(l, v, w, \tau)$ of sort System to represent a PCA state. It has four arguments, because the behavior of PCA is represented with timed states of (l, v, w, τ). For simplicity, v and w refer to values of a clock and a weight, respectively, although they are valuations in the formal definitions.

11.5.2.2 State Transitions

We show how we encode state transitions. An edge of a PCA, $l_1 \xrightarrow{\varphi, E_1, r} l_2$, where φ is a guard condition on the transition source, is interpreted as an event-trigger discrete transition, $(l_1, v_1, w) \xrightarrow{e} (l_2, v_2, w)$. This transition is translated into an instantaneous rule of Real-Time Maude. We have two alternative rules. The first is a transition, along which the clock is not changed, and the second is a transition with a clock reset.

$$E_1\ S(l_1, v_1, w, \tau) \Longrightarrow S(l_2, v_1, w, \tau)\ \text{if}\ \ \varphi(v_1) \wedge Inv(l_2)(v_1)$$
$$E_1\ S(l_1, v_1, w, \tau) \Longrightarrow S(l_2, 0, w, \tau)\ \text{if}\ \ \varphi(v_1) \wedge Inv(l_2)(0).$$

A delayed transition is encoded as a tick rule of Real-Time Maude. The time-dependent behavior needs two functions δ and mte. Value d is chosen in a nondeterministic manner as long as $d \leq mte(S(l, v, w, \tau))$ is satisfied.

$$1 : \{\ S(l, v, w, \tau)\ \} \Longrightarrow \{\ \delta(S(l, v, w, \tau), d)\ \}$$
$$\textbf{in time}\ d\ \textbf{if}\ \ (d \leq mte(S(l, v, w, \tau))) \wedge Inv(l)(v + d)$$

$$\delta(S(l, v, w, \tau), d) = S(l, v + d, M^l \times d + w, \tau + d).$$

When the amount of time that has passed is d, the clock v becomes $v + d$ and the weight is updated to be $M^l \times d + w$ for a given constant M^l at the state l.

Function mte just returns infinity (`INF`) for stable state l_{stable}, in which the system waits for an input symbol to initiate a discrete transition. Conversely, for state l_{change} to have a guard condition on the clock, mte returns an amount of time that the PCA will remain in the l_{change} state. Here, X_c is assumed to be a given time-out constant representing an upper limit that the system will remain in the state.

$$mte(S(l_{stable}, v, w, \tau)) = \text{INF}.$$
$$mte(S(l_{change}, v, w, \tau)) = X_c\ \text{monus}\ v.$$

The `monus` is a built-in operator that returns the difference $(X_c - v)$; it returns 0 if the calculated value is negative.

11.5.2.3 Time-bounded Search

The bounded reachability checking method is conducted with the time-bounded search command of Real-Time Maude.

tsearch [N] ⟨term⟩ ⇒* ⟨pattern⟩ **such that** ⟨cond⟩ **in time** ≤ B.

It searches for N numbers of targets of the form ⟨pattern⟩ in a breadth-first manner starting from ⟨term⟩. The ⟨cond⟩ is an additional predicate that must be satisfied by valid target terms, matching ⟨pattern⟩. The search is conducted within a time bound of B.

Now, we investigate how we use this time-bounded search command to check the above-mentioned energy consumption property. First, we assume that an initial state σ_s is a system term $S(l_s, 0, 0, 0)$. We then let $S(\_, \_, w, \_)$ be the state σ_e, which satisfies the proposition P_e. Here, the symbol $\_$ stands for *do-not-care*. The command below can ensure that a given PCA has at least one transition sequence from σ_s to σ_e.

tsearch [1] { $S(l_s, 0, 0, 0)$ } $\Rightarrow^*$ { $S(\_, \_, w, \_)$ } **such that** P_e **in time** $\leq$ B.

We inspect the value w to see whether $(w \leq C_{max})$ is satisfied [28]. Alternatively, we issue a similar command again with a modified condition $(P_e \wedge (w > C_{max}))$.

tsearch [1] { $S(l_s, 0, 0, 0)$ } $\Rightarrow^*$ { $S(\_, \_, w, \_)$ }
      such that $(P_e \wedge (w > C_{max}))$ **in time** $\leq$ B.

If this command returns successfully, the constraint $(w \leq C_{max})$ is violated. Note that this method is possible only for cases in which we can choose an appropriate state l_s so that we can construct initial state term $S(l_s, 0, 0, 0)$ manually.

11.5.3 LTL Model Checking

We now consider how we translate the model checking problem of an fWLTL formula and a PCA into an LTL model checking problem in Real-Time Maude.

11.5.3.1 LTL Model Checker in Real-Time Maude

Real-Time Maude extends the LTL model checker of Maude. The model checker makes use of the sampling abstraction to choose appropriate time points to check nondeterministically. The analysis method here, in particular, relies on the time-bounded search. The method checks whether the system behavior satisfies a given temporal logic formula ϕ up to a certain time bound B.

mc ⟨initState⟩ $\models^t$ ϕ **in time** $\leq$ B.

Atomic propositions, to which a formula ϕ refers, must be defined as equational specifications of the Maude built-in operator $\models$. A sort symbol `Prop` is predefined in Maude to represent those atomic propositions.

    $\_ \models \_$: `System Prop` $\rightarrow$ `Bool`.

A system state in Real-Time Maude is a `GlobalSystem` term, and state propositions are defined with respect to this term.

$\_ \models \_ : \text{GlobalSystem Prop} \rightarrow \text{Bool}$
$a : \{\, T\, \} \models p(A) = \text{true } \textbf{if } \varphi.$

This is a definition of parameterized proposition $p(A)$. Its truth value is dependent on condition φ, which may refer to the argument A, as well as to the term representing the entire system T.

11.5.3.2 Removing Freeze Quantifiers

We show how freeze quantifiers are removed to obtain LTL formulas to employ the LTL model checker of Real-Time Maude [26]. In the translation, we assume restrictions enabling the over-approximation method.

Encoding a binding environment Γ is straightforward. We introduce, as a subsort of System, a new sort *Bindings* that has pairs of *Var* and Real as its elements. The sort *Var* refers to quantified variables.

$\Gamma : Var \ \text{Real} \ \rightarrow \ Bindings.$

Updating an environment requires appropriate rewriting rules, because *freezing* a variable is synchronized with the state transitions of a PCA. We denote such an *updating* rule as $\xrightarrow{\Gamma}$.

In the operational semantics of a PCA, event-trigger discrete transitions and delayed transitions are interleaving. The updating transition must be fired after a delayed transition, so that newly changed time-dependent values are seen. The sequence of transitions is expected to be $\xrightarrow{e1}; \xrightarrow{d1}; \xrightarrow{\Gamma1}$. In Real-Time Maude, however, the updating rule is an instantaneous rule, and thus it is enabled and fired together with discrete transitions of the PCA. Now, imagine we have two consecutive discrete transitions where $\xrightarrow{e1}$ is followed by $\xrightarrow{e2}$, namely, $\xrightarrow{e1}; \xrightarrow{d1}; \xrightarrow{e2}$. We merge the updating rule ($\xrightarrow{\Gamma1}$) to $\xrightarrow{e2}$ to obtain the instantaneous rule $\xrightarrow{\Gamma1,e2}$. Then, we can have the intended execution order.

Let l be a location, let τ be a time point, and let $S(l, v, w, \tau)$ represent a PCA state. The discrete transition $\xrightarrow{e2}$ is represented as a single instantaneous rule if we ignore the Γ. E_2 stands for the input alphabet involved in $\xrightarrow{e2}$.

$E_2\, S(l_1, v, w, \tau) \Longrightarrow S(l_2, v, w, \tau) \quad \textbf{if } \varphi_2.$

We consider how we define $\xrightarrow{\Gamma1,e2}$, where Γ is updated at l_1 ($\xrightarrow{\Gamma1}$). Since the tick rules are defined so that they are fired after $\xrightarrow{e1}$, the time-dependent values are kept in the term for which $\xrightarrow{e2}$ is to be fired. The above instantaneous rule must be modified to include the updates in Γ. As a concrete example, we consider the case of $\exists^D x.(p^g \wedge \phi^x)$ where D is either m or τ, p^g is a guard state proposition, and *fresh* is a special value denoting that the value of a variable is undefined. In the first rule, b is either $(v \cup w)(m)$ or τ.

$$E_2 \; S(l_1, v, w, \tau) \; \Gamma[x := fresh] \implies S(l_2, v, w, \tau) \; \Gamma[x := b]$$
$$\quad \textbf{if } \varphi_2 \wedge (p^g \in Lab(l_1))$$
$$E_2 \; S(l_1, v, w, \tau) \; \Gamma[x := b] \quad \implies S(l_2, v, w, \tau) \; \Gamma[x := b]$$
$$\quad \textbf{if } \varphi_2 \wedge ((p^g \notin Lab(l_1)) \vee (b \neq fresh)).$$

In addition, because Γ is an environment for frozen variables, it is not time dependent. We define two functions, δ and mte, such that $\delta(\Gamma, \tau) = \Gamma$ and $mte(\Gamma) = \text{INF}$.

11.5.3.3 Model Checking Command

Now, a formula to be checked does not have freeze quantifiers and thus is written in LTL. We put it in a negation normal form (NNF).

$$\phi := C \mid p \mid \neg p \mid \phi_1 \wedge \phi_2 \mid \phi_1 \vee \phi_2 \mid \Diamond \phi \mid \Box \phi.$$

Atomic propositions, to which a formula ϕ refers, must be defined as equational specifications for the built-in operator $\_\models\_$. Atomic propositions $p \in Prop$ and cost constraints C are defined as such equations.

$$\{ S(l, v, w, \tau) \; \Gamma \} \models p \quad = \texttt{true} \quad \textbf{if } p \in Lab(l)$$
$$\{ S(l, v, w, \tau) \; \Gamma \} \models \neg p = \texttt{true} \quad \textbf{if } p \notin Lab(l)$$
$$\{ S \; \Gamma[x := b] \} \quad \models C \quad = \texttt{true} \quad \textbf{if } C[b/x].$$

Given an initial global state with appropriate terms, we conduct time-bounded LTL model checking in Real-Time Maude as follows. The command returns either a constant `true` of sort `Bool` successfully, or a counterexample trace if the property is violated.

$$\textbf{mc } \{ S(\ell_0, 0, 0, 0) \; \Gamma_0 \} \models^t \Diamond(P_s \wedge \Diamond(P_e \wedge C)) \textbf{ in time} \leq B.$$

11.5.3.4 Correctness of Translation

After the translation to Real-Time Maude, the formula for checking does not have freeze quantifiers, but in the Real-Time Maude, the updating rules are merged with discrete transitions in addition to delay transitions. We must consider properties regarding the correctness of the translation. The property 1 was originally introduced in Reference 29, and we followed the proof method there.

Property 11.1. *An instantaneous rule to encode an updating rule does not have any effect on (a) the instantaneous rules used to encode the event-trigger discrete transitions, nor on (b) the tick rules corresponding to the delayed transitions.*

Proof Outline:

a. The rule does not have any additional effect on S of the instantaneous rules.

b. Two functions ensure that Γ does not have any effect on the tick rules ($\delta(\Gamma, \tau) = \Gamma$, $mte(\Gamma) =$ INF).

The next property is concerned with the formula to be checked. We will first consider the reachability case ϕ^{ReA}, $F^{\langle \tau; P \rangle}(Q_1 \wedge F_C^{\langle \tau; P \rangle} Q_2)$. Q_1 and Q_2 are guard state propositions.

Property 11.2. *LTL model checking by Real-Time Maude returns a counterexample* $\Longleftrightarrow \phi^{ReA}$ *returns a counterexample.*

Proof Outline:

a. The case without the constraint: ϕ^{ReA} can express $\Diamond(Q_1 \wedge \Diamond Q_2)$ because it is obtained by inserting *true* in $F_C^{\langle \tau; P \rangle}$ as $F_{true}^{\langle \tau; P \rangle}$. In the Real-Time Maude translation, the propositions Q_1 and Q_2 are defined in terms of equations for an operator symbol $\_\models\_$. Therefore, both can generate counterexamples for the formula $\Diamond(Q_1 \wedge \Diamond Q_2)$ if the property is violated.

b. The case with the constraint: We consider the case in which $\Diamond(Q_1 \wedge \Diamond Q_2)$ is satisfied because of the case (a). A counterexample of ϕ^{ReA} consists of the indices j and k such that $\langle \rho^j, \Gamma \rangle \models Q_2$, k=$min^{(j)}(i \mid \langle \rho^i, \Gamma \rangle \models Q_1$ and $(i \leq j))$, and C is violated for the frozen values obtained at ρ^k and ρ^j. In the Real-Time Maude translation, Γ is updated only for the j and k because of the translation of $\xrightarrow{\Gamma 1, e2}$. C is also shown to be violated against the values obtained from such j and k by the LTL model checker of Real-Time Maude. Therefore, both can generate counterexamples that consist of the same timed state sequences.

The proof for the response property ϕ^{ReS} is essentially the same as the method for the reachability because of the following:

$$
\begin{aligned}
&\Box \exists^\tau x. \exists^P u.(Q_1 \Rightarrow \Diamond \exists^\tau y. \exists^P v.(Q_2 \wedge C(x, y; u, v))) \\
&= \Box \exists^\tau x. \exists^P u.(\neg Q_1 \vee Q_1 \wedge \Diamond \exists^\tau y. \exists^P v.(Q_2 \wedge C(x, y; u, v))) \\
&= \Box(\neg Q_1 \vee \underline{\exists^\tau x. \exists^P u.(Q_1 \wedge \Diamond \exists^\tau y. \exists^P v.(Q_2 \wedge C(x, y; u, v)))}).
\end{aligned}
$$

Since $\neg Q_1$ is independent of any constraints, we only consider the underlined part of the subformula, which is similar to the formula of the reachability case.

11.6 SAT-BASED ANALYSIS METHODS

Because the problem of duration-bounded cost constraints is solved by scope-bounded search, this section focuses on SAT-based analysis methods. The term *SAT method* refers to both pure Boolean satisfiability and satisfiability modulo theories (SMT) methods.

11.6.1 Basic Concepts

We first introduce the basic concepts that are essential to understand discussions in this section. The precise definitions are in the standard literature (e.g., [30]).

11.6.1.1 Bounded Model Checking Problem

The *SAT method* is the basis of various automatic analysis methods, such as bounded model checking (BMC) [31]. Given a system, we encode the potential execution paths of the system in a trace formula (TF) φ_{TF}. We also have an assertion (AS) or a property to be checked as φ_{AS}, and $\varphi_{BMC} = \neg(\varphi_{TF} \Rightarrow \varphi_{AS})$, which illustrates failure situations. Because $\varphi_{BMC} = \varphi_{TF} \wedge \neg\varphi_{AS}$, the BMC problem is to determine whether $\varphi_{TF} \wedge \neg\varphi_{AS}$ is satisfiable. If φ_{BMC} is satisfied, then the obtained assignments constitute a counterexample to demonstrate that system (φ_{TF}) violates the property (φ_{AS}).

11.6.1.2 Fault Localization Problem

Let a formula φ_{FL} be $\varphi_{EI} \wedge \varphi_{TF} \wedge \varphi_{AS}$ for the above-mentioned φ_{TF} and φ_{AS}, and φ_{EI}, which encodes error-inducing input data values. We construct φ_{EI} by extracting from the counterexample a set of input data values that lead the system to such a failing execution.

The φ_{FL} is unsatisfiable, and the fault localization problem is to find clauses in φ_{TF} that are responsible for this unsatisfiability. Clauses in the identified unsatisfiable core constitute a *conflict*, which is an erroneous situation containing the root causes of the failure. Both φ_{EI} and φ_{AS} are supposed to be satisfiable because they encode the input data values and the property, respectively. This is exactly the problem in which we search for root causes of the faulty system (φ_{TF}).

Below, C refers to a set of clauses that constitute formula φ in conjunctive normal form (CNF). We use C and φ interchangeably.

11.6.1.3 Minimal Unsatisfied Subset

A set of clauses M, $M \subseteq C$, is a minimal unsatisfiable subset (MUS) *iff* M is unsatisfiable and $\forall c \in M : M \setminus \{c\}$ is satisfiable.

11.6.1.4 Maximal Satisfiable Subset

A set of clauses M, $M \subseteq C$, is a maximal satisfiable subset (MSS) *iff* M is satisfiable and $\forall c \in (C \setminus M) : M \cup \{c\}$ is unsatisfiable.

11.6.1.5 Minimal Correction Subset

A set of clauses M, $M \subseteq C$, is a minimal correction subset (MCS) *iff* $C \setminus M$ is satisfiable and $\forall c \in M : (C \setminus M) \cup \{c\}$ is unsatisfiable. By definition, an MCS is a complement of an MSS.

11.6.1.6 Hitting Set

Let Ω be a set of sets from some finite domain D. A hitting set of Ω, H, is a set of elements from D that covers every set in Ω by having at least one element in common with it. Formally, H is a hitting set of Ω *iff* $H \subseteq D$ and $\forall S \in \Omega : H \cap S \neq \emptyset$. A minimal hitting set is a hitting set from which no element can be removed without losing the hitting set property.

11.6.1.7 Partial Maximum Satisfiability
A maximum satisfiability (MaxSAT) problem for a CNF formula is finding an assignment that maximizes the number of satisfied clauses. Partial MaxSAT (pMaxSAT) is a variant of MaxSAT, in which some clauses are marked *soft* or relaxable, and other clauses are marked *hard* or nonrelaxable. A pMaxSAT problem is finding an assignment that satisfies all the hard clauses and maximizes the number of satisfied soft clauses.

11.6.1.8 Model-Based Diagnosis Framework
In the model-based diagnosis (MBD) framework [32], fault localization involves finding a subset of clauses, called *diagnosis*, in the unsatisfiable formula φ_{FL} so that removing the clauses in this subset makes the formula satisfiable. A *conflict* is an erroneous situation and a *diagnosis* refers to the root causes. A faulty system usually contains multiple conflicts and diagnoses. Formally, diagnoses are a set of MCS elements (or MCSes), and conflicts are a set of MUS elements (or MUSes). MCSes and MUSes are related by a hitting set relation [33]. The formula-based approach [34,35] calculates an MSS to obtain an MCS by complementing the MSS, and repeats this process to collect MCSes. Any solution to a MaxSAT problem is also an MSS. However, every MSS is not necessarily a solution to MaxSAT [36]. Various algorithms for efficient enumeration of MCSes (cf. [36,37]) have been developed for this.

The fault localization problem requires a way to represent the fact that φ_{EI} and φ_{AS} are satisfiable and that some clauses in φ_{TF} are suspicious. The pMaxSAT approach is well suited to satisfying this requirement. The clauses in φ_{EI} and φ_{AS} are marked *hard*. Suspicious clauses in φ_{TF} are *soft*. The other clauses in φ_{TF} that are assumed to be bug-free are *hard*. This decision, namely which clauses are marked *soft* in φ_{TF}, is dependent on a failure model.

11.6.2 Boolean Encoding
BMC and fault localization require Boolean encoding of a trace formula φ_{TF}. We follow the encoding method of TAs in Reference 38 and extend this to PCAs. For simplicity, we assume here that a PCA has one clock variable and one weight variable.

11.6.2.1 Trace Formula
The state of a PCA $\mathcal{A}$ is characterized by a location variable (at), a clock variable (x), a weight variable (w), and a variable act referring to an input symbol. The input alphabet is $\Sigma \cup \{\epsilon, delay\}$, where a special symbol $delay$ indicates that the automaton takes a delay transition. We introduce a set, $S = \{at, x, w, act\}$, and a PCA $\mathcal{A}$ is defined as $\langle I, T \rangle$ over S.

1. *Initial state*: Clock and weight variables are initialized to zero.

$$I = (at = \ell_0 \wedge x = 0 \wedge w = 0).$$

2. *Discrete transition step*: $T(e)$ is a relation on S and S', where $e = l \xrightarrow{g,a,r} l'$, and $z = 0$ if $x \in r$; $z = x$ otherwise.

$$T(e) = (at = l \land at' = l' \land act = a \land g \land x' = z \land w' = w \land Inv(l')(x')).$$

3. *Delay transition step*: $D(\delta, \ell)$ is a relation on S and S' at a location $\ell \in Loc$, where δ is a positive real variable of delay, and M^ℓ is a nonnegative real-valued constant given at each location. In particular, the weight variable is updated in accordance with the flow dynamics.

$$D(\delta, \ell) = (Inv(\ell)(x') \land at' = at \land at = \ell \land act = delay \land x'$$
$$= x + \delta \land w' = w + M^\ell \times \delta).$$

4. *Inactivity transition step*: F is a relation on S and S' to encode that a PCA performs neither a discrete nor a delayed transition step.

$$F = (at' = at \land x' = x \land w' = w \land (\bigwedge_{\alpha \in (\Sigma \cup \{delay\})} act \neq \alpha)).$$

5. *Transition relation*: $T = (\bigvee_{e \in Edg} T(e)) \lor (\exists\, \delta > 0 . \bigvee_{\ell \in Loc} D(\delta, \ell)) \lor F$.

6. *K-step unfolding of N numbers of PCAs*: A total of N PCAs $A^{(1)} \ldots A^{(N)}$ are composed, where $A^{(i)}$ is $\langle I^{(i)}, T^{(i)} \rangle$. Here, $I^{(i)}$ and $T^{(i)}$ define the initial state and transition relation of an ith automaton, respectively. $T_j^{(i)}$ is the jth transition of the ith automaton.

$$\varphi_{TF}^K = \bigwedge_{i=1..N} (I^{(i)} \land (\bigwedge_{j=1..K} T_j^{(i)})) = I \land (\bigwedge_{j=1..K} T_j)$$

where

$$I = \bigwedge_{i=1..N} I^{(i)} \text{ and } T_j = \bigwedge_{i=1..N} T_j^{(i)}.$$

11.6.2.2 LTL-Bounded Model Checking

We consider LTL-BMC, not fWLTL-BMC, in which we check the property $\Box(E \leq C_{max})$ in a duration specified by the guide constraint φ_G. As mentioned in Sections 11.4.1 and 11.4.2, this property is written in fWLTL as $(I \land \Diamond_d^P v. (P_e \land (v \leq C_{max})))$. The BMC problem here is considered to work on a small subset of fWLTL.

Recall that a PCA is defined over a set of atomic propositions *Prop*, and that has a labeling function *Lab* from each location to a set of atomic propositions, $Lab : Loc \rightarrow 2^{Prop}$. The set *Prop* here is defined over state variables U such that $U = \{x, w\}$. Specifically, a proposition p ($p \in Prop$) takes the form of clock constraints for clock variable (x), and the form of weight constraints for weight variable (w). These constraints are represented in the LRA theory. Below, we use a j-suffix state such that $U^j = \{x_j, w_j\}$.

Although a BMC problem is usually defined for arbitrary formulas of LTL [31,38], we consider safety properties only. We let $\psi(u)$ be a propositional formula constructed from *Prop*, and u_j be state variables in U^j. The safety property is expressed in LTL as $\Box\psi$. Recall a BMC problem is checking the satisfiability of $\varphi_{TF} \wedge \neg\varphi_{AS}$. Because $\neg(\Box\psi) = \Diamond\neg\psi$, the entire formula for a K-unfolded case is

$$\varphi_{safety} = I(u_0) \wedge \left(\bigwedge_{j=1..K} T_j(u_{j-1}, u_j) \right) \wedge \left(\bigvee_{j=1..K} \neg\psi(u_j) \right).$$

This is searching for an erroneous state, that satisfies $\neg\psi(u_j)$, within K transition steps from an initial state.

As the energy consumption analysis is a duration-bounded cost-constraint problem, we need a way to specify a duration in which the numerical constraints on the weight variables are checked. Such a duration is defined by LTL formula φ_G, a *guide constraint*. Then, the entire formula for the BMC problem is $\varphi_{TF} \wedge \neg(\varphi_G \Rightarrow \varphi_{AS})$, which is $(\varphi_{TF} \wedge \varphi_G) \wedge \neg\varphi_{AS}$. Particularly, we consider the case in which $\varphi_{AS} = \Box(w_E \leq C_{max})$ and $\varphi_G = \Diamond(P_s \wedge \Diamond P_e)$. Thus, for the K-unfolded case, the formula becomes

$$I(u_0) \wedge (\bigwedge_{j=1..K} T_j(u_{j-1}, u_j))$$
$$\wedge (\bigvee_{j=1..K-1} (P_s(u_j) \wedge (\bigvee_{i=j..K} P_e(u_i)))) \wedge (\bigvee_{j=1..K} (w_j > C_{max})).$$

We assume here that the value of the weight variable w_E is zero at a state that satisfies $P_s(u_S)$ for a certain S ($1 < S < K - 1$) so that w_j ($S < j < K$) represents the total consumed energy up to the jth state.

11.6.3 Fault Localization

11.6.3.1 Fault Localization Problem for a PCA

We consider that a whole system consists of multiple PCAs. Parallel compositions of PCAs are *closed* in that a sender has a matching receiver. Although the original fault localization problem refers to the error-inducing input φ_{EI}, the problem here does not need to consider this condition. Because the set of PCAs is closed, no input data are given externally. However, as discussed above, we must take into account the guide constraint φ_G. Therefore, the formula φ_{FL} used in the fault localization problem for PCAs takes the form $\varphi_{FL} = \varphi_{TF} \wedge \varphi_G \wedge \varphi_{AS}$. The problem is searching for MCS elements in φ_{TF} of the unsatisfiable φ_{FL}.

11.6.3.2 Failure Model

We assume that faulty behavior in PCAs originates from some faults in using clock variables. PCAs are state-transition systems that control the enabling of transitions by nonnegative real-valued clocks. The weight variables are *observers* that do not affect the enabling of transitions. When multiple PCAs are composed, synchronization can be considered to introduce further constraints on the transition sequences of the constituent PCAs. This synchronization indirectly affects possible transition sequences controlled by the clock variables.

PCAs refer to clocks in invariants, or transition guards and resets, $Inv(l)$, or g and r on an edge of $l \xrightarrow{g,a,r} l'$. We consider the possibilities that the elements referring to clocks may contain root causes. These are suspicious elements in faulty systems. The fault localization problem now involves checking the set of clock constraints collected from a failing trace with respect to the given property φ_{AS}.

In the fault localization problem using the pMaxSAT approach, these suspicious elements are marked *soft*. Since the initial state is usually definite, all of the elements in the initial state formula (I) are *hard*. The inactivity transition F, in parallel compositions of automata, is also definite because it encodes situations in which a constituent automaton does not take any transition.

Next, we consider discrete transition step $T(e)$ and delay transition step $D(\delta, \ell)$. Elements, related to invariants, or guards and resets in the transition step, are marked *soft*. Below the notation p^H (or p^S) indicates that p is *hard* (or *soft*).

1. *Discrete transition step*:

$$T(e) = (\, (at = l)^H \wedge (at' = l')^H \wedge (act = a)^H \wedge (g)^S$$
$$\wedge (x' = z)^S \wedge (w' = w)^H \wedge (Inv(l')(x'))^S \,).$$

2. *Delay transition step*:

$$D(\delta, \ell) = ((Inv(\ell)(x'))^S \wedge (at' = at)^H \wedge (at = \ell)^H \wedge (act = delay)^H$$
$$\wedge (x' = x + \delta)^H \wedge (w' = w + M^\ell \times \delta)^H).$$

11.6.3.3 Nondeterministic Transitions

PCAs have nondeterministic transitions, namely, more than one transition can be enabled simultaneously. A PCA may have multiple edges e_i, such that $e_i = l_s \xrightarrow{g_i, a_s, r_i} l_i$. The edges share a common source location l_s and enabling conditions, but have a different reset r_i and a destination location l_i, where the enabling conditions are described in terms of a common input symbol a_s and overlapped guard conditions g_i. Furthermore, the PCA has a delay transition $D(\delta, l_s)$ at the same source location l_s, which is competing with the discrete transitions. We have a set of transitions that are enabled nondeterministically at l_s, $\{D(\delta, l_s), T(e_1), \ldots, T(e_N)\}$.

Nondeterministic transitions complicate the fault localization problem. The formula-based fault localization method relies on the fact that MCSes are calculated from the unsatisfiability of φ_{FL}. However, if the system has nondeterministic transitions, it can take transitions other than those in the failing execution and some of the paths may be successful. Consequently, φ_{FL} may be satisfiable and have empty MCSes.

The above observation implies that we cannot use the full flow-sensitive trace formula, which is successful in the case of imperative programs [34]. The trace formula used in the BMC is full flow-sensitive because the formula encodes all potential execution paths and thus contains nondeterministic transitions when we consider PCAs. Note that this issue does not come up because imperative programs are deterministic.

11.6.3.4 Trace Formula for Fault Localization

Since φ_{FL} is $\varphi_{TF} \wedge \varphi_G \wedge \varphi_{AS}$, the φ_{TF} in the unsatisfiable formula φ_{FL} can be restricted to capture conflict situations only. These are in the subset of all potential transition sequences, and φ_{TF} does not need to encode all the sequences. Furthermore, the conflict situation is related to the counterexample trace that the K-scoped BMC procedure returns. The trace contains a transition sequence leading to the violation of φ_{AS} and other information. The sequence is actually a mixture of discrete steps ($T(e_j^i)$) and delay transition steps ($D(\delta^i, \ell^i)$). Here, $T(e_j)$ is a formula for a discrete transition of e_j, and $T(e_j^i)$ is an instance of $T(e_j)$ when it is taken in the ith step in the sequence. If the length is K, the sequence to be encoded is $\bigwedge_{i=1..K}(T(e_j^i)$ or $D(\delta^i, \ell^i))$.

Recall that the formula φ_{AS} is a safety property ($\Box\psi$). If it is violated, then an index L ($L \leq K$) exists, such that L is the state at which φ_{AS} violates at the first time, that is, a minimum of such indices. Since the transition sequence up to L contains enough information leading to the violation, the sequence is encoded in the formula $\bigwedge_{i=1..L}(T(e_j^i)$ or $D(\delta^i, \ell^i))$. We can ensure that $I(u_0) \wedge (\bigwedge_{i=1..L}(T(e_j^i)$ or $D(\delta^i, \ell^i))) \wedge \neg\psi(u_L)$ is satisfiable.

11.6.3.5 Sliced Transition Sequence

We use a *sliced transition sequence* as the trace formula (φ_{TF}^{sliced}) for the fault localization. It is a conjunction of the transition steps in an L-scoped counterexample trace.

$$\varphi_{TF}^{sliced} = I \wedge (\bigwedge_{i=1..L}(T(e_j^i) \text{ or } D(\delta^i, \ell^i))).$$

The formula involves *soft* clauses referring to clock variables. Because *soft* clauses are relaxable, we can find MCS elements that make the formula $\varphi_{TF}^{sliced} \wedge \varphi_{AS}$ satisfiable. The formula φ_G does not appear because φ_{TF}^{sliced}, by construction, satisfies the guide constraint.

11.6.3.6 Fault Localization Method

The fault localization steps are described below.

1. Execute BMC of $\varphi_{TF} \wedge \varphi_G \wedge \neg\varphi_{AS}$.

2. Construct φ_{TF}^{sliced} from the counterexample trace.

3. Use pMaxSAT for $\varphi_{TF}^{sliced} \wedge \varphi_{AS}$ to enumerate MCSes.

We conducted some example cases [39] that used Yices-1 [8], a pMaxSAT solver supporting the LRA theory. Below, we will present some findings collected from the experiments.

Because diagnoses of a fault in the formula-based fault localization method are MCS elements, we have two aspects to discuss in regard to the precision of the identified root causes. First, an MCS element may not have a minimum number of clauses and may have extra clauses. This fact implies that a diagnosis may not be accurate enough to pinpoint a specific fault location. However, in our experience of either PCAs or imperative programs [34], an MCS element consists of a single clause or at most a few. Thus, being minimal is not an issue in practice. Second, the number of MCS elements in the obtained MCSes is sometimes large, which means that many root cause candidates exist. Therefore, human insight is required to determine which candidate MCS is a real root cause to repair. The fault localization method is not concerned with such differences. The experiments in Reference 39 reported that the obtained MCSes contained a real root cause and a spurious one as well.

The failure model is based on the assumption that PCA elements referring to clock variables are suspicious. Here, φ_{TF}^{sliced} is considered as a sequence of transition steps and each transition step has a conjunction of conditions on clock variables as its subformula. The sliced trace formula is essentially a set of conditions on clock variables that is collected along the counterexample transition sequence. We denote ϕ_{clk} as the formula encoding such constraints. Imagine that a given property ϕ_{AS} uses clock variables and weight variables. Because the values of the weight variables are dependent on time, formula ϕ_{AS} can be considered essentially as clock constraints. A violation of a given property is indeed an unsatisfiability of clock constraints, $\phi_{clk} \wedge \phi_{AS}$. These clock constraints in PCAs are expressed in the LRA theory, and thus decidable.

The failure model currently does not consider possible structural bugs. We can, however, extend our method to consider such bugs, but in restricted cases only. Recall that the method relies on finding MCS elements, which is finding clauses that must be removed. If a new failure model considers the case in which a certain transition step, or an edge in PCA, is suspicious, then an edge, or actually a clause corresponding to the edge, can be a candidate for removal. Contrarily, if a bug appears because of a missing edge, we cannot find any root cause. No MCS element accounts for missing edges. This limitation is unavoidable in the MBD approaches.

11.7 CONCLUSION

The work reported in this chapter began from an observation that Android-based smartphones occasionally became *hot* as the result of e-bugs [2]. Although smartphones are small, the underlying Android infrastructure is multilayered and complex. A key to understanding the observed phenomena is modeling at an appropriate level of abstraction that, at the same time, is detailed enough to delineate the problems at hand.

We first formulated a problem by studying the measured results of an energy profile [15], as well as international standard recommendations [16]. It turned out that the problem was captured by duration-bounded cost constraints on a state-transition model, called the PCA. Then, we looked for appropriate formal models to represent the PCA, and studied extensively the classical work on LHAs [5]. Although LHA was certainly a formal basis of our problem, we found that a restricted model, a variant of WTA [19], was expressive enough. The theoretical results on LHAs are valuable to further understanding the problem. However, many other issues must be considered to enable automatic analyses, which are mandatory in software engineering for establishing a model-based analysis method of energy consumption behavior. We studied such technical solutions from two viewpoints, one using Real-Time Maude [6,7] and another using SAT-based verification methods [30,31]. We also showed that the SAT-based methods could automatically localize faults. Although the subject matter appeared only recently, the solutions are proposed on the basis of classical results reported since the 1990s. However, various technical work on model checking of fWLTL still remains to be done.

We believe that the work reported here can be a starting point for further research to enable *model-based analysis of energy consumption behavior* in practice, which is increasingly important in developing *trustworthy cyber-physical systems*.

ACKNOWLEDGMENTS

This work is partially supported by JSPS KAKENHI Grant Numbers 24300010 and 26300095.

REFERENCES

1. E.A. Lee and S.A. Seshia. *Introduction to Embedded Systems—A Cyber-Physical Systems Approach*, California, LeeSeshia.org, 2011.
2. A. Pathak, Y.C. Hu, and M. Zhang. Bootstrapping energy debugging on smartphones: A first look at energy bugs in mobile devices, In *Proc. Hotnets'11*, pp. 5:1–5:6, Cambridge, MA, 2011.
3. J.M. Wing. Cyber-physical systems, *Computer Research News*, 21(1), p. 4, 2009.
4. Android. http://developer.android.com.
5. R. Alur, C. Courcoubetis, N. Halbwachs, T.A. Henzinger, P.-H. Ho, X. Nicollin, A. Olivero, J. Sifakis, and S. Yovine. The algorithmic analysis of hybrid systems, *Theoretical Computer Science*, 138(1), pp. 3–24, 1995.
6. P.C. Ölveczky and J. Meseguer. Semantics and pragmatics of real-time maude, *Higher-Order and Symbolic Computation*, 20(1–2), pp. 161–196, 2007.
7. P.C. Ölveczky and J. Meseguer. Abstraction and completeness for real-time maude, *ENTCS*, 176(4), pp. 5–27, 2007.
8. B. Dutertre and L. de Moura. The Yices SMT Solver. http://yices.csl.sri.com.
9. A. Pathak, Y.C. Hu, and M. Zhang. Where is the energy spent inside my app? Fine grained energy accounting on smartphones with Eprof, In *Proc. EuroSys'12*, pp. 29–42, Bern, Switzerland, 2012.
10. L. Zhang, M.S. Gordon, R.P. Dick, Z.M. Mao, P. Dinda, and L. Yang. ADEL: An automatic detector of energy leaks for smartphone applications, In *Proc. CODES+ISSS'12*, pp. 363–372, Tampere, Finland, 2012.
11. J.L. Hennessy and D.A. Patterson. *Computer Architecture: A Quantitative Approach (5 ed.)*, Morgan Kaufmann, Massachusetts, 2011.
12. S. Nakajima and M. Toyoshima. Behavioral contracts for energy consumption, *Ada User Journal*, 35(4), pp. 266–271, 2014.
13. S. Nakajima. Model-based power consumption analysis of smartphone applications, In *Proc. ACES-MB 2013*, pp. 5:1–5:10, Miami, Florida, 2013.
14. S. Nakajima. Model checking of energy consumption behavior, In *Proc. 1st CSDM Asia*, pp. 3–14, Singapore, 2014.
15. J. Manweiler and R.R. Choudhury. Avoiding the rush hours: WiFi energy management via traffic isolation, In *Proc. MobiSys'11*, pp. 253–266, Washington, DC, 2011.
16. IEEE Standard 802.11, Wireless LAN Medium Access Control (MAC) and Physical Layer (PHY) Specifications, 1999.
17. S. Nakajima. Everlasting challenges with the OBJ language family, In *Proc. SAS 2014*, pp. 478–493, Kanazawa, Japan, 2014.
18. A. Brekling, M.R. Hansen, and J. Madsen. MoVES – A framework for modeling and verifying embedded systems, In *Proc. ICM2009*, pp. 149–152, Malaga, Spain, 2009.
19. R. Alur, C. Courcoubetis, and T.A. Henzinger. Computing accumulated delays in real-time system, In *Proc. CAV 1993*, pp. 181–193, Boulder, CO, 1993.

20. R. Alur and D.L. Dill. A theory of timed automata, *Theoretical Computer Science*, 126(2), pp. 183–235, 1994.
21. R.J. Wieringa. *Design Methods for Reactive Systems*, Morgan Kaufmann Publishers, San Francisco, 2003.
22. J. Ouaknine and J. Worrell. Some recent results in metric temporal logic, In *Proc. FORMATS 2008*, pp. 1–13, Saint Malo, France, 2008.
23. R. Alur and T.A. Henzinger. A really temporal logic, *Journal of the Association for Computing Machinery*, 41(1), pp. 181–204, 1994.
24. P. Bulychev, A. David, K.G. Larsen, A. Legay, G. Li, D.B. Poulsen, and A. Stainer. Monitor-based statistical model checking for weighted metric temporal logic, In *Proc. LPAR-18*, pp. 168–182, Merida, Venezuela, 2012.
25. S. Demri, R. Lazic, and D. Nowak. On the freeze quantifier in constraint LTL: Decidability and complexity, *Information and Computation*, 205(1), pp. 2–24, 2007.
26. S. Nakajima. Using real-time maude to model check energy consumption behavior, In *Proc. FM 2015*, pp. 378–394, Oslo, Norway, 2015.
27. M. Clavel, F. Duran, S. Eker, P. Lincoln, N. Marti-Oliet, J. Meseguer, and C. Talcott. *All About Maude—A High-Performance Logical Framework*, Springer, Berlin, 2007.
28. S. Nakajima. Formal analysis of android application behavior with real-time maude, In *Proc. CPSNA 2015*, pp. 7–12, Hong Kong, 2015.
29. D. Lepri, P.C. Ölveczky, and E. Abraham. Model checking classes of metric LTL properties of object-oriented real-time maude specification, In *Proc. RTRTS 2010*, pp. 117–136, Longyearbyen, Norway, 2010.
30. A. Biere, M. Heule, H. Van Maaren, and T. Walsh (eds.). *Handbook of Satisfiability*, IOS Press, Amsterdam, 2009.
31. A. Biere, A. Cimatti, E.M. Clarke, and Y. Zhu. Symbolic model checking without BDDs, In *Proc. TACAS 1999*, pp. 193–207, Amsterdam, 1999.
32. R. Reiter. A theory of diagnosis from first principles. *Artificial Intelligence*, 32(1), pp. 57–95, 1987.
33. M.H. Liffiton and K.A. Sakallah. Algorithms for computing minimal unsatisfiable subsets of constraints. *Automated Reasoning*, 40(1), pp. 1–33, 2008.
34. S.-M. Lamraoui and S. Nakajima. A formula-based approach for automatic fault localization of imperative programs. In *Proc. ICFEM'14*, pp. 251–266, Luxembourg, 2014.
35. S. Safarpour, H. Mangassarian, A. Veneris, M.H. Liffiton, and K.A. Sakallah. Improved design debugging using maximum satisfiability, In *Proc. FMCAD'07*, pp. 13–19, Austin, Texas, 2007.
36. A. Morgado, M. Liffiton, and J. Marques-Silva. MaxSAT-based MCS enumeration. In *Proc. HVC'12*, pp. 86–101, Haifa, Israel, 2013.
37. M.H. Liffiton and A. Malik. Enumerating infeasibility: Finding multiple MUSes quickly, In *Proc. 10th CPAIOR*, pp. 160–175, Yorktown Heights, New York, 2013.
38. M. Sorea. Bounded model checking for timed automata, *ENTCS*, 68(5), pp. 116–134, 2002.
39. S. Nakajima and S.-M. Lamraoui. Fault localization of energy consumption behavior using maximum satisfiability, In *Proc. CyPhy 2015*, pp. 99–115, Amsterdam, 2015.

CHAPTER 12

A Formal DSL for Multi-Core System Management

Alexei Iliasov

CONTENTS

12.1 INTRODUCTION

With the end of processor frequency growth, scaling into the direction of multi- and many-core systems has become the primary way to translate advances in manufacturing technology into increased computer performance. This change has already started to affect operating systems and application software design. Clearly, we have to plan now for the many-core systems of tomorrow. The problem of scheduling and core management is an example of a challenging Cyber-Physical System (CPS) where operating environment (temperature and background radiation), core physical states (temperature, age, frequency, voltage, and heat dissipation) and software load and deadlines are all intertwined in a complex fashion requiring an intricate balance for a smooth, sustained operation.

In this work we propose a method and toolkit that has grown around an effort to build a database of knowledge related to multi-core systems [1]. The exercise started with a simple compilation of facts in a natural language. The result turned out to be inadequate due to many ambiguities, hidden assumptions, and subtle interrelations between concepts. A decision then was made to transfer into a formal specification language. We have chosen the Event-B modeling language [2] mainly based on our previous experience with it. The formal specification of the knowledge database resolved the known ambiguities and exposed many omissions and inconsistencies. The effort took several months and resulted in a substantial model with hundreds of properties. It was assumed that the model would be consulted when designing or writing OS-level software for multi-core systems. It turned out that a formal model of this scale is a difficult read and cannot readily serve as a blueprint for software design. Moreover, there was a lack of confidence in the model completeness and liveness properties—conditions that are impractical to address with static theorem proving and, in this model, impossible, due to scale, to delegate to a model checker.

To address these weaknesses while staying in the domain of formal reasoning, it was decided to build an extension to the developed Event-B model that would "drive" (to take the term from a conceptual inspiration—the CSP∥B technique [3]) it in a sense of restricting event enabling. The pursuit of this goal has led to the development of the DSL-Kit tool. DSL-Kit offers a unique approach to the construction of a formal imperative-style domain-specific language where the heart of the language—its domain-specific parts, such as commands, variables, and constants—is designed in a separate formal notation (in our case Event-B), while the "glue" part (control flow constructs like *if* and *while*) are generic and reused by all DSL instances.

The most profound implication is the facilitation of iterative design: design prototypes realized in a DSL would often highlight deficiencies in the DSL itself (lack of progress, missing concepts); this may be addressed by going back to Event-B specifications of the domain, doing necessary changes, proving them correct and, *automatically*, with the help of a translation tool, transferring the result into DSL-Kit to construct a new instance of domain DSL (Figure 12.1). DSL-Kit itself offers fairly rich reasoning facilities. On top of this, it can execute or animate (with many limitations) DSL specifications and a custom state visualization may be plugged in to facilitate debugging and design comprehension.

The first part of this chapter (Section 12.3) is a narrative build around an Event-B model that attempts to capture the essence of multi-core systems primarily from the software and

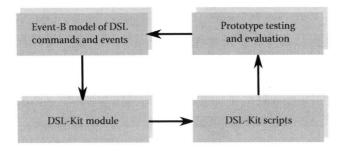

FIGURE 12.1   Iterative development process with the DSL-Kit.

OS perspectives. The model does not include any notions of control logic. Instead, it gives a concrete definition of the subjects of the prospective control system and attempts to explain their interrelationships; it also describes the general life cycle of a multi-core system. The presumption is that any design of an OS-level load distribution, power management, or reliability control may be defined by *refining* this model; that is, constraining its behavior and, possibly, introducing some new concepts.

The second part (Section 12.2) describes the transition in the DSL-Kit environment and relates the initial experience to the design of an adaptive run-time controller. We show how the domain model may be turned into a formal virtual machine over which control logic executes. The reasoning style also changes from refinement proofs and inductive verification of safety invariant of Event-B to the verification of imperative programs in the style of Floyd-Hoare logic [4].

12.2 DSL-KIT

The developed Event-B model is a useful source of domain knowledge and reference material. However, it is hard to utilize directly for design prototyping, testing, and verification. The scale of the model and the complexity of its parametrization made model animation and model checking impossible, although we have successfully used ProB [5] for the few initial refinement steps.

The ability to do design prototyping with the developed formal domain model was perceived to be one of the more interesting directions. To this end, we have developed an extension to the Event-B Rodin [6] modeling framework that adds a capability of rapidly defining a custom DSL on the basis of an Event-B specification. The extension, called DSL-Kit, works by combining a general-purpose imperative specification language with a custom set of "commands" defined by an Event-B specification. The imperative layer provides control flow constructs that allow one to write scenarios or imperative programs in the terms of Event-B events. From the viewpoint of Event-B, the imperative layer is an extra refinement step strengthening event guards and declaring new hidden states.

The core of DSL-Kit is a formal specification language based on the following principal structuring units:

- System: The top-level unit defined as a parallel composition of several *actors*

- Module: A self-contained unit providing definitions of *actors* and *actions*

- Actor: A unit of concurrency; its body is sequential code guarded by rely/guarantee conditions

- Action: A function-like entity

- Control flow statements: Sequential composition, *if*, *while*, *for* and auxiliary variable declarations

- Expressions and predicates, expressed in the Event-B mathematical language: This make it possible to relate DSL-Kit to Event-B without logic mapping

12.2.1 Semantics

DSL-Kit glue language offers common control flow constructs of an imperative language to define sequential models and an actor-based framework for reasoning about concurrency.

The core part of DSL-Kit notation is the statement-level control flow language. It is detailed in Figure 12.2.

In Figure 12.2, pred and expr are exactly Event-B predicates and expressions of the Event-B mathematical language. Also, type are all valid Event-B type expressions and bapred are Event-B predicates with primed variables.

It is assumed that the state model comes from a *domain model* so in the glue language one may only define auxiliary variables. The aux operator introduces a list of fresh identifies nondeterministically bound to some values by a predicate over known state and initial values of the new identifies.

stmt	:=	**skip**	skip statement
	|	**aux** *var*:type :| bapred	auxiliary variable declaration
	|	*var* :| bapred	auxiliary variable assignment
	|	stmt; stmt	sequential composition
	|	{	
		(**invariant** pred)∗	
		stmt	
		}	statement block
	|	[pred]	assertion
	|	*action*((expr(, expr)∗)?)	action invocation
	|	*action*(pred)	action invocation
	|	**if** (pred) stmt (**else** stmt)∗	conditional statement
	|	**while** (pred)	
		variant expr	
		(**invariant** pred)+	
		stmt	loop statement

FIGURE 12.2 Summary of DSL-Kit statement notation.

The assignment also takes a predicate to constrain variable states. The old states, however, are not forgotten when executing a model (symbolically) as they, in general, may be necessary to carry out a proof. As an example, if one writes the following sequence of assignments,

$$\text{aux } n : \text{int} :| \ n' > 0;$$
$$n :| \ n' == 2 * n + 1; n :| \ n' == -n;$$

a proof context will contain an impression of each assignment with new identifies generated automatically to distinguish among current, parent, and grandparent versions of n. When state is complex and there is a long sequence of statements, such accumulation of historical trace may significantly complicate the proof process. There are three ways to combat this: use deterministic assignment; introduce assertions to compress history and filter out unnecessary details; and/or use invariants to highlight essential facts.

There are two forms of invariant. A block invariant applies within the body of a statement block and helps to reason about the properties of statement sequence making up the block body. Such an invariant may reference variables declared within the block.

A loop invariant serves the obvious purpose of semantically folding a terminating loop statement while (c) invariant I s into $[\neg c \wedge I]$. A loop may be nonterminating, which is indicated by omitting a loop variant. A variant expression must be of an integer type; set variants are not supported. Unlike a loop invariant, a block invariant does not propagate outside of a block: semantic folding of a block must be done by explicitly writing an assertion immediately after the block.

The following is a small example using auxiliary variables and the assertion. Variable declarations use before-after predicates to express relations defining possible values for a new variable on the basis of known state.

$$\text{aux } n, m : \text{int} :| \ n' \text{ mod } 6 == 0 \wedge m' == n' * n';$$
$$[m \text{ mod } 12 == 0];$$
$$n :| \ m == n' * 12.$$

When assignment is deterministic one may use syntactic sugar $v = E$ equivalent to $v :| \ v' == E$. Thus, for instance, the first line of the example above may be also written as

$$\text{aux } n : \text{int} :| \ n' \text{ mod } 6;$$
$$\text{aux } m : \text{int} = n * n.$$

The example defines three steps where helper variable m is first initialized to a value divisible by 6, then m is set to square of n and finally new value of n divides m by 12. The intermediate assertion helps to show that assignment to n is feasible: some such n' satisfying predicate $m == n' * 12$ may be found.

Central to the DSL-Kit language is the notion of *actions*. These may be likened to procedures of a programming language. In fact, one may define actions with executable bodies that look and behave much like a programming language procedure. However, in most scenarios an action is defined indirectly by defining the way it affects state model. Formally, this is done with a pair of pre- and postconditions. Action preconditions are predicates over state; postconditions are before-after predicates stating how state may be updated upon action execution. Whenever possible pre- and postconditions are split into subconditions and clustered by the segment of state space they constrain or update.

An action may be referred to in the style of a procedure call by specifying values of its formal parameters. If action has a body (i.e., a statement defining its behavior) and the body is deterministic then an action invocation is also deterministic and, generally, is executable.[*]

An action may also be invoked by constraining all or some of its formal parameters with a predicate. Certain naming conventions are used to disambiguate between state variables and formal parameters.

The primary use of a DSL-Kit script is symbolic execution. We use forward style (as opposed to the original Hoare logic) where known state model H, made of a set of predicates over state, is transformed and extended by each successive statement. For instance, a variable assignment statement to a variable renames all instances of the variable in model H to some fresh identifies and then inserts before-next predicate as an extra bit of knowledge.

Some statement incur side conditions that check for properties like assignment feasibility, precondition enabledness, invariant satisfaction, and so on. These are collected as conjectures during the process of symbolic execution. Each such conjecture must be proven correct for a script to be considered well formed.

The definition in Figure 12.3 defines the interpretation rules for the statement part of DSL-Kit grammar. Here H, T, C, I and V are state model, typing context, conjectures, invariants, and variants. Each element of the grammar transforms one or more of these elements although, in the end, we care only about state model H and conjectures C. The set of predicates making up a state model may be given to a constraint solver to find a solution to a part of a script making up an action which is an atomic step of execution in DSL-Kit. Conjectures C are delegated to automated theorem provers. Operator $H_1 \uplus H_2$ conjoins two state models by renaming in model H_1, to some fresh names, all the identifiers common to the models. After renaming, it combines model predicates into a single set defining a new combined model. Operator $H \uparrow T$ projects state model H on state definition T so that any identifier occurring in H but not T is removed from H by rewriting the affected predicates using an existential quantifier. Finally, operator $c \rightarrow H$ rewrites every predicate p in H to $c \Rightarrow p$.

An action declaration has the following syntax:

$$
\begin{aligned}
\text{actn} \quad := \quad & \text{action } action((var(, var)*)?)(:type)? \\
& (\text{variant } expr)? \\
& (\text{pre } pred)* \\
& \text{stmt} \\
& (\text{post } pred)*
\end{aligned}
$$

An action may return a value and calls to such actions may appear in expressions. Alternatively, an action is a statement and its call is a separate step of a statement block. When

---

[*] It is not always possible to execute even a deterministic model as it might depend on abstract constants that may be too complex to instantiate in a constraint solver.

$$\llbracket \textbf{skip} \rrbracket (H, T, C, I, V) \quad := \quad (H, T, C, I, V)$$

$$\llbracket \textbf{aux } v{:}t :| \ p \rrbracket (H, T, C, I, V) \quad := \quad (H \cup p[v' \mapsto v])) \},$$
$$T \cup \{var \mapsto \text{type}\},$$
$$C \cup \{H \vdash \exists v' \cdot (p \wedge I[v \mapsto v']), V)$$

$$\llbracket v :| \ p \rrbracket (H, T, C, I, V) \quad := \quad (H \uplus \{p\}, T, C \cup$$
$$\cup \{H, I \vdash \exists v' \cdot (p \wedge I[v \mapsto v'])\}$$
$$\cup \{H, I, p \vdash \text{prime}(V, v)\}, V)$$

$$\llbracket s_1; s_2 \rrbracket (H, T, C, I, V) \quad := \quad (H_1 \uplus H_2, T_1 \cup T_2, C_1 \cup C_2, V)$$
$$\text{where } (H_1, T_1, C_1, I_1, V) =$$
$$= \llbracket s_1 \rrbracket (H, T, C, I, V)$$
$$\text{and } (H_2, T_2, C_2, I_2, V) =$$
$$= \llbracket s_2 \rrbracket (H_1, T_1, C_1, I, V)$$

$$\llbracket \{\textbf{invariant } i \ s\} \rrbracket (H, T, C, I, V) \quad := \quad (H_1 \uparrow T, T, C_1, I, V)$$
$$\text{where } (H_1, T_1, C_1, V) =$$
$$= \llbracket s \rrbracket (H, T, C, I \cup \{i\}, V)$$

$$\llbracket [p] \rrbracket (H, T, C, I, V) \quad := \quad (H, I \uplus \{p\}, T, C \cup \{H \vdash p\}, I, V)$$

$$\llbracket action(a) \rrbracket (H, T, C, I, V) \quad := \quad (H \uplus (\{\text{post}(action)(a)\} \uparrow T),$$
$$T, C \cup \{H, I \vdash \text{pre}(action)(a)\},$$
$$I, V)$$

$$\llbracket \textbf{if } (c) \ s_1 \textbf{ else } s_2 \rrbracket (H, T, C, I, V) \quad := \quad ((H_1 \cap H_2) \uplus$$
$$\uplus (c \rightarrow H_1 \uplus \neg c \rightarrow H_2) \uparrow T,$$
$$T,$$
$$C_1 \cup C_2, I, V)$$
$$\text{where } (H_1, T_1, C_1, I_1, V) =$$
$$= \llbracket s_1 \rrbracket (H \uplus \{c\}, T, C, I, V)$$
$$\text{and } (H_2, T_2, C_2, I_2, V) =$$
$$= \llbracket s_2 \rrbracket (H \uplus \{\neg c\}, T, C, I, V)$$

$$\llbracket \textbf{while } (c)$$
$$\textbf{variant } v$$
$$\textbf{invariant } I$$
$$s \rrbracket (H, T, C, I, V) \quad := \quad (H \uplus \{\neg c \wedge I\}, T, C_1, I)$$
$$\text{where } (H_1, T_1, C_1, I_1, V) =$$
$$= \llbracket s \rrbracket (H \uplus \{c\}, T, C, I \cup \{i\}, V \cup \{v\})$$

FIGURE 12.3   DSL-Kit statement symbolic execution and proof semantics.

returning a value, the return is used; it has the semantics of assigning to a special *result* as well as the termination of a symbolic execution.

The pre- and postconditions of an action specify its behavior to potential action users; if an action body is defined it must be checked against the postcondition contract.

The variant part allows one to have simple recursive definitions guarded by an action variant.

The following is an example of an action declaration. It computes the sum of two numbers in a contrived manner:

$$
\begin{aligned}
&\text{action } add(a : \text{int}, b : \text{int}) : \text{int} \\
&\text{variant } a + b; \\
&\text{pre } a \geq 0 \wedge b \geq 0; \\
&\{ \\
&\quad \text{if } (a > b) \\
&\qquad \text{return } add(a - b, b) + b; \\
&\quad \text{else if } (b > a) \\
&\qquad \text{return } add(a, b - a) + a; \\
&\quad \text{else} \\
&\qquad \text{return } 2 * a; \\
&\} \\
&\text{post } result == a + b;
\end{aligned}
$$

Despite an unusual internal behavior, the external contract of the action is very simple: it adds any two positive numbers. Since an action is semantically atomic, any recursive action must define a variant expression to rule out runaway recursion.

The ability to define and reason about recursive definitions is extremely helpful, as many algorithms lend themselves more naturally to a recursive style. An equivalent behavior specification in Event-B would lead, at best, to a cryptic and laborious model.

Statements and actions are grouped into one or more *actors*. When there is just one actor it is still a sequential program as discussed above. However, when there are two or more actors a new phenomenon arises: Executions of atomic statements of actors may be interleaved in an arbitrary order and one must demonstrate that any possible interleaving results in a correct behavior. It is no longer possible to consider a verification constraint solely in its immediate context. For instance, an assertion referencing variables updated in two different actors may not be proven by analyzing just the state model of its parent actor.

The technical solution we employ is the rely guarantee model [7] where an actor defines a number of before-after *rely* predicates describing the cumulative effect of all other actors on relevant global variables, and before-after *guarantee* predicates defining the boundaries of possible changes in the global state changes performed by the current actor. The guarantee of an actor must be satisfied by every actor statement, while any two atomic statements of an actor are separated by an implicit assignment induced by the rely predicate.

12.2.2 DSL Instantiation and Domain Model

The custom part of a DSL comes in the form of a *module* made of a number (often substantial) of actions, variable declarations, and axioms. Unlike normal script, a module always comes in a binary form (via translation from an Event-B model) and cannot be created directly in a textual notation.

Axioms and actions of a module are defined exclusively over its variables. A module cannot reference other modules or import actions and variables from other places. All these precautions are attempts to ensure that axiomatic statements are in agreement with module actions. When a module is constructed from an Event-B model, invariants become module axioms and events become body-less actions. During the translation phase, an engineer may remove some invariants so that they do not appear in axioms of a target module.

Module variables may not be updated directly with an assignment statement. They are manipulated exclusively through the module actions.

12.2.3 Tool Support

One point we have discovered in experimentation with the DSL-Kit is that a domain model frequently requires small changes. Such changes must be in the source Event-B model and proven correct. Reconstruction of an entire DSL from a new domain model was initially a fairly slow process requiring a number of manual steps. The current version of the tool, however, can replace the module of a domain model in a running instance of DSL without requiring recompilation or even stopping the execution. The swap of a domain is accomplished literally in a few milliseconds and automatically triggers recomputation of the state model and conjectures.

There is, in fact, no great complication involved in construction of a module from an Event-B model. Since the mathematical language is exactly the same and there is no heavy-weight mapping between logics the translation is purely structural: Event guards and actions are mapped into pre- and postconditions, model invariants and axioms become module axioms, and constants and variables become module variables (the constant property is respected simply via the fact that no module action alters a constant).

A module is generated in a binary abstract syntax tree (AST) format that is convenient to read without an intermediate parsing stage. The module file includes a path to a source module and an MD5 hash over a combination AST object and source .bum (unchecked machine) XML file. This allows the toolkit to automatically detect out-of-date module files.

Module generation is accomplished by means of a dedicated Rodin Platform plug-in. A generated module file is placed within the Rodin project folder and, if there is a DSL instance dependent on the module, the module is automatically reloaded.

The screenshot in Figure 12.4 shows the DSL-Kit editor; it does syntax highlighting (customized automatically to highlight domain model actions, variables, and types), symbolic execution, and animation via constraint solving of a symbolic solution.

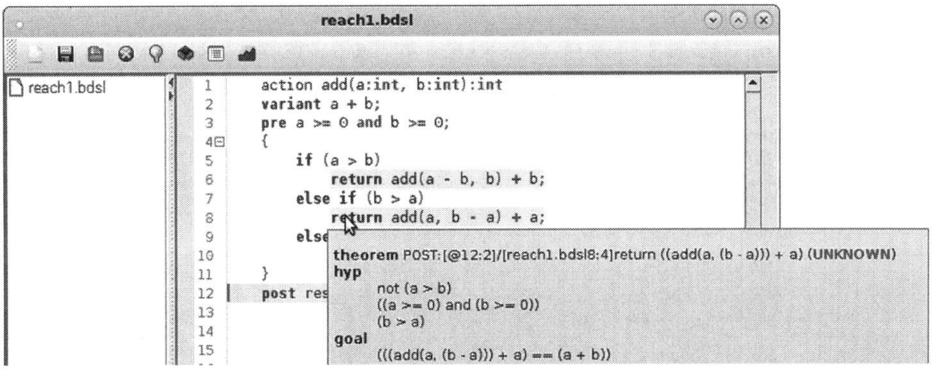

FIGURE 12.4 DSL-Kit support tool generated for a particular DSL from template code. An engineer may choose which parts of the toolkit (editor, evaluator, prover bridge, state plotting, etc.) to include. Highlighted in yellow are parts of the script that have undischarged conjectures. The tooltip annotates a conjecture showing its goal and some of its hypotheses.

12.3 DOMAIN MODEL

Although the concepts of a multi-core system do not pose any challenges in isolation, it is still a formidable task to try and compile a consistent and thorough description of multi-core system concepts in a natural language. Many revisions and cross-checks are necessary and some form of informal executable prototype would be essential to give a degree of confidence. A formal specification achieves the same goal at an arguably smaller price: Terse mathematical notation aids in precision and delegates a large portion of consistency checks to purely mechanical logic manipulation. A formal specification itself may be employed as an executable artefact to help understand the system and reconfirm intuitions.

12.3.1 Event-B

The basis of our discussion is a formalism called Event-B [2]. It belongs to a family of state-based modeling languages that represent a design as a combination of state (a vector of variables) and state transformations (computations updating variables).

An Event-B development starts with the creation of a very abstract specification. A cornerstone of the Event-B method is the stepwise development that facilitates the gradual design of a system implementation through a number of correctness-preserving *refinement* steps. The general form of an Event-B model (or *machine*) is shown in Figure 12.5. Such a model encapsulates a local state (program variables) and provides operations on the state. The actions (called *events*) are characterized by a list of local variables (parameters) vl, a state predicate g called *event guard*, and a next-state relation S called *substitution* or event *action*.

Event guard g defines the condition when an event is *enabled*. Relation S is given as a generalized substitution statement [8] and is either a deterministic ($x := 2$) or non-deterministic update of model variables. The latter kind comes in two notations: selection of a value from a set, written as $x :\in \{2, 3\}$; and a relational constraint on the next state v', for example, $x :| \, x' \in \{2, 3\}$.

The INVARIANT clause contains the properties of the system, expressed as state predicates, that must be preserved during system execution. These define the *safe states* of a system. In order for a model to be consistent, invariant preservation is formally demonstrated. Data types, constants, and relevant axioms are defined in a separate component called *context*.

$$
\begin{aligned}
&\textbf{MACHINE } M \\
&\quad \textbf{SEES } Context \\
&\quad \textbf{VARIABLES } v \\
&\quad \textbf{INVARIANT } I(c, s, v) \\
&\quad \textbf{INITIALISATION } R(c, s, v') \\
&\quad \textbf{EVENTS} \\
&\qquad E_1 \quad \triangleq \quad \textbf{any } vl \textbf{ where } g(c, s, vl, v) \textbf{ then } S(c, s, vl, v, v') \textbf{ end} \\
&\qquad \cdots \\
&\textbf{END}
\end{aligned}
$$

FIGURE 12.5   Event-B machine structure.

12.3.2 Event-B Domain Model Specification

The domain model is a plain Event-B specification fully proven and structured into nine refinement steps. Rodin Platform version 2.8 or a later compatible version is recommended for work with the model. A number of proof obligations (POs) were discharged with the SMT plugin; this must be installed to replay the proofs. The model cannot be animated directly due to the generalized summation operators, which are hard to solve.

The following discussion loosely follows refinement steps from the Event-B specification. To improve model legibility, model parts in this document are formatted to highlight the semantic class of each free identifier, as follows:

x	a quantifier-bound identifier
par	a local parameter of an Event-B event
var	an Event-B machine variable
SET	a carrier set, except predefined sets like $\mathbb{Z}$ and $\mathbb{N}$
CST	a constant
evt	an event label

12.3.2.1 Cores

A system contains a number of cores; at any moment a core may be operating or switched off. The set of all cores in a system is defined by a finite and nonempty set CORES. Current core status (on or off) is given by function variable *status*:

$$status \in \text{CORES} \rightarrow \text{STATUS}$$

where STATUS $= \{\text{ON}, \text{OFF}\}$. Initially, all the cores are switched off,

$$status := \text{CORES} \times \{\text{OFF}\}$$

At this level, we may observe a core being switched on or off, as captured by the following two events:

$$on \triangleq \textbf{any } c \textbf{ where } status(c) = \text{OFF} \textbf{ then } status(c) := \text{ON } \textbf{end}$$
$$off \triangleq \textbf{any } c \textbf{ where } status(c) = \text{ON} \textbf{ then } status(c) := \text{OFF } \textbf{end}$$

The diagram below illustrates switching a core on and off in the case of a four-core system CORES $= \{1, 2, 3, 4\}$.

12.3.2.2 Frequency and Voltage

Two essential characteristics of a running core are the voltage of its power supply and the clock frequency. These are the principal attributes used to control core performance, power, and reliability. Recall that the overall heat dissipated by a core is nearly equal to consumed power, which can be estimated as

$$P = cfV^2$$

where c is the average logic capacitance switch in a core cycle, f is the clock frequency in Hz, and v is the voltage. There is an interrelation between f and V that must be respected at all

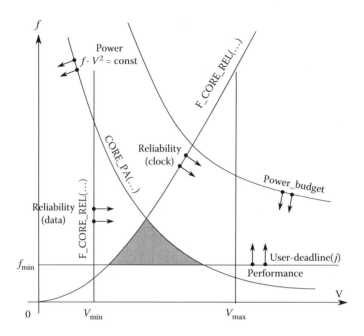

FIGURE 12.6   The meaning of operational constraint functions.

times; roughly, higher frequency requires higher voltage (see also Figure 12.6). The dynamic voltage/frequency scaling (DVFS) subsystem of a core allows a core or an operating system to react to load demands by adjusting frequency and voltage.

The following partial functions define frequency and voltage of a core; these are undefined for cores that are switched off.

$$\begin{aligned} freq &\in \text{CORES} \nrightarrow \mathbb{N} & vdd &\in \text{CORES} \nrightarrow \mathbb{N} \\ \text{dom}(freq) &= status^{-1}[\{ON\}] & \text{dom}(vdd) &= status^{-1}[\{ON\}] \\ freq &:= \varnothing & vdd &:= \varnothing \end{aligned}$$

Since all the cores are initially off, voltage and frequency functions are initially empty. Voltage and frequency attributes are changed dynamically by event *core_dvfs*:

> *core_dvfs* ≜
> **any** *c*, *f*, *v* **where**
>   *status* (*c*) = ON
>   . . .
> **then**
>   *freq* (*c*) := *f* ∥ *vdd* (*c*) := *v*
>   . . .
> **end**

The dots stand for omitted clauses; in a refined version, the event checks whether frequency/voltage pairs are correct for a given core, and also whether power budget and temperature constraints are satisfied for the new settings.

Switching a core on also necessitates setting some initial voltage and frequency values; this is accomplished by a refined version of event *core_on*.

12.3.3 Heat Generation

An operating core must dissipate heat generated through resistive heating. Using Fourier's law, heat transfer due to radiation is given as

$$Q_d = U_d c_d \Delta_d T$$

where Q_d is the heat flux in direction d, $\Delta_d T$ is the temperature gradient in direction d; and U_d and c_d are the directional conductance and conductive area. It is reasonable to assume that $\Delta_d T$ is constant within area c_d so that the amount of heat transferred during a unit of time is determined as

$$\frac{\partial Q_d}{\partial t} = U_d c_d^2 \Delta_d T$$

where c_d^2 and $\Delta_d T$ are time invariant; as a first approximation, we assume U_d is time invariant as well. We assume heat flux to be distributed evenly over core surface and thus *heat rate* is an adequate characterization of heat dissipation requirements.

Heat exchange happens between a core and the environment and a core and its neighboring cores. The environment is assumed to possess infinite heat capacity so that its temperature is not affected by heat exchange.

For some core c, the amount of heat generated per unit of time is determined by core frequency f and core voltage V, according to the following law:

$$\frac{\partial H_c}{\partial t} = C_c f_c V_c^2$$

where C_c is a core-specific constant. Heat rate $\partial H / \partial t$ and heat exchange laws define core temperature delta over a time period. For any given core, given its heat capacitance, we can determine the time duration necessary to increase core temperature by one Kelvin. Alternatively, considering some fixed time period, we can determine temperature change caused by the heat rate during the period.

The following event controls core temperature change. At this step, it is untimed and time will be added in a later refinement step.

$$
\begin{aligned}
&core\_temp \triangleq \\
&\textbf{any } c \textbf{ where} \\
&\quad status(c) = \text{ON} \\
&\textbf{then} \\
&\quad temp(c) := \max(\{temp(c)+ \\
&\qquad \text{HRATE}(c \mapsto vdd(c) \mapsto freq(c))+ \\
&\qquad \textstyle\sum_{n \in \text{NHB}(c)} \text{NRATE}(n \mapsto c \mapsto vdd(c) \mapsto freq(c))- \\
&\qquad \text{ERATE}(c), 0\}) \\
&\textbf{end}
\end{aligned}
$$

In the above HRATE(...), NRATE(...), and ERATE(...) are the heat rates times time delta for the heat gained through resistive heating, heat exchange with neighbor cores, and heat loss to the environment. When a core overheats, the system immediately (we will clarify the meaning of immediacy with the introduction of time) switches off the core.

$$core\_shutdown \triangleq$$
$$\textbf{any } c, t \textbf{ where}$$
$$status(c) = \text{ON}$$
$$temp(c) > \text{CORE\_TEMP\_CRIT}(c)$$
$$\dots$$
$$\textbf{then}$$
$$status(c) := \text{OFF}$$
$$\dots$$
$$\textbf{end}$$

We will see later that shutting down a core cancels all the jobs and stops all the running threads. All the computation progress is irrecoverable.

12.3.3.1 Threads

A thread is a basic concurrency and program structuring unit in our model. By "thread" we understand a potentially infinite sequence of commands that are continuously or intermittently executed by a core, perhaps switching between cores during its lifetime. At any given time, there is some number (potentially zero) of threads in the system:

$$threads \subseteq \text{THREADS}$$

where THREADS is the universe of threads. A thread may be assigned to a core and then it is said to be running or *scheduled*. The thread/core association is functional and partial in the domain:

$$affinity \in threads \rightarrowtail \text{CORES}.$$

It is only possible to schedule a thread on a running core:

$$\text{ran}(affinity) \subseteq status[\{\text{ON}\}].$$

Note that *affinity* is functional in one direction only (i.e., it is not injective); indeed, it is possible to map several threads onto the same core, for instance,

$$t, h \in threads \wedge t \neq h \wedge affinity(t) = affinity(h)$$

describes a situation where distinct threads t, h are running on the same core. In our model we choose to fix the lowest time resolution at the scale of ~ 1 millisecond; thus for a sequence of events with overall time smaller than this limit, we see all the events happening concurrently. This is a standard abstraction technique and it allows one to conduct a formal refinement to a higher time resolution. The following illustrates a system with three threads scheduled over two cores:

A new thread may be added to the system and, at some point, it may be destroyed:

$$th\_start \triangleq \textbf{any } t \textbf{ where } t \notin threads \textbf{ then } threads := threads \cup \{t\} \textbf{ end}$$
$$th\_stop \triangleq \textbf{any } t \textbf{ where } t \in threads \textbf{ then } threads := threads \setminus \{t\} \textbf{ end}$$

An existing thread may be scheduled to run on some operating system core c. An already running thread may be unscheduled and left in a dormant state to be scheduled again:

$$thread\_schedule \triangleq$$
any t, c **where**
$\quad\quad t \in threads \setminus \text{dom}(affinity)$
$\quad\quad status(c) = \text{ON}$
then
$\quad\quad affinity(t) :=c$
end

$$thread\_unschedule \triangleq$$
any t, c **where**
$\quad\quad t \in \text{dom}(affinity)$
$\quad\quad affinity(t) = c$
then
$\quad\quad affinity := \{t\} \lhd affinity$
end

12.3.3.2 Application

Several threads are grouped into an application. The primary role of an application is to define a workload type shared by a number of threads. Applications may be created and destroyed during the system's lifetime. The set of current applications is defined by set

$$apps \subseteq \text{APPS}.$$

Each thread belongs to an application and each application owns at least one thread. This is captured by the following (surjective and total) relation

$$app\_threads \in apps \longleftrightarrow\!\!\!\twoheadrightarrow threads.$$

For each thread there is just one owning application

$$app\_threads^{-1} \in threads \rightarrow apps.$$

Application/thread association may be depicted in a diagram as follows.

When an application is created, it appears together with one thread of its own (this event *refines* event *thread_start*):

$$app\_start \triangleq$$
any a, t **where**
$\quad a \notin apps \wedge t \notin threads$
then
$\quad apps := apps \cup \{a\}$
$\quad threads := threads \cup \{t\}$
$\quad app\_threads := app\_threads \cup \{a \mapsto t\}$
end

Destroying an application cancels all the running application threads and removes them from the system:

$$app\_stop \triangleq$$
any t, a **where**
$\quad t \in threads$
$\quad a \mapsto t \in app\_threads$
$\quad app\_threads[\{a\}] = \{t\}$
then
$\quad threads := threads \setminus \{t\}$
$\quad affinity := \{t\} \lhd affinity$
$\quad apps := apps \setminus \{a\}$
$\quad app\_threads := app\_threads \setminus \{a \mapsto t\}$
end

12.3.3.3 Workload

The purpose of an application is to provide a computation service. It accomplishes this by assigning incoming *workload* to application threads. The unit of a workload is a *job*. Every job is specific to an application so that only threads of a certain application may process a given job.

The pending and executing jobs of a system are defined by variable *jobs*:

$$jobs \in \text{JOBS} \rightarrow\!\!\!\rightarrow apps$$

where JOBS is the universe of all jobs. If a job is to run, it must be allocated to a thread. At any given time a thread executes at most one job. The following partial injection captures this relationship:

$$job\_alloc \in dom(job\_alloc) \rightarrowtail threads.$$

A job allocation must agree with the ownership of a thread to which the job is assigned:

$$jobs^{-1}; job\_alloc \subseteq app\_threads.$$

Here $jobs^{-1}$ is the converse of function *jobs* and $f; h$ denotes forward functional composition.

To reason about the computational complexity of a job, we define the number of *steps* comprising a given job. A step is a normalized complexity measure independent of core properties and shared by all the jobs:

$$job\_steps \in dom(jobs) \rightarrow \mathbb{N}.$$

A step execution time varies from core to core and with core frequency. The following illustrates two jobs assigned to threads and a thread suspended and ready to accept a job allocation.

A new job may appear at any moment; it must be assigned to an existing application.

> $job\_create \triangleq$
> **any** j, a, w **where**
> $j \notin dom(jobs) \wedge a \in apps \wedge w > 0$
> **then**
> $jobs := jobs \cup \{j \mapsto a\} \| job\_steps(j) := w$
> **end**

In the above, w defines the job complexity in terms of steps. An existing but as-yet unassigned job may be allocated to a thread of the job application. The thread in question must be mapped to a core but not already executing any other job:

> $job\_allocate \triangleq$
> **any** $j, t,$ **where**
> $j \in dom(jobs) \setminus dom(job\_alloc)$
> $t \in (app\_threads[\{jobs(j)\}] \setminus ran(job\_alloc)) \cap dom(affinity)$
> **then**
> $job\_alloc := job\_alloc \cup \{j \mapsto t\}$
> **end**

When a job is assigned to a scheduled thread, the job "runs" in a sense; it goes through the steps, one by one, until all the steps are done:

$$job\_run \triangleq$$
any j, c **where**
$$j \mapsto c \in job\_alloc; \textit{affinity} \wedge job\_steps(j) > 0$$
then
$$job\_steps(j) := job\_steps(j) - 1$$
end

At some point, a job finishes and vanishes from the system:

$$job\_finish \triangleq$$
any j **where**
$$j \in \mathrm{dom}(job\_alloc)$$
then
$$job\_alloc := \{j\} \lhd job\_alloc$$
$$jobs := \{j\} \lhd jobs$$
$$job\_steps := \{j\} \lhd job\_steps$$
end

12.3.3.4 Timing

A number of already defined phenomena require some form of timing. There is no native support for clocks and timers in Event-B but there are a number of established approaches to time modeling. The one we use is closely related to timed automata with integer clocks. To keep track of time progress and time various activities of a system we introduce the number discrete *timers*. A timer functions like a stopwatch—it counts down to zero from the initial integer value. All the system timers count in synchrony (that is, driven by one global clock) and there is no limit on the number of timers in the system. Once any timer reaches zero, time freezes to allow for a subsystem interested in this timer to react to a deadline; to enable further progress, the subsystem must also reset the timer or delete it. The set of all clocks is defined by function τ:

$$\tau \in TA \rightarrow TIME, \quad TIME = \mathbb{N} \cup \{DISABLED\}.$$

$\tau(x)$ gives the current value of clock x which is > 0 before deadline, 0 on the deadline, and DISABLED if the clock is not used. To avoid complicated progress arguments, we assume there is a plentiful supply of clocks. At this level of abstraction it suffices to require that set TA is infinite.

As one example, let us consider the timing of a core off state. The aim here is to ensure that a core, once switched off, is not switched on too soon afterward. The core off event sets up a timer:

$$off \triangleq$$
any $\ldots, ta$ **where**
$$\ldots$$
$$\tau(ta) = DISABLED$$
then
$$\ldots$$
$$\tau(ta) := T\_CORE\_OFF(c)$$
end

where $T\_CORE\_OFF(c)$ is a delay between switching core c off and on. When a core is started, event *on* checks that there is no off-state clock for this core. The reaction on the clock deadline is expressed in the following new event:

$$core\_reset\_off\_timer \triangleq$$
any c **where**
$$c \in \mathrm{dom}(time\_core\_off)$$
$$\tau(time\_core\_off(c)) = 0$$
then
$$time\_core\_off := \{c\} \lhd time\_core\_off$$
$$\tau(time\_core\_off(c)) := \mathrm{DISABLED}$$
end

Strictly speaking, event *core_restart* is superfluous as its effect is always the sum of *core_reset_off_timer* and *core_on*. However, it may be interesting to know when a system had to actively wait before restarting a core and the occurrence of this event highlights such situations.

All the clocks are synchronously updated by event *time*:

$$time \triangleq$$
any $p1, p2$ **where**
$$0 \notin \mathrm{ran}(\tau)$$
$$p1 = \tau^{-1}[\{\mathrm{DISABLED}, 0\}]$$
$$p2 = \mathrm{dom}(\tau) \setminus p1$$
$$\forall c \cdot c \in status [\{\mathrm{ON}\}] \Rightarrow temp(c) \leq \mathrm{CORE\_TEMP\_CRIT}(c)$$
then
$$\tau := (p1 \lhd \tau) \cup \{x \cdot x \in p2 | x \mapsto \tau(x) - 1\}$$
end

Notice the disabling condition $0 \in \mathrm{ran}(\tau)$ which stops the timer until the deadline of a clock is processed. The last guard prevents clock progress when there is an overheated core: *it makes reaction to core overheating immediate.*

12.3.3.5 Job Deadlines

One application of timing is the definition of job deadlines and, to make the concept meaningful, the notion of job execution time. The latter is linked to the notion of job steps defined above.

A user deadline is set at the point of job allocation as opposed to the point of job creation. We do not yet model job queues (it is instead a set in this model) so there is no fairness property for job allocation. An extended version of the *job_allocate* event sets a user deadline for a job.

$$job\_allocate \triangleq$$
any $j, t, ta, ta\_user, udln$ **where**
$$\ldots$$
$$\tau(ta) = \mathrm{DISABLED}$$
$$\tau(ta\_user) = \mathrm{DISABLED}$$
$$ta \neq ta\_user$$
$$uldn \in \mathbb{N}1$$
then
$$\ldots$$
$$job\_step\_time(j) := ta$$
$$user\_deadline(j) := ta\_user$$
$$\tau := \tau \Leftarrow \{$$
$$\qquad ta \mapsto \mathrm{F\_STEP\_TIME}(affinity(t) \mapsto freq(affinity(t))),$$
$$\qquad ta\_user \mapsto udln\}$$
end

The user deadline clock is stored in *user_deadline(j)*; another clock times the execution of job steps and is stored in *job_step_time(j)*. The assignment to τ initializes these two clocks. Note that the step running time is defined by the kind and frequency of a core on which a thread executing the job is scheduled. User deadline clock *user_deadline(j)* runs down without pauses from the point of job allocation irrespectively of whether a thread processing the job is running or not.

Job steps are timed according to the respective core performance. Event *job_run* is extended to time itself with a job step timer:

$$
\begin{aligned}
&job\_run \triangleq\\
&\textbf{any } j, t, c \textbf{ where}\\
&\quad \cdots\\
&\quad \tau(job\_step\_time(j)) = 0\\
&\textbf{then}\\
&\quad \cdots\\
&\quad \tau(job\_step\_time(j)) := \text{F\_STEP\_TIME}(c \mapsto freq(c))\\
&\textbf{end}
\end{aligned}
$$

Job step execution time may change with the change of core frequency:

$$
\begin{aligned}
&job\_cancel \triangleq\\
&\textbf{any } j \textbf{ where}\\
&\quad j \in \text{dom}(job\_alloc)\\
&\quad \tau(user\_deadline(j)) = 0\\
&\quad job\_steps(j) > 0 \lor \tau(job\_step\_time(j)) > 0\\
&\textbf{then}\\
&\quad job\_alloc := \{j\} \lhd job\_alloc\\
&\quad jobs := \{j\} \lhd jobs\\
&\quad job\_step\_time := \{j\} \lhd job\_step\_time\\
&\quad \tau := \tau \lhd \{job\_step\_time(j) \mapsto \text{DISABLED},\\
&\qquad\qquad\qquad user\_deadline(j) \mapsto \text{DISABLED}\}\\
&\quad user\_deadline := \{j\} \lhd user\_deadline\\
&\textbf{end}
\end{aligned}
$$

Should a job fail to meet the user deadline, it is canceled even before all the job steps are done. This is a situation a management system should try to avoid. In some situations, a viable alternative is a dynamic extension of a user deadline. This allows a slightly delayed job to run until completion:

$$
\begin{aligned}
&job\_deadline\_extension \triangleq\\
&\textbf{any } j, ext \textbf{ where}\\
&\quad j \in \text{dom}(job\_alloc)\\
&\quad \tau(user\_deadline(j)) = 0\\
&\quad job\_steps(j) > 0 \lor \tau(job\_step\_time(j)) > 0\\
&\quad ext \in \mathbb{N}1\\
&\textbf{then}\\
&\quad \tau(user\_deadline(j)) := ext\\
&\textbf{end}
\end{aligned}
$$

Finally, if a job completes before a user deadline expires, it is accepted as a successful job execution:

$$
\begin{aligned}
&job\_finish \triangleq \\
&\textbf{any } j \textbf{ where} \\
&\qquad j \in \text{dom}(job\_alloc) \\
&\qquad job\_steps(j) = 0 \\
&\qquad \tau(job\_step\_time(j)) = 0 \\
&\qquad \tau(user\_deadline(j)) \geq 0 \\
&\textbf{then} \\
&\qquad \ldots \\
&\textbf{end}
\end{aligned}
$$

12.3.3.6 Scheduling

As defined above, a core runs several threads at the same time. In our model, we do not go into the minute details of scheduling within a group of threads assigned to a core, but rather assume that the core indeed runs all the mapped threads in parallel (that is, the *time band* in which the core model is given does not allow us to distinguish between instances of individual thread executions and the whole picture is blurred to give an illusion of a multi-threaded core). There are limits to the number and kind of threads that may be run on a single core before the execution times of individual threads are affected. Intuitively, if a thread is computationally intensive and never has to wait for data, it cannot share a core with another thread without sacrificing performance. In practice, many applications require data retrieval or do blocking system calls which make a thread idle and hence free a core to run another thread. The purpose of the scheduling refinement step is to bundle threads from various applications into thread groups that may be assigned to the same core without hindering thread performance.

Variable *thread_load* characterizes a thread in terms of a lower bound of operations per second (or a comparable normalized measure) necessary to run the thread at full speed.

$$thread\_load \in threads \nrightarrow \text{LOAD}.$$

Thread load must be defined for all the running threads, $\text{dom}(affinity) \subseteq \text{dom}(thread\_load)$. The crucial property is that the overall load of threads assigned to a core does not exceed the core computational capability at the current core frequency:

$$
\begin{aligned}
&\forall c \cdot c \in status^{-1}[\{\text{ON}\}] \Rightarrow \\
&\quad \text{sum}(affinity^{-1}[\{c\}] \lhd thread\_load) \leq \text{CORE\_LOAD\_MAX}(c \mapsto freq(c)).
\end{aligned}
$$

Thread load affects the way threads are mapped on to cores:

$$
\begin{aligned}
&thread\_schedule \triangleq \\
&\textbf{any } t, c \textbf{ where} \\
&\qquad \ldots \\
&\quad \text{sum}(affinity^{-1}[\{c\}] \lhd thread\_load) + \\
&\qquad\quad thread\_load(t) \leq \text{CORE\_LOAD\_MAX}(c \mapsto freq(c)) \\
&\textbf{then} \\
&\qquad affinity(t) := c \\
&\textbf{end}
\end{aligned}
$$

12.3.4 Power Budgeting

Power budgeting limits core voltage settings and, possibly, the overall number of operational cores. An overall power budget is set externally for the whole system

$$pwr\_b \in \mathbb{N} \wedge pwr\_b > 0.$$

The power consumption of a running core is determined by a (constant) function CORE_PA defined for each triple of core type, frequency, and voltage. In normal operation, the power drawn by the individual cores does not exceed the power budget:

$$c\_pwr \leq pwr\_b$$
$$c\_pwr = \{c \cdot c \in status^{-1}[\{ON\}] \mid c \mapsto CORE\_PA(c \mapsto freq(c) \mapsto vdd(c))\}$$
$$\cup \{c \cdot c \in status^{-1}[\{OFF\}] \mid c \mapsto 0\}.$$

The power budget may be adjusted at any moment:

$$power\_set \triangleq \textbf{any } pb \textbf{ where } pb \in \mathbb{N}_1 \textbf{ then } pwr\_b := pb \textbf{ end}$$

Power budgeting constraints control switching cores on and changing core frequency and voltage.

$$core\_on \triangleq$$
any c, f, v **where**
$$\ldots$$
$$\text{sum}(c\_pwr) + CORE\_PA(c \mapsto f \mapsto v) \leq pwr\_b$$
then
$$\ldots$$
end
$$core\_dvfs \triangleq$$
any c, f, v **where**
$$\ldots$$
$$\text{sum}(\{c\} \lhd c\_pwr) + CORE\_PA(c \mapsto f \mapsto v) \leq pwr\_b$$
then
$$\ldots$$
end

For a short period of time, the power budget may be lower than the amount of power already drawn. To emphasize the immediacy of a reaction, the system timer is stopped until the power budget constraint is resolved either by shutting cores or adjusting their frequency and voltage.

$$time \triangleq \textbf{any} \ldots \textbf{ where } \cdots \wedge \text{sum}(c\_pwr) \leq pwr\_b \textbf{ then } \ldots \textbf{ end}$$

12.3.4.1 Reliability Control

The reliability control matches user reliability expectations against the overall reliability of a job execution (which may be spread over several cores and affected by DVFS). As a reliability measure we employ mean time to failure (MTTF). The objective is to schedule threads in such a way that the resultant MTTF figure is above certain job-specific thresholds.

MTTF defines the mean duration of time from the start or repair of a system to the next failure. Let $f(t)$ be the time density function of system failures:

$$\int_0^\infty f(t)dt = 1$$

then MTTF is defined to be

$$\text{MTTF} = \int_0^\infty f(t)t\,dt$$

replacing integration with summation over nonempty time intervals $P \subseteq \mathbb{P}(\mathbb{R}_+)$ partitioning half-interval $[0, \inf)$, we have

$$\text{MTTF} = \sum_{p \in P} f(p)\text{diam}(p)$$

where $f(p) = f(t)$ for some $t \in p$ and $\text{diam}(d)$ is the length of interval p. Let $r(p)$ be the MTTF value for a constant failure density $f(p)$ in interval p: $r(p) = f(p) \sum_{p \in P} \text{diam}(p)$. The formula above may thus be written as

$$\text{MTTF} = \frac{\sum_{p \in P} r(p)\text{diam}(p)}{\sum_{p \in P} \text{diam}(p)}.$$

This value is computed for each job and is matched against a user-defined job reliability threshold:

$$user\_rel \in \text{dom}(jobs) \rightarrow \text{REL}$$

which is initialized at the point of job creation. The actual reliability figure is computed as a weighted sum of core MTTF values, one per each period p of constant core type, voltage, frequency, and temperature:

$$jobmttf(j) = \frac{\sum_p mttf(p,j)duration(p,j)}{\sum_p duration(p,j)}.$$

Should job execution fail to meet its reliability requirement, the job is said to be rejected (event *job_reject*). A smarter controller may preemptively cancel a job that is unlikely to meet its reliability requirement. The bulk of reliability computations happen in the refined version of *job_run* event.

12.3.4.2 Interpretation of Some Constant Functions
The bounding curves and lines of the diagram in Figure 12.6 are the uninterpreted functions characterizing a core. Specifically,

1. Constant function CORE_PA defines the power drawn by a core at a given voltage and frequency. The summation of all such values must be under the power budget curve, defined by variable *power_budget*

2. Constant function F_CORE_REL defines core reliability for the given core frequency, voltage, and temperature. This accounts for the vertical "noise" boundary (bit flips due to thermal noise and radiation) and the clock reliability (latency) curve (registers latching to incorrect values). Variable *user_rel* defines how closely a core may approach these curves. These two curves are in fact bands defining failure probability

TABLE 12.1 Proof Effort Summary

Component	Total POs	Automatic	Manual
(Contexts)	16	14	2
Cores	5	5	0
Cores1	30	26	4
Affinity	16	16	0
Workload	42	41	1
Workload1	11	10	1
Application	11	6	5
Timing	208	132	76
Scheduling	56	42	14
Power	35	21	14
Reliability	89	62	27
Total	**582**	**422**	**158**

3. Horizontal QoS line is defined by the user-defined job deadline *user_deadline*

4. Normalized core performance is given by constant function *job_step_time*

The summary of the proof effort for the domain model is presented in Table 12.1.

12.4 CORE-DSL

We construct a DSL for reasoning about multi-cores systems by translating the final refinement of the domain model into a DSL-Kit module. Tables 12.2 and 12.3 show environment and system actions translated from the model. We call the result Cores-DSL and use it to construct some interesting models of scheduling and load balancing.

Not all scenarios defined on top of the Event-B domain model correspond to behaviors permitted by the domain model. For instance, the following defines a scenario made of a sequence of two events:

$$on(c, v, f); off(c);$$

The first event switches a core on and the second switches the same core off (assuming all identifiers are locally bound). If, instead, we write

$$on(c, v, f); on(c, v, f);$$

it appears that the second event cannot be truly enabled and the scenario must halt prematurely. In fact, even in the first example, we could not know that $on(c, v, f)$ starts in a state where the guard of *on* instantiated with arguments c, v, f is enabled. To collect all such conditions automatically, we need to know the kind of states in which an event (or, for the sake of generality, any imperative statement) is enabled and also the kind of states it produces upon termination.

To stay close to the Event-B semantics we not only reuse Event-B mathematical language but also persist with the notion of atomic events. In the DSL-Kit events are rendered as *actions* with explicit identification of arguments; guards become preconditions; actions are written as postconditions.

TABLE 12.2   A Summary of the Cores-DSL System Commands

$on(c, f, v)$	$c \in$ CORES, $f \in \mathbb{N}$, $v \in \mathbb{N}$
switch on a core with given initial frequency and voltage settings	
$off(c)$	$c \in$ CORES
switch off a core	
$core\_dvfs(c, f, v)$	$c \in$ CORES, $f \in \mathbb{N}$, $v \in \mathbb{N}$
adjust core Vdd and frequency (idle core case)	
$core\_dvfs\_running(c, f, v)$	$c \in$ CORES, $f \in \mathbb{N}$, $v \in \mathbb{N}$
adjust core Vdd and frequency while running threads	
$thread\_start(a, t, l, p)$	$a \in$ APPS, $t \in$ THREADS, $l \in$ LOAD, $p \in$ CTYPE
create a new thread of a given type	
$thread\_stop(t)$	$t \in$ THREADS
destroy a thread	
$thread\_schedule(c, t)$	$c \in$ CORES, $t \in$ THREADS
schedule thread execution at a particular core	
$thread\_unschedule(t)$	$t \in$ THREADS
unschedule a thread	
$app\_start(l, p)$	$l \in$ LOAD, $p \in$ CTYPE
create a new application with its own thread of a given type and load	
$app\_stop(a)$	$a \in$ APPS
destroy an application	
$job\_allocate(j, r, t)$	$j \in$ JOBS, $r \in$ REL, $t \in$ THREADS
assign a job to a thread of an application	
$job\_add(a, udln, w)$	$a \in$ APPS, $udln \in \mathbb{N}1$, $w \in \mathbb{N}1$
add a job to an application job queue and start the deadline clock	
$job\_deadline\_extension(j, e)$	$j \in$ JOBS, $e \in \mathbb{N}1$
extension of user deadline for job completion time	
$power\_set(b)$	$b \in \mathbb{N}1$
change of the overall system power budget	

As an example, event *on* takes the following form in DSL-Kit notation:

```
action on(c : CORES,  v : int, f : int)
pre status(c) == OFF;
pre f ≥ 0;
pre v ≥ 0;
pre temp(c) ≤ CORE_TEMP_CRIT(c);
pre c ∉ dom(time_core_off);
pre sum(core_power) + CORE_PA(c, f, v) ≤ power_budget;
post status' == status ⊲ {c ↦ ON};
post freq' == freq ⊲ {c ↦ f};
post vdd' == vdd ⊲ {c ↦ v};
post core_power' == core_power ⊲ {c ↦ CORE_PA(c, f, v)};
```

All such translated events are put into a *module*; in this module DSL commands (see Table 12.3) are the exported actions (i.e., callable functions) and environment events are the actions wrapped into two actors. One actor is solely responsible for time progress; the other automatically executes all the housekeeping events related to core and job management.

TABLE 12.3 A Summary of the Cores-DSL Environment Events

Control Commands

core_reset_off_timer(c)	$c \in$ CORES
	reset core off-state clock
core_powerdown(c)	$c \in$ CORES
	automatically shut down a core due to power undersupply
core_shutdown(c)	$c \in$ CORES
	automatically shut down a core due to overheating
core_heat(c)	$c \in$ CORES
	core heat exchange event
job_cancel(j)	$j \in$ JOBS
	job cancellation due to overrunning user deadline
job_reject(j)	$j \in$ JOBS
	job is rejected due to violation of reliability expectations
job_finish(j)	$j \in$ JOBS
	successful job completion
job_run(j)	$j \in$ JOBS
	job execution and update of reliability indicators
time	
	synchronous update of all the clocks

The module containing the Cores domain model is made available for new actor and system definitions. Module variables (translations of Event-B variables) are accessible as read-only. Their state may only be modified via module actions. Event-B invariants thus become module axioms (Event-B has already established that events preserve invariants). This protects high-level design from small changes in the domain model specification, regenerated mechanically from an Event-B model.

DSL-Kit uses a variant of Floyd-Hoare logic to compute the strongest postcondition given some current state and a statement. On the basis of the strongest postconditions, the verification conditions generation procedure constructs a number of theorems. Among them are conditions establishing that every scenario runs until completion. To facilitate reasoning about design prototypes, DSL-Kit offers invariant, auxiliary variables, iteration, and recursion variants, and rely/guarantee conditions for reasoning about concurrency and shared state.

Action *on* defined above is purely a specification statement. In its current form, also completely deterministic, it cannot be executed. DSL-Kit also supports definitions of actions with bodies that strongly resemble conventional functions, that is,

```
action fact(n : int) : int
variant n;
pre n > 0;
    if (n == 1) return 1;
        else return n * fact(n − 1);
post n == 1 ⇒ result == 1;
post n > 1 ⇒ result == mul(n);
```

In general, actions constructed from Event-B events are not executable. Specifically, events containing nondeterministic substitutions are assumed to be nonexecutable. However, nondeterminism arising due to event local variables is handled differently. The Event-B to DSL-Kit translation tool generates Java code responsible for solving given constraints either through iterating by means of candidates if their numbers can be statically determined to be small; or by calling an external constraint solver [5]. Quantifiers and set comprehension are treated in the same fashion. There may be runs when a constraint solver fails and then a script execution is aborted. It took some time and effort to exclude such cases and make the Cores domain model executable for all the practically relevant situations (it is still nondeterministic). Note that this notion of execution differs from running deterministic programs: specification execution explores only one possible trace out of many. Where possible, we use randomization to hit different traces on each run.

The following script is an example of a complete DSL actor. The actor switches of all the cores that are currently on (set $status^{-1}[\{ON\}]$) but do not have any threads mapped (minus set **ran**($affinity$)) and the core frequency may not be set any lower (the *if* statement condition):

```
actor alloff
rely invariant status, affinity;
        for(c in status⁻¹[{ON}] \ ran(affinity))
            if (not exists d : int
                    enabled core_dvfs(c, vdd(c), freq(c) − d))
                        off (c);
        guar status'⁻¹[{ON}] ⊆ status⁻¹[{ON}];
        guar nothing timer, time_core_off;


    actor dvfs(c : CORES)
    rely . . .
    {
        if (status(c)! = OFF){
        core_threads : set(THREADS) = affinity⁻¹[{c}];
        ml : int = CORE_LOAD_MAX(c, freq(c));
        if (core_threads! = {}){
            total_load : int = sum(thread_load[core_threads]);
            if (ml < total_load + STEP)
                if (enabled core_dvfs_running(. . .))
                    core_dvfs_running(c, vdd(c) + CH, freq(c) + CH);
            else if (ml > total_load + STEP and ml > MIN_LOAD)
                if (enabled core_dvfs_running(. . .)
                    core_dvfs_running(c, vdd(c) − CH, freq(c) − CH);
        } else
            #nothing is running, run down frequency gradually
            if (enabled core_dvfs(. . .) − 1) and ml > MIN_LOAD)
                core_dvfs(c, vdd(c) − 1, freq(c) − 1);
        }
    }
    guar . . .
```

FIGURE 12.7   Cores-DSL script for a simple on-demand frequency/voltage governor.

The actor in the script demands that no other actor may switch off or on a core and map threads to a core; with this assumption it guarantees that the only globally observable effect during and after actor execution is the decrease in the number of running cores. Such a design may be used to test the shutting down of cores that have been idling for a while; the frequency check assumes a DVFS governor clocking down idle cores. The script generates five verification conditions, all of which are discharged automatically by a collection of automated provers managed by the Why3 tool [9].

As one example, the satisfaction of the actor guarantee requires proving theorem

$$c \in status^{-1}[\{ON\}] \setminus \mathbf{ran}(affinity)$$
$$\vdash$$
$$(status' \lhd\!\!\!- \{c \mapsto OFF\})^{-1}[\{ON\}] \subseteq status^{-1}[\{ON\}].$$

The script in Figure 12.7 shows a simple yet functional on-demand frequency governor. The performance of the script is shown in Figure 12.8. The plotted graph is discrete and it depicts a time series of domain model variables *freq* and *core_temp*.

There is an important relation between DSL specifications and the original Event-B domain model.

Theorem 12.1. *For every DSL specification there exists an Event-B model refining the Event-B specification of the domain model.*

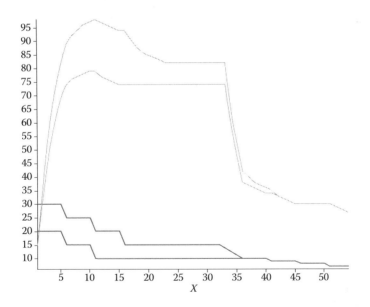

FIGURE 12.8 Time graph of core frequencies (dark gray, lower) and temperatures (light gray, higher); the graph shows the performance of an on-demand DVFS governor (see the script in Figure 12.7) in a scenario where two jobs are scheduled simultaneously for both cores and cancelled after running for 33 s. The thread mapper script ramps up initial frequencies and voltages and these are gradually lowered by the DVFS script. Idling produces rapid cooling starting at the 33 second mark.

The proof, which we omit here, consists of exhibiting a translation procedure converting a DSL specification to the Event-B model. Such a translation is trivial in principle but there does not seem to be a practical reason to implement it.

12.5 CONCLUSION

The chapter presents a formal DSL for multi-core systems. It is based upon an Event-B specification with nine refinement steps. Much of the final specification and "debugging" (that is, proving statistically that omissions and progress problems for safety and correctness have been corrected) was accomplished by building design prototypes in the developed DSL. Fifteen revision cycles were done over the course of 5 months since the initial Event-B domain model was developed.

The DSL is still growing; it is missing, for instance, such concepts as thermal cycling and silicon aging. Application lifetime modeling is also quite rudimentary. There is no model of catastrophic failures or noise in sensor reports. These would be necessary for conducting stochastic experiments with design prototypes to establish stability and convergence properties. We also plan to construct a separate domain model for application and thread lifetime that would detail an API through which software and OS may exchange information about resources management. The two domain models would be united under the same modeling DSL.

Finally, we are exploring options to generate production code from DSL scripts (Figure 12.9).

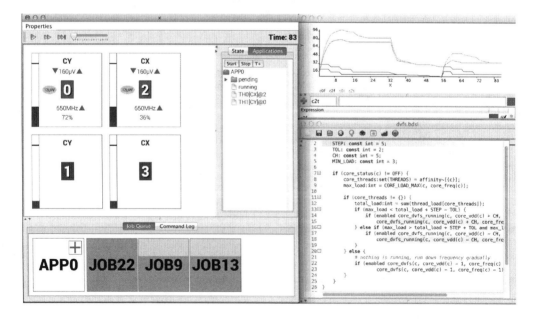

FIGURE 12.9   A screenshot of a simulation environment running (animating) a Cores-DSL script.

ACKNOWLEDGMENTS

This work has been supported by the EPSRC/UK PRiME program grant on Power-Efficient, Reliable, Many-Core Embedded Systems, as well as the EPSRC/UK TrAmS-2 platform grant on Trustworthy Ambient Systems: Resource-Constrained Ambience.

REFERENCES

1. A. Iliasov, A. Rafiev, F. Xia, R. Gensh, A. Romanovsky, and A. Yakovlev. A formal specification and prototyping language for multi-core system management. In *Parallel, Distributed and Network-Based Processing (PDP), 2015 23rd Euromicro International Conference on*, Turku, Finland, 2015.

2. J.-R. Abrial. *Modelling in Event-B*. Cambridge University Press, 2010.

3. H. Treharne, S. Schneider, and M. Bramble. Composing specifications using communication. In *ZB 2003: Formal Specification and Development in Z and B, Third International Conference of B and Z Users, Turku, Finland, June 4–6, 2003, Proceedings*, volume 2651 of Lecture Notes in Computer Science. pp. 58–78. Springer, 2003.

4. C. A. R. Hoare. An axiomatic basis for computer programming. *Communications of the ACM*, 12(10):576–580, 1969.

5. M. Leuschel and M. Butler. ProB: A model checker for B. In A. Keijiro, S. Gnesi, and M. Dino, editors, *Formal Methods Europe 2003*, Vol. 2805, pp. 855–874. Springer-Verlag, LNCS, Pisa, Italy, 2003.

6. The RODIN platform. Online at http://rodin-b-sharp.sourceforge.net/.

7. C. B. Jones. *Systematic Software Development Using VDM*, 2nd edition, Prentice-Hall, Inc., Upper Saddle River, NJ, 1990.

8. J.-R. Abrial. *The B-Book*. Cambridge University Press, 1996.

9. F. Bobot, J.-C. Filliâtre, C. Marché, A. Paskevich. Why3: Shepherd your herd of provers. In *Boogie 2011: First International Workshop on Intermediate Verification Languages*, pp. 53–64, Wroclaw, Poland, August 2011.

CHAPTER 13

New Standards for Trustworthy Cyber-Physical Systems

Alan Wassyng, Paul Joannou, Mark Lawford,
Thomas S. E. Maibaum, and Neeraj Kumar Singh

CONTENTS

C YBER-PHYSICAL SYSTEMS (CPSs) are extremely complex systems that combine components with both physical and cyber interfaces and potentially complex interactions between these parts. They are also often both security and safety critical if the physical system being controlled can harm people. It is imperative that these systems be developed and certified to be safe, secure, and reliable—hence the focus on *Trustworthy Cyber-Physical Systems*. The current safety-critical or high-integrity standards primarily set out objectives on the process, as is typical in much of software engineering. Thus, the acceptance criteria in these standards apply to the development process much more than to the product being manufactured. Having manufacturers use these "good" processes is, indeed, advantageous. However, their use does not guarantee a "good" product, except in a statistical sense. We need to evaluate the quality of the product, not only the process by which it was built [1].

Regulators, for instance, the U.S. Food and Drug Administration (FDA), have been dissatisfied with the frequency with which many of the products evaluated in such process-based regimes have been recalled. Of course we can, and should, make our standards specify the product-focused evidence that is required, as well as the acceptance criteria for this evidence. However, software engineering does not have a good track record in this regard. One approach that we can take is to identify the critical properties of a system that are necessary to achieve tolerable risk regarding the safety, security, and reliability of that system. *Assurance cases* have been gaining traction as a means of documenting such claims about these critical properties of a system, together with the evidence and associated reasoning as to why the claims are valid. Not surprisingly, the FDA has turned to assurance cases to help improve the quality of products submitted for approval [2]. Assurance cases provide one way of documenting a convincing argument regarding the trustworthiness of the resulting system—built on the identification of specific items of evidence, and the satisfaction of explicit acceptance criteria involving both the product and the process.

We have been exploring the use of *assurance case templates* that can be used to drive the development of a software-intensive critical system. Such a template will also provide explicit guidance on an effective assurance case for a specific product within the template's identified product scope. We believe that a *product-domain-specific template* can serve as a standard for the development and certification of a CPS in that specific *product domain*.

We have two distinct user groups in mind in this endeavor:

- The developers and certifiers of a safety-critical CPS.

- Standards developers in this community.

We further believe that assurance case templates can be much more effective than the standards we have now, for both of these groups. It is probably quite easy for readers to understand why this should be true for the first group—developers and certifiers. However, it may not be so readily obvious for the second group—standards developers. If we are correct, and it is possible to build product-domain-specific assurance case templates that are suitable for use as standards, then, the structure of the assurance case template will help standards developers in a number of crucial ways, described more fully later in this chapter. A distinct benefit to standards developers is that the assurance case structure actually guides the development of the standard such that we (can) have a better understanding as to why compliance with the standard will result in a high-integrity system. This is the way in which we envisage standards developers using such templates. There is another way that would, perhaps, be more palatable politically, but less satisfying technically. This would be to use an assurance case template as a guide in creating a more traditionally organized standard.

This chapter presents the motivation for the use of assurance case templates as standards by first describing the role of standards (Section 13.1.1), and the shortcomings of current standards (Section 13.1.2). It also includes a discussion of the basic concepts involved in assurance cases (Section 13.2), what an assurance case template is (Section 13.3.1), essential components and concepts of such a template (Section 13.3.2), and the problems (Section 13.4.2) and advantages of an assurance case template-based standard (Section 13.4.3).

13.1 STANDARDS-BASED DEVELOPMENT

13.1.1 The Role of Standards

The construction of a standard is a community effort that uses state-of-the-practice knowledge to describe requirements on the development process and (hopefully) the product that will be developed through an application of the process. In any industrial domain, we can typically find a number of international standards that fulfill this role. In addition, if the domain is regulated, such as the nuclear and medical device domains, a government-appointed regulator will likely issue regulatory requirements that may refer to international standards. In many regulatory jurisdictions, compliance with relevant standards and regulatory guidelines is a necessary prerequisite to obtaining approval to market that device.

A useful definition of a standard is given by the International Organization for Standardization (ISO):

> Standards are documented agreements containing technical specifications or other precise criteria to be used consistently as rules, guidelines, or definitions of characteristics, to ensure that materials, products, processes, and services are fit for their purpose.

There are at least two important points to note about this definition. The first is that it is an "agreement," and the parties to the agreement are members of the community involved in manufacturing, developing, and certifying the products, processes, services, or notations governed by the standard. The second point is that the goal of the standard is to ensure that the entities governed by the standard are "fit for purpose." This is especially pertinent to our discussion when "fit for purpose" includes safety and security attributes.

There are many different kinds of standards, and for the remainder of this chapter we are going to limit our discussion to standards that govern the development of safety-critical CPS/products. We have seen that standards have the potential to help developers/manufacturers by documenting the best practice as determined through community consensus. Of course, the assumption here is that the standard is "good," truly reflects best practice, and is written so that it is reasonably easy to understand and follow. Additionally, standards promote a common understanding among the various stakeholders— manufacturers, suppliers and consultants, and certifiers/regulators. It is true that many excellent products are built through compliance with the current standards.

However, to be truly useful, a standard should include acceptance criteria on the process and/or product that will help to reduce the variance in conformance to the critical properties of the product, such as safety, security, and reliability. Our observation is that this is woefully lacking in most current standards. In general, there are additional problems and limitations with current standards that affect how useful they are in governing the development and certification of CPSs and other safety-critical software-intensive systems. In the following section, we discuss the potential weaknesses in current standards as a motivation for doing something different.

13.1.2 Challenges with Standards

Wassyng et al. [3] described eight potential problems with the current standards. These were

- Most software-related standards are primarily process based

- Outdated standards [4]

- Complex and ambiguous standards

- Checklist-based completeness

- Lack of design and technical detail [5]

- Excessive documentation

- Not adopted by small companies [6]

A similar list was compiled by Rinehart et al. [7]

- Unknown assurance performance (same as "primarily process based")

- Lack of flexibility

- Development cost

- Maintenance cost

All these potential deficiencies impact our ability to consistently develop, manufacture, and certify CPSs that are safe, secure, and reliable. However, in our opinion, by far the most important of these is "*Most software-related standards are primarily process based*," and it is on this point that we now focus our attention.

There have been a number of papers written on product-versus process-based certification [1,8], and the role each plays in the confidence we have in the quality of the resulting system. The process requirements embedded in standards are not a guarantee that the resulting product will exhibit the desired level of reliability, safety, and security. The reasons for choosing those requirements are statistical, in that we have historical evidence (mostly anecdotal) to show that a certain percentage of products developed using those process requirements were successful. We think of these requirements as being derived from principles, such as stepwise refinement, but such principles are still only supported by a statistical argument as to their effectiveness for any one specific system. At best, therefore, we can hope that there is a reasonable/good chance that the product will be satisfactory. Of course, there are required processes that should result in safe and secure products—for example, hazard analysis. The fact that our process-based standards are also not very prescriptive usually diminishes the reliance we can place on such expectations.

If we want to use process as a basis for our standards, then, there are a couple of approaches that we can take to make the outcomes more predictable.

- We can specify more acceptance criteria on aspects of the process, and more precise acceptance criteria on the product.

- We can specify more detailed process steps. Vague or overview process requirements will never be enough to affect the outcome of the process to the degree we need.

Without requirements on the product as well as the process, we are really not able to claim with any certainty that our developed systems are reliable, safe, and secure. In practice, therefore, successful companies sometimes impose their own outcome-based requirements on specific processes in the development life cycle. An obvious example of one area in which we have done a little better is testing. Some standards, at least, have requirements not only on the process, but also on the outcome of the process. The civil aviation standard DO-178B (and now DO-178C) is a good example in this regard, with its mandating of the use of *Modified Condition/Decision Coverage* [9,10]. This may not have been by intent, but it turns out that this test coverage criterion itself also imposes requirements on the outcome of the process.

Other benefits of specifying requirements on the outcome of a process are that

- It informs developers of how the acceptability of their products will be judged.

- It results in consistency of an assessment when different assessors evaluate the outcomes, so the acceptability is less prone to be a result of the views of an individual assessor.

In the end, we need to identify what it is specifically about process-based standards that is hampering us in achieving more consistent and higher-quality outcomes.

We believe that there are two essential elements that are missing in most current standards:

1. Identification of *evidence* that must be produced (the evidence has to relate to the product as well as process entities).

2. *Acceptance criteria* for the evidence produced to be regarded as valid, enabling us to use the evidence to argue that the CPS has the desired properties.

13.2 ASSURANCE CASES

Assurance cases are based on safety cases [11], and safety cases have been in use for over 50 years in the United Kingdom. An assurance case presents a *structured argument* in which the developer of a product makes a claim regarding the critical properties of the product (e.g., safety, security, and reliability), and then systematically decomposes the claim into subclaims so that, eventually, the lowest-level subclaims are supported by evidence. There are a number of notations and tools for assurance cases, the most popular notation being the *goal-structuring notation (GSN)*, developed by Tim Kelly [12]. Figure 13.1 shows what an assurance case may look like, represented in GSN. The major components in a GSN diagram [13] are

- *Assumptions*—identified by "A" followed by a number

- *Contexts*—identified by "C" followed by a number

- *Goals*—represent claims and are identified by "G" followed by a number

- *Justifications*—explain why a strategy was chosen and is appropriate, and are identified by "J" followed by a number

- *Solutions*—represent evidence, identified by "Sn" followed by a number

- *Strategies*—explain why and how a claim was decomposed into specific subclaims, and are identified by "S" followed by a number

A recent technical report by Rushby [14] is an excellent description of the history of assurance cases, current assurance case practice, and also delves deeply into the essential characteristics of assurance cases.

13.2.1 The Role of an Assurance Case

The traditional assurance case was designed to document a structured argument that some critical properties of a system, product, or process are satisfied to a tolerable level, for the purpose for which they were constructed. Its primary role was thus seen to be as providing a believable demonstration of compliance. An assurance case that assures safety for a product, for example, must demonstrate that the product, when used as intended, in its intended environment, will function with tolerable risk of harm to anyone, over its lifetime. That role is likely to remain as a crucial role for assurance cases for many years to come. In the past few

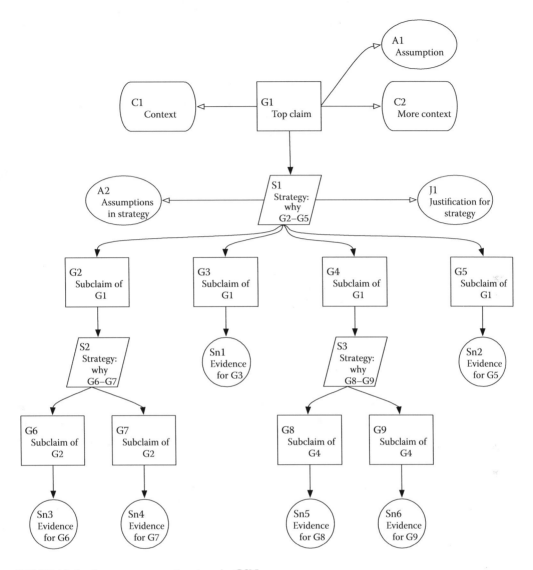

FIGURE 13.1   Assurance case structure in GSN.

years, an additional role has slowly been emerging—that of driving the development of the product, for example, in such a way that the resulting product will satisfy, to a required level, the critical properties of interest, and this assurance is documented by way of the assurance case [15].

13.2.2   State-of-the-Practice

One of the problems researchers have in the field of assurance cases, is that many of the real, industrial assurance cases are proprietary. This makes it difficult to ascertain exactly what the state-of-the-practice is. It also makes it difficult to back up any claims we make regarding the state-of-the-practice. We have seen real industrial assurance cases and formed an opinion about the state-of-the-practice. We have also taken note of what our colleagues have said about the cases they have been able to examine and evaluate. However, we still cannot cite the specifics and cannot show extracts from these proprietary assurance cases.

We take "State-of-the-Practice" to mean what most practitioners do, day in and day out. It is not likely to be the "best practice," especially in a relatively new field, since there will be relatively large variations in practioners' knowledge and skills. An excellent description of current practice for assurance cases in aviation is given in Reference 7.

Our impression of industrial-level assurance cases is that

- *Improved over the past 5 years:* The concepts and basic idea behind an assurance case seem to be better understood and more widely spread now. In earlier years, we even saw one submitted assurance case that had a top-level claim equivalent to "*we followed an accepted process.*"

- *Developed for certification:* In spite of the fact that some companies and many researchers recognize the need to develop the assurance case in tandem with system development [15], the cases we have seen were clearly constructed late(r) in the development cycle. In discussion with manufacturers, the development of the assurance case is seen as a necessary evil that consumes time and effort, and its purpose is to convince a certifier that the system is of sufficient quality—and safe.

- *Structured to mimic hazard analysis results:* Most assurance cases, since all of them had safety concerns, have a top-level structure based on hazard mitigation. We do not believe that this is the best way to structure assurance cases. Some of our reasons are presented in Section 13.3.7.

- *It is not clear whether the precise nature of the evidence is predetermined:* In most assurance cases, the evidence that supports a terminal claim is simply presented, and it is not clear as to why that specific evidence is used, or why it supports the claim.

- *No acceptance criteria:* Presumably because the cases that we have seen or heard about were created post development, we are not aware of any attempt to define the acceptance criteria for evidence. We have seen the discussion, after the fact, as to why the evidence is applicable and "good enough." There is significant attention paid to the *confidence* we have in the evidence and other aspects of the assurance case. Acceptance criteria could have a positive effect on some aspects of this confidence.

- *No explicit argument:* It seems that the currently accepted definition "explicit argument" is

  - The tree structure of the claim, subclaim, and evidence makes the argument structure explicit.

  - The claim decomposition rationale (i.e., described in the strategy in GSN) explicitly informs us as to the nature of the argument.

Both of these aspects are useful, maybe even essential, but the actual "reasoning" involved in the argument is not explicit. Our experience is that it is never provided.

13.2.3 Improving the State-of-the-Practice

As mentioned above, there has been a rapid improvement in many aspects of the state-of-the-practice as regards assurance cases. However, there are some things we can do that will improve it even more—and they can be done reasonably easily and thus, soon.

- *Prepare the assurance case before developing the system:* The critical properties that we want in our system must be designed in—they cannot be added successfully after the development of the system [16]. There are many reasons as to why the assurance case should not be created after system development (or late in the system-development process). The two primary reasons are

  - Safety, security, and dependability have to be built in to the system from the beginning, they cannot be bolted on after the fact.

  - Companies complain about the cost and effort involved in constructing the assurance case. They mistakenly believe that they construct the assurance case to get whatever certification they seek. It is true that developing the assurance case after the fact is costly and time-consuming; it is also a duplication of effort. If the company does its work well, it will have had to plan the development of the system in a way that involves many aspects of what will be repeated in the assurance case. Combining the planning and the documented assurance case right from the beginning is much more cost-effective. This is explicitly recognized and developed in the work on *Assurance-Based Development* (ABD) [15]. As mentioned earlier, some companies already do this, but our experience is that companies considering assurance cases, or who have recently started using assurance cases, are still considering them to be certification hurdles.

- *Use a robust structure that facilitates defense-in-depth:* Defense-in-depth is a long-standing principle used to ensure the safety of a system. Its use predates the use of software in critical systems. The idea is that safety must not be dependent on any single element of the design, implementation, maintenance, or operation. The decomposition structure that we use for the assurance case impacts the ease and stability with which we can make changes to the assurance case, and also influences how easy it will be to plan and document a defense-in-depth strategy for the development process in addition to the system. This is discussed in slightly more detail in Section 13.3.7, and Figure 13.5 illustrates a top-level decomposition that we believe will lead to a much more robust structure, as compared with basing the top-level decomposition on hazard mitigation.

- *Decide and document ahead of time what evidence must be obtained:* Creating the assurance case early—preferably before system development starts—enables us to plan what items of evidence must be produced to support predetermined claims. This has a number of advantages, the main one being that it provides direction to the developers. They know ahead of time what their process should produce, whether it be documentation or system components. This, together with acceptance criteria as discussed below, provides a direction that process alone simply cannot achieve. It is true that we will come across situations in which the "desired" evidence cannot be obtained. In these cases, alternative evidence must be substituted. The substitution needs to be noted (just as we would note a deviation from a standard). If the reasoning (argument) is really made to be explicit (see below), then, we have a tool whereby we can evaluate whether or not the alternative evidence is sufficient to support the specific claim.

- *Determine and document the acceptance criteria for evidence:* Determining what evidence to produce is one thing. We also need to specify what must be true of that evidence for it to be sufficient in supporting a specific claim. This is discussed in more detail in Section 13.3.3.

- *Make the reasoning explicit—not just the claim decomposition structure:* One of the original claims for assurance cases is that their argument structure would be explicit. Technically, that is true. The tree structure showing the claim, subclaim, and evidence is explicit, at least it is in the graphical notations for assurance cases. However, the structure leaves out the essential aspects of the argument, and we believe that the reasoning aspects of the argument help us to evaluate the validity of the claims in a way that the decomposition structure on its own cannot do. Currently, there is no real agreement on how to perform and structure the reasoning, so we are concerned that in terms of a relatively quick benefit, if we wait until there is a consensus on how to perform the argumentation, then we will have squandered an opportunity to improve assurance cases in the short term. We propose that, at a minimum, whenever a claim is decomposed into subclaims or is directly supported by evidence, the assurance case must include a strategy (why and how the claim was decomposed into specific subclaims), a justification (why the strategy was chosen and is appropriate), and the reasoning (that the claim follows from its premise—the subclaims or direct evidence). A partial example showing the suggested strategy and reasoning is shown in Figure 13.5 in Section 13.3.7.

13.2.4 Aiming for State-of-the-Art

We believe that the items above can be put into general practice now, with a corresponding improvement in the state-of-the-practice for assurance cases. There are two additional items that we see in the longer term, that will raise the state-of-the-practice to tomorrow's state-of-the-art. The first of these is to reach some degree of consensus on how to perform argumentation in assurance cases in such a way that we can dramatically improve the soundness of the argument, that is, all the claims are valid and all the premises are true. The second is the evaluation of confidence in the assurance case.

13.2.4.1 Argumentation

Arguments supporting assurance should ideally be formal reasoning in some appropriate formalism defined for the purpose. This is because only when the argument is formalized in this way can we be sure that the reasoning is sound and that we have made no mistakes or omitted any necessary details. Of course, we may not have to or even want to present the argument entirely formally, but instead rely on a style similar to that used by mathematics: a rigorous approximation to a completely formal proof, that can, if necessary, be turned into a formal proof.

Then, there arises the question of what the appropriate formalism looks like. It is commonly accepted that some aspects of such reasoning in assurance cases will not be expressible in conventional logics, such as first-order logic. The ideas of Toulmin [17] have been used as a departure point for characterizing the kind of reasoning that might be appropriate. Toulmin's argument falls into that group of reasoning known as *inductive reasoning*, in which we cannot provide "*proofs*" in the same way that we can in *deductive reasoning* [14]. Toulmin proposed a scheme for what rules in an argument might look like. That is, he proposed a template that should be general enough to represent the form of any rule used in an argument. So, a logic appropriate for reasoning in assurance cases would be populated with reasoning rules that may conform to Toulmin's scheme.

13.2.4.2 Confidence

One important aspect of Toulmin's scheme is the uncertainty associated with the elements of the rule scheme. In conventional logics, if one accepts the premises of some inference/argument step and applies a rule to produce a conclusion, one can then have absolute confidence in that conclusion. Confidence is not talked about as there is no uncertainty associated with any of the premises, the applicability of a rule of inference, or the resulting conclusion. Toulmin argues that in certain domains, such as law, science, and, we would claim, in assurance cases, all three elements above have levels of uncertainty associated with them, which we conventionally refer to as confidence.

Over the past few years *confidence*, as it applies to assurance cases, has become an important and much-discussed topic [18–24]. What confidence are we, as a community, attempting to define, estimate, and even quantify?

The answer is *confidence in the validity of the top-level claim*. This is clearly the primary concern of the assurance case (if it is not, we have the wrong top-level claim), and we want to be able to evaluate, sensibly and as objectively as possible, how certain we are that the claim is valid. Note that confidence in the top-level claim will typically be governed within the context of the appropriate *safety integrity level*, and its associated *integrity-level claim* [25]. To this end, we can try and evaluate our confidence in two items as shown below.

- *Confidence in the argument*: The argument is supposed to be explicit, but even if it is not, the claim, subclaim decomposition, and evidence do describe an argument— we must simply discover for ourselves what it is, rather than be told explicitly in the assurance case itself.

- *Confidence influenced by evidence:* The purpose of an item of evidence is to support the parent subclaim directly connected to the evidence. Confidence is thus related to how well the evidence supports the claim. For example, if the terminal subclaim is *"The system behavior as implemented is adequately[*] similar to the documented required behavior,"* what can we say about our confidence in the assurance case in the two examples of evidence that follow? (1) System test reports and (2) A mathematical verification report.[†] Confidence here will be influenced by the coverage criteria used for the tests; the specific test coverage achieved; the number of failed tests at the time of installation; the fact that a mathematical verification was performed; and the number of residual discrepancies that remain at the time of installation.

However, even if we can establish our confidence in the two items above, we have no generally accepted theory or any other means of using this confidence to ascertain the confidence we have in the top-level claim.

13.3 ASSURANCE CASE TEMPLATES

In the discussion that follows, readers who know something about assurance cases may conclude that we simply mean an *assurance case pattern* when we talk about an *assurance case template*. We do not. At least, we do not believe that we do. The real difference between a *pattern* and a *template* is one of completeness. Our observation of assurance case patterns is that they are assurance case *design artifacts* that can be instantiated for a specific situation, and that are used within an overall assurance case. The assurance case template is a complete assurance case in which claims are specialized for the specific situation and the evidence for terminal claims is described together with acceptance criteria for the evidence and, as development progresses, the evidence is accumulated and checked against its acceptance criteria. The concepts are, indeed, very similar—but these templates are not the same as current patterns. Patterns could prove to be useful within assurance case templates.

13.3.1 What Is an Assurance Case Template?

Probably the best way of defining or describing an *assurance case template* is to show an overview example of the process that led us to believe that assurance case templates could be a productive and important step in the development of a reliable, safe, and secure CPS.

To begin with, let us assume that we have successfully developed a product (system) and also documented an effective assurance case for the product. This product is creatively called *"Product 1,"* and the assurance case decomposition structure is shown in Figure 13.2. In the interests of clarity, assumptions, contexts, strategies, etc., are not included in the figure, and the "1" in the top-level claim box simply indicates that this is the top-level claim and assurance case for *Product 1*.

Now, let us further assume that we follow up our success by developing another product, which we call *"Product 2." Product 2* is a product different from *Product 1*, but is related in

---

[*] We assume that "adequately" is defined for each specific use of the word, within the assurance case.
[†] In general, we expect the evidence to be much more finely grained than that presented in this example.

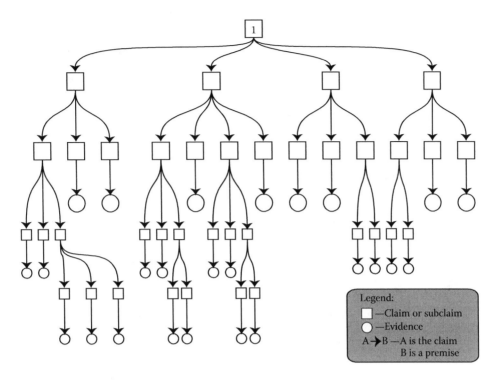

FIGURE 13.2   Assurance case structure for Product 1.

that it is within the same *product domain* as *Product 1*. For instance, perhaps *Product 1* is an insulin pump, and *Product 2* is an analgesic infusion pump. Or, perhaps *Product 1* is a car and *Product 2* is a minivan. We again document an assurance case for *Product 2*, and we are so expert at developing assurance cases that the new assurance case differs from that for *Product 1* only where absolutely necessary. The assurance case for *Product 2* is shown in Figure 13.3.

Figure 13.3 also highlights the differences between the two assurance cases by explicitly showing which components have been added, removed, or modified in the assurance case for *Product 2* as compared with that for *Product 1*. Now, consider what the figure would look like if we developed *Product 2* before *Product 1* and then highlighted the differences in the assurance case for *Product 1*. It should be clear that components added in the case for *Product 2* would be shown as components removed in the case for *Product 1*. In other words, the difference between "added" and "removed" is one of time—it depends on whether (the assurance case for) *Product 1* is developed before or after (the assurance case for) *Product 2*.

So, now we get to the heart of the idea of an assurance case template. We want to develop an assurance case for both *Product 1* and *Product 2*—before we actually develop the products themselves. Of course, since an assurance case must be product/system specific, we cannot actually build a single assurance case that precisely documents the assurance for both products. What we can do is build a template that has a structure that will cater to both the products, but in which the content of components will need to be different for the two products. If our template has to handle only the two products, *Products 1* and *2*, a template that

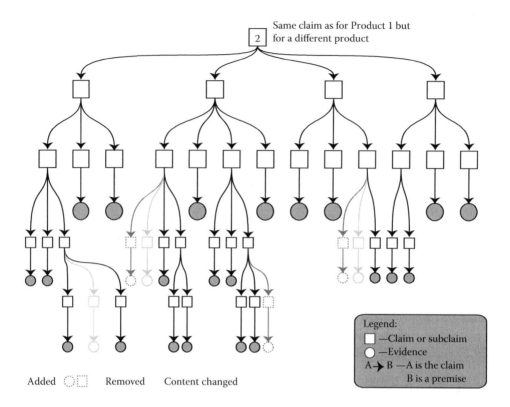

FIGURE 13.3   Assurance case structure for Product 2.

may achieve this is shown in Figure 13.4. The idea here is that claims and subclaims can often be parameterized, in a manner similar to that done in *assurance case argument patterns* [26,27], so that the claim can reflect the specific product that the assurance case deals with. The details are not visible in the figure, but we will discuss this aspect of the template later in this chapter.

Figure 13.4 now needs some explanation. There are two specific aspects of the template that we will discuss at this stage.

- *Optional paths:* The paths shown in gray in the figure are optional, depending on the specific product for which the assurance case is being instantiated. The numbers next to the optional paths show the *multiplicity* of the paths.

  - *Optional 0–1:* This is a single path that may or may not be required for a specific product.

  - *Exclusive-Or 1:* One of the paths (there can be more than 2) must be instantiated for a specific product.

  - *Non-exclusive-Or 1–n:* One or more of the paths can be instantiated for a specific product.

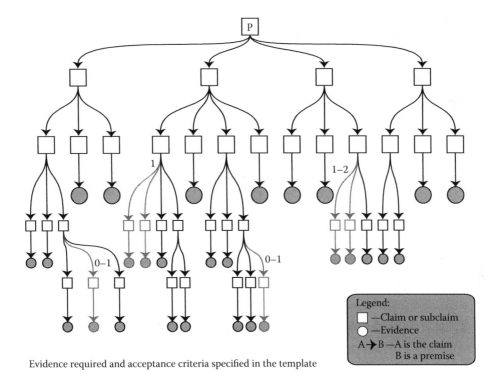

Evidence required and acceptance criteria specified in the template

FIGURE 13.4   Assurance case template structure.

- *Evidence nodes:* The actual evidence for products will differ from product to product. That is why all the evidence nodes in Figure 13.4 are shaded. If this was not true, we would not need to develop an assurance case at all! Since this is a template, the content of the nodes in the template will not be the actual evidence. What will it be? The answer is simple and reflects one of the benefits of building an assurance case template. Each evidence node must contain:

  - A description of the required evidence.

  - Acceptance criteria for that item of evidence, that is, what must be true of that evidence to raise the level of confidence that the critical properties that the system must have are true.

Not surprisingly, the GSN community, in their work on assurance case patterns, had to deal with exactly this type of optionality. GSN now includes a specific notation to deal with what they have termed "Multiplicity" and "Optionality" [13]. A small solid ball together with a number representing the cardinality is used for multiplicity, and a small hollow ball is used for optionality (in this case 0 or 1). The notation was further extended to use a solid diamond together with the cardinality number as a general *option* symbol. What seems to be missing from GSN is the *non-exclusive-Or* option. An example of an assurance case template using this notation is provided in Figure 13.6 in a later section of this chapter. We have used the option symbol together with an "at least n" descriptor to describe the *non-exclusive-Or*.

13.3.2 Characteristics of an Assurance Case Template

The essential characteristics of an assurance case template must also include characteristics of state-of-the-art assurance cases. Why state-of-the-art? Primarily, to reap the benefits of a template.

13.3.2.1 Characteristics of State-of-the-Art Assurance Cases

We believe that the following characteristics/components are essential for any state-of-the-art assurance case:

- *Explicit assumptions:* An assurance case applies to a particular system, and we know that we cannot claim to assure any critical property in that system without precisely describing what has to be true about the system. Assurance cases have improved the general approach to system assurance by highlighting the fact that assumptions are first-class components of such assurance. They are useful/essential because

  - Reviewers can determine whether or not the assumptions are practical

  - They help to constrain the system, and without constraints, usually, we will not be able to achieve the required quality

  - They provide an easy opportunity to check that the system does not violate any assumptions

  - They enable us to evaluate changes in the system to determine whether or not the assurance case will have to be modified to account for the changes (even if the changes do not violate the stated assumptions, we may need to modify the argument component of the assurance case)

- *Explicit context:* An area where current assurance cases do reasonably well is in the documentation of context for the primary system components. This is important since one of the typical entities described in the context is the definition of terms. This may seem trite, but becomes crucial the more precise we try to be—and precision/formality is something we strive for in the case of safety-critical systems.

- *Explicit argument:* One of the benefits "claimed" for safety and assurance cases is that they make the assurance argument explicit. Typically, this is not true in extant examples that we have seen. Most assurance cases present an explicit decomposition of the top-level claim(s) into subclaims, and eventually, at the end of each path of subclaims, there is an evidence node (or nodes) that "supports" the subclaim to which it is attached. The decomposition of subclaims is not an argument—it does not provide a reasoning that uses the premise(s) to support the (sub-)claim (see Figure 13.4 above). Actually, there is no agreement at this time as to what an assurance case argument should be. Some researchers believe that we can provide localized arguments as described above. Others believe that there are some properties, safety being the prime example, that are global to the system, and local arguments (the premises "imply" the claim) ignore emergent behavior. In this case, the argument would have to be global to the assurance case. Over the years, there have been a number of attempts at defining how to perform argumentation (reasoning) in assurance cases [28–32]. The report by Rushby [14], mentioned earlier, presents an excellent discourse on the different types of arguments for assurance cases. At this time, any explicit argument, of whatever level of formality, even if we find out in the future that it has shortcomings, will be a huge improvement. The sad truth is that, in practice, there is almost no attempt to provide an explicit reasoning in assurance cases. Since so many successful projects have been completed without it, do we really need it? Our answer is "yes"—especially in the long term. The size and complexity of CPS is stretching our ability to produce systems/products of sufficient quality and reliability. The complexity and distributed nature of these systems makes it even more difficult to identify emergent behaviors and protect against hazards that arise because of these emergent behaviors. Evaluating the safety, security, and reliability of these systems is assuming more and more importance, whether it be for external certification (licensing and approval), or internal certification (quality control). Making the argument explicit, so that we can evaluate the soundness of the argument more easily and accurately, will help us build systems with adequate quality.

- *Explicit strategies:* Although the strategy is not an argument, it is still useful and important. The idea of the strategy is to explain how the parent claim is decomposed into subclaims and subclaim paths that explicitly characterize the safety design aspects of the artifact.

- *Explicit justifications:* Justifications are usually regarded as "nice to have," but they are more than that. While the strategy describes the "how" of the decomposition, the justification describes the "why." The structure and rationale for the decomposition is akin to the module guide in Parnas's *Rational Design Process* [33], and this is precisely what the strategies and justifications, together, should deliver.

13.3.2.2 Characteristics Essential for an Assurance Case Template
- *Adequate descriptions of evidence to be provided:* An important difference between an assurance case and an assurance case template is that, in a template, the specific evidence for a system cannot be included, because the specific system is not known at the time the template is constructed. However, we do (should) know what evidence will be required to support the claims that we want to make—we know the claims as well. Why should this be true? The simple fact is that any manufacturer of safety-critical systems sets out to build a safe (reliable and secure) system, and an incredible amount of work goes into planning how to build the system so that it will be safe! This planning uses knowledge accumulated by practitioners through the ages and documented in books and research articles in appropriate fields, as well as in corporate memory. It may not have been framed or even recognized as claims and evidence, but it is— or can be viewed as such. So, what we need for an assurance case template is explicit

documentation of the nature of the evidence that is required for each "bottom-level" subclaim. This description can include alternatives, since it is sometimes possible to support a claim in a number of ways. In fact, it may be possible to provide more than one item of evidence for a single claim—and that certainly helps to build confidence in the assurance case.

- *Acceptance criteria for evidence:* A description of the kind of evidence required for each terminal subclaim is necessary, but not sufficient. We also need to describe, as precisely as possible, what must be true of the documented evidence to be sufficient to support a specific claim. The fact that this is not routinely required in assurance cases reflects the fact that most assurance cases are developed after the fact (or, at best, during development) to document why a particular system satisfies a specific criteria. They are not developed to drive development so that those criteria will be satisfied. There are academic papers that exhort practitioners to build the assurance case as early as possible [34,35]. However, those papers do not discuss the acceptance criteria for evidence. Our ideas regarding an assurance case template were prompted by the work of Knight and colleagues. Their idea of *ABD* [34] was intriguing, and seemed to reflect many of the principles in which we also believe. However, this body of work also does not promote the idea of acceptance criteria.

- *Arguments that cope with optional paths:* As we saw in Section 13.3.1 and Figure 13.4, an assurance case template requires arguments that deal only with logical conjunction. An assurance case template requires arguments that can deal both with logical disjunction and logical conjunction. This should not constitute a real problem, but it is worth noting that the situation has to be dealt with.

- *Why explicit arguments will be crucial:* As stated earlier, we strongly believe that explicit arguments are essential if we are to realize the true potential of assurance cases to help improve the safety, security, and reliability of a complex CPS. Explicit arguments (reasoning) are even more important in the construction of assurance case templates, simply because the template is predictive rather than reflective. We can consider two different cases:

  - What if we want to use evidence that is not explicitly included in the description of evidence in the template? In this situation, it is possible that the template erroneously omitted evidence that we now want to include. Or, it may be that the developers of the template did consider whether or not to include that evidence but, for whatever reason, decided not to. In both situations, if we have an explicit argument in the template, then we will have able to determine what effect our potential change in evidence will have on the validity of our claims.

  - What if we want to modify a claim or subclaim? Since our current knowledge does not adequately inform us as to whether or not it is at all possible to perform incremental safety assurance, modifying claims in a way that requires a change in the reasoning/argument appears to be risky. Well, it may not be quite as bad as it seems.

First of all, if the original argument is "bottom up" and global (as opposed to a set of localized arguments, each constrained in scope to a specific set of subclaims that support the directly connected parent claim), then we should be in a position to evaluate the effect of the change in terms of how it affects the "top-level" claim(s), and what would be necessary to make the argument "sound" again. Second, if the argument consists of a set of localized arguments, then we must have decided that these localized arguments are sufficient for our purpose—at least at this time and stage of the knowledge in argumentation for assurance cases. Then, we should be able to use the localized argument that encompasses the change to evaluate the effect of the change in terms of how it affects the "parent" claim, and what would be necessary to make the argument "sound" again.

13.3.3 Acceptance Criteria and Confidence

The notion of *confidence* in assurance cases was discussed in Section 13.2.4.2.

Why is it that, with all the research on *confidence* in assurance cases, we hear so little about *acceptance criteria*? The answer to this is quite simple, but sometimes hotly contested. The focus on *confidence* as compared with *acceptance criteria* is primarily due to the perception that assurance cases are developed after the system/product has been built—or, at best, while the system is being built.

Why are we concerned about this? There is significant research effort on confidence related to evidence. We believe that in dealing with this aspect of confidence, our task would be better defined if we could assume that the acceptance criteria for evidence have been defined. The more we can reduce uncertainty in this regard, the better off we will be. A consensus of expert opinion used to define acceptance criteria would be of definite benefit.

13.3.4 Dealing with Evidence

We have already stated that, in terms of evidence in assurance case templates, the two most important attributes we must document in the template for each evidence node are a description of the evidence required to support the parent claim, and acceptance criteria for the evidence. Related to the first of these, we need to discuss the nature of evidence in an assurance case node. It has become prevalent to simply reference documents produced during development as the relevant evidence. We believe that evidence in assurance case nodes should be much more specific. If we include specific sections of a document as part of the reference, then, this can still be achieved by referring to development (or other) documents. This is sometimes difficult when done "after the fact," since the section referred to may contain more than the specific item to which we wish to refer. In the case of a template, this should be less of a problem since we will know ahead of time that we will want to refer to the specific item.

As far as the evidence is concerned, one important thing to note is that the evidence and assurance case argument are strongly linked, and both are essential to successful assurance cases. Researchers at the University of York have a saying regarding this: "*An argument without evidence is unfounded whereas evidence without argument is unexplained*" [36]. You cannot necessarily change one without making changes in the other. This is always true in any assurance case. For an assurance case template, this dependence is both a potential challenge and extremely useful. It is a challenge in that changing the evidence will incur a reexamination of the argument, and reexamining the argument can be costly in terms of time and effort. One mitigation in this regard is that the evidence is at the bottom of the argument tree, and our intuition is that a change in evidence is more likely to have a local impact on the argument. The dependence can also be useful in helping developers understand why that particular evidence is necessary.

13.3.5 Product-Domain Specific

The template that we envisage will be product-domain specific. It has to be. A major assumption in this chapter is that there are classes of products/systems for which an assurance case structure (claims and arguments) exists that handles all the products within a class without *too many* differences. The only possible way that this assumption could be valid would be if we were to restrict our assurance case templates to product-domain-specific systems. We do not have a proof yet that this is possible. We do have some preliminary results from the consideration of infusion pumps—and also preliminary ideas with respect to the automotive domain.

13.3.6 Different Development Processes

In Section 13.3.1, and in Figures 13.2 through 13.4, we presented motivation for assurance case templates by describing how we could handle differences caused by product/system differences. There is another kind of difference that we have not yet discussed—differences in the development process. Many differences in the development process can be handled without any change in the structure of the assurance case. For many years, we have argued that even when we promote "product-focused" certification, there is always an assumption of an "idealized development process," as described in Reference 37. Typically, it is these elements of an idealized process that make their way into an assurance case. However, there is one situation that is definitely different, and may result in requiring a slightly different assurance case structure. If one system is developed using a traditional development life cycle (especially as regards the software), and another system is developed using a model-driven engineering approach [38], then we can expect differences in the assurance case structures. In addition, the use of two substantially different development processes can/will occur even within a product domain. Thus, an assurance case template must be able to handle these differences, and we believe that a template with optional paths as shown in Figure 13.4 will be able to handle them well.

13.3.7 Suggested Structure

Researchers have been trying to work out how to achieve incremental assurance for complex systems, preferably using assurance cases. The difficulty here is that some properties, such as safety, are global to the system, and emergent behaviors have implications for the safety of the system. We do not know how to isolate portions of the system in such a way that each portion being "proven" to be safe would imply that the entire system is safe. We do not even know if this is possible. We surmise that it may not be possible without revisiting the argument in the assurance case. At the same time, we cannot afford for that to be the final answer, since we will not be able to redo complete assurance case arguments every time we make changes to a safety-critical CPS.

In some ways, software-engineering research has always grappled with problems like this, though not always concerning the safety, and not always with regard to global properties. Throughout the past 50 years, we have had a number of challenges in our pursuit to build more dependable software-intensive systems. One distinct challenge was how to cope with change in software-intensive systems. In the 1970s, Parnas suggested that "*information hiding*" [39] would help us to cope with anticipated changes, and that we could then build systems in which dependability would not be degraded when we made these anticipated changes. Information hiding helps us to structure software designs so that they are robust with respect to anticipated changes. It is time that we put serious effort into determining how to design assurance case structures that are robust with respect to change. This seems to be particularly apropos with regard to assurance case templates. We will use the top-level decomposition of an assurance case for a medical device, as shown in Figure 13.5, to begin this discussion.

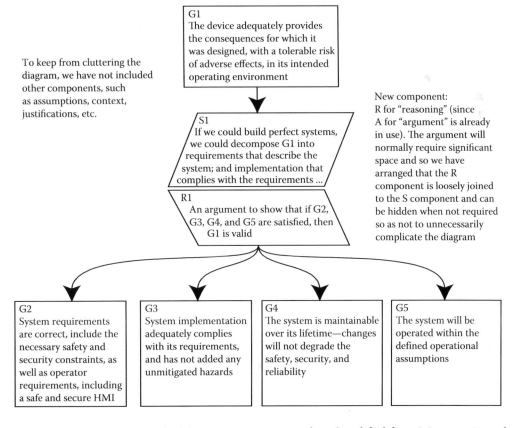

FIGURE 13.5   Robust top level of the assurance case template. (Modified from Wassyng, A. et al., *Des. Test, IEEE*, 32(5), 1–11, 2015.)

The first thing we notice (probably) about the claims G2, G3, G4, and G5 in Figure 13.5 is that they should not have to be changed no matter what system/product the assurance case template is developed for. Not only is this level robust with respect to change, but it also reflects our experience of how we have successfully planned safety-critical systems for more than 20 years. One principle that has occurred to us for designing robust structures for assurance cases, and that is supported by what we see here, is to move product-specific components of a case as low down in the structure as possible. This is an immediate difference from what we see in many current assurance cases. In most cases that we have looked at, the assurance case structure is very heavily based on hazard mitigation, with the hazard analysis providing pertinent evidence. This not only violates the principle that we have just espoused, but it also seems to be faulty for a number of other reasons.

- *Completeness:* It is impossible to "prove" that any hazard analysis is complete with respect to the hazards in the system. There are many aspects of assurance in which we cannot prove completeness. The way we deal with this is usually twofold: do as good a job as we possibly can, and have more than one way to achieve a critical outcome. Basing the structure of the assurance case on the structure of the hazard mitigation makes this more difficult. One of the touted benefits of a hazard analysis [40] is that it avoids *confirmation bias.* Ironically, structuring an assurance case primarily by documenting mitigation or elimination of hazards, seems to inject an aspect of confirmation bias into the assurance case.

- *Iterative nature of hazard analysis:* As we progress through the development life cycle we introduce more hazards, and the simple act of mitigating a hazard may introduce even more new hazards. Since we intend to use the assurance case template to drive development, it is important to be able to convey the iterative aspect of hazard analysis and mitigation.

In our suggested structure, the system hazard analysis is included in the claim–evidence path leading off from subclaim G2, and it is not the only means whereby we assure safety and security. The actual wording of the subclaim in G3, "*…and has not added any unmitigated hazards,*" has been worded that way explicitly to help convey the idea that any hazards introduced in the development process must be mitigated. The way in which the usual claim is typically phrased, "*all hazards have been mitigated,*" does not convey the same imperative to the developers. Finally, the product-specific aspects of the hazard analysis are much lower down in the structure.

Another principle that we believe facilitates building assurance case structures that are robust with respect to change is to try and make claim–argument–evidence paths as independent as possible. "Independence" in this context means that, at each "level" of the structure, for each claim at that level, there are no argument-specific dependencies between the lower paths that support that claim. This is clearly difficult to do, sometimes not possible, and it may be difficult to know when you have really achieved that independence. However, we believe that it is worthwhile keeping this in mind while deciding on the structure for the assurance case.

These are very basic and obvious ideas for improving the robustness of the structure of an assurance case. What we are lacking right now is something equivalent to the elegance and effectiveness of information hiding as applied to software designs.

13.3.8 Example: Infusion Pumps

To illustrate the different kinds of optional paths that we foresee for assurance case templates, we have three brief examples drawn from our experience with insulin pumps. The three cases that we identified earlier (Section 13.3.1) were

- *Optional 0–1*: Some insulin pumps include a built-in light that helps the user to read the pump's screen in the dark. This has an impact on the assurance case in a number of ways, including access to data from the pump, and battery usage. These paths in the assurance case would not exist at all if the pump did not have this feature.

- *Exclusive-Or 1*: Some pumps use a standard Luer connector for their infusion sets, while others do not. The pumps definitely need to use some sort of connector for the infusion set, and there are different pros and cons depending on what connector is used. The use of a Luer connector is not mandatory. The assurance case has paths that depend on the connector since it affects both the safety and effectiveness of the delivery of insulin to the patient. However, these paths are different because of the different pros and cons of the connectors. Therefore, the template will include a number of paths, depending on the number of connectors likely to be used in commercial pumps. Only one of those paths will apply for a particular instantiation of the template. (In the following section, we will discuss how to deal with the situation where a new connector becomes available that is not included in the current assurance case template.)

- *Non-exclusive-Or 1–n*: Some pumps allow the input of glucose readings directly from an associated meter, or those readings can be entered manually. The assurance case then has to handle the situation where both options are present in a single pump, as well as the situations where only one of the options is available for a specific pump. Note that this is not a situation where redundancy is being used to increase the reliability. The pump that allows both modes of input does so for ease of use—only one mode is used at any one time. The arguments for safe, secure, and effective are different in the two modes. An easy difference to note is that there is (probably) less chance of a security problem arising when the manual mode is used.

13.4 ASSURANCE CASE TEMPLATES AS STANDARDS

There are two major reasons that we considered using assurance case templates as development standards. The first reason was that it seemed to be an obvious extension of using an assurance case template to drive the development of a system, and then use the resulting assurance case as a basis for certification. The second reason was the research over the past few years in which multiple efforts have been made to discover the assurance case that is implicit in an existing standard [41,42].

While exploring assurance case templates and their potential to replace standards, we learned about a talk given by Knight [43], in which he outlined the use of an assurance case as an alternative to the avionics standard DO-178B. This reinforced our belief that we are on the right track. Our assurance case template is different from the *fit-for-purpose assurance case patterns* used in ABD. It does have much in common with Knight's ideas expressed in Reference 43.

The assurance cases we build are targeted at assuring the safety, security, and reliability of the complete system, and not only the software, for instance. There are likely to be aspects of existing standards that do not currently fit easily within the type of assurance case template we envisage. Since at this stage we do not know what those aspects are, we do not know whether or not they are really needed. Our intent is to explore how they can be incorporated. In the meantime, we need to realize that the assurance case template may need to be supplemented by "extra" information that is currently included in various standards.

13.4.1 Using an Assurance Case Template as a Standard

If we have done our work effectively, using the assurance case template as a standard should be reasonably straightforward—most of the time. The template will naturally dictate certain aspects of the development process, and system development and instantiation of the template proceed in tandem. As instantiation proceeds, the resulting assurance case documents the specific system being built. As usual, project planning determines how to comply with relevant standards, including how to comply with the relevant assurance case template. During this period, we may find that the template does not cover a situation that we want to include. In such a situation, we can modify the assurance case template that we will use for our specific development. It is important though, that we carefully document how we have deviated from the community-developed template, just as we would currently note any deviations from any other standard. Figure 13.6 shows how we can achieve this, by using a notation that highlights entities that have been modified, and entities that are completely new, that is, were not included in the template at all.

Figure 13.6 also shows how the optional paths included in Figure 13.4 are now described using GSN, as discussed in Section 13.3.1. For example, the two optional paths are now described by the inclusion of small hollow circles in the arrows. Also, the exclusive-Or path is shown with a solid diamond and a "1 of 2" descriptor, while the non-exclusive-Or is shown with the solid diamond and an "at least 1" descriptor (our modification of the GSN "option").

13.4.2 Problems in Constructing Assurance Case Templates

There is work needed to really *demonstrate* that we can construct effective assurance case templates at all, and especially for CPSs.

13.4.2.1 General Problems

- *Avoiding a check box mentality:* It is crucial that the template cannot be treated in any way as something where we mindlessly "fill in the blanks," or simply note that we have

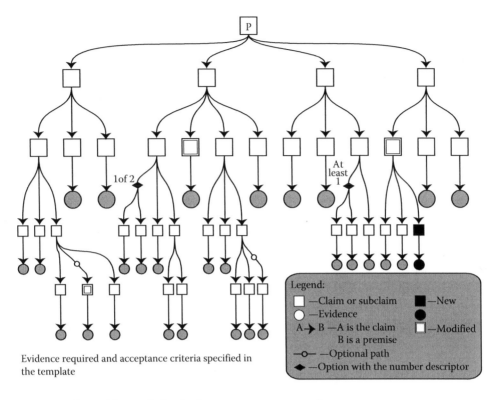

Evidence required and acceptance criteria specified in the template

Legend:
☐ —Claim or subclaim   ■ —New
○ —Evidence   ●
A➔B —A is the claim
        B is a premise   ☐ —Modified
—○— —Optional path
◆ —Option with the number descriptor

At least 1

1 of 2

FIGURE 13.6   Describing variation in the assurance case template.

provided what was specified in the template. Claims, subclaims, assumptions, contexts, strategies, justifications, and reasoning must be worded, as much as possible, in a way that makes the developer seriously consider each claim and how it relates to their specific situation. The evidence nodes in the template do not contain evidence as such until the developers document the relevant evidence. What is available in the node is the nature of the evidence required, and its acceptance criteria. The format used for the acceptance criteria should be such that it requires thoughtful interaction on the part of the developer in ascertaining whether or not the specific evidence generated meets the acceptance criteria and the related argument, and the developer has to document as to why it is sufficient.

- *Coping with omissions in the template:* No matter how well constructed our template is, there will be omissions and "errors." Related to the point above, we need to develop a mechanism for recognizing "holes" in the argument structure for a specific situation that was not predicted.

- *Essential items that do not fit within an assurance case template:* There are likely to be items currently included in standards that do not have an obvious place in an assurance case. We do not yet know what these are, and we have merely speculated that they exist. We need to find out what they are, and how we can handle them.

13.4.2.2 Problems Especially Related to CPS

- *Interrelated cyber, physical, and hybrid components:* Many CPSs include both cyber and physical components, and even hybrid components. It would seem that this must pose a distinct problem for the development of assurance case templates, since it is likely that the claims and evidence associated with these components must reflect the nature of the relevant component. The development of such a system is incredibly demanding, and regularly stretches our current capabilities. The essential requisites for constructing an assurance case template are

  - Predicting the claims associated with the development and certification of a specific system/product

  - Predicting the nature of the evidence required to support a specific (terminal) claim

  - Specifying the acceptance criteria for each set of evidence

  - Knowing enough about how to build the system to be able to predict the required claims and evidence

  So, although structuring the assurance case template may appear to be complicated by the hybrid nature of CPS, developing the claims and evidence is not more difficult than for non-CPS—assuming that we know enough up front about how the system will be developed. For safety-critical systems, we cannot afford to embark on "*radical design*" [44]. So, even in the case of a complex CPS, we should know how we are going to build the system so that it is dependable, safe, and secure.

- *Open systems:* A much more serious problem to deal with is that of constructing an assurance case template for an "*open system*," that is, a system constructed by combining existing systems (perhaps with other systems/components that are specially constructed for the new system). Such Systems of Systems pose unique and extremely difficult challenges for dependability, safety, and security. The current concept of an assurance case template will not cope adequately with such systems. This does not mean that we will not be able to extend the concept in the future; it is simply an acknowledgment that right now, we do not know how to achieve this. This is not surprising. We believe that assuring the dependability, safety, and security of such open systems is not adequately dealt with in general at this time.

13.4.3 Benefits of Using an Assurance Case Template as a Standard

It is one thing to say that we could replace some existing standards by assurance case templates and it is another thing to explain why we should do so. Without significant benefit, it would be pointless to attempt this. This section presents the major perceived benefits.

- *Explicit, consistent, and understandable guidance:* An assurance case template provides guidance to both the developer and certifier in a way that is much more structured than is a textual standard. Yes, the template is likely to be complex and large. However, so are modern standards. The structure does have some inherent benefits that help us to contain and control the complexity.

- It is easier to make the standard internally consistent

- Given the strategies used to decompose claims, the arguments that explain why the premises support the claims, what evidence is required, and the acceptance criteria for the evidence, users of the template understand how items depend on each other, what evidence to document, the rationale for the evidence, and what should be true of the evidence to satisfy the claims

  - There is a consistency to the way in which these items are presented to the user. This aids understandability, and it also aids completeness since it is more difficult to ignore/forget items

  - The need to confirm and/or enter assumptions and contexts also helps to develop a deep understanding of the system. The fact that these are supposed to hold consistently across the product domain makes it easier for developers and certifiers to understand whether anything is "out of the ordinary" for the system at hand

- *An opportunity for incremental safety:* In Section 13.3.7, we discussed incremental safety, and pointed out that it may not be possible—and that we could not afford for that to be true. We may have an opportunity with an assurance case template to make some inroads in this regard. We have built the template so that the facts that the argument is explicit (to reinforce the strategy), that the description of the evidence to support a claim is explicit, and that the acceptance criteria for the evidence is explicit, make it possible to create an assurance case that will eventually apply to a variety of systems. If this is achievable, we will have created a methodology for incremental assurance— if the modified system still fits within the assumptions and contexts that govern the assurance case template.

- *Easier to develop an expertise in evaluating the safety, security, and reliability:* One of the first things that occurred to us when we first started evaluating assurance cases for regulated systems was that if every assurance case has a one-off structure, then it will be extremely difficult for regulators to develop expertise in evaluating assurance cases submitted to them [45]. A product-domain-specific assurance case template will help to alleviate this problem.

- *Valid arguments built by experts:* The argument in the assurance case template is the state-of-the-art component that will help us tame the complexity in our modern CPS, and thus make it possible to develop and certify adequately safe, secure, and reliable systems. For this to happen, the argument has to be both valid and sound. Unfortunately, the argument is one of the most difficult elements in the creation of assurance cases, and the number of people with the expertise to do this is rather limited. The fact that the assurance case template we envisage will be created by a community of experts is of crucial importance. Among those experts, we can make sure that some of them have the requisite expertise in argumentation. Hopefully, the number of people with this expertise will gradually increase.

- *Avoid confirmation bias:* The very fact that the assurance case template is developed prior to the development of a system should definitely help in avoiding confirmation bias—a potential problem postulated by Leveson [40]. In this regard, there is a tremendous difference between developing the assurance case, including the acceptance criteria on evidence, before developing the system, and developing the assurance case after or during development of the system—and then justifying the acceptability of the evidence.

- *Publicly available examples of good assurance cases:* Assurance case templates, developed by a community of experts, will provide something we do not have at all at this time—examples readily available so that more and more practitioners can develop a thorough understanding of, and a facility with, assurance cases.

13.5 CONCLUSION

We propose that we use a product-/system-domain-specific assurance case template as a standard for the development and certification of products/systems within that product domain. Assurance cases have been growing in acceptability as an excellent way of determining the confidence we have that a system fulfills the need for which it was intended, and possesses the critical properties that it is required to possess. The assurance case template we envisage has the structure of a complete assurance case with (optional) subclaims and requirements on generated evidence. The template is to be produced, much like a standard, through a consensus of experts in the relevant domain. Such an approach promises to deliver significant benefits, as itemized in the sections above.

Although it may seem that our discussion in this chapter is largely independent of CPS, one of the guiding motivations of this chapter is to be able to cope with the complex, multidisciplinary, and connected nature of today's and tomorrow's CPS. Developing and certifying such systems, so that they are dependably safe, secure, and reliable, calls out for a significant change in the way we guide people through the development and certification of these systems. We believe that a concerted effort in understanding how to produce effective assurance case templates for this purpose will be one effective step in meeting the challenges of the (near) future. Luckily, they should be just as useful for less complex systems as well.

REFERENCES

1. T. Maibaum and A. Wassyng. A product-focused approach to software certification. *Computer*, 41(2):91–93, 2008.
2. U.S. Food and Drug Administration. Infusion pumps total product life cycle: Guidance for industry and FDA staff, December 2014. OMB Control Number: 0910-0766.
3. A. Wassyng, N. Singh, M. Geven, N. Proscia, H. Wang, M. Lawford, and T. Maibaum. Can product specific assurance case templates be used as medical device standards? *Design and Test, IEEE*, 32(5):1–11, 2015.

4. R. Bloomfield and P. Bishop. Safety and assurance cases: Past, present and possible future—An Adelard perspective. In C. Dale and T. Anderson, editors, *Making Systems Safer, Proceedings of the Eighteenth Safety-Critical Systems Symposium*, pp. 51–67, 2010. Springer, London, UK.

5. M. Huhn and A. Zechner. Arguing for software quality in an IEC 62304 compliant development process. In *Leveraging Applications of Formal Methods, Verification, and Validation—4th International Symposium on Leveraging Applications, ISoLA 2010*, Heraklion, Crete, Greece, October 18–21, 2010, *Proceedings, Part II*, pp. 296–311, 2010.

6. S. Basri and R.V. O'Connor. Understanding the perception of very small software companies towards the adoption of process standards. In A. Riel, R. O'Connor, S. Tichkiewitch, and R. Messnarz, editors, *Systems, Software and Services Process Improvement*, volume 99 of *Communications in Computer and Information Science*, pp. 153–164. Springer, Berlin, Heidelberg, 2010.

7. D.J. Rinehart, J.C. Knight, J. Rowanhill, and D. Computing. *Current Practices in Constructing and Evaluating Assurance Cases with Applications to Aviation*. National Aeronautics and Space Administration, Hampton, VA, 2015.

8. I. Habli and T. Kelly. Process and product certification arguments: Getting the balance right. *ACM SIGBED Review*, 3(4):1–8, 2006.

9. RTCA. Software considerations in airborne systems and equipment certification. *RTCA Standard DO-178B*, 1992.

10. RTCA. Software considerations in airborne systems and equipment certification. *RTCA Standard DO-178C*, 2012.

11. P. Bishop and R. Bloomfield. A methodology for safety case development. In F. Redmill and T. Anderson, editors, *Industrial Perspectives of Safety-Critical Systems*, pp. 194–203. Springer, London, 1998.

12. T. Kelly. Arguing safety—A systematic approach to managing safety cases. PhD thesis, University of York, September 1998.

13. GSN Community. GSN community standard, 2011.

14. J. Rushby. Understanding and evaluating assurance cases. (SRI-CSL-15-01), 2015.

15. P.J. Graydon, J.C. Knight, and E.A. Strunk. Assurance based development of critical systems. In *Dependable Systems and Networks, 2007. DSN '07. 37th Annual IEEE/IFIP International Conference on*, Edinburgh, UK, pp. 347–357, June 2007.

16. J. Hatcliff, A. Wassyng, T. Kelly, C. Comar, and P. Jones. Certifiably safe software-dependent systems: Challenges and directions. In *Proceedings of the on Future of Software Engineering*, pp. 182–200. ACM, Hyderabad, India, 2014.

17. S.E. Toulmin. *The Uses of Argument*. Cambridge University Press, Cambridge, Edinburgh, 2003.

18. R.E. Bloomfield, B. Littlewood, and D. Wright. Confidence: Its role in dependability cases for risk assessment. In *Dependable Systems and Networks, 2007. DSN'07. 37th Annual IEEE/IFIP International Conference on*, Banff, Canada, pp. 338–346. IEEE, Edinburgh, 2007.

19. E. Denney, G. Pai, and I. Habli. Towards measurement of confidence in safety cases. In *Empirical Software Engineering and Measurement (ESEM), 2011 International Symposium on*, pp. 380–383. IEEE, Banff, Canada, 2011.

20. J. Goodenough, C.B. Weinstock, and A.Z. Klein. *Toward a Theory of Assurance Case Confidence*. 2012. Software Engineering Institute, Technical Report, http://repository.cmu.edu/sei/679/.

21. J.B. Goodenough, C.B. Weinstock, and A.Z. Klein. Eliminative induction: A basis for arguing system confidence. In *Proceedings of the 2013 International Conference on Software Engineering*, pp. 1161–1164. IEEE Press, San Francisco, 2013.

22. S. Grigorova and T.S.E. Maibaum. Argument evaluation in the context of assurance case confidence modeling. In *Software Reliability Engineering Workshops (ISSREW), 2014 IEEE International Symposium on*, pp. 485–490. IEEE, Naples, 2014.

23. S. Grigorova and T.S.E. Maibaum. Taking a page from the law books: Considering evidence weight in evaluating assurance case confidence. In *Software Reliability Engineering Workshops (ISSREW), 2013 IEEE International Symposium on*, pp. 387–390. IEEE, Pasadena, 2013.

24. C.B. Weinstock, J.B. Goodenough, and A.Z. Klein. Measuring assurance case confidence using Baconian probabilities. In *Proceedings of the 1st International Workshop on Assurance Cases for Software-Intensive Systems*, pp. 7–11. IEEE Press, San Francisco, 2013.

25. P. Joannou and A. Wassyng. Understanding integrity level concepts. *Computer*, 47(11):99–101, 2014.

26. E. Denney and G. Pai. A formal basis for safety case patterns. In *Computer Safety, Reliability, and Security*, pp. 21–32. Springer, Toulouse, 2013.

27. S. Yamamoto and Y. Matsuno. An evaluation of argument patterns to reduce pitfalls of applying assurance case. In *Assurance Cases for Software-Intensive Systems (ASSURE), 2013 1st International Workshop on*, pp. 12–17. IEEE, San Francisco, 2013.

28. V. Cassano and T.S.E. Maibaum. The definition and assessment of a safety argument. In *Software Reliability Engineering Workshops (ISSREW), 2014 IEEE International Symposium on*, pp. 180–185. IEEE, Naples, 2014.

29. R. Hawkins, T. Kelly, J. Knight, and P. Graydon. A new approach to creating clear safety arguments. In C. Dale and T. Anderson, editors, *Advances in Systems Safety*, pp. 3–23. Springer, Southampton, 2011.

30. J. Rushby. Formalism in safety cases. In C. Dale and T. Anderson, editors, *Making Systems Safer*, pp. 3–17. Springer, Bristol, 2010.

31. R. Weaver, G. Despotou, T. Kelly, and J. McDermid. Combining software evidence: Arguments and assurance. In D. Budgen and B. Kitchenham, editors, *ACM SIGSOFT Software Engineering Notes*, vol. 30, pp. 1–7. ACM, St. Louis, 2005.

32. R. Weaver, J. Fenn, and T. Kelly. A pragmatic approach to reasoning about the assurance of safety arguments. In *Proceedings of the 8th Australian Workshop on Safety Critical Systems and Software—Volume 33*, pp. 57–67. Australian Computer Society, Inc., Canberra, 2003.

33. D.L. Parnas and P.C. Clements. A rational design process: How and why to fake it. *Software Engineering, IEEE Transactions on*, 12(2):251–257, 1986.

34. P.J. Graydon, J.C. Knight, and E.A. Strunk. Assurance based development of critical systems. In *Dependable Systems and Networks, 2007. DSN '07. 37th Annual IEEE/IFIP International Conference on*, pp. 347–357, Edinburgh, June 2007.

35. T.P. Kelly and J.A. McDermid. A systematic approach to safety case maintenance. *Reliability Engineering and System Safety*, 71(3):271–284, 2001.

36. G. Despotou, S. White, T. Kelly, and M. Ryan. Introducing safety cases for health IT. In *Proceedings of the 4th International Workshop on Software Engineering in Health Care, SEHC '12*, pp. 44–50, Piscataway, NJ, 2012. IEEE Press.

37. A. Wassyng, T. Maibaum, and M. Lawford. On software certification: We need product-focused approaches. In *Foundations of Computer Software. Future Trends and Techniques for Development*, pp. 250–274. Springer, Berlin, Heidelberg, 2010.

38. E. Jee, I. Lee, and O. Sokolsky. Assurance cases in model-driven development of the pacemaker software. In T. Margaria and B. Steffen, editors, *Leveraging Applications of Formal Methods, Verification, and Validation*, pp. 343–356. Springer, Heraklion, 2010.

39. D.L. Parnas. On the criteria to be used in decomposing systems into modules. *Communications of the ACM*, 15(12):1053–1058, 1972.

40. N. Leveson. The use of safety cases in certification and regulation. *Journal of System Safety*, 47(6):1–5, 2011.

41. A.B. Hocking, J. Knight, M.A. Aiello, and S. Shiraishi. Arguing software compliance with ISO 26262. In *Software Reliability Engineering Workshops (ISSREW), 2014 IEEE International Symposium on*, pp. 226–231. IEEE, Naples, 2014.

42. C.M. Holloway. Making the implicit explicit: Towards an assurance case for DO-178C. In *Proceedings of the 31st International System Safety Conference (ISSC)*, 11 pages, Boston, 2013.

43. J. Knight. Advances in software technology since 1992. In *National Software and Airborne Electronic Hardware Conference*, 46 slides, Denver, FAA, 2008.

44. W.G. Vincenti. *What Engineers Know and How They Know It: Analytical Studies from Aeronautical History*. The Johns Hopkins University Press, Baltimore, MD, 1990.

45. A. Wassyng, T. Maibaum, M. Lawford, and H. Bherer. Software certification: Is there a case against safety cases? In R. Calinescu and E. Jackson, editors, *Foundations of Computer Software. Modeling, Development, and Verification of Adaptive Systems*, volume 6662 of *Lecture Notes in Computer Science*, pp. 206–227. Springer, Berlin, Heidelberg, 2011.

CHAPTER 14

Measurement-Based Identification of Infrastructures for Trustworthy Cyber-Physical Systems

Imre Kocsis, Ágnes Salánki, and András Pataricza

CONTENTS

14.1 CLOUD-CPS: EXTRA-FUNCTIONAL DESIGN CHALLENGES[*]

Embedded systems are present everywhere in our everyday life. They interface with the physical environment via sensors and actuators—jointly referred to as transducers. Embedded systems have an extremely large spectrum of applications and, correspondingly, fall into a wide range of functional and architectural complexity. The simplest are smart autonomous transducers; at the other extreme we can find interconnected critical infrastructures under joint supervision and control (like the networked energy production and distribution system).

The earliest applications already served such networked embedded systems. For instance, in the late '60s, autonomous measurement devices connected to a central computer (calculator) via a dedicated information exchange bus (e.g., HP-IB, the Hewlett-Packard Interface Bus, standardized as IEEE-488). This hierarchy has clear advantages: autonomous units perform general-purpose measurement/control tasks via well-defined service and physical interfaces, and the application-specific parts of the composite problem are best solved by software solutions that are easy to change. Naturally, the potential scope of applications was limited by the properties of the measurement arrangement, such as the transducers included into the configuration, and the processing speed limits stemming from communication and program execution (instead of using dedicated hardware).

This service integration principle became the fundamental paradigm for building complex embedded systems in the last decade. One remarkable observation was that the centralized intelligence in the supervisory computer dominantly determined the scope and quality of the service provided to the end user. In addition, modernization can be carried out by upgrading the central intelligence and its underlying platform, as the useful operational time of the transducers tends to be very long.

This trend continues in the Internet age and with the proliferation of cloud computing—the availability of computing power as utility. The emerging new class of Cyber-Physical Systems (CPS) aims to seamlessly integrate the cyber and physical worlds. The physical world is still observed and influenced via embedded systems and transducers. The cyber world provides the (elastic) resources for sophisticated intelligence—and the possibility for integrating external knowledge. For instance, an intelligent home control solution may rely on remote services for centralized supervisory control, which in turn may exploit weather forecasts provided by a professional meteorological service provider.

The overall intelligence of the entire application depends on the computing power that is available for data processing and control. Cloud computing meets the increasing demand in

---
[*] This chapter was partially supported by research scholarships on cloud performance evaluation, awarded to the authors by the Pro Progressio Foundation and funded by Ericsson Hungary Ltd.

an affordable and agile form—if no specific timeliness guarantees are needed. Thus, CPSs are increasingly becoming amalgamations of classic, embedded in-field devices and various flavors of cloud computing.

While cloud platforms bring immense computational power into a CPS, they also introduce system-level resilience [1] assurance challenges that are entirely new for embedded systems as well as CPS. One core problem to be addressed is that the "central computer" is now a resource shared across applications and different users, with limited user control—a stark contrast to the dedicated systems used in traditional embedded systems.

Traditional embedded systems establish guarantees for the fulfillment of extra-functional requirements during design time, by selecting an appropriate architecture and deployment. Each instance of a particular embedded system simply reimplements these design decisions. In contrast, CPSs incorporating cloud services lack such guarantees; the general-purpose cloud platforms provide at most loose and incomplete Service Level Agreements (SLAs) on platform service quality. The deployment of several system instances cannot rely on the infrastructure consistently having the same service quality; thus, identifying service quality errors and monitoring for them becomes necessary. Even the service quality of a single platform instance can change over time—for example, due to insufficient performance separation of the users and their applications. It follows that extra-functional requirements can be guaranteed only by using runtime assurance mechanisms.

However, runtime monitoring and control based on it need an underlying runtime state modeling methodology for rigorous assurance design (a topic that the next chapter will cover in detail). IT system management (and general-purpose cloud application management) typically uses expert-determined thresholding of "best practice" system metrics for that. However, this cannot be adapted to the cloud-CPS context, due to a combination of increased criticality, dynamic system composition, timeliness criteria, and a range of hard-to-anticipate faults in the platform behavior. Such a level of complexity needs a model-based approach in order to keep track of all the aspects and their changes. Also, to make adaptation to different runtime environments and changes feasible, a general model that summarizes the key properties of the underlying phenomena and their management is needed.

Engineering thinking is typically of a qualitative nature. For instance, for Ethernet-based communication a network engineer differentiates between the qualitative utilization domains of "*low traffic,*" "*overloaded,*" and "*saturated.*" Each has its own characteristics and (use-case driven) management responses. The domains are separated by *landmarks*; for instance, one may state that "proper" operation is assured until we switch from the first to the second domain at, say, 40% workload. Note that management policy is entirely independent of the actual quantitative value inside a particular domain. Also, if management rules are based on the relative workload metric value of 40%, adaptation to the particular bandwidth of a specific system becomes a simple task.

Qualitative Reasoning (QR), a well-established technique in artificial intelligence, fits this approach of describing system behavior and its control. Classically, QR was primarily used for modeling physical systems [2]. Lately, it has seen applications in other fields, too (e.g., ecology [3]). Here, we propose using QR for describing and managing the extra-functional behavior of cloud platforms and applications that form part of a CPS.

A qualitative model that, using qualitatively different operational domains, reflects the main behaviors and phenomena in a system at a highly abstract level is basically the "core knowledge" needed for system management. It can be applied to a particular runtime environment and its actual state by mapping the landmarks into concrete quantitative values in the system. This way, a QR model is actually a generalized form of the most widely used approach in system monitoring: making quantitative measurements, quantizing them into qualitative domains, and detecting *"situations"* from domain combinations.

In this chapter, we focus on the modeling problem. First, we provide an overview of the most typical metrics to be used for system identification in order to assure compliance of trustworthy CPS applications to their respective extra-functional requirements. Subsequently, we introduce QR and show how it can form the basis of monitoring the "cloud part" of a CPS using a simple case study. In the following, we will refer to CPSs that significantly depend on cloud-deployed applications as *cloud-CPS*, when we want to emphasize this property.

14.2 CLOUD SERVICE QUALITY

A cloud-CPS has overall extra-functional qualities: trustworthiness and resilience properties form a major part of these qualities. The quality of the included cloud-deployed services and applications influences these directly. Ensuring the quality of cloud-deployed applications focuses on the runtime control of application Key Quality Indicators (KQIs). Here, we use the definition of KQIs [4] being a subset of KPIs (Key Performance Indicators) across the customer facing service boundary that characterize key aspects of the customer's experience and perception of quality. In this context, KPIs mean "technically useful service measurements" in a broad, but technical sense. Cloud platform behavior evidently influences application KQI; this way, observing and predicting the KQIs of the "cloud platform service" forms a major part of the control problem.

Note that managing the effect of deviations in platform KQIs is much more important in a cloud setting than in classic, nonvirtualized environments. The next chapter includes an evaluation of platform failure modes in Infrastructure as a Service (IaaS) [5] settings. "Hard" failures will form a distinct minority; most failures are actually deviations from the application-expected value of a *cloud service* KQI.

This section provides an overview of cloud-platform quality modeling with runtime application quality control as the intended application. Recognizing the deficiencies of the state of the art, we introduce a new runtime quality model for IaaS cloud platforms. We also touch on the basic cloud application KQIs for cloud-CPS.

These deliberations are largely descriptive; their main goal is to establish the quantitative domains over which we propose the use of qualitative models in the next section.

14.2.1 Application KQIs

What are, then, the primary KQIs for cloud-hosted applications that are part of a cloud-CPS? We do not yet have a definite answer for this question; however, we do have established best practices for telecommunications (telco) applications. Telco applications are very similar to CPS in the sense that they provide critical, interactive or (soft) real-time functionality and they do not tolerate network impairments well without additional design for dependability. For virtualized telco applications, Reference 6 defines the following (not application-specific) KQIs—in addition to the generic aspects of availability, latency, reliability, and throughput:

- Service accessibility: Largely as a synonym for availability on demand.

- Service retainability: For session-oriented services, the probability of a session remaining fully operational until user-initiated termination.

- Service timestamp accuracy: The ability of the application to accurately timestamp events.

For CPS, we can complement this list with *service timeliness*: the ability of the service to perform operations with the exact deadlines expected (and possibly specified) by its users.

14.2.2 Existing Models for Cloud Platform Quality

As in any context where business critical functionality has to rely on externally sourced services, the need for defining and evaluating service quality naturally arose for IaaS, too. Established IT Service Management approaches for SLA and availability management—for example, those included in the IT Infrastructure Library (ITIL)—have long recognized that for externally sourced services, many key activities should be based on selecting and prioritizing service attributes and measuring the metrics that best express these activities. Such activities include selecting a service, agreeing on SLAs with the provider, and tracking compliance or runtime control of the services used upon service impairments. This way, the effects of various disruptions in the externally sourced services become much more tractable and the risks threatening the overall system more manageable.

In recent years, industry and academia have made considerable progress in capturing notions of cloud service quality for general-purpose usage of IaaS. For performance and availability, benchmarking IaaS, (virtualized) resource capabilities, and service availability tracking are established fields by now, with readily usable commercial offerings. For other service aspects—including for example, usability, reliability, maintainability, or security—various quality models have been proposed in research papers as well as in standardization efforts. Still, none of the frameworks seem to be able to address the runtime KQI modeling needs of CPS application design and management.

14.2.2.1 Quality Modeling Concepts

First, let us introduce a number of basic quality modeling concepts that we will use. Regardless of the subject system or service, the main goal of their quality modeling is either to control their quality or to support their comparison. Software quality modeling has long recognized that addressing both requirements has to rely on multiple types of quality models. A useful categorization is the definition, assessment, and prediction (DAP) model [7] that distinguishes quality models by purpose. With some simplifications:

Definition models: Establish quality factors (usually hierarchically) and provide their meaning. Their main goal is to define (usually through decomposition) what quality means; the appropriate corresponding measures are usually so context-dependent that it is only logical to define the basic concepts separately.

Assessment models: Map directly (and numerically) assessable metrics and measures to quality factors, including hierarchical aggregation.

Prediction models: Use quality factor metrics and measures of quality drivers to predict quality.

Note that the separation between definition and assessment model is not always clear in practice. Notions that are closer to being a metric (not much has to be added for a full measurement specification) are sometimes part of a definition model, and conversely, properties specifying the exact means of measurement can be represented as metrics.

14.2.2.2 Standard Models

For general-purpose as well as telco-oriented cloud computing, a number of standard and quasi-standard quality models have been already published. In this subsection, we briefly review these.

The International Organization for Standardization (ISO) released a software product quality model standard in 2011: ISO/IEC 25010 [8] (a successor of the standard 9126). ISO/IEC 25010 is a *definition* model with accompanying standards and guidelines for assessment and prediction. It establishes a hierarchy of quality factors (*characteristics*, *subcharacteristics*, and *sub-subcharacteristics*) for software products and computer systems that may be directly calculable by a measure (note that quality factor is not a concept in the standard). If direct evaluation is not possible, then so-called *quality properties* (that have measures) can be attached to the factors.

The standard tries to model the quality of software and systems in development as well as "software in use"—that is, deployed, running, and used by clients. Reference 9 presents a systematic evaluation of current web service quality model proposals (including standards), with an analysis of the coverage of the ISO/IEC 25010 quality factors. Even more importantly for our purposes, [10] sets up a "performance measurement framework" for cloud computing services with a quality model part that heavily uses ISO/IEC 25010; however, the factor selection is only cursorily justified. (The authors of this work have other reservations regarding this quality model, too; for example, treating dependability essentially as a subcharacteristic of performance is a curious choice.) Reference 11 provides an example for setting up an assessment model for the "timeliness" characteristic of a private IaaS service.

The Service Measurement Index (SMI) is "*a framework of critical characteristics, associated attributes, and measures that decision-makers may apply to enable comparison of non-cloud based services with cloud-based services or cloud services available from multiple providers.*"[*] It provides seven characteristics and subcharacteristics of these; additionally, a

---

[*] Source: http://csmic.org.

number of measures are also provided (meaning that partial assessment model are available for the definition model). References 12 and 13 attempt to apply SMI for IaaS.

Network Function Virtualization (NFV) is the ongoing effort in the telco domain to enable hosting a wide range of telecommunications functions—from home gateways to VoIP "switching"—on mostly IaaS clouds. The European Telecommunications Standards Institute (ETSI), the most important standardization body for NFV, has released a specification for NFV Service Quality Metrics [14]. That metric taxonomy is not exhaustive; it focuses on the select few aspects that are the most important in the industry right now. That said, the specification includes a key methodological aspect: it arranges the metrics in a service/life-cycle category and quality criterion matrix. The quality criteria of speed, accuracy, and reliability are simply a subset of the standard eight telco service quality criteria.[*] For cloud services, the hierarchical decomposition and refinement of quality factors necessarily has to happen along multiple aspects; at the very least, the service-type hierarchy and various service quality attributes (Reference 15 employs the same technique specifically for IaaS performance capacity).

Although mainly a research effort and not a standard, we will build on multiple insights of Reference 16. Reference 16 sets up a VM-focused, abstract performance capacity feature model in the earlier introduced terminology, a performance capacity quality model for IaaS. Accompanied with References 15 and 17, an assessment model is also available for performance assessment through planned benchmark experiments. For our purposes, these works stand out from the very rich literature of cloud performance benchmarking exactly due to the establishment of an explicit quality factor and assessment model. (A relatively recent, very extensive, systematic review of experiment-based cloud evaluation can be found in Reference 18.) Reference 19 sets up a conceptual cloud SLA framework, including specific metrics for IaaS, PaaS, and SaaS.

14.2.2.3 Critical Applications

The apparent abundance of IaaS quality models—and to a lesser extent, assessment models—is misleading, if the goal is designing and controlling resilient critical applications that are deployed on clouds. Reference 6 demonstrates that for timeliness-sensitive and critical applications, the main known threats towards delivering application service quality are to be found in the *fine grained* and *runtime* behavior of the cloud platform. Characterizing this behavior leads to a need for a quality model that focuses on cloud service *runtime* KQIs.

From the above-cited quality modeling approaches, neither ISO/IEC 25010 nor SMI is readily usable for cloud service KQI modeling. For both approaches, specializing the set of their very general quality characteristics (and the refinement of these) down to the runtime KQI level is nontrivial and involves a very high level of ambiguity and guesswork. We hasten to add that broad-scope quality modeling with deep hierarchies is a fully justified approach for other purposes. A prime example is ranking the *overall* quality of cloud services from financial aspects through usability to (average) performance for hosting general-purpose applications.

---

[*] See the specification ETSI EG 202 009-1 V1.2.1 (User Group; Quality of telecom services; Part 1: Methodology for identification of parameters relevant to the users).

In contrast, Reference 15 as well as the ETSI NFV QoS recommendation provide quality models that not only specify runtime KQIs, but also acknowledge that taxonomizing these has to happen necessarily along multiple, mostly independent aspects. Reference 15 uses two or three dimensions for service type, capacity type and potentially statistical behavior; for example, [*Computation; Throughput; Variance*] is the time-invariance of the computational throughput-capacity of a virtual machine. Notable here is the "structural" decomposition along service type—although only Virtual Machines (VMs) already in the running state are considered—and the introduction of statistical behavior (for evolution in time, across instance populations and under resource scaling, respectively).

The ETSI NFV QoS specification introduces a categorization that is based on the life-cycle phases of leased resources (in addition to the property to be expressed). This way, for example, VM Provisioning Latency is a KQI (and at the same time, also a metric) expressing the "*Speed*" of "*Orchestration Step 1.*" One of our main ideas for setting up our KQI model will be to use *both* decompositions with standard-driven taxonomies (in that respect, both proposals are lacking). At the time of its publication, Reference 19 provided a usable and practical SLA framework (that by its nature has to include quality definition and assessment modeling); however, it is not applicable for critical applications and does not provide a usable multi-aspect factor framework.

14.2.3 A Runtime Quality Model for Cloud-CPS

We propose a new runtime quality model for critical applications. The model has the following base characteristics:

KQI modeling through attribute vectors: A number of KQI attributes are defined and (base) taxonomies for each attribute are established. The attributes are orthogonal to a great extent, meaning that in a given technical and/or business context, the various attribute-combinations can be enumerated and their importance (or whether they make sense) can be determined in a structured way.

Resource, service, and quality type attributes: As we will demonstrate, for our purposes all aspects of a tenant using an IaaS offering can be interpreted as he or she using a service defined on a resource. Using the capabilities of a leased resource and requesting actions to be performed on one both fit this model. These two attributes are accompanied by a quality characteristic, providing our primary KQI attribute set.

Secondary attributes: The intended (default) meaning of a triple of our primary attributes is the expression of the chosen characteristic at a given point in time, for a single resource service usage engagement (however, that can be continuous, not only transactional—see later) on a single resource instance. Following Reference 15, we lift out the spatial and temporal scope of the attribute into an explicit secondary attribute and the quality defined over a scope (e.g., variability over time, or central tendency over population) into a further secondary attribute.

Quasi-flat structure: While the taxonomies given for the individual attributes are hierarchical, only leaves of the trees are valid attribute values. This is to facilitate the capture of those properties that are directly measurable (simplifying assessment modeling to mapping the in the given context important KQIs to metrics) and directly influence application-level nonfunctional behavior. In the following subsections, we introduce our taxonomies for the three primary KQI attributes and for the three secondary attributes.

14.2.3.1 Resources and Services

The same way as "infrastructure" is an umbrella term for computing, communication, and storage services of on-premise environments, IaaS offerings consist of a wide range of services. When IaaS quality is investigated in the context of a deployed application, this composite nature cannot be readily abstracted away and quality defined on "the" service; the important quality aspects will depend on the usage of the component services by an application, the service impairment sensitivities of the application and the implemented control mechanisms.

By today, IaaS has become so well-established that functionally, the expected services and capabilities can be enumerated in a provider-independent way with a good coverage of the actual offerings. Importantly, the International Telecommunication Union (ITU) has even released a functional requirement definition document for IaaS as an official ITU-T recommendation [20]. We chose to use this as a basis for our IaaS service taxonomy.

As can be expected, the standard defines *resources* and *tenant capabilities* provided in the context of instances of each resource type. Multiple capabilities are generic in the sense that they are defined for more than one resource type; for example, *reservation* is recommended to be supported for storage resources as well as VMs. This basic insight gives rise to the generalized IaaS taxonomy presented in Figure 14.1. We subdivide tenant capabilities into the following service types:

Resource service types: Service types that are provided in the context of a resource instance.

Resource service properties: Properties of services that in themselves do not justify introducing a new service (e.g., the ability to instantiate a resource from a template).

Nonresource services: Services that are not defined in the context of a resource type instance.

In this generalized framework, the actual *resource services* that may be offered by an IaaS solution are pairings of resource types and resource service types. We hasten to add that not all combinations are technically meaningful.* Note that we also abstracted away the required, recommended or optional nature of the individual requirements, as specified by the standard.

---

* Due to lack of space we have to omit presenting our analysis of the combinations that are (a) standard-defined; (b) not defined by the standard, but potentially meaningful; and (c) technically nonsense.

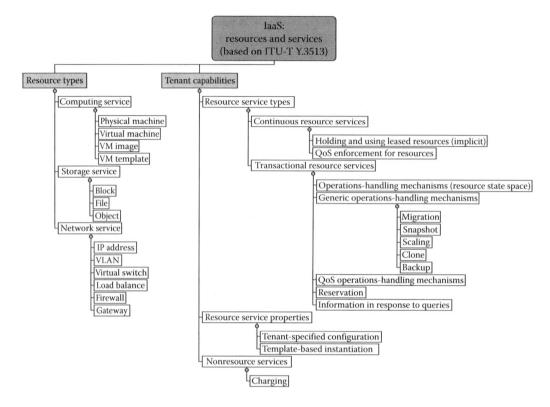

FIGURE 14.1   A generalized IaaS taxonomy based on ITU-T Y.3513.

Unsurprisingly, resource service types (in the above sense) very clearly dominate this definition of IaaS by functional capabilities. It is also apparent that resource service types fall into two fundamentally different categories: *continuous* and *transactional*. With rough simplification: the ability to use acquired resources and the ability to *change* these are both service types provided for the tenant, but the first is of "ongoing" nature, while we expect service requests of the second category to finish as soon as possible.

Numerous service types in our taxonomy are categories themselves, with contents differing from one resource type to another. Most importantly, the specific operations handling mechanisms (effectively the transitions of the state space of a resource) can be established only after specifying the resource type. Table 14.1 displays those prohibited or suggested by the recommendation.

TABLE 14.1   IaaS Resource Operations Handling Mechanisms (ITU-T Y. 3513)

Resource Type	Operations Handling Mechanisms
Physical machine	Start, shutdown, hibernate, wakeup
Virtual machine	Create, delete, start, shutdown, suspend, restore, hibernate, wakeup
VM Image	Add, import, store, register, deregister, query, update, delete, export
VM template	Upload, update, disable, enable, query, delete
Storage	Create, attach, detach, query, delete
IP address	Apply, bind, unbind, query, release

We reasoned that resource type and resource service type can be to a great extent treated as largely orthogonal dimensions and actual resource services being defined by combinations of these. (There are certainly technically meaningless combinations.) Notice that from the point of view of KQI modeling, they are conceptually very much the same (requesting an operation to be performed on a resource); therefore, we opt to simply further annotate with an additional dimension these KQIs—one with a domain depending on the valuation of the other dimensions.

Additionally, certain resource types behave as *resource type bundles* from the point of view of influencing the QoS of deployed applications; this way, can and arguably, should be decomposed further. This is most apparent for the computing services; that is, physical machines and virtual machines in our taxonomy. For instance, a virtual machine is equipped at least with (virtual) CPU cores and (virtual) memory, even if we treat VM block storage devices and network interface cards as part of other services.

We propose to handle these resource aggregates in a similar way to operational handling mechanisms (i.e., with an "extra" dimension, the value set of which depends on the valuation of the resource type and resource service type and other dimensions). VMs are decidedly the most problematic and most important from this point of view for IaaS. We propose to decompose them into virtual CPU or virtual CPU core, virtual memory, virtual block storage device, virtual network interface, virtual clock, and virtual GPU.

Note that this taxonomy does not explicitly distinguish a VM potentially using physical resources directly (as allowed by PCI pass-through); however, conceptually these cases *are* covered by these categories.

14.2.3.2 Quality Characteristics

Figure 14.2 represents our quality type taxonomy. Here we omit defining the quality types, as they are well known and/or self-explanatory.

This is not by accident. We aim at modeling KQIs—quality characteristics that can be directly measured or estimated by technical means; if a quality factor needs an elaborate explanation and its assessment sophisticated computation over observations, then chances are it is not a KQI. (Or it is not a good one.)

First and foremost, the taxonomy aims to capture *dependability*, *operational precision* (i.e., timing, motivated by Reference 6), and *performance capacity* attributes [17]. Notice that the last category is not simply called "performance." As we talk about cloud platform KQIs, the truly interesting question is that at any given point in time, what is the capacity of the platform that a tenant can potentially utilize.

Semantically, actual performance observations of speed, latency, and throughput form a subset of this notion—the operation throughput capacity of a virtual storage at a given moment is at least as much as what we observe. *Security* is present as a category, but without leaves; we are still investigating whether similar qualities—that almost directly translate to well measurable technical metrics—can be gathered for it at all.

These categories all express platform service *trustworthiness*. Using the framework of Reference 21, we also decided to include attributes from the other major aspect of resilience: *challenge tolerance*. Again, these concepts are widely known, with multiple competing

FIGURE 14.2   A taxonomy for quality types in IaaS KQI modeling.

definitions at the abstract level. For KQI modeling, this is not a major problem—binding all other dimensions for a given KQI usually constrains their meaning satisfactorily enough. Also note that for cloud applications, these are platform qualities that actually do change at runtime (e.g., the remaining fault tolerance in a redundant cluster versus the initial fault tolerance) and the application may want to respond to these changes.

There is one major problem with these quality types: in general-purpose IaaS, they simply cannot be measured. In today's general-purpose clouds, these aspects are treated as KPIs by the providers and not shared with the tenants. The reason they are present is that exactly for the cloud platforms that will most likely host cloud-CPS applications this is expected to change; the NFV standards already mandate sharing various aspects of cloud substrate runtime status with the tenants. Finally, let us emphasize that we do not treat this taxonomy as final; as CPS design and operation methodologies mature it will need revisiting. However, to our knowledge it is a first in gathering the relevant runtime quality factors of IaaS in a structured way for the CPS application context.

The *secondary* attributes (see Figure 14.2) establish the temporal-spatial scope of a quality defined over a resource service and the statistical behavior of a quality that a KQI expresses over that scope. For instance, the well-known CPU steal time reported by the kernel in Linux virtual machines measures the CPU time "stolen" from a VM by the other VMs co-hosted with it. Continuously measuring it, this metric can be used to express the *variability* of the operational speed of a virtual CPU over a time interval; that is, how stable is the CPU performance of a VM. Notice that we do not name specific aggregation functions that belongs to metric selection and not quality modeling. Rather, we establish the most important (descriptive) statistical categories.

14.2.3.3 KQI Modeling with the Framework

In our framework, an IaaS KQI category is a vector of attributes:

$$k = \left[rt, rst, qt, op, prt, ts, ss, stat \right],$$

where

1. rt is a resource type

2. rst is a resource service type

3. qt is a quality type

4. op is an operational management mechanism name, if applicable

5. prt is a "part" name in a resource bundle, if applicable

6. ts is a temporal scope

7. ss a spatial scope

8. $stat$ is a statistical aspect of quality over scope kind

Let K be a set of KQI categories accepted to be technically meaningful and potentially important in a given technical context. For each $k_i \in K | i = 1 \dots |K|$, the set s_i carries the names of the KQIs that represent that combination in a given context. Let us call the set containing all s_i-s S.

This way, we can provide simple synonyms for attribute combinations instead of the long-winded vector representation. Also, we have to concede to the fact that depending on the context, a combination may not translate directly to a single KQI, but cover multiple ones (thus the term KQI category).

We say that an IaaS runtime KQI model is specified as $M = [K, S]$, that is, by a set of permitted attribute combination vectors and specific KQI names (possibly multiple) mapped to these.

14.2.4 Applications of the Quality Model

The quality model we defined here is primarily intended to serve as a generic basis for QR modeling, discussed in the next chapter. In order to keep the discussion easily tractable, our example will focus on modeling performance characteristics in the absence of platform deviations, connecting VM CPU performance capacity, application load, and application latency. Note, however, that QR modeling can easily incorporate all platform and application KQI types identified in the current section. Notably, the variability of VM CPU performance capacity can be modeled either directly, or incorporated into the example model of the next section via platform-induced CPU usage changes. After all, if the usable CPU capacity diminishes (e.g., is "stolen"), then user-mode CPU utilization will increase, if the load remains the same. Beyond this primary application, the KQI modeling approach we set up has other potential uses, too. We close this section by outlining these.

Configuring monitoring: A KQI model can be used to define the KQIs that have to be monitored for an application and inform the provider that these are the KQIs that are important for the application. The tenant can check whether all necessary KQIs are monitored; informed by the priorities of the application, the provider can opt to configure its platform so that applications are only (or mostly) exposed to such threats they are not sensitive for and also track the important measures.

    On the monitoring synthesis side, the quality model enables modeling sensors as well as monitoring requirements in the same structured framework, thus enabling a cost-optimizing search for monitoring sensor deployment that covers all requirements.

Application sensitivity assessment: Application design models can be traversed along the resource type and resource service type dimensions and a rough overapproximation of the KQI model for high-impact KQIs can be established. Following that, targeted fault injection measurements can help to further reduce the set of KQIs that have to be treated as threats.

    Due to the large KQI space, this can lead to large gains in rigorous sensitivity modeling. For instance, a single VM performing signal processing (e.g., sound transcoding) will be very sensitive for CPU cycle allowance stability (a continuous resource service inside a VM); on the other hand, for a continuously "right-scaled" web server cluster the availability, reliability and consistent performance of specific VM operations handling mechanisms will be far more important. CPU allowance instability might even be a nonissue due to QoS-based load balancing and the sheer number of hosts. Notice that in the first case, beginning with the question "do we depend on operations handling mechanisms?" leads to discarding a huge number of possible KQIs. We plan to investigate such "variable-ordering" problems in future work.

Application-aware platform evaluation: The state of the art of in-cloud platform evaluation focuses on application-independent performance benchmarking and dependability benchmarking for private clouds; specifically for NFV, see [22]. In reality, an application is truly sensitive only for a subset of the cloud services; it follows that platform evaluation, which can focus solely on these aspects. KQI models expressing application sensitivity can drive such targeted evaluation efficiently.

Application-aware platform ranking and selection: The state of the art for cloud platform selection is performing multiple criteria decision making (MCDM) on cloud quality characteristics [23]. We propose translating KQI models expressing application sensitivities as requirements into MCDM parameterization; for example, the relative priorities of quality factors in an Analytic Hierarchy Process.

Semantically enriched measurement storage and processing: Especially in the CPS field, describing and storing observations together with their metadata is rapidly becoming a fundamental technique. Such data includes, for example, the observed object, the observed property of the object, the sensor performing the observation, and the basic attributes of the quality (in the measurement theory sense) expressing the property (e.g., dimension). There are even established standards supporting this process (see [5,7]). Standards-based semantic observation data representation facilitates data exchange, but maybe more importantly, data stored in databases with semantic technology support (e.g., modern triple stores) enables truly agile and self-contained exploratory data analysis, too. Our ongoing work indicates that approach carries over with the same benefits to IT observation data, too. In this context, having a KQI "ontology" for expressing quality properties is a fundamental prerequisite.

14.3 QUALITATIVE REASONING OVER CLOUD-CPS BEHAVIOR

Separation of the invariant logic from the varying quantitative domains to which they apply is actually how an engineer tends to *think* and informally *reason* about the behavior of a system. A typical system manager supervising a computing system thinks primarily in qualitative terms: "if a particular machine is near to overload and its workload is increasing, it is time to take a corrective action by adding more resources to it or moving a part of its workload to another platform."

This small example represents a simple form of proactive control that assures the prevention of service-level faults by predicting faults from platform-level symptoms. Specific quantities play a role only in deciding whether a particular VM is in an "overloaded" state or not (i.e., situation recognition). Further reasoning uses only qualitative, that is, quantized discrete values of the VM load.

The resolution of the quantization in terms of the cardinality of the qualitative value domain may depend on the set of control objectives to be executed. If operation continuity of this VM is the only objective, then a simple binary not overloaded—in other words overloaded categorization—is sufficient. However, if the VM is a potential task migration target candidate for other, overloaded VMs, then it is reasonable to also introduce a category corresponding to the "underutilized" state.

14.3.1 Motivation for QR Modeling

Model-based CPS monitoring necessitates the observation, evaluation and checking of a large number of continuous entities. A core problem is the selection of a model that is reliable, but also simple enough to serve as a central entity in the supervisory monitoring and control system.

Pure discrete system models: Are unable to express the continuity of the variables; this way, their interfacing to observation and control functions is problematic.

Pure continuous models: Are able to capture the behavior of the continuous characteristics of the system, but they need a long elaboration/learning preparatory phase especially if fault impacts are anticipated as well. They cannot serve as a central entity in dependability and resilience management.

Accordingly, we propose the use of qualitative reasoning (QR) based models as the central models orchestrating supervisory monitoring and control. The main principle of QR is to focus on the qualitatively different operational domains in a continuous system. The core idea of the abstraction provided by QR models is the assumption that all states within of a particular operational domain behave identically.

For instance, all overloaded states in a VM are expected to have the same potential logic consequences like dropping requests, violating time limits, etc. This assumed equivalence of the behavior of all states within a particular operational domain allows the representation of all members in the domain by a single state variable in the abstract QR model. This way, the infinite continuous domain of the state variables in the continuous model is mapped into a single abstract state in the QR model indicating the actual operational domain.

A discrete event model describing the operational domain changes will be constructed over the set of abstract states. Such models exactly correspond to the representation of different situations in traditional supervisory systems of IT infrastructures. The major difference from the industrial practice is that we propose a systematic way to establish the QR model itself and determine its corresponding quantitative parameters for its instantiations. Also, QR models carry transitions, enabling formal reasoning about the possible system trajectories in the qualitative domain. QR models are capable of forming a static model of the supervised system by clearly identifying the system-level operational domains and expressing the existence and main characteristics of relations between individual and composite state variables.

14.3.2 Elements of QR Models

QR offers several notions (see Table 14.2 for the basic definitions and concepts of QR modeling) to map the main characteristics of a continuous model into a discrete one without an extreme level of complexity (Table 14.2). The QR paradigm introduces notions like monotonicity or proportional dependence among state variables. Similarly, the speed of change (the first derivative of a continuous entity) and occasionally the second derivative of it can be introduced into a model, as well. For further details on QR modeling, see [26,2].

TABLE 14.2 Basic Definitions and Concepts of QR Modeling

Name	Definition	Name	Definition
Entity	Physical objects or abstract concepts	Value assignment	Assignment of a qualitative to a magnitude or derivative value
Agent	External processes or actors	Influence	Source magnitude → target-derivative relation
		I+	
		I−	
Configuration *Configuration*	Structural relation between entities and agents	Proportionality	Source derivative → target-derivative relation
		P+	
		P−	
Point value	A particular quantitative value	Value correspondence	Simultaneous occurrences of two qualitative values
		V	
Interval value	A set of quantitative values	Quantity space correspondence	Simultaneous occurrences of quantity space values
		Q	
Quantity space	Ordered, alternating set of point (landmark) and interval (domain) values	Inequality	Relation between quantitative values
		≥	
Quantity	Changeable features of entities during simulation		
Magnitude	Current value of a quantity		
Derivative δ	Current changes of quantity value		

QR supports the creation of dynamic models involving different operation domains, as well. Transitions between qualitative states represent domain changes after a continuous variable crosses the boundaries (called landmarks) between two neighboring domains (continuity constraint).

14.3.3 Advantages in the CPS Context

QR has several advantages for modeling CPS

1. The general-purpose mathematical modeling paradigm is able to uniformly describe cyber and physical entities and their interrelations.

2. The QR model is well portable among different environments. For instance, the QR model of a saturation effect refers only to the existence of a landmark below which there is no saturation and above which the system is saturated. All the different systems generated by instantiating the same application under different environments may share the same model of logic, and only the actual values of the landmarks have to be fitted to the concrete system.

3. QR models can incorporate partial models and information and process them (thanks to the underlying paradigm of constraint programming). For instance, if some values of state variables in a QR model are unknown (because they are not measured or unmeasurable in a particular configuration) QR performs exhaustive simulation delivering all the potential behaviors of the model by taking the values of the known variables and traversing the solution phase over the entire qualitative domain of the unknown variable.

4. Finally, the QR model is simply a nondeterministic finite state model of a system. The entire spectrum of formal methods can be applied to answer a variety of common correctness and dependability-related questions or prove their correctness.

14.4 QR MODELING CASE STUDY

Using a simple case study, we show in this section how the QR model of an application can be established by analyzing runtime observations of its quality characteristics and the runtime platform of it.

14.4.1 Application and Experimental Environment

The pilot setup represented a multiclient service system performing several operations upon user request. This is a typical model used in a variety of applications, like those in telco, external sensor data processing in CPS, etc.

In the pilot workflow

i. Devices register themselves

ii. Send connection requests

iii. They terminate the connection after a specific time representing communication and processing

In the simulation model, these steps are refined into fine granular action sequences in order to truly represent the operation and its sensitivity to the different parametrization of individual phases. The capacity limitation in terms of requests concurrently serve models, typically effects related to resource saturation triggering unsuccessful execution in steps (i) and (ii).

Measurements are performed both at the service level (aggregated success rate of the requests and end-to-end service latency) in the clients serving as workload generator and at the technology level (fine-granular measurements on the progress of service request processing at the application level and on VM utilization as resource metrics). However, neither the details of the internal application nor the beyond VM level platform metrics are used in order to faithfully model the typical use case by and end-user of a public cloud. A series of campaigns was executed sweeping over the entire workload domain.

The extraction of a QR model on the base of measurements is composed by the following main steps:

  i. Exploration of the performance characteristics of the system

  ii. Determination of the operation domains

  iii. Estimation of interdomain transitions

  iv. Searching for hypotheses justifying these transitions

14.4.2 Visual Exploratory Data Analysis

Visual exploratory data analysis (EDA), a well-established and fast method for analyzing empirical data [1] was used to create the first QR model.*

The initial system identification follows typically the following main steps:

  1. Estimation of the initial set of operation domains and landmarks separating them from the individual observations.

      a. Distribution analysis searching for well-separated value domains delivers the initial partitioning of operational domains and their high-level characteristics.

      b. Analysis of the distribution of the first derivates of the state variables in their respective individual domains in the initial partitions prepares the inclusion of elements describing the dynamics of change impacts into the QR model.

  2. Time series analysis searches for potentially relevant platform metrics and parameter candidates subject of further analysis to reveal hidden timing parameters of the system.

  3. Exploration of multivariate relations extracts the interactions among different entities and characteristics in the system. For instance, relations among service and platform metrics estimated by correlation analysis can substantiate an initial monitoring model.

---

* Note that the usual follow-up phase of confirmatory data analysis in realistic experiments validating the hypotheses gained during EDA can perform simultaneously the checking the QR model, as well. The elaboration of details is subject of ongoing research work.

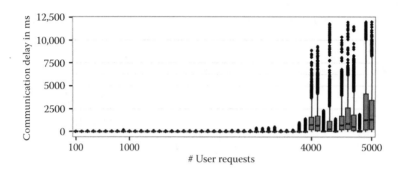

FIGURE 14.3 Distribution of communication delay along increasing user request number.

14.4.2.1 Landmark Hypotheses

The distributions of request processing times and the success rates indicate a change in the operational domain both at the same level of workload by around 4000 (Figure 14.3) thus Step 1 partitions their domains by this landmark. In the current case, the analysis of the derivate of the observations did not result in further domain refinements.

The same landmark appeared in the distribution of other observations like platform metrics, for example, the CPU usage of the virtual machine. Such a coincidence of landmarks of different state variables indicates a potential joint root of differences in the operation. All the individual state variables sharing landmarks are objects of later relation analysis between them.

14.4.2.2 Operational Domain Hypotheses

Multivariate relation analysis (Step 2) is illustrated by the heatmap in Figure 14.4. The two-dimensional distribution exposes a finer structure of domain 1 splitting it into subdomains 0, 1.1, 1.2, and 1.3, respectively in order to have the subdomains of a homogenous behavior. Observations in 1.1 suggest a linear CPU usage—number of user requests relation. Samples in 1.3 are rare, presumably being produced by system maintenance manifested as noise.

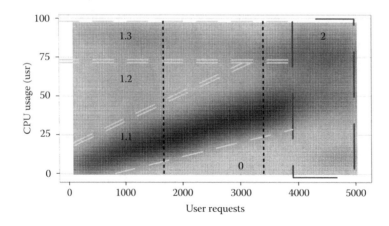

FIGURE 14.4 Heatmap of CPU usage (%) observations along increasing user request number.

Note that the domain partitioning can be further refined along the two dotted vertical lines in Figure 14.4. Here, the continuity constraint indicates that by around 3000 there is another landmark. Below this landmark, some external factor can cause 1.1 ↔ 1.2 and 1.2 ↔ 1.3 transitions as they are neighboring domains. Above this landmark only 1.1 ↔ 1.3 can occur.

14.4.2.3 Exploring Dynamics

The normal domain has a constantly low service delay while these values increase extremely after the landmark of 4000 (note the values in Figure 14.3). The most probable cause for the observed phenomenon is that near to saturation the client aiming to generate a constant workload resends the requests after a connection failure caused by capacity limitations of the server (Figure 14.5). This feedback loop leads to an oscillation.

In Step 3, a strong correlation was found between service quality and resource (CPU and network related) metrics. Due to the nonuniform strength of the correlation, the influence between individual platform metrics should be modeled as well.

14.4.2.4 The Impact of Experiment Design

Even this simple example indicates the role of careful design of the measurement campaigns for system identification.

- Visual EDA typically uses only a moderate subsample of logs. In the subsequent phase of confirmatory data analysis the initial landmarks have to be fitted to the entire pattern set together with the checking of the correctness of the clustering.

- The continuity constraint assumes that the state variables change gradually along a continuous trajectory. However, if measurements are based on undersampling, a sequence of transitions can appear as a direct state transition. For instance, if the intermediate state 1.2 escapes sampling, a 1.1 ↔ 1.3 direct transition can be observed.

- A special problem is exposed by the domain marked by 0 with no samples at all, meaning there is no representation in the QR model. An occurrence of a value from this domain during runtime monitoring indicates an error in the QR model or in the measurement.

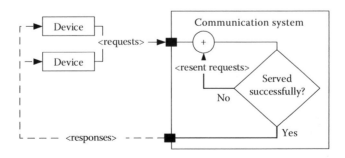

FIGURE 14.5   Dynamic model of the overloaded communication system.

- The incompleteness of the QR model may indicate an insufficiently exhaustive benchmark. In typical critical CPS applications this is intolerable as the designated high-level availability necessary to cope with rare events as well. At least theoretically, the extension of benchmarks with fault-injection can provide in the future some solution for this representativeness problem. Insufficient parametrization of the benchmark model can cause such problems as well. For instance, if the pilot VM is overpowered, potentially no saturation will occur thus resulting an insufficient representativity of the measurement campaign.

- Measurement errors are relatively frequent. In this case, redundant monitoring may help to overcome the problems by adapting a design for diversity approach.

14.4.3 Abstraction Generalization: Parametric QR Models

All the analysis described above was performed over a single setup in the terms of runtime platform parameters (e.g., maximal number of requests simultaneously served). However, this capacity bound changes with the size of the virtual machine. It can be estimated only during instantiation and deployment time knowing a platform to business capacity metrics mapping (e.g., defining the upper bound of concurrently served requests). The current best practice uses curve fitting over measurement results in different settings.

The creation of a proper model is performed typically by a domain expert, who assigns an interpretation to the different derived observations. For instance, in our running example the generated workload is represented exactly on the left side of Figure 14.6a to the observations in Step 1 and introduces "saturation boundary" as a quantitative value representing the landmark at 4000. Similarly, the different operational states get an assigned quantitative variable, Domain 1 is splitted into "Green" corresponding to the proper functioning in domain 1.1., "Yellow," (1.1 and 1.2) and malperforming "Red" (oscillations in domain 2).

The correspondence of the values of the generated workload in the over the saturation boundary interval and "Red" represents the fact that workloads above the landmark bring the system into "Red" (domain 2).

This initial model can be extended by relations describing the proportionality between workload and resource usage in Domain 1.1, the random occurrence of some OS activity moving the system to Domain 1.3, etc. (Figure 14.6b).

This way, QR helps to create reliable models between different continuous entities and their changing dynamic at a high level of abstraction, and at a price of only a low complexity. A well-elaborated process model on the elaboration of QR models is illustrated in Table 14.3 based on Reference 27.

QR-based models (especially when deriving from observations) are able to incorporate fault impacts in a phenomenological, assuming that the input observation data set consists of related samples. However, trusted solutions need a much more organized approach by generating mutations of the QR model of normal, fault-free operation according to the fault anticipated. In the next section, we address the problem of modeling typical platform failures (and their application effect) for cloud-integrated CPS in order to find a technological foundation for fault modeling.

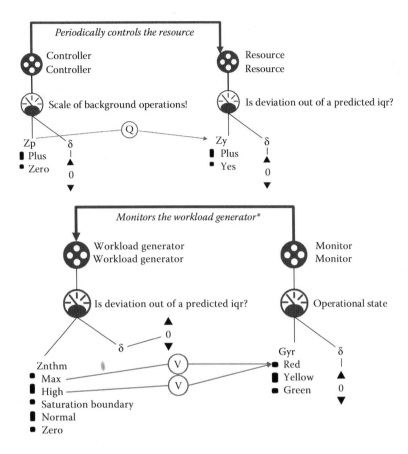

FIGURE 14.6   QR models of the (a) internal system maintenance process and (b) overload.

TABLE 14.3   Outline of Qualitative Modeling Steps

Modeled Feature	Terms Used	Introduced Simulation Result
1. System concept	Concepts, relations	—
2. Distinction of structure and behavior	Entities, configurations. quantities, causal relations, value assignments	Single state simulation
3. System behavior	Quantity spaces, correspondences	State graph, value history
4. Causes of changes	Influences, proportionalities, inequalities	—
5. Conditional knowledge	Conditional fragments	—
6. Generic and reusable knowledge	Model fragments, scenarios	Expression fragment, conditional fragments

14.5 SUMMARY

Today, a CPS is more and more likely to contain cloud-hosted application components. CPS trustworthiness (and more generally, resilience) is fundamentally influenced by the run-time quality of these cloud platforms and our ability to efficiently capture the response of applications to disturbances in these. This knowledge can then efficiently drive runtime resilience assurance—by monitoring platforms and applications and performing actions that are planned at runtime.

For this sake, we proposed Qualitative Reasoning as the underlying modeling formalism for application state and sensitivity modeling. To facilitate structured QR model determination we also defined an Infrastructure as a Service cloud quality model (that has other eminent uses in the cloud-CPS context, too). We did not touch on questions of *planning* assurance mechanisms: that is the purpose of the next chapter.

REFERENCES

1. A. Pataricza, I. Kocsis, Á. Salánki, and L. Gönczy. Empirical assessment of resilience. In A. Gorbenko, A. Romanovsky, and V. Kharchenko (Eds). *Software Engineering for Resilient Systems*, pages 1–16. Springer-Verlag, Berlin Heidelberg, 2013.

2. B. Kuipers. *Qualitative Reasoning: Modeling and Simulation with Incomplete Knowledge*. MIT Press, Cambridge Massachusetts, USA, 1994.

3. P. Salles and B. Bredeweg. Building qualitative models in ecology. In: L. Ironi (Ed.), *Proceedings of the 11th International Workshop on Qualitative Reasoning (QR97)*, Italy, June 3–6, pages 155–164, 1997.

4. TM Forum. Multi-Cloud Service Management Pack: Service Level Agreement (SLA) Business Blueprint, TR 197, V1.3., 2013.

5. P. Mell and T. Grance. The NIST Definition of Cloud Computing, 2011.

6. E. Bauer and R. Adams. *Service Quality of Cloud-Based Applications*. John Wiley & Sons, Hoboken, New Jersey, 2013.

7. S. Wagner. *Software Product Quality Control*. Springer-Verlag, Berlin Heidelberg, 2013.

8. International Organization for Standardization. IEC25010: Systems and software engineering—Systems and software quality requirements and evaluation (SQuaRE)—System and software quality models. 2011. Available at https://www.iso.org/obp/ui/#iso:std:iso-iec:25010:en. Accessed on September 14, 2015.

9. M. Oriol, J. Marco, and X. Franch. Quality models for web services: A systematic mapping. *Information and Software Technology*, 56(10):1167–1182, 2014.

10. L. Bautista, A. Abran, and A. April. Design of a performance measurement framework for cloud computing. *Journal of Software Engineering and Applications*, 5(2):69–75, 2012.

11. A. Ravanello, J.-M. Desharnais, L. E. B. Villalpando, A. April, and A. Gherbi. Performance measurement for cloud computing applications using ISO 25010 standard characteristics. In *Software Measurement and the International Conference on Software Process and Product Measurement (IWSM-MENSURA), 2014 Joint Conference of the International Workshop on*, Rotterdam, The Netherlands, October 6–8, pages 41–49. IEEE, 2014.

12. S. K. Garg, S. Versteeg, and R. Buyya. SMICloud: A framework for comparing and ranking cloud services. In *Utility and Cloud Computing (UCC), 2011 Fourth IEEE International Conference on*, Melbourne, Australia, pages 210–218. IEEE, 2011.

13. S. K. Garg, S. Versteeg, and R. Buyya. A framework for ranking of cloud computing services. *Future Generation Computer Systems*, 29(4):1012–1023, 2013.

14. ETSI. Network Functions Virtualisation (NFV); Service Quality Metrics, December 2014.

15. Z. Li, L. O'Brien, H. Zhang, and R. Cai. On a catalogue of metrics for evaluating commercial cloud services. In *Proceedings of the 2012 ACM/IEEE 13th International Conference on Grid Computing*, Beijing, China, pages 164–173. IEEE Computer Society, 2012.

16. Z. Li, L. O'Brien, R. Cai, and H. Zhang. Towards a taxonomy of performance evaluation of commercial Cloud services. In *Cloud Computing (CLOUD), 2012 IEEE 5th International Conference on*, Honolulu, Hawaii, USA, pages 344–351. IEEE, 2012.

17. Z. Li, L. O'Brien, H. Zhang, and R. Cai. On the conceptualization of performance evaluation of IaaS services. *Services Computing, IEEE Transactions on*, 7(4):628–641, 2014.
18. Z. Li, H. Zhang, L. O'Brien, R. Cai, and S. Flint. On evaluating commercial Cloud services: A systematic review. *Journal of Systems and Software*, 86(9):2371–2393, September 2013.
19. M. Alhamad, T. Dillon, and E. Chang. Conceptual SLA framework for cloud computing. In *Digital Ecosystems and Technologies (DEST), 2010 4th IEEE International Conference on*, Dubai, United Arab Emirates, pages 606–610. IEEE, 2010.
20. ITU-T. Y.3513: Cloud computing—Functional requirements of Infrastructure as a Service, 2010.
21. Resilience Metrics and Measurements: Technical Report ENISA. https://www.enisa.europa. eu/activities/Resilience-and-CIIP/Incidents-reporting/metrics/reports/metrics-tech-report. Accessed on September 14, 2015.
22. D. Cotroneo, L. De Simone, A. K. Iannillo, A. Lanzaro, and R. Natella. Dependability evaluation and benchmarking of network function virtualization Infrastructures, *1st IEEE Conference on Network Softwarization (NetSoft)*, Seoul, Korea, 2015.
23. L. Sun, H. Dong, F. K. Hussain, O. K. Hussain, and E. Chang. Cloud service selection: State-of-the-art and future research directions. *Journal of Network and Computer Applications*, 45:134–150, 2014.
24. M. Compton, P. Barnaghi, L. Bermudez, Raúl García-Castro, O. Corcho, S. Cox, J. Graybeal, M. Hauswirth, C. Henson, A. Herzog et al. The SSN ontology of the W3C semantic sensor network incubator group. *Web Semantics: Science, Services and Agents on the World Wide Web*, 17:25–32, 2012.
25. M. Botts, G. Percivall, C. Reed, and J. Davidson. OGC® sensor web enablement: Overview and high level architecture. In S. Nittel, A. Labrinidis, and A. Stefanidis (Eds), *GeoSensor Networks*, pages 175–190. Springer-Verlag, Berlin Heidelberg, 2006.
26. J. Liem. Supporting conceptual modelling of dynamic systems: A knowledge engineering perspective on qualitative reasoning, University of Amsterdam, 2013.
27. A. J. Bouwer. *Explaining Behaviour: Using Qualitative Simulation in Interactive Learning Environments*. Ponsen & Looijen, Wageningen, Netherlands, 2005.

CHAPTER 15

MDD-Based Design, Configuration, and Monitoring of Resilient Cyber-Physical Systems

László Gönczy, István Majzik, Szilárd Bozóki, and
András Pataricza

CONTENTS

15.1 INTRODUCTION*

As described in Chapter 14, one of the core challenges in the evolving category of cloud-based cyber-physical systems (CPS) is the assurance of a proper service over dynamic, and affordability of common-quality clouds.

Traditional embedded system design relies on a strict design time resource allocation in providing a sufficiently dependable platform for critical applications. In contrast, the dynamic nature of cloud platforms necessitates not only the deployment of functional components of applications, but also for an extension with appropriate measures to maintain the required level of service during the operation phase. In this way, the core problem addressed in the current chapter deals with the creation of a virtually "carrier grade" platform even if the underlying physical cloud is not of a distinguishably high service level. Provisioning a high Service Level Agreement (SLA) over inexpensive computing platforms is one of the problems addressed previously in the context of autonomic computing.

Kephart and Chess [1] identifies four main self-* strategies (self-configuration, self-healing, self-protection, and self-optimization) to achieve this goal:

> *Self-configuration* refers to the complex activities of fitting the requirements of applications to the characteristics of resources provided by the cloud. Note that one particular problem is originated by the very principle of resource sharing: the availability of free resources becomes a changing characteristics in contrast to the fixed resource allocation performed during the final phase of traditional embedded system design. Moreover, the isolation between independent applications in the performance domain is less perfect than the usual intertask isolation in the logic domain. The imperfect assurance of the exclusivity of the resources allocated to a particular application may lead to parasitic interferences, like the noisy neighbor problem potentially leading to catastrophic outages in soft real-time systems [2].

> *Self-healing* assures the early detection of errors and mitigation of faults in order to guarantee fault tolerance and resilience of the application. This means that runtime verification techniques both in monitoring the infrastructure and the application level are turning into a fundamental component in critical applications. While one of the main promises of cloud computing originates in the feasibility of using rough granular redundancy schemes, thanks to the availability of low-cost computational power, timeliness criteria require fast failover processes in order to reduce downtimes.

> *Self-protection* is used to limit and preferably avoid the impact of external incidents including security-related intrusions.

* This chapter was partially supported by the Pro Progressio scholarship for research on cloud performance evaluation, funded by Ericsson Hungary Ltd.

Self-optimization deals with the maintenance of an effective infrastructure utilization under changing tasks, both in terms of the set of applications to be served and changes in the workload induced by these in the cloud. One important new characteristic of cloud infrastructures is that the size of the problem to be managed excludes a global one-step optimization method (like design time optimization in traditional embedded system design [3]), and has to be substituted by a continuous and incremental (change-driven) optimization approach [4].

The requirements described above are valid for all kinds of cloud infrastructures. However, the evolving class of cloud-based CPS implies such specific needs and benefits of new opportunities that extend beyond the usual requirements for data centers processing pure data related applications. CPS applications by their very nature underlie at least soft real-time requirements. However, the use of public communication and computation resources is unavoidable in many cases and is unable to deliver hard timeliness guarantees [5].

A better solution is to use private cloud as the core computational engine for processing data delivered by the underlying network of embedded systems, including when they are as simple as intelligent sensors. Note that this concentration of computational power may compensate the strong limits of simple intelligent sensors in terms of both of computational power and energy consumption (consider battery-powered devices). The practical relevance of such private cloud-based CPS is well emphasized by the recent appearance of micro-clouds, small dedicated private clouds serving exclusively a limited amount of mobile communication and/or CPS applications. Naturally, self-optimization plays an even more important role in such limited capacity clouds as compared to large scale data centers, as proper utilization is a primary design requirement for the feasibility of the deployment during such tasks.

At the same time, it can easily be predicted that similar problems and solutions will appear in intelligent embedded systems and transducers. Virtualization as a means of separating the physical platform and the task to be solved also started to appear at the lowest level of CPS in the form of real-time virtualization. The benefits of resource sharing lead to better utilization in integrated distributed embedded systems; moreover, there are natural consequences for the widespread use of credit card-sized cheap computers at the lowest level of embedded systems as well.

A proper service level at the lowest level of the CPS architecture will lead inevitably to the introduction of the self-* properties in some downscaled form. Timeliness requirements imply the necessity of fast error detection and fault mitigation in order to assure proper resiliency after faults. Note that the set of faults to be anticipated fundamentally exceeds those in traditional data centers. CPS-based cloud applications are sensitive to all of the faults in traditional data processing applications; moreover, the central intelligence deployed over the cloud also has to take care of the faults of transducers.

This means that a large number of components may appear faulty even as a consequence of their normal operational characteristics. For instance, moving mobile sensors may pass out of communication range when leaving the environment of physical phenomena of interest. Energy supply in battery-powered devices is a major source of causing states of inactivity. Moreover, transducers distributed in the environment have frequently only a limited physical protection, thus, they are at risk for potential physical damages as well.

Accordingly, the core intelligence has to assure proper handling of such faults as an integral part of the operation. This means that faults in the central cloud platform play a dominant role in the assurance of dependability and resilience for the entire application as such faults can lead to service level violations. A major part of the anticipated faults is identical with those in pure data centers. For instance, self-protection against intrusions attacking the core cloud is a joint requirement for all cloud applications. These aspects will not be detailed further in this chapter. At the level of application, this means that depending on the particular purpose, high-level algorithm-based fault tolerance and architectural self-healing solutions are needed both in the cloud and in the embedded parts of the system.

The current chapter focuses on some core problems and solutions that are common for the dominating part of cloud-based CPS:

Modular Redundancy based fault tolerance is one of the standard solutions in fault tolerant computing. However, its use was confined during the last decades to ultracritical applications, like nuclear plants or space applications, due to the extreme overhead related to hardware replication. We have reached the stage of cloud computing at which replicas become cheap and can rely on the same mechanisms as standard virtual machine replication and migration, so this technique is one the most promising in trustworthy CPS applications.

Software-Defined Networks are new means for fast configuration in a variety of operations, including self-configuration after launching or terminating applications in a cloud. In this way, they serve as a basis for resilience after changes in the workload. However, the very same mechanism can be exploited for resilience after faults well complementing the principle of modular redundancy.

Local monitoring supported by *runtime verification* helps to filter out critical events at the "physical" side of a cloud-based CPS. By introducing distributed, context-sensitive monitors (systematically generated from the formalized properties of the component), the cloud-based processing unit does not necessarily have to know the internal behavior and quantitative parameters of embedded physical devices.

Complex Event Processing forms the foundation of early error detection both at the application and infrastructure levels. Here, the chapter examines how a model driven approach can be applied by providing a solid basis to overcome problems related to the a priori unknown deployment and operational characteristics of the application and its runtime environment.

Our proposal is to put a qualitative model in the core of the system contains a compact logic view on the characteristics of the application and the platform, and to focus on the main phenomena in the logic domain facilitates the resolution for the majority of scalability and portability problems by separating the main logic of the quantitative attributes.

15.2 CPS AND CLOUD COMPUTING FAULT TOLERANCE

The high-level overview of cloud augmented CPS consists of at least two distinct components: the embedded (sensory) systems and the cloud. In this setup, some degree of computation and intelligence is removed from the embedded systems and centralized in the cloud, providing multiple benefits. For example, if network communication is a viable alternative to local computation then the complexity and power consumption of the embedded system part could be reduced. Meanwhile, cloud computing is posing new challenges. For instance, while the exclusive nature of private clouds could provide laboratory-like isolated environments, public clouds might not due to resource sharing and multitenancy. Cloud, by its nature, is a dynamic environment since the task deployment is usually optimized in order to improve characteristics, such as utilization, energy efficiency, and availability. Consequently, task migration is a key element of cloud computing. Virtualization alone is an enabler technology for task migration because it decouples the application from the hardware, making it redeployable. Software-Defined Network (SDN) is also a key technology in efficient task migration because it allows quick reconfiguration of the network for a migrated task by making the classically static network of nodes and edges more flexible [6,7].

In this section, fault-induced resiliency mechanisms are investigated with a focus on the new attributes and mechanisms of cloud computing with a keen eye on new possibilities for critical embedded systems (as summarized in Figure 15.1). First, cloud and SDN are discussed further in order to highlight new aspects relevant to fault tolerance. Then the idea of deploying a critical service (application) on the CPS cloud is explored to present some fundamental requirements and failure modes. Finally, a pattern applicability analysis is conducted.

Cloud computing has been reshaping IT for many years. With the rapid advancements of technology, migration to the cloud has become a viable alternative for manifold use cases. For example, the structure of cloud computing not only allows for design deployment diversity, but also for avoidance of correlated faults, greatly enhancing resiliency, an attribute important for CPS. Meanwhile, SDN has been changing the decades-old networking paradigms, creating huge potential for the data center of the future [6,7]. For example, the swift network level reconfigurability of SDN promises fast failover, downtime reduction, and increased reliability, which are all deemed important for CPS. Moreover, the freedom

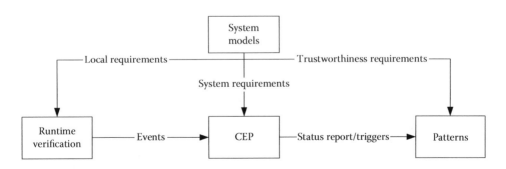

FIGURE 15.1   Model-based methods for CPS.

SDN provides in networking increases the adaptability of CPS to ever present a plethora of changes in the physical world, necessary for CPS. Furthermore, cloud computing and SDN together could provide even more leverage to meet the various requirements of CPS, especially in the areas of fault tolerance and resiliency. For example, the incorporation of scalable cloud computing resources into CPS could be enhanced through the flexibility of SDN.

Mission critical, high-value, strict SLA-bound applications are, as usual, a step behind in adapting to new technologies due to the massive constraints and requirements. However, there is a clear industry trend toward carrier grade cloud platforms. For example, the telecommunications industry is on the brink of adopting cloud computing through Network Function Virtualization (NFV), effectively creating a platform capable of real-time duplex communication and high availability [8].

The often present hardware constraints of CPS devices make the deployment of rich, complex, and resource intensive services difficult. In contrast to this, cloud computing is said to offer virtually unlimited resources, making it a possible remedy for CPS constraints. However, cloud computing was originally developed for batch-like jobs without time limits or for soft real-time applications with greatly relaxed limits, a far from ideal situation for CPS. In order for CPS to embrace cloud computing, significant groundwork is needed to satisfy the highest priority requirements of CPS while working within the performance limitations that are characteristic of cloud computing. For example, the physical cloud topology could affect not only intercloud network propagation time, but also the availability of cloud services due to correlated failures. Additionally, the varying domain size of the different cloud service models has an obvious impact on the scope and granularity of fault management and resiliency. In the case of Infrastructure as a Service (IaaS), the cloud user might have the means to rollback and roll forward between different Virtual Machine (VM) images or reinitialize a VM from an image. Platform as a Service (PaaS) may provide similar means for application bundles, even supporting design per implementation diversity through providing different runtime containers. On the other hand, Software as a Service (Saas) typically elevates the high-level services provided by the application framework (middleware), which might have built-in fault tolerant mechanisms that could mask different kinds of failures.

Cloud platforms introduce new infrastructure component (and infrastructure) service failure modes [9], for example, the virtualization usually underlying the VM service itself can cause such "virtualized infrastructure impairments" [10] as nondelivery of configured VM capacity. The affected components encompass the virtualized resources as well as the operational services defined on them, for example, a slowdown in the speed of spinning up of new VM instances. The various failure modes, in turn, not only can influence the services provided by the tenant via error propagation, but also the capability of the tenants to perform corrective or preventive actions.

15.2.1 Physical Cloud Layout

In this section, the physical cloud layout is further evaluated because both availability and networking are greatly affected by it. The physical layout of cloud platforms is extremely

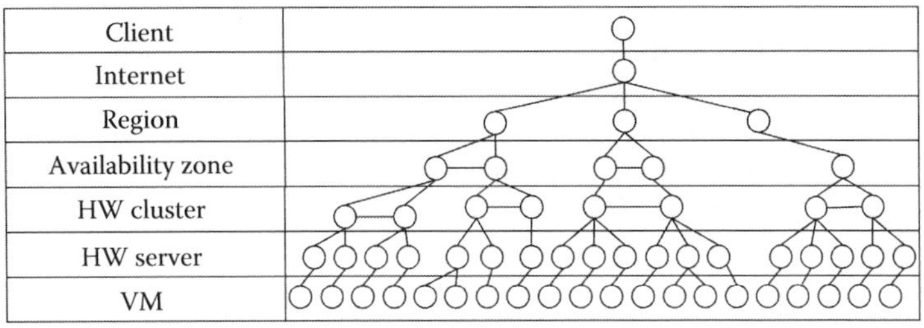

FIGURE 15.2   Physical cloud layout with IaaS example.

diversified, due to the various real life scenarios. For example, private cloud deployments could range from micro private clouds in a student laboratory to scientific-grid-turned-research-institution-scale multilocation based clouds. However, the physical layouts of global public cloud providers follow a similar pattern, where cloud resources are hosted in multiple locations, reachable via internet (data centers, a.k.a regions), providing geographic diversity (Figure 15.2) [11]. Availability zones (AV) are physically isolated locations within regions interconnected through low-latency links. Interregion communication is done via the Internet, implying higher cost, and lower security and quality.

With a fault tolerant mindset in focus, the hierarchical structure of the physical cloud layout contains intrinsic correlated failures. Moreover, the cost and complexity of replacement units decrease by region through the AV zone to the provisioned cloud resource (e.g., VM), which could impact Mean Time To Repair (MTTR). Regarding storage, there is storage associated with AVs and virtual machines as well. Due to the networking implications of the physical cloud layout, storage access works the same way as a network, making interregion storage access less reliable, and more vulnerable, which has to be factored in when mission critical systems are designed.

15.2.2 Software-Defined Network

Dynamicity of CPS infrastructure is a key requirement for the on-demand fulfillment of various functional requirements to provide an adaptive performance management in order to assure the designated timeliness and throughput characteristics and as a main enabling factor for redundancy-based fault mitigation. CPSs typically deal with a large amount of data simultaneously due to the complexity of the functions implemented. Traditional reconfiguration methods like those used in high-availability servers need the transfer of these data for failover. The related significant data transfer times may lead to excessive and intolerable service downtimes.

Typically, networks belong to the most vulnerable parts of a system as they are exposed to a variety of unintentional faults and intentional attacks. Moreover, they are even physically less protected than the usual server farm operating in an isolated environment. In this way they form a primary source of faults to be mitigated.

Both reconfiguration and mitigation of network related faults can benefit if the network topology can be directly reconfigured in a relatively short time. SDNs offer the opportunity for fast alteration of the communication fabric. In this way, traditional data and task migration activities can be substituted by real-time network level reconfiguration. Accordingly, SDNs may serve as a core resiliency enabling mechanism for CPS, especially in highly distributed systems. Rerouting the internode communication and/or involving new computing/transducer nodes into the deployment structure of the CPS is a natural way for self-healing after performance or fault related errors.

Networking is made up of two distinct components (planes): (1) the control plane, responsible for decision making about setting up traffic rules (where to send what traffic) and (2) the data plane, responsible for executing these rules (actually sending traffic according to the rules).

Traditionally, these two planes were colocated in networking nodes, making them tightly coupled, complex, and expensive, etc. Compared to this, in essence, SDN relies on two core ideas: separation of the control and data planes, and abstraction of network control. These two concepts allow higher-level programming (even model based design) of networks and give greater flexibility over traditional networking implementations [12].

For example, traditional control mechanisms are mostly distributed because the control plane is deployed along with the data plane. This alone makes networks with traditional control mechanisms challenging to quickly reconfigure for at least two reasons: independent decision making and limited network view. In contrast to this, SDN control mechanisms are centralized and work on a substantially better (practically complete) central network view, making SDN networks easier to quickly reconfigure. Moreover, due to the central network view, complex distributed routing protocols could be replaced by simpler graph algorithms.

From the point of view of dependability, the innovative SDN concept greatly enhances the agility of network management, a trait that fundamentally complements on-demand self-service and rapid elasticity, providing even greater freedom in resource configuration within the infrastructure [6]. This enables better, more delicate fault tolerance schemes to be deployed and utilized by CPS.

However, the benefits of using SDN differs in the individual parts of a typical cloud-based CPS. As intraregion communication is fully controlled by the cloud provider, the benefits of SDN can be applied end to end, enabling customized, unique, and more stringent requirements to be met. On the other hand, the Internet serving as an intercloud and cloud to transducer medium, is not under the control of the cloud provider, therefore it cannot be entirely covered by a single SDN.

Moreover, network reconfiguration facilitated by SDN is a new efficient candidate for swift failover. In this case, the network can execute the failover by rearranging the network routing topology. A concept with the similar idea, known as network topology transparency, can be found in Reference 13.

15.2.3 Critical Services of the CPS Cloud

In this section, important aspects of the critical applications utilizing the cloud are introduced.

Telco applications need a carrier grade infrastructure due to the strict reliability, availability, throughput, and timeliness requirements. Interestingly, Telco applications are very similar to CPS in the sense that they provide critical, interactive or (soft) real-time functionality. They also do not tolerate network impairments well without additional design for dependability, making them a well founded specimen to start with. It has to be noted that regarding timeliness, most telco application are somewhat more permissive than CPS in general; breaches of Service Level Agreements on latencies lead "only" to financial penalties and not to safety violations, as can happen with, for example, missing computation deadlines in CPS control loops (although this is not necessarily something that should be performed in the cloud backend of a CPS anyway). Still, due to the above reasons cloud-deployed Telco applications are expected to be a very good initial model for cloud-deployed CPS components, especially considering comprehensive platform failure (and error propagation) models [13,14]. It can even be argued that NFV actually forms a class of CPS by directly providing building blocks for CPS systems.

With this introduction, let us enumerate the major platform failure modes for NFV, as identified by Reference 14

- Latency variation of VM services (e.g., resource accesses, real-time notifications, system calls, instruction execution, and interruption handling)

- Variability of resource service performance

- VM failure

- Nondelivery of configured VM capacity

- Delivery of degraded VM capacity

- Changes in the tail latency of IaaS constituent services

- (VM) clock even jitter

- (VM) clock drift

- Failed or slow allocation and startup of VM instance

Almost trivially, these failure modes all can be valid threats for CPS applications. Even when the cloud-deployed CPS application is not directly part of a hard real-time control loop that influences the physical world, breaches of soft real-time requirements and inaccurate timestamping of data submitted to the in-field elements can translate to reduced quality, or even erroneous functionality. This, in turn, can lead to unsafe situations. An example is computationally intensive situational analysis with knowledge integration (that is at least partially offloaded to the cloud) for autonomous vehicles. Note that this list of significant failure modes is mainly based on industrial experience behind NFV [13,14]; the structured multidimensional cloud quality model we introduce later generates a significantly larger failure space, enabling a more rigorous analysis of application dependability and resilience.

15.2.4 Fault Tolerant Design Patterns

Traditional fault tolerance uses well established patterns for error detection, error recovery, error mitigation, and fault treatment [9,15]. In the subsequent discussion we will use a representative subset of these items to indicate the adaptation, reinterpretation, and reevaluation process when applying them in the context of cloud-based CPS. The remaining patterns can be adapted in a similar way.

Redundancy was usually considered expensive in traditional systems. However, with the advent of new technologies like cloud computing and SDN, the age-old, proven, general purpose engineering principles of resiliency need to be reevaluated in order to broaden their usability in the new era [16].

Fault tolerance aims to maintain the continuous service operation, even in the presence of faults. Patterns of fault tolerant software offer general purpose solutions applicable at different levels of the architecture to ensure improved resiliency and undisrupted processing [15]. Being general in nature, there is a wide range in their applicability.

For example, the same redundancy architectural pattern is usable in several layers. On one hand, hardware redundancy resides mostly in the cloud provider domain, offering a huge number of physical platforms to host virtual machines in a way completely hidden from the user. On the other hand, the replication of VMs is viable for both the cloud user and cloud provider, showcasing drastically different cases of the pattern.

For instance, the *Failover error recovery* pattern assures fault masking by replacing a faulty resource with a fault free alternate. The whole process of substitution should be seamless and opaque to the services using the corresponding resource. There are several ways to fail over: resource reallocation, task mitigation, or the rearrangement of the system topology. For instance, an external entity could act as a proxy or load-balancer, initiating a task reallocation based failover upon detecting an error. On the other hand, using SDN, the network could be rearranged without the need of an external entity, as previously mentioned.

15.2.4.1 Phases of Fault Tolerance

Fault tolerance has four phases: error detection, error recovery, error mitigation, and fault treatment [9].

Error detection aims at finding errors within a system state. Considering a public cloud from the cloud user's point of view, the system boundary hides the details of the provider, somewhat thwarting the error detection capability. On the other hand, if the system boundary is well specified, along with a proper SLA, then it might be enough for the cloud user to trace the error to a proof of the violation of the corresponding SLA.

Error recovery aims at returning an erroneous system to an error-free state by applying corrective actions that eliminate the error. The layered architecture of the technologies usually involved in cloud computing enables layered error recovery. Moreover, due to the flexibility of cloud, error recovery could leverage on rapid provisioning, where new fault-free cloud resources are instantaneously created upon failure and deployed to replace a faulty resource.

Error mitigation aims at adapting a system to maintain operation in the presence of errors. The layered architecture of cloud technologies usually allow some degree of error mitigation

TABLE 15.1   Overview of IaaS and CPS Pattern Applicability

Pattern Group	Pattern Name	Public IaaS	Private IaaS	CPS Part
Architectural	Units of Mitigation	OK	OK	OK
	Redundancy	OK	OK	OK
Detection	Fault Correlation	Difficult	OK	OK
	System Monitor	Difficult	OK	OK
Recovery	Restart	OK	OK	Difficult
	Rollback	OK	OK	Difficult
	Roll-forward	OK	OK	Difficult
	Return to reference point	OK	OK	Difficult
	Failover	OK	OK	OK
	Checkpoint	OK	OK	Difficult
	What to save	OK	OK	Difficult
	Remote storage	OK	OK	Difficult
	Data reset	OK	OK	Difficult
Error Mitigation	Final handling	OK	OK	Difficult
Fault Treatment	Reproducible error	Difficult	OK	Difficult

within the layers, however, if cross-layer error mitigation is needed, the separation between provider and user in public clouds could pose difficulties.

Fault treatment aims at fixing the system by removing the faults (the causes of errors). With the harder to modify infrastructure residing at the public cloud provider, the cloud user is then free from such burden. Moreover, if the fault resides in the provider domain, migration to another provider could easily solve the problem for the cloud user. On the other hand, if the cloud user and provider cooperation is required, fault treatment could become more complex.

15.2.5 Infrastructural Patterns

In this section, a representative subset of commonly used FT patterns will be discussed with cloud augmented CPS (Table 15.1). For the rest of the chapter, IaaS will be in focus, as infrastructure resources are easier to comprehend and align with the CPS world than the various services provided by SaaS and PaaS providers.

15.2.5.1 IaaS Aspects

In the case of IaaS the cloud serves basic infrastructure resources. VMs are primary cloud resources in IaaS, therefore they represent a natural candidate for the role of unit of fault tolerance (*Units of Mitigation* architectural pattern). There are manifold faults that can be mitigated in this way, especially from the single point of failure and environment related faults. For example, overload situations could be resolved through allocation of new virtual machines, migration of VMs to hardware with a lesser workload, or by simply increasing the resource share (CPU cycles, I/O) of a VM. Moreover, design per implementation diversity could be used to defend against guest OS, host OS, and VMM failures by creating several redundant VMs with a different configuration in order to tolerate failures of one specific component by having alternative ones. However, protection against application logic failures is not guaranteed.

Operating multiple VMs concurrently (*Redundancy* architectural pattern) and switching between them (failover error recovery pattern) is also an obvious idea; SDN could play a vital role in conducting the fast failover. These patterns protect against the failure of VMs; if the VMs reside in different AV zones or regions (deployment diversity), then even AV zone or region unavailability could be mitigated. Even failure from the provider domain could be mitigated by running VMs at another provider [17–19]. Obviously, this pattern does not provide protection against application level failures.

Storing information independently from the VMs is a regular cloud service with high reliability within the same AV zone, owing to the high quality of internal networking, thus, the *Remote Storage* error recovery pattern is also easily applicable. SDN could support error recovery, through the optimization of the available network resources during error recovery. The pattern protects against data loss due to failure, however, dirty write is not negated.

Moreover, since the complete state of a VM can be captured, stored, or altered. The following error recovery patterns are easily applicable: *Restart; Rollback; Roll-Forward; Return to Reference Point; Checkpoint; What to save*; and *Data Reset*. All VM state related failures can be recovered. Moreover, the Final Handling error mitigation pattern is also applicable due to the flexibility of the VM state. However, systematic provider state failure is not mitigated.

Since the logic interface hides the majority of details of the underlying cloud platform from the cloud user, error detection patterns using large chunks of information are hindered: *Fault Correlation, System Monitor* [20]. Essentially, these patterns would protect against complex failures. However, without the whole spectrum of information available, multilayer correlated faults of both the cloud user and provider domain are not negated. Private clouds do not have such problems because of available observability and controllability in the cloud provider domain.

In addition, the *Reproducible Error* fault treatment pattern is also difficult to implement on a public cloud, as both the cloud user and the cloud provider lack the whole system state, which might be required in the error reproduction during fault treatment. However, some degree of subsystem stubbing is achievable [21]. Private clouds do not suffer from such difficulties.

15.2.5.2 CPS Aspects

Due to the long history of embedded system design, the Units of Mitigation architectural pattern is not difficult to apply. Moreover, operating on multiple resources concurrently (*Redundancy* architectural pattern) and switching between them (*Failover* error recovery pattern) is a natural idea even in the CPS domain, effectively preventing SPOF (Single Point of Failure). For example, a drone with multiple cameras is protected against the failure of a single camera.

The quality and availability of network CPS connectivity might not allow for critical fault tolerant logic to rely on remote resources, thus, the applicability of the *Remote Storage* error recovery pattern is problematic. For example, a flying drone might fly into spot without any network connectivity; even in such a scenario, the drone must be able to autonomously operate properly, avoiding crashing into obstacles.

While the software state in a CPS could be saved, altered, restarted, or set to a reference point, the physical world surrounding it, (the physical state), might not be savable or changeable in the same way, which makes the applicability of the following error recovery patterns difficult: *Restart, Rollback, Roll-Forward, Return to Reference Point, Checkpoint, What to save,* and *Data Reset*. Moreover, the *Final Handling* error mitigation pattern could also be difficult to apply to CPS for the same reason: physics, such as inertia.

On the other hand, CPSs are not limited by a logic interface like that on a public cloud, thus, as much as the instrumentation and measurement support, error detection patterns are applicable: *Fault Correlation, System Monitor*. The *Reproducible Error* fault treatment pattern is also difficult to implement because the physical environment might be difficult to reproduce.

15.2.5.3 Pattern Context

Table 15.1 presents the overview of the applicability of the patterns. Based on the overview, it can be concluded that public clouds are at a disadvantage, making them an ill-advised place for deploying functionality not proven to be fail safe. Moreover, critical cloud augmented CPS should be able to tolerate the loss of public cloud resources and run solely on private cloud resources.

15.3 MODEL-DRIVEN COMPLEX EVENT PROCESSING FOR CPS

While traditional IT systems had methodologies for managing system assets, changes in system topology and configuration, in the case of CPS have proven that these methods and tools often reach their barriers. For example, ITIL (Information Technology Infrastructure Library) is a prominent method for a large enterprise but fails when it comes to managing systems where a large number of resources are changing and topology is often reconfigured. Therefore, it is crucial to apply methods capable of capturing large amounts of incoming (monitored) data and retrieve relevant information in a flexible and scalable way.

Complex Event Processing (CEP) refers to a set of languages and technologies that aim at capturing the behavior of distributed systems and detecting situations which are relevant for the operators of the complex system. In the context of CPS, this task is crucial by the very nature of such interconnected, heterogeneous systems. While the information collection/retrieval is typically a monitoring task, actions in large systems are often triggered by the output of event processing built upon monitoring events. Therefore, CEP should be configured properly to collect all relevant information. This needs a comprehensive modeling approach as described in Reference 22. Here we first define CEP concepts and tasks in a CPS environment, and then we concentrate on how to apply model-driven principles in CEP synthesis.

CEP is a term converged from approaches of different domains: embedded systems, sensor networks, enterprise software, fraud detection, etc. Main characteristics of CEP systems include the following [23]:

- The aim of the system is to capture behavior from *heterogeneous information* flow of a*synchronous events*.

- *Timing* plays an important role, therefore event processing operators support the description of timeliness. Time can be represented as single points or intervals and measured as logical (e.g., number of ticks) or physical (exact timestamp) time.

- Events are evaluated over *sliding windows* which are handled by the processing environment and help to wipe out unnecessary events from the memory (this is usually supported by static analysis of the active queries/patterns).

- Quality of event processing is measured by *preciseness* (whether all occurrences of complex events were detected), *throughput* (number of events processed per time unit), *latency* (the processing time needed to detect a complex event counted from the arriving time of the last event needed for identification of the situation). As CEP itself is a cloud-based service, its qualitative properties can be described by following the approach of the previous Chapter 14.

Let us take the example of a request–response system with possible event sources from multiple levels of system architecture as shown in Figure 15.3. In this system, service level and resource/infrastructure level metrics are shown. The values/changes (delta) of these metrics are handled as events, coming from either the monitoring agent or system logs. Note that CEP can be applied at different levels in a CPS (from application level to resource monitoring).

Typical CEP (event processing) language concepts include the following:

An *event* is the occurrence of something that corresponds to a change in the state of an entity or an action/trigger performed by an entity or coming from the environment.

Everything that is observable can be an atomic event. An event is represented by an object which is characterized by different *properties*, and is identified by its *source* and a *timestamp*. In different CEP systems this timestamp can represent either a physical or logical time, and

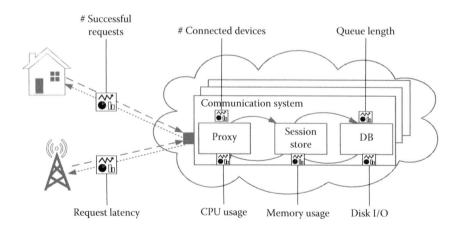

FIGURE 15.3   Example: monitored data in a CPS.

can be attached to the event either by the sender entity (e.g., a sensor sending temperature data) or by the engine receiving the event.

A *complex event* is an event which corresponds to the occurrence of multiple atomic events complying with predefined *timing* and *ordering* requirements and fulfilling constraints formulated on event properties.

Operators are used to define complex events. Semantics of these operators are typically determined by some automaton, defined by, for example, a process algebraic representation. One of the first definitions of such operators is the paper of Allen from 1983 [24] where 13 possible cases are identified to cover possible relationship between points and intervals. Most CEP engines work with a subset of these operators, for example, References 25 and 26.

One key challenge is modeling the relevant combinations and maintaining the model. While for two events the number of combinations is 13, for three it is 409, while the possible combinations of four intervals can be 23,917, which is complex to be modeled correctly if one is using explicit operators for all combinations, also considering the property space of the events. Therefore, regarding timing, arithmetic and logic operators are also used to break down complex expressions

Before an incoming event can be processed by the CEP engine, it passes a preparation phase. In most cases, it must be transformed into a self-contained format which can be processed during the queries. This means that raw data (alerts sent by sensors, component-level monitoring information, application level protocol-specific messages, etc.) have to be transformed into an object-like representation, following the semantics of the particular event processing language. Model-driven development can facilitate this transformation by defining mappings and enrichments at the data model level instead of writing code, for example, by adding metadata to the events.

The next step of event processing is *filtering data* to sort out the relevant information from the information flow. This step is supported by SQL-like queries (e.g., in EPL [27]) by keeping only basic events that may contribute to the identification of a complex event. This is followed by *aggregation* where certain properties (e.g., the average during a time window) are calculated over a number of events.

Figure 15.4 shows the typical procedure for event processing based on the terminology of Reference 23. Benefits of a model-driven approach are summarized below.

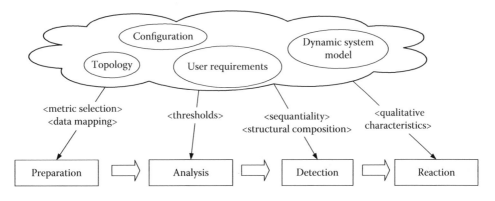

FIGURE 15.4   Steps of Complex Event Processing.

During the design of analysis rules (e.g., scoring, rating, and classification of incoming events), combined queries over topology, configuration and requirement models support the processing phases. Event detection consists of aggregation and compositions steps where the definition of the domain-specific logic is supported by information coming from the system model like sequential (e.g., process model) or structural (de)-composition of event sources (e.g., a class diagram-like representation). In the case of CPS, these models are in most cases runtime. Triplestores are good candidates for these as they are scalable and offer a uniform data retrieval interface.

In the reaction phase, qualitative characteristics of the system help to identify the relevant action, combining information captured by system models and event processing patterns. Such corrective actions implement the Monitoring FT pattern, and, depending on when the phase event is detected, it can be used for mitigation or prevention as well. Following the system configuration, event definition may have a hierarchy which is reflecting the hierarchy of the event sources as well (see Event patterns section).

15.3.1 Property Definition

During the definition of domains for event properties, the qualitative model of the system is taken as a basis. Quantity spaces are encoded as possible values while the event processing rules monitor transitions among these quantity spaces. While in QR concrete values (time windows, alert thresholds) are abstracted, and as per CEP rules these must be reflected, therefore a systematic mapping (depending on the concrete system configuration) is needed.

Consider the example of the communicating devices. In this case, the following complex events can be defined:

- Monitoring device messages: Certain message sequences can refer to situations which might need intervention, and therefore must be identified (application level monitoring, single event source).

- Monitoring device communication: Certain communication patterns correspond to suspicious/dangerous events detected by the devices (application level monitoring, multiple event sources).

- Performance monitoring at the server side. In this case, multiple layers of the server-side architecture provide information in order to evaluate the overall performance of the system, for example, how user workload and server-side metrics are changing together. These include application, container, virtual machine, host (in the case of private/hybrid cloud), and network (resource level monitoring, multiple event sources, etc.).

Model-driven techniques can be used in all of the above cases, hereby we focus on how MDD can help system level performance and availability monitoring. Note that in the case of one single VM, a few hundred metrics can be monitored which results in a practically unmanageable quantity of data even in the case of a mid-sized system of a few dozen nodes, considering that data is provided typically 10–60 times per minute. Even if such an amount can be effectively stored in memory/distributed storage, it is difficult to formulate the queries which cover all important cases (i.e., are *complete*) and are not contradicting (*correct*) by not identifying a situation that is both dangerous and normal at the same time.

15.3.2 Model-Driven Rule Definition

CEP rules usually follow an IF-THEN structure where the IF part refers to a number of preconditions which should be met in order to apply the actions specified in the THEN branch. (See e.g., [22] for a detailed metamodel of a rule language.) These actions can also create new events (e.g., alerts or *pseudo-events* which correspond to situations derived from real events).

In order to explain how modeling can help the definition, let us take the example of the QR model of a cloud-based service shown in Figure 15.5 (see Chapter 14 for a detailed explanation on QR).

In this abstract model of monitoring, a resource (e.g., CPU) implicitly determines system level characteristics, which is represented by the increased need of background operations, therefore the throughput of the system will be decreased. This will be revealed by the monitor component which, at the same time, collects data about application workload.

The quality model is used to generate a hypotheses (i.e., rules) for system diagnostics which are then validated on historical logs and operational data. In the above case, it depends on the concrete implementation whether and how the background operations (buffer cleaning, initialization of sockets/connection pools, garbage collection, etc.) will reach a level where the system irreversibly gets into the critical state.

In this case, the complex event to be detected will be the resource utilization and the level of background operations growing by at least X% for at least Y time units (X and Y here refer to concrete, configuration-specific quantitative parameters which can be determined by data analysis during the system identification).

Event processing, on the other hand, can provide feedback for maintaining a correct qualitative model of the system. In the above case, the first model may not cover the behavior when the system gets into a dangerous state because of its internal operations (without further increase in the user workload). This will be reflected by the unexpected rule/pattern

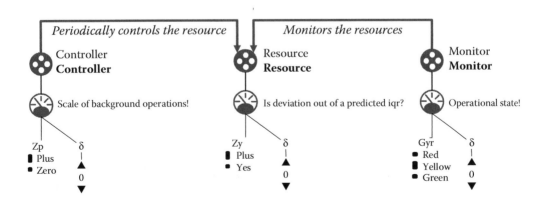

FIGURE 15.5   Qualitative model of monitoring.

matches. The system model in this case will be iteratively refined to cover this kind of self-inducing behavior.

15.3.3 Event Patterns

Besides information processing and rule definition, MDD can also help in creating rules which provide relevant information by following event processing patterns well-known in IT systems monitoring [28]. Such patterns include root-cause analysis, event suppression, and topological correlation.

To illustrate the application of such patterns based on the structural (topology) and behavioral (dynamic) model of the system, one can consider the situation when one of the database components of the client-server architecture is down. This will cause error messages at the level of the database, session store, proxy, the entire application, and the client-server communication. These could refer to faults at the level of individual components, but a carefully designed CEP configuration will detect that all these messages are caused by the database failure.

15.4 RUNTIME VERIFICATION FOR CPS COMPONENTS

Runtime verification is a technology for the dynamic monitoring and checking of system or component behavior with respect to precisely specified properties. Accordingly, the program code of the monitors that observe and analyze the behavior can be synthesized from the property specifications in an automated way by utilizing MDD technologies.

The monitor synthesis technique described in this section is suitable for local monitoring of CPS components that are characterized by real-time, event-driven, and context-aware behavior (for example, the control components of autonomous robots). Basically, the monitors check safety properties ("something bad never happens") and liveness properties ("something good will eventually happen"). Accordingly, the monitored properties are specified using linear temporal logic extended with timing and context related expressions.

Using these component-level monitors, a monitoring hierarchy can be formed. Local monitoring offers a fine granular checking that results in short detection latency, moreover, deploying the monitors close to the monitored components reduces the communication overhead. In the case of detecting errors or witnesses of specified local properties, these monitors are able to provide events for checking *system level properties and interaction scenarios* by Complex Event Processing (see in the previous section). Note, however, that monitors synthesized from temporal specifications can also be applied at the system level if the classic problems of ordering the observed events and providing a common time base are solved.

A monitor can observe the behavior of the monitored component in two ways. Monitoring the externally observable behavior (i.e., noninvasive monitoring) means that the monitor observes the timed sequence of inputs and outputs (events) on the interface of the component, together with context and configuration related information, and then decides whether the runtime trace of these events is conformant with the specified properties (that define a set of allowed traces). Tracking (or another appropriate synonym) the internal behavior (i.e., when the monitor is able to observe the variables and the control flow of the component) is possible if the component is instrumented to send to the monitor relevant information (events) which allows the checking of the related properties.

15.4.1 Specification of Properties

The monitor synthesis technique was inspired by the classical code generation approaches that have been proposed to support the synthesis of monitors and executable assertions on the basis of properties given in temporal logic languages. Our language, which is used to unambiguously express temporal properties for checking context-aware real-time behavior, supports the following features:

- *Explicit context definitions*: Context may appear in the properties an a condition for a given behavior. As an example, for an autonomous robot, in the context of a nearby obstacle a slowdown command is required.

- *Timing*: Timing related criteria can be expressed. For example, the slowdown command shall occur in 10 time units after the observation of the nearby obstacle.

- *Modality*: Properties may define mandatory or optional behavior.

- *Requirement activation*: Ordering between the required properties is supported. For example, violation of the slowdown property given above shall activate the subsequent monitoring of an emergency stop action.

In our monitoring approach this language may be used in two ways. First, it is available as a direct property specification language to formalize properties. Second, it can be used as an intermediate language, when its expressions are mapped from higher-level graphical languages like extended Message Sequence Charts or Property Sequence Charts. In any case, the resulting properties form the input of the monitor synthesis.

Our language is an extension of the Propositional Linear Temporal Logic (PLTL) [27] which is particularly popular in runtime verification frameworks. PLTL expressions can be evaluated on a trace of observed steps, in which each step can be characterized by *atomic propositions*, that is, local characteristics of the step. Below we call these atomic propositions in general as *events*, and the trace of steps is the *trace of events*. In the same way as in the case of the CEP terminology, the concept of the event includes all elements of an observed execution that are relevant from the point of view for the specified properties: input/output signal, sent/received message, function call/return, started/expired timer, entered/left state, change of context, change of configuration, predicate on the value of a variable, etc.

Besides the usual Boolean language operators, basic PLTL has the following temporal operators:

- *X*: "Next" (*XP* means that the next step shall be characterized by event *P*).

- *U*: "Until" (*PUQ* means that a step characterized by the event *Q* shall eventually occur, and until that occurrence all steps shall be characterized by event *P*).

- G: "Globally" (*GP* means that each step in the trace shall be characterized by *P*).

- F: "Future" or "Eventually" (*FP* means that eventually a step shall occur in the trace that is characterized by *P*).

To be able to interpret PLTL expressions on finite traces, so-called *finitary semantics* is used which allows the evaluation of properties at any time. A three-valued logic with "true," "false," and "inconclusive" values is applied, where the evaluation is "inconclusive" if (especially in case of a partial observation trace) no final verdict can be given. The semantics is impartial (a final verdict must not change when a new step is evaluated) and anticipatory (verdict is given as early as possible) [29].

To handle data in the properties, PLTL can be combined with first-order logic and theories (e.g., to check sequence numbers, the theory of natural numbers with inequality) [30]. In this framework PLTL atomic propositions are substituted with first-order data expressions that are evaluated with respect to a particular step (applying direct data processing or SMT solvers to reason about data), and the temporal operators describe the ordering of these steps.

To support the expression of context dependence and real timing, we defined a new extension of PLTL, the *Context-aware Timed Propositional Linear Temporal Logic* (CaTL). In the following subsections these time and context-related extensions are detailed.

15.4.2 Timing Extensions

The basic PLTL cannot specify real-time requirements as it is interpreted over models which retain only the temporal ordering of the events (without precise timing information). Therefore, PLTL cannot specify requirements like "An alarm must be raised if the time difference between two successive steps is more than 5 time units." To tackle this issue, various approaches can be found in the literature. The *Explicit Clock Temporal Logic* formulas [31] contain static timing variables and an explicit clock variable (referring to the current time). In the *Metric Temporal Logic* [32], instead of having timing variables, there are time bounded temporal operators. The *Timed Propositional Temporal Logic* [33] utilizes the so-called freeze quantification which means that each variable can be bound to the time of a particular state. We apply the *Timeout-based Extension of PLTL* [34] which uses an explicit global clock and both static and dynamic timing variables, which makes it flexible and expressive. Using this extension, the property example above is expressed as $G((x = t_0) \rightarrow X((x > t_0 + 5) \rightarrow alarm))$, where x is the clock variable and t_0 is a timing variable.

15.4.3 Context Related Extensions

In our approach, the context is captured in the form of a context model that describes the environment perceived and internally represented (as a runtime model) in the checked component to influence its behavior. The *static part* of the context model supports the representation of the environment objects, as well as their attributes and relations, creating a scene of the environment (e.g., the furniture of a room with their colors, sizes, and positions). The objects are modeled using a type hierarchy. The *dynamic part* of the context model includes the concepts of changes with regard to objects as well as their properties and relations. Changes are represented as *context events* (e.g., appears, disappears, and moves) that have attributes and relations to the changed static objects and their relations (depending on the type of the context event).

The abstract syntax of the context model is defined in the form of a *context metamodel.* Note that this type of hierarchy metamodel can be systematically constructed on the basis of existing domain ontologies.

The metamodel is completed with *well-formedness constraints* (that define conceptual rules) and *semantics constraints* (that are application-specific preconditions or expectations).

In properties to be monitored, the contextual condition is referenced in the form of *context fragments*, which are (partial) instance models of the context metamodel (for example, an instance of *LivingBeing* is in a *tooClose* relation with a *Robot* instance).

15.4.4 The Elements of CaTL

The basic vocabulary of the Context-aware Timed Propositional Linear Temporal Logic (CaTL) consists of a finite set of propositions, static timing variables, and static context variables (these static variables are implicitly quantified with a universal quantifier). The value of a context variable is an instance of the context metamodel (e.g., a context fragment, which contains objects with unique identifiers, attributes, and links). Moreover, two dynamic variables are used to represent the *current time* (clock variable t) and the *current context* (observed context e). The set of *atomic formulas* consists of the following elements:

- *Propositions* are events in the observed trace (each step may include multiple events). Each proposition can be evaluated as true or false in each step.

- *Property constraints* are predicates over properties of a context object.

- *Timing constraints* are defined as inequalities on the timing variables, constants, and the dynamic clock variable.

- *Context constraints* are defined using a compatibility relation $\approx$ between context definitions and the current context. Context definitions can be created from context variables and operators as object exclusion, object addition, relation exclusion, and relation addition. A context definition e_1 is compatible with the current context e (denoted as $e_1 \approx e$) if, and only if, there exists a bijective function between the two object sets e_1 and e, which assigns to each object in e_1 a compatible object from e. Two objects are compatible, if and only if, both have the same type and both have the same relations to other compatible objects.

To form CaTL expressions, these atomic formulas can be connected by using Boolean operators and PLTL temporal operators. Note that for each atomic formula, a modality can be assigned, where *hot* means a mandatory formula, and *cold* means an optional one.

In summary, PLTL atomic propositions are extended with context constraints (to be evaluated with respect to the observed context) and timing constraints (evaluated with respect to the current clock). These expressions are evaluated with respect to a particular step, without affecting the evaluation of the temporal operators. The precise semantics of CaTL is described in Reference 35.

Let us demonstrate the use of CaTL through, the following examples:

- It is always true that if the checked component is in the connected state, then it will eventually become disconnected: $G(connected \rightarrow F(disconnected))$

- It is always true that if the component is connected, then it will be disconnected in 5 time units: $G(connected \wedge t = t_0 \rightarrow F(disconnected \wedge t < t_0 + 5))$

- It is always true that if the component is connected and it is in the e_1 context, and it will be disconnected in the next step, then eventually it will be in the e_2 context: $G(connected \wedge e_1 \approx e \wedge X(disconnected) \rightarrow F(e_2 \approx e))$

15.4.5 Synthesis of Checking Automata

There are several approaches in the literature to synthesize specific monitoring code from PLTL specifications, based on rewriting (e.g., [36,37]) or automata theory (e.g., [38]).

For the finitary semantics of the extended PLTL, the monitor can be synthesized as a finite-state machine [39]. The PLTL formula is mapped to a Bchi automaton with state labeling to give an evaluation ("true," "false," or "inconclusive") in each step of the observed trace.

Another approach is an optimized tableau-based synthesis of monitors [40]. In this case a direct iterative decomposition of PLTL formulas is used. The main idea is that the evaluation of a PLTL expression in a given step of the trace depends on (1) the events observed in a given step, evaluated on the basis of the Boolean operators on the events and the current-time part of temporal operators and (2) the events observed in the future, related to the next-time part of temporal operators. Accordingly, the PLTL formula (with its temporal operators) is rewritten to separate subexpressions that are evaluated in a given step (*current-time expressions*) from subexpressions to be evaluated in the next step (*next-time expressions* with the X operator). On the basis of this rewriting, so-called *evaluation blocks* (EB) are constructed. The internal logic of an EB is the current-time expressions, where inputs are the events observed in the current step and the results of the next-time expressions. Here the three-valued logic according to the finitary PLTL semantics is used (more precisely, if there are *cold* atomic formulas, then a fourth "undecided" value is used).

In the case of next-time expressions, the X operator is removed and the resulting expression is the basis of forming the next EB in a similar way. Accordingly, as the results of the next-time expressions come as the outputs of the EB belonging to the next step, a chain of EBs is formed. If there are "inconclusive" values then these are resolved at the end of the trace at the very latest.

In Figure 15.6 an EB belonging to the expression $G(r \rightarrow (pUd))$ is presented where r, p, and d are atomic propositions. The next-time expressions are at the bottom interface of the

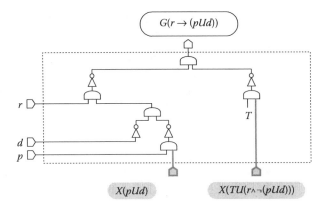

FIGURE 15.6 An evaluation block.

EB. Note that the hardware analogy is just a straightforward illustration of the idea: from a software engineering point of view the logic circuits correspond to data types and functions.

The synthesis of checking automata includes (1) the construction of EB types on the basis of the PLTL expression (there are a finite number of EB types in the case of a finite PLTL expression), (2) generating source code for the internal logic of EB types, and (3) generating an execution context for runtime instantiation of the EB types (according to next-time subexpressions). Memory needs of monitors are most linear with the length of the trace (repetition of blocks can be reduced). In comparison with other implementations, it turned out that this monitoring approach is faster than online symbolic term-rewriting and checking by alternating automaton [40]. We detail the solution applied for the evaluation of the special features of CaTL below.

15.4.6 Time and Context Related Expressions

The handling of time and context require the adaptation of the evaluation block approach. The handling of the current time and context is done through the clock and the context variable. At each EB, the clock variable will be substituted with the current time at the evaluated step, and the context variable will be replaced with the current observed context of the component.

However, the handling of static timing or context variables requires a bit more care. Two states of a variable must be distinguished: free (no value has been assigned) and bound (when it has a value). At first, all static variables are free. If a current-time expression of an EB contains a $var = value$ atomic formula, where var is a free static variable, then var will be immutably bound to $value$ (i.e., it has to be true in all succeeding EBs). For this reason two new ports are added to the EBs: the input and the output valuation port. The EBs receive previous valuations, which will be used at the local evaluation (the valuations cannot be altered, but new valuations can be added). When creating new EB for the succeeding state, the received set of valuations expanded with locally created valuations are forwarded to the next EB.

While the evaluation of clock constraints with the newly introduced valuations is relatively straightforward, the efficient evaluation of context constraints is a more complex context matching task, which can be traced back to the problem of graph matching. An efficient context matching solution, based on [41], was presented in Reference 42. The general idea is to map the contexts to graphs consisting of vertices (labeled with the type and identifier of the object) and edges (labeled with the relation between the objects). The graph matching process involves multiple context fragments (as the typical use case of monitoring addresses not only one, but multiple properties), thus, the result of the simultaneous matching process is a set of valuations between the context fragments and the observed context. Before the matching is calculated, all context fragments (from all properties) are represented in a so-called *decomposition structure*, which is optimized for storing multiple similar graphs in a compact form by storing the common subgraphs only once (as context fragments refer to the same environment, these are often similar).

The context matching algorithm based on this decomposition structure can efficiently search for all possible valuations between an observed context and the context fragments. The complexity of matching is discussed in Reference 42, while more details about the implementation of monitor prototypes can be found in References 35 and 40.

15.5 CONCLUSIONS

In the development of trustworthy CPS, the integration of embedded components into IT systems needs a careful design. While traditional embedded systems and the algorithms behind embedded applications are, in most cases, already designed to cope with uncertainty and faults in the environment, this is not true for IT infrastructures. Therefore, we presented methods to enhance the trustworthiness of CPS built on the combination of embedded/networked devices and cloud-based services.

The application of fault tolerant patterns, complex event processing, and runtime verification together help to create trusted applications. While embedded components comply with critical requirements both from algorithmic and implementation aspects, even in the case of private nanoclouds, there is no guarantee of availability, timeliness, or integrity requirements. The methods we propose show that these can be met using cheap redundancy patterns, verification, and monitoring techniques at the level of platform or application. This means that redundancy schemes and monitoring configuration have to be developed together with the application, and should utilize parametrized deployment time. Model-driven methods support the systematic development and parametrization of these. Such methods can be used even if the observability and controllability of the system is user level, that is, restricted, which leads to a reinforced trust.

REFERENCES

1. Jeffrey O. Kephart and David M. Chess. The vision of autonomic computing. *Computer*, 36(1):41–50, 2003.
2. Imre Kocsis, Andras Pataricza, Zoltan Micskei, Andras Kovi, and Zsolt Kocsis. Analytics of resource transients in cloud-based applications. *IJCC*, 2(2/3):191–212, 2013.
3. Andras Balogh, Andras Pataricza, and Judit Racz. Scheduling of embedded time-triggered systems. In *Proceedings of the 2007 Workshop on Engineering Fault Tolerant Systems*, p. 8. ACM, ESEC/FSE '07 Joint 11th European Software Engineering Conference (ESEC) and 15th ACM

SIGSOFT Symposium on the Foundations of Software Engineering (FSE-13) 2007, Cavat, Croatia, September 03-07, 2007.

4. Zoltan Adam Mann. Allocation of virtual machines in cloud data centers—A survey of problem models and optimization algorithms. *ACM Computing Surveys*, 48(1):11:1–11:34, August 2015.

5. Anatoliy Gorbenko, Vyacheslav Kharchenko, Seyran Mamutov, Olga Tarasyuk, and Alexander Romanovsky. Exploring uncertainty of delays as a factor in end-to-end cloud response time. In *Dependable Computing Conference (EDCC), 2012 Ninth European*, pp. 185–190. IEEE, Sibiu, Romania, 2012.

6. Open Networking Foundation. Software-defined networking: The new norm for networks, Palo Alto, CA, 2012.

7. Sakir Sezer, Sandra Scott-Hayward, Pushpinder-Kaur Chouhan, Barbara Fraser, David Lake, Jim Finnegan, Niel Viljoen, Mary Miller, and Neeraj Rao. Are we ready for SDN? Implementation challenges for software-defined networks. *Communications Magazine, IEEE*, 51(7):36–43, 2013.

8. Margaret Chiosi, Don Clarke, P Willis, A Reid, J Feger, M Bugenhagen, W Khan, M Fargano, C Cui, H Deng et al. Network functions virtualisation introductory white paper. In *SDN and Open Flow World Congress*, Darmstadt, Germany, 2012.

9. Algirdas Avizienis, Jean-Claude Laprie, Brian Randell, and Carl Landwehr. Basic concepts and taxonomy of dependable and secure computing. *Dependable and Secure Computing, IEEE Transactions on*, 1(1):11–33, 2004.

10. Eric Bauer and Randee Adams. *Service Quality of Cloud-Based Applications*. Wiley-IEEE Press, Piscataway, NJ, 2013.

11. Amazon AWS documentation. http://docs.aws.amazon.com/AWSEC2/latest/UserGuide/using-regions-availability-zones.html. Accessed: 14-Sep-2015.

12. Jan Medved, Robert Varga, Anton Tkacik, and Ken Gray. Opendaylight: Towards a model-driven SDN controller architecture. In *2014 IEEE 15th International Symposium on*, pp. 1–6. IEEE, Sydney, Australia, 2014.

13. ETSI. NFV, Network Functions Virtualization and Use Cases, Resilience Requirements V1.1.1. http://www.etsi.org/deliver/etsi_gs/NFV-REL/001_099/001/01.01.01_60/gs_nfv-rel 001v010101p.pdf, Sophia Antipolis Cedex–France, 2013.

14. ETSI. Network Functions Virtualisation (NFV); Service Quality Metrics—ETSI GS NFV-INF 010. http://www.etsi.org/deliver/etsi_gs/NFV-INF/001_099/010/01.01.01_60/gs_NFV-INF 010v010101p.pdf, Sophia Antipolis Cedex–France, 2014.

15. Robert Hanmer. *Patterns for Fault Tolerant Software*. Wiley-IEEE Press, Naperville, IL, 2013.

16. Bill Wilder. *Cloud Architecture Patterns: Using Microsoft Azure*. O'Reilly Media, Inc., Sebastopol, CA, 2012.

17. Szilárd Bozóki, Gábor Koronka, and András Pataricza. Risk assessment based cloudification. In *Software Engineering for Resilient Systems*, pp. 71–81. Springer, Berlin-Heidelberg, 2015.

18. Andreas Polze, Peter Troger, and Felix Salfner. Timely virtual machine migration for pro-active fault tolerance. In *Object/Component/Service-Oriented Real-Time Distributed Computing Workshops (ISORCW), 2011 14th IEEE International Symposium on*, pp. 234–243. IEEE, Newport Beach, CA, 2011.

19. Topology and Orchestration Specification for Cloud Applications (TOSCA). http://docs.oasis-open.org/tosca/TOSCA/v1.0/os/TOSCA-v1.0-os.html. Accessed: Sep. 14, 2015.

20. Giuseppe Aceto, Alessio Botta, Walter De Donato, and Antonio Pescapè. Cloud monitoring: A survey. *Computer Networks*, 57(9):2093–2115, 2013.

21. Linghao Zhang, Xiaoxing Ma, Jian Lu, Tao Xie, Nikolai Tillmann, and Peli De Halleux. Environmental modeling for automated cloud application testing. *Software, IEEE*, 29(2):30–35, 2012.

22. István Dávid and László Gönczy. Ontology-supported design of domain-specific languages: A complex event processing case study. In Vicente García Díaz, Juan Manuel Cueva Lovelle,

B. Cristina Pelayo García-Bustelo and Oscar Sanjuán Martinez (Eds), *Advances and Applications in Model-Driven Engineering*, p. 133, IGI Global, NY, 2014.

23. Event Processing Technical Society. Event processing glossary. http://www.complexevents.com/2011/08/23/event-processing-glossary-version-2/, 2011. Accessed: Sep. 14, 2015.

24. James F Allen. Maintaining knowledge about temporal intervals. *Communications of the ACM*, 26(11):832–843, 1983.

25. Drools Business Rule Management System. http://www.drools.org/. Accessed: 14-Sep-2015.

26. Esper Event Correlation Engine. http://www.espertech.com/products/esper.php. Accessed: Sep. 14, 2015. Wayne, NJ.

27. Amir Pnueli. The temporal logic of programs. In *Foundations of Computer Science, 1977., 18th Annual Symposium on Foundations of Computer Science*, pp. 46–57. IEEE, Providence, Rhode Island, 1977.

28. IBM Redbook SG246094. *Event Management and Best Practices*. IBM (online), 2004.

29. Andreas Bauer, Martin Leucker, and Christian Schallhart. Comparing LTL semantics for runtime verification. *Journal of Logic and Computation*, 20(3):651–674, 2010.

30. Normann Decker, Martin Leucker, and Daniel Thoma. Monitoring modulo theories. In Erika Ábrahám and Klaus Havelund (Eds), *Tools and Algorithms for the Construction and Analysis of Systems*, pp. 341–356. Springer, Berlin-Heidelberg, 2014.

31. Eyal Harel, Orna Lichtenstein, and Amir Pnueli. Explicit clock temporal logic. In *Logic in Computer Science, 1990. LICS'90, Proceedings, Fifth Annual IEEE Symposium on Logic in Computer Science*, pp. 402–413. IEEE, Philadelphia, PA, 1990.

32. Ron Koymans. Specifying real-time properties with metric temporal logic. *Real-time systems*, 2(4):255–299, 1990.

33. Rajeev Alur and Thomas A Henzinger. A really temporal logic. *Journal of the ACM (JACM)*, 41(1):181–203, 1994.

34. Janardan Misra and Suman Roy. A decidable timeout based extension of propositional linear temporal logic. *arXiv preprint arXiv:1012.3704, Journal of Applied Non-Classical Logics*, 2010.

35. Gergo Horanyi. Monitor synthesis for runtime checking of context-aware applications (MSc thesis). *Budapest University of Technology and Economics*, Budapest, Hungary, 2014.

36. Howard Barringer, David Rydeheard, and Klaus Havelund. Rule systems for run-time monitoring: From Eagle to Rule R. *Journal of Logic and Computation*, 20(3):675–706, 2010.

37. Allen Goldberg and Klaus Havelund. Automated runtime verification with Eagle. In *MSVVEIS*, Workshop on Verification and Validation of Enterprise Information Systems (VVEIS'05), INSTICC Press, ISBN 972-8865-22-8, Miami, FL, May 24, 2005.

38. Andreas Bauer, Martin Leucker, and Christian Schallhart. Runtime verification for LTL and TLTL. *ACM Transactions on Software Engineering and Methodology (TOSEM)*, 20(4):14, 2011.

39. Andreas Bauer, Martin Leucker, and Christian Schallhart. Monitoring of real-time properties. In S. Arun-Kumar and Naveen Garg (Eds), *FSTTCS 2006: Foundations of Software Technology and Theoretical Computer Science*, pp. 260–272. Springer, Chennai, India, 2006.

40. Gergely Pintér and István Majzik. Automatic generation of executable assertions for runtime checking temporal requirements. In *High-Assurance Systems Engineering, 2005. HASE 2005. Ninth IEEE International Symposium on High-Assurance Systems Engineering*, pp. 111–120. IEEE, Heidelberg, Germany, 2005.

41. Bruno T. Messmer and Horst Bunke. Efficient subgraph isomorphism detection: A decomposition approach. *Knowledge and Data Engineering, IEEE Transactions on*, 12(2):307–323, 2000.

42. Gergo Horanyi, Zoltan Micskei, and Istvan Majzik. Scenario-based automated evaluation of test traces of autonomous systems. In *SAFECOMP 2013-Workshop DECS (ERCIM/EWICS Workshop on Dependable Embedded and Cyber-physical Systems) of the 32nd International Conference on Computer Safety, Reliability and Security*, pp. 181–192, Toulouse, France, 2013.

CHAPTER 16

Education of Scientific Approaches to Trustworthy Systems for Industry

After 10 Years

Fuyuki Ishikawa, Nobukazu Yoshioka, and Yoshinori Tanabe

CONTENTS

16.1 INTRODUCTION

Software-intensive systems have played an essential role in human and social activities, with ever-increasing complexity that engineers need to tackle. The emergence of Cyber-Physical Systems (CPS) further promotes this trend. CPSs are systems that involve computational elements as well as elements that interact with the physical environments [1,2]. CPSs thus have a stronger, more direct impact on the users and surrounding environments, which means that ensuring trustworthiness of CPS is challenging and the demand for it is high. The actors here are a wide range of engineers who are required to tackle the difficulties in emerging CPS.

One of the key directions is the use of scientific methods and tools with solid theoretical foundations (usually mathematics) for strong confidence and trustworthiness. For software-intensive systems, a lot of methods and tools based on computer science have been leveraged for modeling, simulation, testing, proofs, and so on. For CPS, novel approaches are emerging that handle both discrete (software) and continuous (physical) aspects. For example, Crescendo is a tool that uses collaborative simulation of discrete models and continuous models by synchronizing simulators for each model [3]. Another example is KeYmaera, a theorem prover for hybrid models that include both discrete and continuous aspects [4]. These tools have demonstrated their usefulness for ensuring trustworthiness of CPSs such as a dredging excavator and air traffic control. While such tools use novel techniques to combine continuous aspects from mechanics or control theory, they are extensions of long-established approaches in computer science. It is therefore necessary to understand foundations for trustworthiness, such as abstraction for discrete models and theorem proving.

This point illustrates the key challenge for engineers because they often have not received strong training in computer science, including areas that serve as the basis for trustworthiness, such as mathematical logic. Another point is the emergence of new techniques, which should be ideally supported by academic-industry collaborations. These points are not really new to CPS and have existed for software-intensive systems. Nevertheless, difficulties imposed by these points have been increasing with the complexity and higher demand for trustworthiness in emerging CPS. For example, one of the key challenges for CPS design is reconsideration of abstraction and semantic models [2], which requires essential advances in both theory and practice.

In this chapter, we address the above issue by focusing on our experience with the teaching of formal methods for the Japanese industry. Formal methods have been strong means for the assessment of trustworthiness, as discussed above: requiring engineers to understand theoretical foundations and tackle new tools (at least the tools seem completely new or unfamiliar to people in the industry). Formal methods have also been discussed in the context of academia–industry or theory–practice gaps. There have been many discussions of these issues, from the viewpoint of myths, or typical misunderstandings [5,6], or obstacles [7]. However, practice with a wide range of engineers (not only top researchers) is limited.

Our intensive experience discussed in this chapter is based on an educational program for the industry called Top SE.[*] The program started in 2004 and has provided a unique place for education of advanced techniques of software engineering to over 300 engineers in the industry.

The lecture classes are carefully designed to involve discussions of practices. Some of the classes focus on practices and are led by lecturers from the industry. This design increases the insights of the participants (engineers) into practical usages of scientific approaches. In addition, the participants also tackle 3- or 6-month studies called graduation studies. In these graduation studies, the engineers identify and tackle their own problems with supervisors. Domains of problems vary depending on the participants' work and interests, including CPS.

The remainder of this chapter is organized as follows. Sections 16.2 and 16.3 give an overview of the Top SE program and formal methods of education in the program. Section 16.4 provides analysis and discussion of the key factors and roles of formal methods education for the industry, including the "side effects" of formal methods education. Section 16.5 gives concluding remarks.

16.2 TOP SE PROGRAM

In the software engineering area, there has been a gap between what is taught in academia and what is required by industry [8]. The Top SE program was developed to bridge this gap by providing a place where academia and industry jointly deliver knowledge and techniques for practical use of advanced scientific methods and tools.

The Top SE program targets engineers in the industry and provides education on a variety of methods and tools by lecturers from academia and industry. The Top SE program began in 2004 [9], with a government sponsorship for 5 years.[†] After that, the program was renewed as a paid course and has its operation has been quite stable. Many companies now consider the program to be a good choice for experience and consistently dispatch participants every year. Thus, the fees are covered by the companies for most of the participants. However, some of the participants pay privately.

Each participant spends one or one and a half years attending lectures as well as a so-called graduation study, which will be explained below in Section 16.2.2. In academic year 2015, the program had 44 lecture classes, 40 participants, and 47 lectures.

The participants are generally engineers from industry, but sometimes include a few graduate students in a given year. Many of the engineers are around age 30. This means that they have experienced difficulties and issues with projects before and thus highly motived to learn more. They are not only eager to learn new techniques to solve problems in their companies, but also often motivated by the opportunity for obtaining systematic knowledge in software

---

[*] http://www.topse.jp/.
[†] By the Ministry of Education, Culture, Sports, Science and Technology, Japan.

engineering. The companies expect these young engineers to obtain more knowledge and experience to lead future activities.

About two-thirds of the lecturers are from industry and many of the lecture classes are operated jointly by lecturers from academia and industry.

16.2.1 Lectures

In the Top SE program, lectures are organized to involve different methods and tools so that participants can compare different approaches while understanding common principles. Each class involves group exercises in which participants jointly tackle practical problems while exchanging ideas on different approaches.

Each lecture class consists of 7 or 15 slots and each slot lasts 1.5 h. As the participants are industry engineers, classes are held in the evenings on weekdays and in the daytime on Saturdays. Participants are required to take at least six lecture classes to graduate. About a quarter of them take this minimum number, but others take many more- in extreme cases more than 20.

The 44 lectures are grouped into seven lecture series. Examples of classes and their target knowledge and methods are shown in Table 16.1. Each lecture series consists of six to ten lecture classes, although only a few are shown in the table.

For software engineering and trustworthy CPS, the key factor is modeling, or how to extract the essences of complex systems for effective analysis, decision-making, verification,

TABLE 16.1   Lecture Series and Examples of Classes

Lecture Series	Examples of Classes and Target Knowledge/Methods
General	Foundational theories (e.g., Hoare-logic and temporal logic) Software testing (e.g., combinational testing)
Requirements engineering	Goal-oriented requirements analysis (e.g., KAOS) Safety requirements engineering (e.g., FMEA)
Architecture	Component-based software engineering (e.g., UML components) Aspect-oriented software engineering (e.g., AspectJ)
Formal specification	Introduction to formal specification (e.g., VDM) Theorem proving (e.g., Coq)
Model checking	Introduction to model checking (e.g., SPIN) Verification and implementation of concurrent systems (e.g., CSP)
Cloud computing	Cloud infrastructure construction (e.g., OpenStack) Distributed processing (e.g., MapReduce)
Project management	Software development scale estimation (e.g., CoBRA) Risk management (e.g., Analytic hierarchy process)

and so on. There are two lecture series that focus on specific phases in which models of the target systems are leveraged (namely the requirements analysis and design phases): the Requirements Engineering and Architecture series. Two other series cover techniques for trustworthiness by formal methods in various phases: the Formal Specification and Model Checking series. The other lecture series cover a wide range of significant topics, including the theoretical foundation of mathematical logic. The Cloud Computing series is a domain-oriented series newly designed to leverage emerging practices that have become widespread.

It is difficult to generally describe the technical level of the lecture content or how the content is advanced from the viewpoint of engineers in industry. However, it is reasonable to say that the lecture content is typically common sense for academic researchers who work in the relevant area. It is also possible to find a few, well-known good books written by some of the leading researchers in the world, but there is often no book in the Japanese language and the content is not handled in the standard industry training or seminars.

Even if there are good books on the target topic, it is necessary to design the lecture class carefully. Acquisition of knowledge and exercises on methods and tools are an indispensable, and thus significant, part of the lecture classes. However, solving each problem further requires discussions on, and decisions regarding, approaches such as modeling scope, the correct abstraction level, and properties for verification. To practice for such aspects, most classes have group exercises in which participants jointly tackle nontrivial problems. The target problems of group exercises should illustrate typical challenges in the industry so that participants can link the exercises to issues they face daily. The design of lecture classes will be further discussed in Section 16.3.1, focusing on the two series for formal methods.

16.2.2 Graduation Studies

Even though group studies in each lecture class are based on practical settings and give deeper insights, they are limited to the target topic of the class. The problems are designed, restricted, and defined to align well with the target topic and techniques in the class. It is therefore desirable for participants to have training with more realistic settings in which approaches to the problems, and even the essence of the problems and requirements, are not known in advance. This is essential for software-intensive systems, including CPS, as such systems by their very nature consist of a variety of requirements and environments, which never lead to the same way of development and thus require tailoring, even if similar approaches are adopted.

Participants may want to go beyond the "training" discussed above. Since they work in industry, they already have difficulties or problems to tackle that have not been solved effectively inside their companies. Thus, the participants often want to tackle their own problems rather than the problems prepared for training. Participants are often motivated to extend or complement the techniques they have learned when they find some limitations or gaps from their practical experience.

The Top SE program provides a place for the activities discussed above, which are called graduation studies. In graduation studies, participants identify and tackle their own problems using scientific approaches they have learned. As discussed above, problems tackled by the participants vary a lot. The duration of the studies is 3 or 6 months, guided by one or more lecturers who act as supervisor(s). The studies are evaluated in terms of validity of problem setting, approach (methods/tools), modeling of target problems, and evaluations.

A few typical topics for graduation studies are presented below.

Case Study: In this type of study, engineers identify problems in development activities at their companies. They tackle these problems by selecting and tailoring scientific methods and tools learned in the course and then evaluate how well the problems are solved. A typical example is the application of a formal verification tool to a typical, but troublesome, problem and the evaluation of cost and effectiveness.

Domain-Specific Finer-Grained Support: In this type of study, engineers identify problems in the application of methods and tools learned in the course. They tackle these problems by providing fine-grained support for the application, often in a domain-specific way. A typical example is a translation tool from spreadsheets currently used for the specific input format of a formal verification tool.

Bridging Gaps between Different Methods/Tools: In this type of study, engineers identify problems in bridging different phases or tasks, with the methods and tools for each often discussed separately. They tackle these problems by providing support for connecting output of one task to the input of another. A typical example is the support of systematic mapping from requirement models to design models.

Development of Methods and Tools: In this type of study, engineers identify problems in the targets or capabilities of current methods and tools. They tackle these problems by constructing new methods and tools. A typical example is tool development to support deeper analysis or visualization of counterexamples generated by a model checker.

16.3 FORMAL METHODS IN THE TOP SE PROGRAM

Among the 44 classes in the program, there are 12 classes on formal methods in the two lecture series, Formal Specification and Model Checking. In this section, notable points are described in greater detail regarding the design of these classes, which illustrate our challenge to disseminate scientific approaches for trustworthiness to the industry.

16.3.1 Lectures

16.3.1.1 Series Design

The class for mathematical logic teaches the set theory, predicate logic, Hoare-logic, and temporal logic, together with brief introductions to their practical applications that motivate the learning. The knowledge is given basically from the viewpoint of "users" of the theories, for example, those who do not prove soundness.

The other 12 classes are divided into two categories: six for formal specification and six for model checking. Each category starts with a basic class that focuses on one method, VDM [10] and SPIN [11], respectively. VDM is a good starting point for formal specification as the method and tools focus on verification by testing (on models) with which engineers are familiar. SPIN is chosen as one of the most popular tools for model checking. In these introductory classes, participants learn the basics and receive experience in modeling, verification, and validation. At the time of writing this chapter, the lecture series covers many methods and tools: VDM, B-Method, Event-B, Coq, SPIN, SMV, LTSA, CSP, UPPAAL, JML, and Java PathFinder. They are selected as representatives for specific topics; for example, Coq for theorem provers and Java PathFinder for software model checkers. Stability of the tools is one of the minimum requirements to enable efficient hands-on experience as well as to not let the participants think they are immature or impractical.

All of the classes have unique designs and do not follow any existing books typically used in university classes. Specifically, general knowledge and exercises are commonly used to introduce the techniques and follow existing books. The classes for the industry need more, such as links to common development processes and deliverables, practical guidelines, pros and cons, and larger exercises. Such content is often designed jointly by academic and industrial lecturers. Actually, several new textbooks have been published based on the developed lecture classes, and many of them are the first books in Japanese about the target topics. Below, we elaborate on the lecture design in terms of group exercises and lecture classes for practice.

16.3.1.2 Group Exercises

Most classes have group exercises in which participants tackle nontrivial problems through discussions with each other. For example, group exercises for formal specification methods use a routing protocol for ad-hoc networking, Optimized Link State Routing Protocol (OLSR) [12]. Each exercise focuses on a specific aspect, such as an algorithm for efficient multicast avoiding duplicate deliveries or distributed management of network information. Discussions include what aspects and properties to focus on, how to remove details irrelevant to the focus, and so on. For example, the protocol includes prevention of multicast flooding by selecting a set of neighbor nodes that forward multicast messages from a node. To analyze and validate this selection algorithm, one may construct a formal model of the whole network to simulate its behavior for multicast messages or one may formalize only the algorithm, which targets reachability of messages to nodes that are two hops away from the sender node. Then, the global behavior can be discussed without simulating the whole network, that is, reachability of messages to all the nodes.

Other examples of group exercises include robots of the classical "dining philosopher" problem. This is an extension of the classical problem since the robots can act in parallel, for example, leading to a state in which two robots touch one fork at the same time. It is also possible to run actual robots by using the models. Discussions for these exercises include how to define abstract and essential state transitions, what properties to verify, and so on.

16.3.1.3 Classes for Practice

Two classes for practice were introduced to formal specification and model checking. The lecturers for these classes are from the industry and have application experience with formal methods.

Both classes teach principles and practice on technical or nontechnical aspects, such as extraction of models from specification as well as communication and understanding within the team. The lecturers also deliver insights drawn from their own experiences, such as the significance behind explaining cost and benefit.

In the practice class for formal specification, participants list questions regarding issues about specification in general and formal methods. Discussions are conducted concerning possible answers and rebuttals. The results for the first few years led to a list of over 100 Frequently Asked Questions. In recent years, the list has been discussed in a shorter time and discussions moved to a new topic: decisions and planning necessary for specification (when natural languages are used and when formal specification languages are used). Experience from application of VDM by a Japanese company is also discussed [13,14], which gives deep insights into principles of specification, team-based work, verification and validation, beyond mere words, such as increased reliability and cost reduction.

In the practice class for model checking, each participant plays the role as an expert of model checking and is asked to support a project given a specification document or source code. He or she is responsible for not only the verification by model checking, but also for presenting the work plan and reporting the result. The plan and result should include analysis of the impacts, including costs and benefits as well as difficulties in the target program and the scope handled by model checking.

16.3.2 Graduation Studies

There have been many graduation studies that constantly use formal methods. Every year, at least a quarter of the participants use some kind of formal methods in their graduation studies.

In this section, two examples of graduation studies are discussed that tackle trustworthiness of CPS by using formal methods. These studies are representatives of activities that go beyond what are learned directly in the lecture classes, but are still based on understanding of the foundations learned.

16.3.2.1 Crescendo Evaluation for Power System

Crescendo [3] is a tool for collaborative modeling and simulation of CPS. The term "collaborative" refers to the point at which continuous models (physical world) and discrete models (software) are combined and simulated. A mechanism is provided to connect languages and tools for each of the aspects. Crescendo is a very new tool developed in a European FP7 project DESTECS and was completed at the end of 2012. The graduation study, after only a few years, was representative of trials on novel tools from academia through academic-industry collaboration.

Crescendo was only briefly mentioned in a lecture class about formal specification in VDM (used for discrete or software modeling in Crescendo), but one participant was interested and decided to try the tool in his graduation study. The participant has worked as a software engineer in the power industry. He has often experienced validation of the software part being delayed or limited. The main reasons are dependency on availability and configuration of hardware environments as well as limited support of software aspects in tools for investigation of continuous aspects. Thus, the participant anticipated potential uses of Crescendo, which allows for validation of software aspects in early phases using high-level tools and connections with the continuous aspects.

In the graduation study, the participant tried to evaluate Crescendo. An existing Pulse Code Modulation system (PCM) system is targeted as an example. In simple terms, PCM allows the amount of electric current to be digitalized and delivered for software-based rich functions of monitoring and control. The study, a short period of 3 months, focused on evaluation of these two points.

- How can Crescendo simulate expected behavior of the system?

- How can the software aspects be handled efficiently?

The first point was carried out by trying various scenarios, including ones with faulty behavior. In all of the scenarios, it was possible to simulate expected behavior of the system. For example, proper (improper) adaptation behavior in response to disturbance of the current leads to convergence in short time (increase) of the disturbance effect. The second point was discussed by comparing different ways to construct the simulation models. Thanks to abstraction of registers and memory maps, the lines of code could be less than one third in VDM compared with the corresponding past data for hardware-based testing with C/assembler prototype coding. Other advantages were also discussed, such as mitigation of risks about configuration cost of hardware devices or deficiencies in hardware devices.

16.3.2.2 Physical Control Validation with Event-B

Another example of graduation studies is validation of physical control. The participant who tackled this study works with a construction company. The target problem is control of two cranes that jointly pull up a burden (e.g., by catching the two ends of a pole). There are several properties that critically affect the trustworthiness, specifically safety, of the system. One example property is to keep the gradient of the burden within a certain range so that the burden does not slip down.

Event-B [15] was chosen to tackle this problem. Event-B is a relatively new formal specification method that targets system modeling. It thus has strong expressiveness for event-based models, including not only software parts, but also various elements of environments. Although there are emerging tools that deal with the continuous aspects involved in the target problem, Event-B was selected with its rich tool support from the methods taught in the Top SE program.

The study went far beyond what was taught in the lecture class. Currently, the class for Event-B is short and only focuses on a few examples after the class for classical B-Method. The challenges the participant was faced with are summarized below.

- The study involved intensive use of tool functions (the RODIN platform for Event-B). The Event-B method considers theorem proving as the core means of verification. This was used in the study as well (to provide assurance of the properties), which also helped clarify assumptions that tend to be hidden. However, it is difficult to have strong confidence by asking the prover and seeing the OK message. Therefore, model checking and simulation are also used to investigate different scenarios by using the ProB plug-in on the RODIN platform. This required investigation of the plug-in during the period of the study.

- It was necessary to carefully design the modeling strategy to handle the system that contains the continuous aspect as well as to encode the target problem into the modeling language (set of events for Event-B). The final model represents each step in a time unit as an event. This was a challenge for the participant and required careful discussions with the experienced lecturer.

- The model with numerical properties required much more effort on the manual (interactive) proving of properties that the automated prover failed to handle. This was also a challenge because given knowledge and experience on the manual proving was very limited in the short time frame of the lecture class.

Finally, the participant succeeded in the construction and validation of the model with good confidence through investigation by different means of validation (theorem proving, model checking, and simulation).

16.3.2.3 Other Studies

The studies presented above illustrate how engineers investigate state-of-the-art techniques beyond what can be learned from well-established lecture classes, seminars, or books.

It is notable that there are other kinds of significant graduation studies that tackle practice of more traditional techniques. A few examples are described below.

- A case study involving colleagues in the participant's company was done to evaluate the cost and benefit of design-related techniques. Different methods are used gradually systematic (informal) analysis of components, formal annotations on components, and validation with formal specification. The cost and improved quality were evaluated in each step to examine the trade-off. This study is a good example that involves other engineers and discusses cost-effectiveness that is often wondered about in the industrial context.

- A mechanism was proposed to verify constraints on time and physical/cyber resources in workflow or business processes. The main case study analyzed processes in a hospital. Nurses support daily activities of patients with some facilities while also being prepared for emergency tasks with strict time constraints. The number of nurses, with a limited number of facilities, should be decided by properly considering various possibilities, such as when delay of an activity with a certain facility blocks activities by other nurses. To support this kind of analysis, first an extension for a standard of business process modeling was proposed to allow for annotations on time and resource constraints (BPMN: Business Process Modeling Notation [16]). Then, a translation mechanism to a timed model checker (UPPAAL) was constructed for verification of properties.

- A tool to support model checking is constructed. By targeting checking between the expected protocol and internal design, the tool provides in-depth support with usable interfacing functions. It accepts input specification in the form of sequence diagrams. If violation of the protocol is found, a counterexample is shown together with an expected behavior. By showing the two models in parallel, with careful adjustment of layout to contrast the difference, it is very easy to see the detected aberration.

16.4 ANALYSIS AND DISCUSSION

In this section, key insights and suggestions are provided along with some data from our experiences presented in the previous sections. The key background of the discussion is that techniques like formal methods are sometimes considered too difficult and only for limited experts. There have also been discussions on academia-industry or theory-practice gaps. However, these obstacles should be overcome in order to tackle increasing difficulties in ensuring trustworthiness of CPS.

16.4.1 Feedbacks to Lecture Classes

Figures 16.1 through 16.3 summarize the comments to the questionnaires distributed after lectures from 2012 to 2014. Only three questions are presented and discussed here: difficulty, availability of practical information, and applicability of the obtained approaches. We used the first basic classes and the practice classes (described in Section 16.3.1) specification (FS) and model checking (MC). In each figure, the "worst" choice is omitted since no one selected "very easy," "no practical information," or "not applicable at all."

16.4.1.1 Difficulty

Figure 16.1 in concerned with difficulty. Although the basic classes are the starting points, many participants felt they were difficult, although not too difficult. Many comments for both classes included students having difficulty due to being unfamiliar with, and new to, logic expressions. Model checking was thought to be more difficult. This is probably

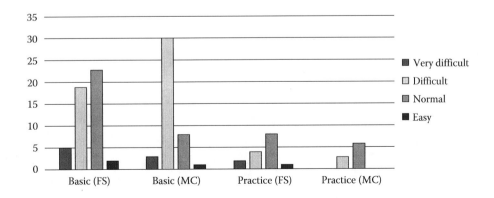

FIGURE 16.1   Difficulty.

because the problems are complex with concurrency, while the basic class for formal specification primarily focuses on sequential specification animation. The practice classes were not thought to be difficult because practical issues, such as management, were the main topics. A few participants thought such issues were the really difficult ones, because they do not have absolutely correct answers and require a wide viewpoint.

Technical difficulties vary from one method or tool to another and also depend on the expected role and tasks. As an instance of tools for CPS, Crescendo allows for separate work on continuous models and discrete models. For an engineer responsible for the discrete models (the software part), the VDM language may be easily accepted as a type of high-level programming language. This is because the purpose is simulation, similar to execution or testing that software engineers are familiar with, and only the software part is extracted and handled. Strictly speaking, it is necessary to jointly discuss the interfaces and collaborative simulations with the continuous models by mechanical engineers. On the other hand, if an engineer is responsible for discharging failed proofs of hybrid models with KeYmaera, then quite different skills from those in common software development are required. Specifically, it is necessary to interactively guide deductive proof with understanding of theories on the target continuous part (e.g., acceleration and velocity). The key point here is that such additional difficulties are not unnecessary difficulties introduced by the tools, but originally come from the difficulties of emerging CPS, the hybrid nature, higher demand on trustworthiness, and so on. The data on the lecture classes in Figure 16.1 show that learning the technical foundation to tackle these difficulties is not hard for a wide range of engineers.

16.4.1.2 Practicality

Figure 16.2 is concerned with the availability of practical information. The fundamental classes involve nontrivial group exercises and give guidelines such as modeling patterns and mapping from UML diagrams to formal models. Although some participants mentioned this was very sufficient, most thought it was not. The practice classes complement this point and most participants seemed satisfied with the practice classes. Most of the positive comments suggested that the participants liked the "real" nature of the exercise problems, guidance for reporting, and so on. A few negative comments on the practice class for formal specification

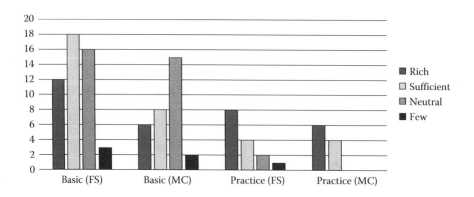

FIGURE 16.2 Availability of practical information.

suggested a need for more details, which were difficult to provide in the limited time of the lectures.

Emerging kinds of difficulties in CPS have led to technical advance, for example, tools for hybrid aspects. However, the sound theoretical foundations and tools based on them are not sufficient even as a starting point to imagine and discuss practical use. Efforts similar to our lecture design described in Section 16.3 are inevitable to promote emerging tools for trustworthiness of CPS.

16.4.1.3 Applicability

Figure 16.3 shows the applicability of the obtained approaches. Even though the participants liked the content of the classes, many had some doubts on the immediate applicability. This point did not change much even with the satisfaction level of the practice classes.

Typical directions of positive comments are given below.

- Positive comments often included specific applications and aspects, for which each commenter felt that formal methods can be (cost-)effective.

- Some comments included "what is left should be done by each project and team," suggesting that there were some gaps, but the participants thought the gaps could not be bridged in a general way.

- Notable comments of nonnegligible numbers were about immediate applicability of "ways of thinking," not the methods or tools themselves. For example, some included the use of principles for specification in natural languages, such as structuring of information and clarification of properties. This point suggests learning formal methods allows the learning of principles significant in handling issues of early phases. On the other hand, these types of comments imply doubt of the applicability of methods or tools themselves, as suggested by negative comments.

These positive results are encouraging, as they show that the participants understood and could imagine how each project and team finally found and tailored solutions for themselves, without any "silver bullets."

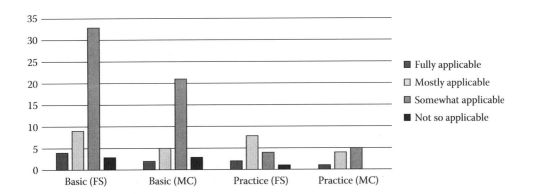

FIGURE 16.3   Applicability.

Typical negative comments were as follows.

- Many comments included that the language and methods seem too specific compared with the currently available practice and skills. The participants could not imagine how to disseminate and use them immediately back at their companies.

- There were comments that suggested the methods and tools should have more useful functions. This point is true compared with popular tools, such as programming IDEs.

- Some comments for the basic classes suggested that the participants could not imagine concrete applications. This type of comment indicates the need for the practice classes and does not appear in the comments for the practice classes.

From the viewpoint of education, it is good to see many negative comments accompany positive comments regarding the experiences and ways of thinking that the participants obtained. However, the negative comments also show long-lasting issues on applicability of formal methods, for example, as discussed in Reference 7.

The first negative comment implies the difficulty of changing the present way of development that has been established and is somehow working. Challenges on CPS with different aspects may serve as a good opportunity to examine and reconsider disciplines, processes, and tools more suitable for tackling difficulties in ensuring trustworthiness.

16.4.2 Graduation Studies

16.4.2.1 Popularity

In Section 16.3.2, we explained that the number of graduation studies has recently been about a quarter of the number of all studies. This number is actually smaller compared with the previous report in which studies on formal methods were about half of all studies [17]. Possible reasons include the following.

- There were strong dissemination activities and thus strong interests in formal methods in Japan at the time of the previous report, such as surveys and guidelines by ministries, public agencies, and private consortia.

- There is now a larger variety of methods and tools taught in the program, for example, large data processing in clouds.

- There is also a larger variety of engineer types, including those who have little programming experience, as the program has become a standard choice in companies rather than reserved for limited elites.

The current ratio of studies with formal methods (a quarter) is quite fair and is almost the same as the ratio of formal methods classes to the total number of classes.

Formal methods are considered significant for trustworthiness of CPS and thus emerging tools have been tried in graduate studies (two of them were described in Section 16.3.2). We believe such trials under academic-industry collaboration are important as the starting point for the industry to investigate new approaches, especially when theoretical foundations are considerably extended.

16.4.2.2 Quality

At least one study on formal methods received the best award every year. This is probably because outputs with formal methods are clearer and more repeatable and are often realized as frameworks or tools with objective rationales (compared to methods that guide human thinking on requirements, risks, or architectures). This point never implies any superiority of formal methods. In any case, clarification of required human efforts is significant, as well as validation of the meaning and impact of the studies.

Nevertheless, formal methods provide good opportunities for participants to discuss and define approaches to be rigorous, repeatable, objective, and general. This point should be very important to tackle the complexity of CPS, which cannot be tackled in an ad-hoc way.

16.5 CONCLUDING REMARKS

Ensuring trustworthiness of CPS requires scientific approaches based on sound theoretical foundations, for example, to decompose the complexity correctly and increase the confidence in trustworthiness. Novel tools have emerged to tackle increasing difficulties in CPS such as the hybrid nature of continuous and discrete aspects. On the other hand, in software engineering there have been gaps between academia and the industry, or between theory and practice.

In order to discuss how such gaps are filled (though not completely), we discussed our experience with formal methods education in the Top SE program. Formal methods are representatives of academia-based or theory-based approaches for trustworthiness that often receive doubt or indifference from the industry.

Although it is difficult to empirically prove any direct or indirect effect of our design for the education, we have reasonable confidence in the following topics:

- Key factors in delivering technology for the industry.
  - Lecturers from the industry.
  - Delivering, working out, and discussing practice, including nontechnical aspects, such as planning and reporting.
- Roles of scientific approaches in software engineering education and training.
  - Opportunities through exercises to discuss and define approaches to be rigorous, repeatable, objective, and general.
  - Clear principles, especially in early phases of development processes, regardless of whether specific tools are used or not.

We also mentioned trials with state-of-the-art tools in graduate studies for trustworthiness of CPS. Even with the short period of lecture classes (2 months for each class) before the studies, participants could obtain foundational understanding of how to think about modeling, verification, and validation. This point illustrates the potential for applications of novel tools that continuously emerge given the new difficulties and increased complexity of CPS.

The Top SE program has been established to be sustainable as a paid course, accepting over 40 participants from companies every year. We would like to have more experience and provide more insights into education as well as academia-industry collaborations.

ACKNOWLEDGMENTS

We would like to thank all of the lecturers and participants involved in the Top SE program.

REFERENCES

1. Manfred Broy. *Complex Systems Design & Management*, chapter Engineering cyber-physical systems: Challenges and foundations, pp. 1–13. Springer, Cham, Switzerland, 2013.
2. Edward A. Lee. Cyber physical systems: Design challenges. In *The 11th IEEE International Symposium on Object Oriented Real-Time Distributed Computing (ISORC 2008)*, Orlando, FL, pp. 363–369, May 2008.
3. John Fitzgerald, Peter Gorm Larsen, and Marcel Verhoef, editors. *Collaborative Design for Embedded Systems: Co-Modelling and Co-Simulation*. Springer, Berlin Heidelberg, 2014.
4. André Platzer. *Logical Analysis of Hybrid Systems: Proving Theorems for Complex Dynamics*. Springer, Los Alamitos, CA, 2010.
5. Anthony Hall. Seven myths of formal methods. *IEEE Software*, 7(5):11–19, 1990.
6. Ciera Jaspan, Michael Keeling, Larry Maccherone, Gabriel L. Zenarosa, and Mary Shaw. Software mythbusters explore formal methods. *IEEE Software*, 26(6):60–63, November 2009.
7. Dines Bjrner and Klaus Havelund. 40 years of formal methods—Some obstacles and some possibilities? In *The 19th International Symposium on Formal Methods (FM 2014)*, Singapore, pp. 42–61, June 2014.
8. Kathy Beckman, Neal Coulter, Soheil Khajenoori, and Nancy R. Mead. Collaborations: Closing the industry-academia gap. *IEEE Software*, 14(6):49–57, 1997.
9. Shinichi Honiden, Yasuyuki Tahara, Nobukazu Yoshioka, Kenji Taguchi, and Hironori Washizaki. Top SE: Educating Superarchitects Who Can Apply Software Engineering Tools to Practical Development in Japan. In *The 29th International Conference on Software Engineering (ICSE 2007)*, Minneapolis, MN, pp. 708–718, 2007.
10. John Fitzgerald, Peter Gorm Larsen, Paul Mukherjee, Nico Plat, and Marcel Verhoef. *Validated Designs for Object-Oriented Systems*. Springer, London, UK, 2005.
11. Gerard Holzmann. *SPIN Model Checker, The: Primer and Reference Manual*. Addison-Wesley Professional, Boston, MA, 2003.
12. Thomas Clausen and Philippe Jacquet. Optimized link state routing protocol (olsr). http://www.ietf.org/rfc/rfc3626.txt, October 2003.
13. Taro Kurita, Fuyuki Ishikawa, and Keijiro Araki. Practices for formal models as documents: Evolution of VDM application to "Mobile FeliCa" IC chip firmware. In *20th International Symposium on Formal Methods (FM 2015)*, June 2015.
14. Taro Kurita and Yasumasa Nakatsugawa. The application of VDM to the industrial development of firmware for a smart card IC chip. *International Journal of Software and Informatics*, 3(2–3):343–355, 2009.
15. Jean-Raymond Abrial. *Modeling in Event-B: System and Software Engineering*. Cambridge University Press, Cambridge, UK, May 2010.

16. OMG. Business Process Model and Notation (BPMN) version 2.0. http://www.omg.org/spec/BPMN/2.0/, January 2011.
17. Fuyuki Ishikawa, Kenji Taguchi, Nobukazu Yoshioka, and Shinichi Honiden. What Top-Level Software Engineers Tackles after Learning Formal Methods—Experiences from the Top SE Project. In *The 2nd International FME Conference on Teaching Formal Methods (TFM 2009)*, Eindhoven, The Netherlands, pp. 57–71, November 2009.

Index